Environmental Cardiology
Pollution and Heart Disease

Issues in Toxicology

Series Editors:
Professor Diana Anderson, *University of Bradford, UK*
Dr Michael D Waters, *Integrated Laboratory Systems, Inc, N Carolina, USA*
Dr Martin F Wilks, *University of Basel, Switzerland*
Dr Timothy C Marrs, *Edentox Associates, Kent, UK*

Titles in the Series:
1: Hair in Toxicology: An Important Bio-Monitor
2: Male-mediated Developmental Toxicity
3: Cytochrome P450: Role in the Metabolism and Toxicity of Drugs and other Xenobiotics
4: Bile Acids: Toxicology and Bioactivity
5: The Comet Assay in Toxicology
6: Silver in Healthcare
7: *In Silico* Toxicology: Principles and Applications
8: Environmental Cardiology: Pollution and Heart Disease

How to obtain future titles on publication:
A standing order plan is available for this series. A standing order will bring delivery of each new volume immediately on publication.

For further information please contact:
Book Sales Department, Royal Society of Chemistry,
Thomas Graham House, Science Park, Milton Road, Cambridge,
CB4 0WF, UK
Telephone: +44 (0)1223 420066, Fax: +44 (0)1223 420247, Email: books@rsc.org
Visit our website at http://www.rsc.org/Shop/Books/

Environmental Cardiology
Pollution and Heart Disease

Edited by

Aruni Bhatnagar
Department of Medicine, University of Louisville, Louisville, KY, US

RSCPublishing

Issues in Toxicology No. 8

ISBN: 978-1-84973-005-1
ISSN: 1757-7179

A catalogue record for this book is available from the British Library

© Royal Society of Chemistry 2011

All rights reserved

Apart from fair dealing for the purposes of research for non-commercial purposes or for private study, criticism or review, as permitted under the Copyright, Designs and Patents Act 1988 and the Copyright and Related Rights Regulations 2003, this publication may not be reproduced, stored or transmitted, in any form or by any means, without the prior permission in writing of The Royal Society of Chemistry, or the copyright owner, or in the case of reproduction in accordance with the terms of licences issued by the Copyright Licensing Agency in the UK, or in accordance with the terms of the licences issued by the appropriate Reproduction Rights Organization outside the UK. Enquiries concerning reproduction outside the terms stated here should be sent to The Royal Society of Chemistry at the address printed on this page.

The RSC is not responsible for individual opinions expressed in this work.

Published by The Royal Society of Chemistry,
Thomas Graham House, Science Park, Milton Road,
Cambridge CB4 0WF, UK

Registered Charity Number 207890

For further information see our web site at www.rsc.org

Printed and bound in Great Britain by CPI Antony Rowe, Chippenham and Eastbourne

Preface

This monograph was assembled to bring together recent developments in the emerging field of environmental cardiology. This new area of research encompasses the study of various environmental factors and their role in the genesis, severity and incidence of heart disease. Although it is widely recognized that environmental factors such as smoking, diet, exercise, and socio-economic status profoundly affect the risk of cardiovascular disease, recent work showing the effects of other environmental factors provides a more complete assessment of the depth and the breadth with which the environment affects heart disease.

This comprehensive view has emerged from three recent developments. First, there has been a relatively sudden explosion in the prevalence of diabetes and obesity, which indicates a strong environmental component. In addition, there has been an accumulation of new evidence suggesting that most cases of heart disease and diabetes could be prevented by healthy lifestyle choices. Finally, extensive studies have shown that exposure to environmental pollutants has a significant effect on heart-disease risk. Among these developments, studies in the area of air pollution provide a more detailed description of how the environment affects heart disease. These studies reveal that cardiovascular tissues are exquisitely sensitive to changes in the external environment, and they broaden the view that cardiovascular health is inextricably linked with natural, social and personal environments. Accordingly, this monograph is devoted primarily to a discussion of pollution and heart disease.

In an attempt to develop a more complete view of the environmental basis of heart disease, assessments of the cardiovascular disease burden of pollutant exposure provide an important missing piece of the puzzle. Putting this piece together with other known environmental effects allows us to see uninterrupted connections between different aspects of the environment and how together they create conditions that promote and sustain heart disease. Studies in particulate

Issues in Toxicology No. 8
Environmental Cardiology: Pollution and Heart Disease
Edited by Aruni Bhatnagar
© The Royal Society of Chemistry 2011
Published by the Royal Society of Chemistry, www.rsc.org

air-pollution research reveal a new "risk factor" for heart disease; but more importantly, they provide a new paradigm for understanding how the environment continuously affects the development of heart disease and how environmental changes adruptly trigger adverse cardiovascular events. Exposure to particulate air pollution is associated with an exacerbation of hypertension and insulin resistance, acceleration of atherogenesis, as well as plaque rupture leading to myocardial infarction. These associations suggest that environmental exposures affect all stages in the development of heart disease. Other environmental factors exert similar effects. Hence, an understanding of environmental influences is likely to be important, not only in the prevention of heart disease, but in its treatment and management as well.

The introductory chapter provides a general view of the field and outlines the effects of different aspects of the environment on heart disease. It provides a context for the discussion that follows, and it maps pollution research within the overall topography of environmental cardiology. Chapter 2 gives an overview of the cardiovascular effects of particulate matter, and Chapter 3 discusses the epidemiological studies supporting this link. In subsequent chapters, the effects of pollution on different aspects of cardiovascular disease – hypertension, stroke, heart failure, ischemic heart disease and atherogenesis – are presented. Because of a close association between diabetes and heart disease, a discussion of the effects of particulate matter on diabetes is included in Chapter 5. Later chapters discuss the effects of individual pollutants such as vehicular emission, metals and aldehydes. A review on manufactured nanoparticles is included because these particles represent an important new threat to cardiovascular health.

Although not exhaustive, this collection provides an inclusive view of research in this area. Like all areas of active investigation, this is a work in progress and therefore subject to modification, elaboration or even revision by future discoveries. Research in this area is progressing at a rapid pace, and therefore it is important to pause and survey how far we have come and to consider where we should go from here. To this aim, the monograph brings together for the first time a broad discussion on the role of the most important environmental factors that affect heart disease.

Many of the studies discussed here suggest that a significant burden of heart disease could be lifted by removing unhealthy environmental influences. These studies show that, for the most part, heart disease does not develop in healthy, unpolluted environments or in individuals who make optimal lifestyle choices and are in synchrony with the primordial rhythms of their natural environment. In addition, it has been shown that the risk of heart disease is rapidly and robustly affected by changes in the environment. Collectively, these facts imply that there is a causative link between the environment and heart disease. While the disease manifests in the individual, its origins frequently lie in the environment. Attributing heart disease to unhealthy environments, however, does not invalidate or deny the role of genetic susceptibility. Genetic and metabolic factors are undeniably important formal and material causes of heart disease. They regulate the forms, the manifestation and the severity of heart disease. But, the environment is usually the efficient cause, as it often engenders the right conditions for the development of heart disease, and in doing so it acts as a primary trigger to which genetic and metabolic processes respond.

While current therapies are aimed at treating pathological responses (blood pressure, cholesterol levels) in the individual, less emphasis is placed on controlling or extinguishing the environmental triggers that elicit these responses. In this regard, the understanding that emerges from this monograph suggests that we must be more alert to the effects of the environment and develop strategies that target not only the diseased individual but the unhealthy, disease-causing environment as well. Because heart disease arises mostly from unhealthy environments, targeting the environment is likely to provide more tangible gains. Although much work is still required to fully redeem the promise of this vision, the research presented here could facilitate and stimulate new investgations and, thereby, encourage the development of a more coherent view of environmental cardiology.

In the last few years, our understanding of the environmental factors that contribute to the risk of heart disease increased significantly. The most rapid growth has been in the area of air-pollution research. This area has attracted wide attention and has been a topic of several commentaries, reviews and symposia. It has also been the subject of a recently updated scientific statement from the American Heart Association. Nevertheless, this monograph fills an important void. It is the first attempt to provide a comprehensive account of the effects of pollutants on heart disease and to integrate this area of research within the overall theme of environmental cardiology. Thus, the publication of this monograph is an important milestone in the development of this field, and the book itself is likely to serve as a valuable resource for both new and established investigators interested in this area of research. The overview and perspectives, as well as the detailed discussions on individual issues, may prove helpful to students and trainees on their path to new discoveries.

The most important element in discovery, however, is the discoverer. All that we know about the environment and its effects on heart disease comes from the work of several creative and committed investigators to whom we remain indebted. In particular, I am thankful to the extraordinary league of scientists who have made key discoveries in this area and who have contributed to this monograph. Their relentless pursuit of truth, even when its path may not be clear or fashionable, is inspirational. I am both proud and humbled to be their colleague and to be able to participate in the discussion they started. I am grateful for the time they took from their hectic research schedules to contribute to this book, and I am convinced that their work will continue to inspire the next generation of scientists.

On behalf of my colleagues, I also wish to express gratitude to the enlightened leadership at the National Institutes of Environmental Health Sciences and the Environmental Protection Agency. They are equal partners on this journey, and their support has been instrumental in the development of this field. Finally, I would be remiss if I did not acknowledge my deep appreciation for members of my family. They have suffered my long absences with extraordinary patience and understanding. But always it was the return home that made it all worthwhile.

Aruni Bhatnagar

Contents

Chapter 1	**Environmental Basis of Cardiovascular Disease**		**1**
	A. Bhatnagar		
	1.1	Introduction	1
		1.1.1 *My Family and Other Animals*	4
		1.1.2 *Peacocks in Siberia*	6
		1.1.3 *Out of Africa*	8
	1.2	Categories of the Human Environment	12
	1.3	Cardiovascular Disease and the Natural Environment	15
		1.3.1 *Cycles of Night and Day*	16
		1.3.2 *Four Seasons*	18
		1.3.3 *I'll Follow the Sun*	21
		1.3.4 *In High Places*	24
	1.4	Cardiovascular Disease and the Plastic Environment	26
		1.4.1 *It Takes a Village*	27
		1.4.2 *Wealth is Health*	29
		1.4.3 *People or Places?*	31
		1.4.4 *With the Help of My Friends*	33
	1.5	Heritability of the Environment	35
	1.6	Pollution and Heart Disease	37
		1.6.1 *Brave New World*	38
		1.6.2 *Weaknesses of the Heart*	39
	1.7	Personal Environment and Lifestyle Choices	40
		1.7.1 *Sum of Our Choices*	42
		1.7.2 *Food for Thought*	43
		1.7.3 *Rolling Stone Gathers no Moss*	46
		1.7.4 *Smoke and Mirrors*	49

Issues in Toxicology No. 8
Environmental Cardiology: Pollution and Heart Disease
Edited by Aruni Bhatnagar
© The Royal Society of Chemistry 2011
Published by the Royal Society of Chemistry, www.rsc.org

1.8	Mechanisms of Environmental CVD	52
	1.8.1 *Risky Business*	52
1.9	Implications of an Environmental Perspective	55
References		59

Chapter 2 Cardiovascular Effects of Particulate-Matter Air Pollution: An Overview and Perspectives 76
J. A. Araujo and R. D. Brook

2.1	Introduction	76
2.2	Air-Pollution Components and Characterization	77
2.3	PM Exposure and Cardiovascular Morbidity and Mortality	78
	2.3.1 Short-Term Exposures	78
	2.3.2 Longer-Term Exposures	82
	2.3.3 Additional Epidemiological Findings	83
2.4	PM Exposure and Clinical and Subclinical Cardiovascular Outcomes	85
2.5	Pathobiological Mechanisms	87
2.6	Conclusions and Perspectives	89
Acknowledgment		90
References		90

Chapter 3 Air Pollution and Atherosclerosis: Epidemiologic Studies 105
V. C. Van Hee and J. D. Kaufman

3.1	Introduction	105
3.2	Atherosclerosis: A Chronic, Inflammatory Disease Leading to Acute Cardiac Events	106
3.3	Subclinical Atherosclerosis: Measurement Methods	106
	3.3.1 Carotid Intima-Media Thickness (CIMT)	107
	3.3.2 Coronary Artery Calcium (CAC)	107
	3.3.3 Aortic Calcium	108
	3.3.4 Ankle-Arm Index (AAI)	108
	3.3.5 Other Methods	109
3.4	Epidemiologic Studies Addressing the Relationship Between Air Pollutants and Atherosclerosis	109
	3.4.1 Particulate-Matter Air Pollution and CIMT in Los Angeles	109
	3.4.2 PM, Traffic-Related Air Pollution, and CAC in Three German Cities	112
	3.4.3 $PM_{2.5}$, PM_{10}, and Multiple Subclinical Measures in Six US Cities	112
	3.4.4 PM, Traffic-Related Pollution, and Aortic Calcium in Six US Cities	113

	3.4.5	PM, Traffic-Related Air Pollution, and ABI in Three German Cities	113
3.5		Consistency between Relationships Observed in Current Studies	114
3.6		Air-Pollution Exposure and Atherosclerosis: Ancillary Epidemiologic Evidence	115
References			117

Chapter 4 Hypertension and Vascular Toxicity of PM 121
Z. Ying and S. Rajagopalan

4.1	Introduction		121
4.2	Current Evidence from Animal and Toxicological Studies		122
	4.2.1	Systemic Oxidative Stress and Endothelial Function	122
	4.2.2	Autonomic Tone and Function	123
	4.2.3	Pulmonary and Systemic Inflammation	124
	4.2.4	Integrated Animal Studies Supporting a Role in Hypertension	125
4.3	Evidence to Support Vascular Effects of Inhaled Particles in Humans		127
	4.3.1	Systemic Oxidative Stress and Endothelial Dysfunction	127
	4.3.2	Autonomic Tone and Function	129
	4.3.3	Systemic Inflammation	130
	4.3.4	Integrated Hemodynamic Studies in Humans	131
4.4	Summary of Biological Mechanisms		132
References			135

Chapter 5 Air Pollution and Diabetes 143
E. H. Wilker and J. D. Schwartz

5.1	Introduction		143
5.2	Evidence from Administrative Data Sources		144
	5.2.1	Mortality	144
	5.2.2	Hospital Admissions and Acute Events	145
5.3	Evidence from Measurements of Physiologic Outcomes		146
	5.3.1	Heart-Rate Variability	146
	5.3.2	Brachial-Artery Diameter and Flow-Mediated Dilation	147
	5.3.3	Evidence from Biomarkers	148
	5.3.4	Toxicology Studies	148
	5.3.5	Potential Mechanisms	149

	5.4 Does Air Pollution Cause Diabetes?	150
	5.5 Conclusions	151
	References	151

Chapter 6 Ambient Particulate Matter and the Risk of Stroke — 159
G. A. Wellenius, D. R. Gold and M. A. Mittleman

6.1	Stroke is a Public-Health Problem	159
6.2	Cardiovascular Health Effects of Ambient Particulate Matter	160
6.3	Effects of Short-Term PM Exposure on Cerebrovascular Hospitalizations	161
6.4	Ischemic Stroke and Transient Ischemic Attack (TIA)	163
6.5	Hemorrhagic Stroke	164
6.6	Effects of Short-Term PM Exposure on Cerebrovascular Mortality	164
6.7	Effects of Long-Term PM Exposure on Cerebrovascular Morbidity and Mortality	165
6.8	Potential Mechanisms	166
6.9	Summary	167
	References	168

Chapter 7 Environmental Pollutants and Heart Failure — 177
S. D. Prabhu

7.1	Introduction	177
7.2	Clinical and Pathological Characteristics of HF	178
7.3	Pollution and Heart Failure: Short-Term Effects	179
	7.3.1 HF Symptoms and Signs	180
	7.3.2 HF Mortality	180
	7.3.3 HF Hospital Visits and Admissions	181
7.4	Pollution and Heart Failure: Long-Term Effects	183
	7.4.1 Particulate Exposure	184
	7.4.2 Motor-Vehicle Traffic Exposure	184
7.5	Pathophysiological Mechanisms of Pollution-Related HF Risk	185
7.6	Aldehydes Impart Significant Cardiotoxic Effects	187
	Acknowledgement	192
	References	192

Chapter 8	**Ultrafine Particles and Atherosclerosis** J. A. Araujo	**198**
	8.1 Introduction	198
	8.2 Particulate Matter of the Smallest Size Has the Greatest Pro-Oxidative Potential	199
	8.3 UFP Activate Proinflammatory Pathways in Vascular Cells	200
	8.4 UFP Exert Largest PM Proatherogenic Effects	203
	8.5 How Do Pro-Oxidative UFP Enhance Atherosclerosis?	207
	8.5.1 Larger Particle Number	208
	8.5.2 Greater Lung Retention	208
	8.5.3 Larger Content of Redox Active Compounds	208
	8.5.4 Greater Bioavailability	209
	8.6 Do UFP Enhance Atherosclerosis in Humans?	209
	8.7 Conclusions	211
	Acknowledgments	212
	References	212
Chapter 9	**Air Pollution and Ischemic Heart Disease** A. Peters	**220**
	9.1 Introduction	220
	9.2 Chronic Exposure to Particulate Matter and the Risk of Ischemic Heart Disease	221
	9.3 Chronic Exposure to Particulate Matter and Atherosclerosis	223
	9.4 Inflammation as a Marker for Increased Cardiovascular Risk	224
	9.5 Evidence for Endothelial Cell Activation and Changes in Coagulation Markers	224
	9.6 ECG Recorded Ischemia	225
	9.7 Acute Exposure to Particulate Matter and the Risk of Ischemic Heart Disease	225
	9.8 Components of the Ambient Air-Pollution Mixture Associated with Ischemic Heart Disease	226
	9.9 Overall Summary and Outlook	226
	References	227
Chapter 10	**Vehicular Emissions and Cardiovascular Disease** M. Campen and A. Lund	**234**
	10.1 Introduction	234
	10.1.1 Vehicle Emissions in the United States: Trends and Policy	235

		10.1.2	Findings from Population Health Studies	235
		10.1.3	Exposure Assessment	237
		10.1.4	Chemistry of Vehicular Emissions	237
	10.2	Toxicological Research Findings		239
		10.2.1	Human Studies	239
		10.2.2	Animal Studies	242
	10.3	Research Needs: Mechanisms, Interactions, and Sensitivities		246
	References			247

Chapter 11 Manufactured Nanoparticles   253
G. S. Kang, P. A. Gillespie and L. C. Chen

	11.1	Nanoparticles and Nanotoxicology		253
	11.2	NP Exposure and Cardiac Toxicity		255
		11.2.1	Direct Cardiac Exposure to NPs	255
		11.2.2	Cardiovascular Effects by Pulmonary NP Exposure	256
	11.3	Study Review – Cardiac Toxicity by NP Exposure		256
		11.3.1	Fullerenes	257
		11.3.2	Carbon Nanotubes	259
		11.3.3	Quantum Dots	261
		11.3.4	Metallic and Metal-Oxide-Based NPs	261
	11.4	A Case Study: Subchronic Effects of Inhaled Nickel Nanoparticles on the Progression of Atherosclerosis in a Hyperlipidemic Mouse Model		262
	11.5	Future Studies		263
		11.5.1	Human Data	263
		11.5.2	Thorough Particle Characterization	264
		11.5.3	Relevant Exposure Scenario	264
	11.6	Summary		265
	Acknowledgment			265
	References			265

Chapter 12 Metals in Environmental Cardiovascular Diseases   272
A. Barchowsky

	12.1	Introduction		272
	12.2	Overview of Metal Exposures		273
	12.3	Mechanisms of Metal Action		275
	12.4	Pathogenic Actions of Metals in the Heart		278
		12.4.1	Metal-Induced Cardiomyopathies	278
		12.4.2	Cardiac Arrhythmias	279
		12.4.3	Ischemic Diseases and Atherosclerosis	281

	12.4.4	Copper	282
	12.4.5	Arsenic	282
	12.4.6	Cadmium	283
	12.4.7	Hypertension	284
	12.4.8	Angiogenesis	288
12.5	Conclusions		289
References			290

Chapter 13 Environmental Aldehydes and Cardiovascular Disease   **301**
*D. J. Conklin, P. Haberzettl, J. Lee and
S. Srivastava*

13.1	Introduction		301
13.2	Epidemiology of Environmental Aldehydes and Cardiovascular Disease		304
	13.2.1	Levels of Environmental Aldehydes	304
	13.2.2	Epidemiology of Aldehyde Exposures	305
	13.2.3	Aldehyde Exposure as a Product of Xenobiotic Metabolism	313
13.3	Cardiovascular Effects and Signaling Mechanisms of Aldehyde Exposure		315
	13.3.1	Enals and Cell Signaling	315
	13.3.2	Cardiovascular Effects of Aldehydes	318
	13.3.3	Role of Protein–Aldehyde Adducts	332
13.4	Aldehyde Metabolism		338
	13.4.1	*In Vitro* Kinetic Studies	338
	13.4.2	Regulation of the Enzyme Activity of Aldehyde-Metabolizing Enzymes by Enals	339
	13.4.3	Cardiovascular Metabolism of Enals	340
	13.4.4	Systemic Metabolism of Enals	341
13.5	Environmental Aldehydes: Detection and Quantitation		341
	13.5.1	Measurements of Aldehydes in Air, Water and Food	341
	13.5.2	Problems and Pitfalls of Aldehyde Measurements in Air, Water and Foods	349
13.6	Conclusions and Future Directions		350
	13.6.1	Environmental and Endogenous Aldehydes: Shared Biological Pathways?	350
	13.6.2	Aldehyde Metabolism as a Modifier of Aldehyde Effects *in Vivo*	350
	13.6.3	Potential Approaches to Reduce and Mitigate Aldehyde Exposure	351

Acknowledgments	352
Abbreviations	352
References	353

Subject Index   **371**

CHAPTER 1
Environmental Basis of Cardiovascular Disease

A. BHATNAGAR

Diabetes and Obesity Center, Division of Cardiovascular Medicine, Department of Medicine, University of Louisville, 580 S. Preston Street, Louisville, KY 40202, USA

1.1 Introduction

The term cardiovascular disease (CVD) refers to a group of illnesses caused by the disorders of the heart, blood vessels and blood flow. The most common cause of cardiovascular diseases is atherosclerosis, which is the hardening of arteries due to the formation of an atheromatous plaque. Abrupt changes in blood flow in atherosclerotic vessels result in acute myocardial infarction and stroke, which are the major clinical manifestations of chronic changes in the vessel wall. In the heart, ischemic injury due to atherosclerotic disease often leads to arrhythmia, hypertrophy, cardiomyopathy and heart failure. Heart disease is accompanied by chronic metabolic and physiological changes that precede and contribute to its clinical manifestations. These include metabolic changes such as high cholesterol (hypercholesterolemia) and insulin resistance and physiological changes such as an increase in blood pressure (hypertension) and changes in cardiac contractility. Although the causes of diabetes are not well understood, diabetes primarily affects the heart and blood vessels and is, therefore, considered to be a major CVD risk factor. Therefore diabetes and obesity are included in this discussion of heart disease.

Significant CVD is also associated with rheumatic disease, which is due to myocardial damage caused by streptococcal bacteria and congenital malformation of the structures of the heart or blood vessels. Several other types of

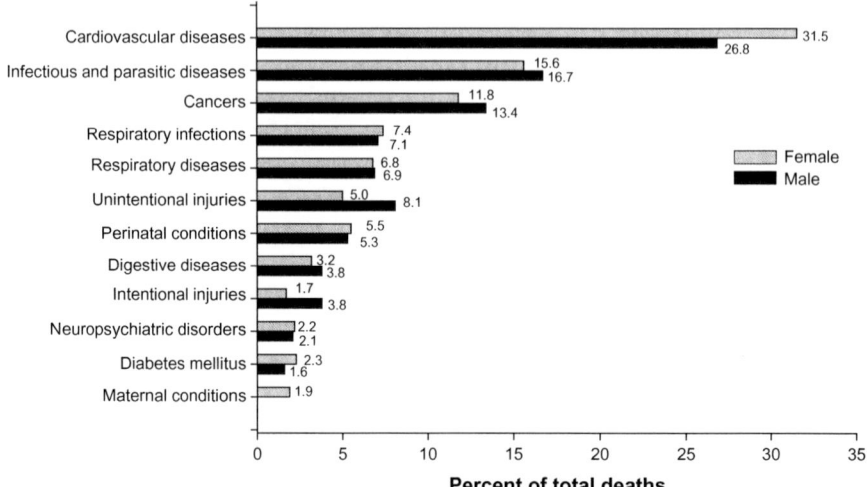

Figure 1.1 Distribution of deaths in the world in 2004 by leading cause groups and gender (WHO report).

congenital CVD are also common. These defects could be overt, resulting from gross malformation of major blood vessels or myocardial tissue in the fetus, or they may be more subtle, leading to an increase in susceptibility to stress or exercise. Congenital defects or prolonged hypertension and infectious diseases could also result in the dilation and rupture of the aorta leading to aortic aneurysm and dissection. Additionally, cardiovascular disorder associated with deep vein thrombosis and pulmonary embolism could result from blood clots in the leg veins that can dislodge and move to the heart and the lungs.

As a group, CVD is the leading cause of death world-wide (Figure 1.1). According to the WHO in 2004, CVD accounted for nearly 30% of all deaths world-wide. It killed twice as many people as infectious and parasitic disease and 3 times as many people as all forms of cancer. Globally, most (43–45%) cardiovascular deaths are due to coronary heart disease (CHD) or ischemic heart disease (IHD), whereas stroke accounts for 33% of CVD. A similar distribution of CVD deaths has been reported for countries such as the US (Figure 1.2).[1]

These statistics suggest that heart disease is the major cause of mortality world-wide. Although the prevalence of heart disease varies considerably (*vide infra*) it still remains a major cause of death in all human populations regardless of their geographic location or ethnicity. It shows no preference for gender or economic status. Both men and women appear to be equally susceptible. World-wide, more women (31.5%) than men (26.8%) die of heart disease. Even in low-income countries (per capita ≤$825) IHD is the number two leading cause of death (9.4%), second only to deaths caused by lower respiratory infections (11.2%), whereas in middle and high income countries ($10,066 or more) IHD and cerebrovascular disease account for 25 to 28% of all deaths

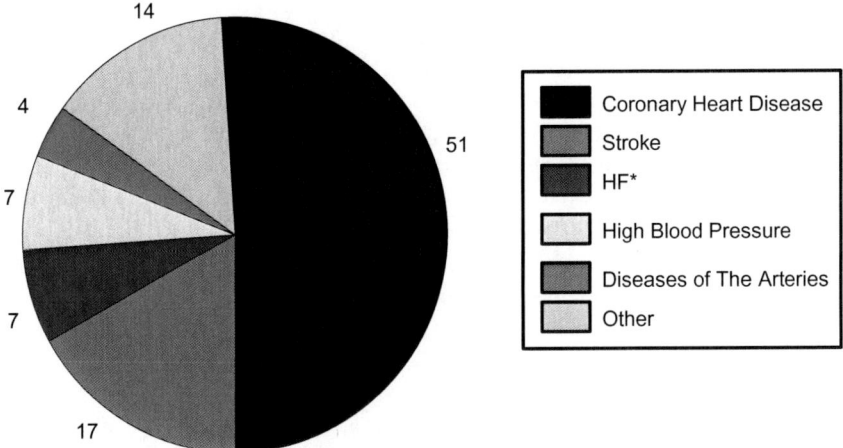

Figure 1.2 Per cent breakdown of deaths due to cardiovascular diseases in the United States in 2006. Data are derived from the 2010 report of the American Heart Association.[1] * Not a true underlying cause. The data may not add to 100% because of rounding.

(WHO, 2005). What is more alarming is that the prevalence of heart disease in increasing. The WHO estimates that 80% of all current CVD deaths are in developing, low- and middle-income countries and it is estimated that by 2010, CVD will be the leading cause of death in developing, low-income countries as well. In developed countries, the emergent epidemics of diabetes and obesity are threatening to erode the pattern of recent gains in health. In the US, the increase in obesity alone has been forecasted to slow down the increase in life expectancy that has been steadily increasing since the early 20th century.[2] Thus, CVD is the most frequent cause of death throughout the world, independent of economic status, gender, or ethnic differences.

The universally high burden of CVD and the extraordinarily high rates of CVD mortality across all communities, suggests that humans as a species are particularly prone to heart disease. It may be argued that CVD is an inevitable consequence of aging, that blood pressure and cholesterol levels inexorably increase with age and that if an individual survives middle age without succumbing to infectious disease, sporadic cancers, accidents or violence, their most likely fate is cardiovascular death. This view is consistent with data showing that the risk of dying from CVD increases with age. In the US, the percentage of population with CVD increases from 14.9 and 8.7% for men and women of 20–39 years of age to 78.8 and 84.7% for men and women more than 80 years of age. It has been suggested that because heart disease develops more often in old individuals, it is not subject to direct selective pressure, *i.e.* that natural selection during evolution is unable to weed out these diseases as they do not affect reproductive success. Natural selection, it is believed, tends to maintain the frequency of genes that increase reproductive success even if the genes have other effects that increase disease susceptibility in older age.[3] However, as we shall see, these arguments do not take into account the important role of the environment, which affects not only the long-range evolutionary susceptibility to disease, but also the proximate causes that lead to the disease development in a specific individual. Moreover, changes in the environment can modify (slow down or accelerate) age-dependent changes in the heart and blood vessels. In addition, a changing environment could continuously alter the context within which the effects of a gene manifest. Thus, a gene could be beneficial in one environment but not the other. As a result, changes in the environment could impart maladaptive predilection to a previously well-adapted genetic variance; thereby significantly and robustly modifying disease susceptibility.

1.1.1 *My Family and Other Animals*

The high prevalence of CVD in human populations suggests shared genetic susceptibility. In comparison with other species, humans are genetically very similar. The low genetic diversity in humans has been linked to a rather small population of ancestors from which modern humans have descended. By some accounts, all modern humans are descendents of a small ancestral family of only 10,000 individuals.[4] As a result, humans are very similar to each other. Moreover, their gene pool has remained shallow because humans spread very quickly over vast expanses of land without acquiring sufficient genetic diversity. Because of their high cognitive abilities and greater capacity to adapt to different environments they migrated to different parts of the planet and segregated into small inbreeding populations, which did not have the time to diverge before significant interbreeding began again. It is therefore not surprising that all humans have similar disease susceptibility and that they succumb to very similar afflictions. But if we take a less parochial view and look outside the human family we might ask – what about other species? Are other animals susceptible to heart disease as well?

In the wild, captivity or domestication, mammals such as dogs, rabbits, rats and mice rarely develop spontaneous atherosclerosis of the type seen in humans. Even when cholesterol metabolism in mice is severely compromised by genetic engineering, they rarely suffer from myocardial infarctions or stroke. This difference may be due to the large evolutionary distance that separates humans from most other mammals and perhaps it is more instructive to look at the great apes, particularly gorillas and chimpanzees. Humans, gorillas, and chimpanzees have descended from a common ancestor that lived 7.3 million years ago.[5] The chimpanzees are our closet living cousins from whom we diverged 5.4 million years ago.[5] Nevertheless, the amino acid sequences of humans and chimpanzees show 99% homology.[6] Hence, it may be expected that, because of their high genetic similarity, humans and chimpanzees would have similar disease susceptibility and might die of similar causes.

Fortunately, several investigators have studied chimpanzee and gorilla mortality both in the wild and in captivity. As expected, the life expectancy of a chimpanzee in the wild is shorter than in captivity. In the wild most chimpanzees live to be around 15 years of age, although occasionally 40- to 50-year-old chimpanzees have been sighted.[7] In the wild they succumb mostly to infectious diseases, most chimpanzees die of respiratory infections[8] while gorillas fall prey to various types of entrocolitis due to viral or fungal infections.[9] In captivity, however, it has been found that cardiac disease is the primary cause of mortality in both gorillas[10] and chimpanzees.[11] Cardiac disease has been reported to be responsible for 41%[11] of deaths of captive adult lowland gorillas and 67.8%[12] of captive chimpanzees. However, the type of heart disease described in chimpanzees and gorillas is not the type commonly seen in humans. In one study most of the heart disease in chimpanzees was reported to be due to an unusual form of cardiomyopathy that was associated with congestive heart failure and the presence of multifocal to coalescing areas of fibrosis, necrosis, mineralization and inflammation[12] and ventricular arrhythmias.[11] Similar findings have been reported by others.[13–15] This type of diffuse cardiac fibrosis leading to congestive heart failure has also been observed in western lowland gorillas.[16] Such pathology is rarely seen in humans and it certainly does not contribute to garden-variety heart disease that kills most humans. In humans, a majority of heart disease is due to atherosclerosis that results in coronary artery disease and stroke. Together, these diseases account for 76% of all cardiovascular deaths world-wide. In contrast, only 2.3% of the captive chimpanzees have been reported to have atherosclerotic disease and although hypertension and hyperlipidemia have been diagnosed in both captive chimpanzees[11] and gorillas,[17] these conditions were found not to be associated with coronary heart disease or with atherosclerosis.

There may be several reasons why chimpanzees are genetically less susceptible to atherosclerotic disease. One of these may relate to the 1% difference in the human and chimpanzee genome. While this does not seem like much, it accounts for the starkly different cognitive, cultural and behavioral differences between humans and chimpanzees. However, this appears not be the case because most the genetic differences between human and chimpanzees are in cortical genes. By contrast, the genes in chimpanzee hearts and livers are nearly identical to humans.[18] Thus, it seems unlikely that humans have recently acquired genes that have increased their susceptibility to metabolic diseases. An alternative explanation is that perhaps during evolution, humans have lost some of the genes that protect chimpanzees from atherosclerotic disease. Indeed, current theories of human evolution suggest that humans have evolved from chimpanzees by loss-of-function mutations (the "less-is-more" hypothesis).[19] It is believed that in many respects, humans are "degenerate apes" who have lost, among other characteristics, much of their muscle strength, hair *etc.*, or their ability to synthesize certain metabolites such as sialic acid.[5] By shedding this excessive baggage, humans have been able to evolve at a more nimble and rapid pace than chimpanzees. Hence, it is conceivable that by losing some genes and acquiring a more retrograde phenotype, humans have become more susceptible to atherosclerotic disease. It is well known that several human diseases such as cystic fibrosis, phenylketouria, and familial breast cancer are due to loss-of-function mutations. Nevertheless, a comparison of human and chimpanzee genomes shows that humans have not lost any of the genes that regulate cardiovascular and hepatic function. On the contrary, several common genetic polymorphisms that are clearly linked to coronary disease and diabetes in humans (*e.g.*, PPARG A12P, PON1 Q192R, and ABCA1 I883M) are ancestral alleles carried not only by chimpanzees but also by out-groups such as macaque.[20] A most likely explanation is that humans and chimpanzees carry the same gene variant and that these ancestral alleles have become human-specific risk factors, not because of a loss of function, but because of a change in the environment. These alleles are natural and widely distributed in living apes. They have evolved and they have assumed the form that they do so that the apes could adapt to their environment. It is likely that they were equally beneficial to humans in their early, ape-like environment, but because the environment in which human live now has changed these genetic variations are no longer beneficial. Instead, they increase disease risk. Thus, a change in the environment has dramatically changed the survival advantage and the disease-risk associated with specific allelic variations.

1.1.2 *Peacocks in Siberia*

Why have the genes that were protective in the ancient environment become maladapted in the current environment? There are several answers to this question.[21] One explanation is that an ancestral gene, which was adaptive in an ancient environment, is no longer beneficial in the modern environment because the environment has changed drastically. In other words, it has become maladaptive in the modern environment. Ancient and modern humans live in very different environments and the gene variants that were important for survival, growth and health under those conditions may not be protective in the modern environment (A to B; Figure 1.3). This model is consistent with the "*thrifty-gene*" hypothesis, which states that genes that favor energy conservation in the wild, food-scarce environment, impart genetic risk for obesity and diabetes in the modern, food-rich environment. Maladaptation between an ancestral allele and modern environment arises because natural selection cannot keep pace with rapid changes in the environment. This is particularly true for human environments, which could change completely within a few generations. However, if after the change, the environment reaches a steady state and if the gene has a survival advantage, the ancestral allele changes to a derived allele under positive selection. The derived allele, once again, confers protection and survival advantage, but only as long as the environment does not change. If the environment changes again, the derived allele may or not may not be beneficial. Because many aspects of human environment are continuously changing, there may be a constant mismatch between environment and gene adaptation. This mismatch could provide a selective pressure for genetic adaptation, but could also account for the temporary persistence of several disease-susceptibility genes in the current environment (*the mismatch hypothesis*).

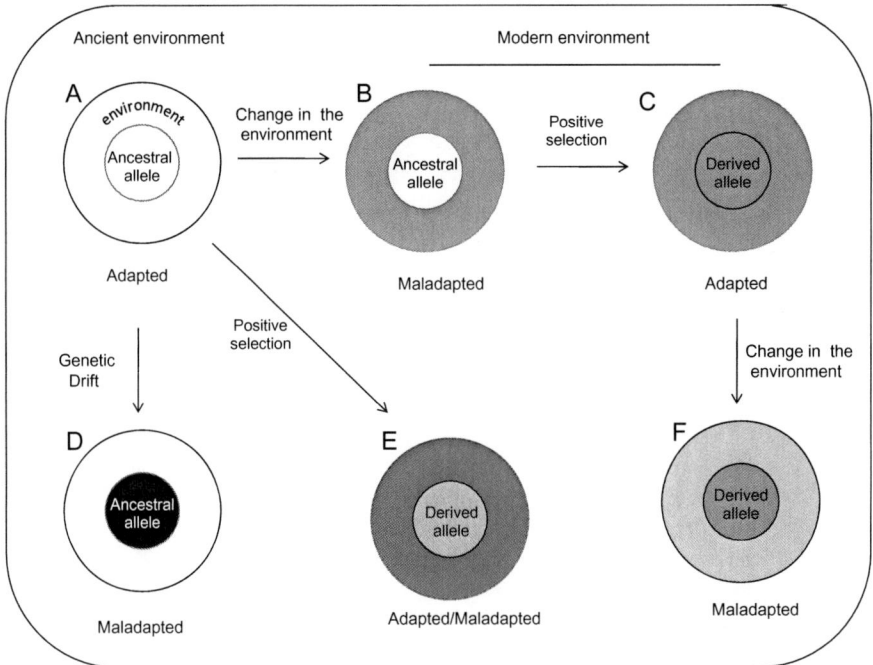

Figure 1.3 Role of environmental changes in disease susceptibility. Ancestral alleles that are adaptive in an ancient environment can become maladaptive in a new environment and thereby increase disease susceptibility. Under positive selection, the ancestral allele may give rise of a derived allele, which is better suited, or more adaptive in the modern environment. The susceptibility of several modern diseases may be high due to a mismatch between the ancestral alleles and the current environment. Adapted from Ding and Kullo.[21]

Another explanation is that the genes that affect CVD susceptibility are retained just by chance (*Neutrality hypothesis*). Because heart disease develops after the reproductive years, it is believed that there are no selective pressures either for or against the genes that regulate CVD susceptibility. When mismatched with the environment these gene variants are not eliminated by a strong purifying selection because they do not impair reproductive fitness. As a result, harmful ancestral genes are retained and tolerated because there is no evolutionary pressure to change them. In addition, new gene variants could arise by genetic drift and these new variants are either beneficial or harmful, but regardless of their effects, these rare variants accumulate because they do not impair reproductive fitness and are therefore not eliminated by strong purifying selection (*the rare-variant-common disease hypothesis*).

A third explanation is that gene variants that drive human evolution by promoting early life survival impair human health in old age. It is believed that one reason that these variants appear and are retained is because they increase reproductive success. Therefore, there is strong positive selection that enriches a derived allele in a population. The derived allele confers a well-adaptive phenotype and promotes reproductive success; however, this victory comes at a high cost because the very gene variants that were advantageous in youth increase disease susceptibility in old age. Thus, early life acclimatization is optimized at the cost of late-life adaptation so that the derived allele is adaptive in youth but not in old age (since this is victory gained at too great a cost, we can call this the *pyrrhic hypothesis*). However, regardless of mechanisms, it is evident that the environment plays a leading role in driving the change and in providing context to the derived or retained alleles because it is only within the framework of the environment that a gene is either adapted or maladaptive.

1.1.3 Out of Africa

There are several notable examples of how a change in the environment affects the genes that affect CVD susceptibility and how alleles adapted to one environment elevate CVD risk in another environment (Figure 1.3; A to B). A particular interesting example, supporting the mismatch hypothesis is apoE. The apoE gene is the code for a plasma protein that is associated with chylomicrons and intermediate density lipoproteins (IDLs). ApoE binds to LDL receptors on the liver and is required for the breakdown of triglyceride-rich lipoprotein constituents. Although several polymorphic forms of apoE have been recognized, the most common alleles are ε2, ε3, and ε4. The proteins coded by these genes differ in their affinity for lipoprotein particles and hepatic receptors.[22] The ε4 allele is the ancestral gene present in chimpanzees and other nonhuman primates[23] but it apparently does not impart excessive CVD risk in them (*vide supra*). In humans, however, the ε4 allele is associated with lower apoE levels, higher cholesterol levels and a higher risk of developing coronary artery disease as well as Alzheimer disease (at least in European or Asian populations[24,25]). Why the ε4 allele has become maladaptive in humans in their current environment is not clear, but it is likely that maintaining higher levels of cholesterol may have been beneficial. Cholesterol is required for the synthesis of steroid hormones and neural function and therefore maintaining high levels of plasma cholesterol in the ancient diet-restricted environment may have been advantageous, however, it has become deleterious in the current diet-abundant environment.

Environmental influences could also foster adaptation by positive selection (Figure 1.3; B to C). For instance, the most common apoE variant in the current human populations is ε3.[24] This variant is believed not to have arisen due to genetic drift but to a positive selection, some 100,000 years ago. Because this change occurred prior to population expansion, it may be reflective of adaptation to a change in diet that accompanied the transition from subsistence to an agricultural economy.[26] This may be because carriers of the ε3 allele were more resistant to the infectious diseases that originate from domestic animals or only become endemic in larger communities that live close together (*e.g.*, smallpox and tuberculosis). The ε3 allele is less strongly associated with CAD and Alzheimer's disease risk than ε4 and it has been suggested to promote reproductive success.[27] Thus, genes can rapidly adapt to new environments and whether in their ancestral form or in their new adaptive form, their overall contribution to disease risk is entirely contingent on environmental influences and subject to environmental modification. In this regard it is interesting to point out that even in present-day humans, the extent to which apoE variations predict cholesterol levels and disease susceptibility varies with environmental factors such as diet and smoking.[28] Hence, environmental changes can render genetic adaptation not only irrelevant or harmful but they can also modify the risk imparted by a specific genotype.

Hypertension is another CVD risk factor associated with mismatch between environment and genetic adaptation. Hypertension results from the inability of multiple compensatory mechanisms involved in the control of blood pressure to maintain pressure within appropriate limits. Because blood pressure is regulated by many interrelated multiorgan control mechanisms, it is considered to be a higher-order emergent function that depends upon, but is not predictable from, the structures and functions of lower levels.[29] The dominating mechanism for systemic regulation of blood pressure is renal-pressure natriuresis, which controls the set point at which the blood pressure is regulated. In addition, blood pressure is also affected by aging and is significantly influenced by both environmental and heritable components. However, the incidence of hypertension varies widely among different geographic and ethnic origins and it has been estimated that 20 to 30% of interindividual variations systolic blood pressure could be attributable to heritable polygenes.[30,31] However, despite the clear genetic component of blood pressure, essential hypertension shows no clear pattern of inheritance.[32]

A leading explanation of the emergence of hypertension in modern humans is the "sodium hypothesis". Salt regulation is a key component of blood-pressure homeostasis and therefore, variable sodium sensitivity could explain the prevalence of hypertension in different human populations. The sodium hypothesis states that when humans first appeared in the hot, dry savannah they adapted to this environment by conserving salt. This adaptation increased their chances of survival in environments where salt was scarce and limiting water loss was advantageous to avoid dehydration. But as humans migrated to more temperate climates, this adaptation became less important. As a result sodium-conserving alleles show strong latitudinal gradients in allele frequency and are more common in Africans than in North Europeans. Many North Europeans retain the sodium-conserving alleles, although others, under positive selection, driven perhaps in large part by climate, have acquired a derived allele, the G(-6)M235 variant in the promoter region of the angiotensinogen gene. In addition to angiotensinogen other genes such as CYP3A5, GNB3, ADRB2 (β2-adrenergic receptor) and SCCN1A also show allelic variation with climate and latitude.[21] The ancestral variants are more prevalent in African populations living near the equator than in those populations that live in northern latitudes. Consequently, rates of hypertension and sodium sensitivity are higher in individuals that carry the ancestral allele in the modern environment (*mismatch hypothesis*) and, therefore, individuals from hot arid climates are more susceptible to hypertension than populations from cold climates.[29] This is borne out by the observation that African-Americans are at a greater risk of hypertension than Americans of European descent.[33] In contrast, in migrant populations that have moved out of Africa a long time ago, acquisition of a derived allele mitigates the harmful effects of the maladapted phenotype (Figure 1.3; B to C). Thus, changes in the environment render an adaptive allele maladaptive if the components of the new environment interact unfavorably with the ancestral trait (Figure 1.3; A to B). Precisely how this happens in not well understood, however, it is clear that whether an allelic variant of a gene results in an adaptive or a maladaptive phenotype depends entirely upon the environment. Hence, the outcome of a specific genetic makeup is nondeterministic because without relevant environmental triggers the presence of a specific genotype does not necessarily or inevitably cause CVD. A similar predominant role of the environment in modulating the outcome of genetic adaptation could be demonstrated for other processes that contribute to cardiovascular disease such as blood coagulation, inflammation, diabetes and obesity.[21]

These examples demonstrate that the genes (apoE, angiotensinogen, PPARG) that regulate the major CVD risk factors such as hypertension, hypercholesterolemia, diabetes change with the environment. While initially maladapted due to a time lag between the change in environment and natural selection, these traits are under positive selection and there is evidence to support the notion that the derived alleles are more compatible or adapted to the new environment. This is not what would be expected if the major variants of CVD susceptibility genes arose by genetic drift and were not subjected to natural selection or purification. If CVD appears late in life and does not affect reproductive success, why are CVD susceptibility genes under positive selection?

One answer is that genes that regulate the major CVD risk factors such as cholesterol, hypertension and coagulation have pleiotropic effects on the general well being of the individual and are therefore likely to be under significant selection pressure. Because the genes involved in cardiovascular function and health affect the general well-being of an individual, any dysfunctional or maladapted variant is removed readily by negative selection or purification and the derived allele increases survival and reproductive success by providing better cardiovascular adaptation to the environment. For instance, as pointed out earlier, the derived ε3 allele of apoE, in addition to decreasing CVD risk, could increase reproductive fecundity, adaptation to dietary changes, facilitate recovery from head injury or decrease susceptibility to lipophylic pathogens.[34] Thus, the derived allele is selected because of its positive pleiotropy. Similarly, the acquisition of the derived angiotensinogen allele by migrants moving out of Africa has been found to be associated with strong positive selection[29] although the environmental factors that favored the selection of the derived allele or selected against the ancestral allele are not known. Regardless, the genetic traits that favor cardiovascular health are favored in all environments and appear to be under strong selective pressure, even though their direct contribution to reproductive success is uncertain. In contrast, the contribution of genetic drift leading to the accumulation of chance variants (rare-variant common disease hypothesis) is less clear.

Genome-wide linkage analyses have shown that there are no loci for coronary artery disease risk, however, there are some mappable loci with modest effects.[3] Significantly, most of the genes that show weak associations are involved in innate and adaptive immunity, consistent with the notion that inflammation is a critical component of cardiovascular disease. Because a highly active immune response is critical for survival and therefore for reproductive success, it has been suggested that the frequency of these genes is maintained by natural selection even though they increase disease susceptibility in older age (*pyrrhic hypothesis*). However, as we have seen, genes that lower CVD risk in old age (cholesterol and hypertension) are under positive selective pressure. In addition, their contribution to CVD is significantly modified by the environment both before and after the reproductive years. For instance, by studying historical data from cohorts born before the 20th century in European countries, Crimmins and Finch[35] have found that increasing longevity and declining mortality in the elderly occurred among the same birth cohorts that experienced a reduction in mortality at younger age. This is consistent with earlier observations that when life expectancy increased, the increases in the elderly began many decades after the increases in younger ages,[36] indicating that being healthy and disease free at a young age delays the onset of age-associated disease. Thus, what happens before or during the reproductive years does not appear to be irrelevant to aging. That healthy children and adults make healthy seniors is a well-understood euphemism – and for good reason; cardiovascular disease begins early (by some estimates in the preteen years) and therefore environmental and genetic factors that promote good cardiovascular health during childhood and adulthood are likely to decrease the CVD burden in old age. Thus, improvements in the environment, better nutrition, lower infection *etc.*, that improve the CVD health in youth also improve health in old age.

Crimmins and Finch[35] also found that the decline in old-age mortality in their 19th century European cohort was promoted by the reduced burden of infections and inflammation during childhood. They hypothesized that reduced infections at young ages delayed the development of atherosclerotic and thrombotic conditions by reducing the lifetime inflammatory burden. This is consistent with the current view of atherogenesis that holds that lesion formation begins early in life and that an increase in systemic inflammation, due to repeated infections, could accelerate the rate of atherogenesis either temporarily or permanently. Several studies show that by chronically elevating the levels of inflammation, persistent infections could increase the risk of atherosclerotic disease.[36] Thus, it seems reasonable to assume that decreasing inflammation early in life could decrease CVD progression and severity. If this is true it would suggest that CVD health before and during the reproductive age could not be optimized to the detriment of health at old age. Atherosclerotic disease is a lifelong process, and its total burden is the record of the entire environmental life history of an individual and it is likely that few, if any, mechanisms that are beneficial only during youth are detrimental in old age.

The recognition that the environment plays a predominant role in modulating genetic disposition for CVD has important practical implications. If we could identify the specific environmental triggers and understand how they interact with specific genetic variants it might be possible to prevent much of the disease by altering the environment or modifying its effects on individual with genotypes. For instance, as pointed out by Willet,[37] although phenylketouria is an entirely genetic disease, it could be completely avoided by eliminating phenylalanine from the diet. So, from another perspective it could be viewed as an entirely environmental disease. This perspective is useful, because it can suggest that simple modifications in the environment could significantly impact the outcome of CVD. Further gains can be made by understanding genetic susceptibilities and gene–environment interactions not only within the context of the current environment but also with the understanding of the evolutionary history of how specific genetic adaptations arose and how they modify CVD risk in the current environment. Historically, medical research has focused on mechanistic or proximate causes of disease, but distal evolutionary causes that determine disease susceptibility (or even normal physiology for that matter) are overlooked. Ideally, a complete explanation must be based on a thorough understanding of both the proximate and distal causes. While much has been learned about the mechanistic, cellular and molecular mechanisms, the importance of environmental influences has been underestimated. This is particularly unfortunate because the environment strongly influences both the proximal and distal causes of disease and it is the link and the context within which to understand both the long-term evolutionary causation and the immediate precipitation of disease in a genetically unique individual. In this regard, heart disease (inclusive of metabolic disease such as diabetes and obesity) is a quintessential environmental disease. Its long-term risk is embedded in the evolutionary history of responses to the ancient environment and its current population and individual risks are largely determined by the modern environment.

1.2 Categories of the Human Environment

All life adapts to its environment. For most animals and plants, the environment is primarily the ecosystem populated by natural geographic features and life forms that coinhabit the niche; however, the human environment is more complex. Given this complexity, how can we understand the effects of the environment on humans? What specific constituents or aspects of the human environment influence health and disease? Do they work independently or synergistically? What types of environmental factors are modifiable? Which ones are nonmodifiable? Which parts of the environment affect human health? And are these effects direct or indirect? To address these questions, it is important to understand specific aspects of the human environment and how they affect individual health, disease-risk and mortality. The major difficulty in understanding environmental influences on human health is that the term "environment" is used currently as a catch-all phrase. It is used to describe all physical, social and cultural surroundings of an individual. In this sense it refers to a host of disparate entities that may be the local climate, food sources or social and economic conditions. Each of these referents, however, has different and unique effects on human health and grouping them together causes these differences to blur. Hence, to delineate the contribution of the human "environment", it is necessary to differentiate among different forms and types of environments.

The term "environment" is derived from the French verb (*en- viron*, circuit) meaning to surround or enclose. If is defined as the set of circumstances or conditions in which a person or community lives, works, and develops. It includes all surroundings, the totality of circumstances and the complex of social and cultural conditions that affect the *nature* of an individual or the society. In its broader sense, it could mean anything that is external to an individual. However, to understand the role of the environment with any degree of specificity we must distinguish between types of environment and its categorically differentiated forms.

From the perspective of an individual, the total environment consists of plastic and aplastic components (Figure 1.4). The aplastic or the nonmodifiable environment is the natural environment that we cannot change significantly. It is the relatively nonmalleable ecosystem that we share will all other living things. This includes the day–night cycle, the season, and our terrestrial rather than aquatic existence – a group of *a priori* conditions that remain relatively constant. There are changes in the natural environment (the seasons, the length of days and nights change), but these are relatively unchangeable by human activity.

In contrast to the natural environment, the *plastic* environment is changeable by human activity. The creation of a plastic environment is the collective work of the community fashioned by its history and culture. The main program of human civilization is to mold the natural environment so as to enhance human safety, comfort, and convenience. However, this self-created, plastic environment completely surrounds and engulfs human lives and it has now become, more so than in the past, the primary domain of our existence. Not only does this environment substitute for our natural ecological environment, it also shields us from nature and it radically modifies our interactions with the natural world. Therefore, to evaluate how the environment affects human disease and health, it is essential to understand not only our interactions with the natural environment, but more importantly, how we interact with the plastic, oysterous environment that we have created around us and how this environment affects our well-being.

Components of the current plastic environment (community, cars, buildings, roads, pollution, *etc.*) have a more profound effect on human health than natural forces or ecological threats (floods, infections, *etc.*). Indeed, the plastic environment has radically changed the natural environment and the current ecosystem itself. While in the past, human gene–environment interactions were primarily driven by changes in the natural environment, ecological changes and geological shifts, the challenge now is to adapt to the ever-changing plastic, man-made environment rather than our natural ecosystem. It is becoming increasing clear that the plastic environment has been a significant agent of natural selection in the past and is likely to be a predominant force in setting our future genetic trajectories.

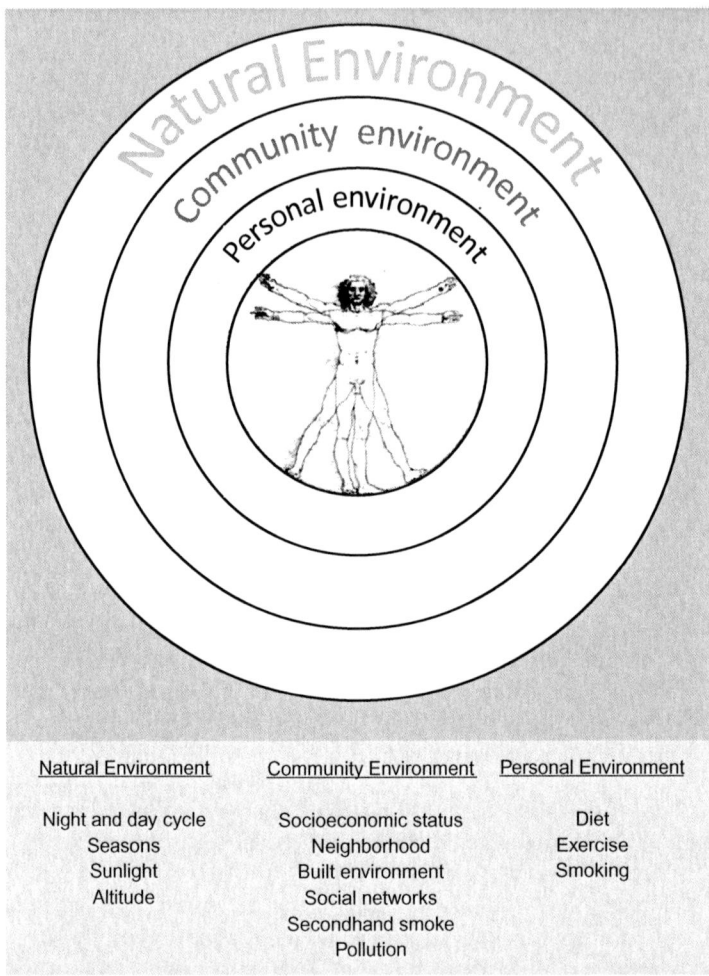

Figure 1.4 Categories of the human environment: Mandala showing the relationship between an individual and the environment. Components of each environment that regulates the risk for cardiovascular disease are listed.

The plastic environment could be further differentiated into a community or societal environment, which is the environment we create as a community and a personal environment that is made up of our own conscious (life-style) and unconscious choices. The community environment includes the social and cultural structures that each generation (like its genes) inherits from their predecessors and contributes to in return. It is the result of the choices that we as a community have made and continue to make. It includes the built environment in which we live (houses, roads, parks) and the environmental conditions (access to healthy food, clean drinking water, *etc.*) that the community (neighborhood, city, country) provides. Recent research has shown that components of the community and built environment have a significant impact on human health and disease, particularly on chronic illnesses such as heart disease and diabetes. Other physical aspects of the community environment such as noise, pollution, food availability, sanitation, *etc.*, also have important health consequences.

The final component of the environment is the personal environment. It is a subset of the plastic environment, but it is not entirely communal. It is populated by the choices that we as individuals make, such as where we live, what we eat, and how we spend our leisure time. These life-style choices have the largest and perhaps the most significant effect on our health. Even though the personal environment is largely created by our own unique choices, these choices are to some extent determined and limited by the community environment. We can only chose from the set of options provided to us by our community. Our freedom to choose is constrained by our community (*e.g.*, peer pressure, fashion, advertisement). Still other choices that we make are unconscious because we lack the information to make the right choices or our choices are forced upon us by the community as a matter of tradition, civic laws, *etc.* Regardless, the personal environment is the most malleable, (because it could be changed by one person) and central to our understanding of the effects of other environments (the natural environment and the community environment) and how they are transmitted to a unique individual.

Despite a clear demarcation between the different environmental categories, their boundaries overlap. Human activity has significantly changed the natural environment and it continues to do so. Although, the effects on human activities such as deforestation, pollution of air and water, deep-sea fishing, *etc.*, are abundantly evident, there is vigorous debate over the extent to which human activity has changed other major aspects of the natural environment such as global climate or weather patterns. In addition, the boundaries between the communal and the personal environment are continually shifting and in most modern societies there are persistent and opposing attempts to enlarge the influence of one over the other.

1.3 Cardiovascular Disease and the Natural Environment

Recent research provides important insights into how each of the categorically differentiated aspects of the environment affects cardiovascular health. The most ancient of these is our natural environment. In common with all other plants and animals, early human adaptation was in response to the natural environment and the ecosystems in which humans evolved. As we have seen, several genes that regulate cardiovascular function, for instance, the angiotensinogen or the apoE gene variants were selected by the conditions in the African savanna. When humans migrated to different part of the world, different variants of these genes appeared in response to positive selection or random genetic drift. However, as human societies became civilized, the direct impact of the natural ecosystem on human health diminished and natural selection favored adaptation to the new urban environment (*e.g.*, retention of the lactase gene;

acquisition of ε4 variant of apoE). It is likely that in the future, the majority of our genetic variation will come from responses to the artificial, plastic environment created by urbanization. Indeed, we are already beginning to see profound effects of urbanization on human health. Nevertheless, the effects of the natural environment cannot be overlooked. We still carry the genes that are adapted to our ancient ecosystem and a disruption of the synchrony between our genes and environment due to differences between the ancient and modern environments is a significant cause of cardiovascular dysfunction and disease.

1.3.1 Cycles of Night and Day

An important aspect of the natural environment that affects cardiovascular function is the day–night cycle. The day–night cycle is a fundamental, aplastic feature of the natural environment. All life is entrained to this cycle, which in turn exerts a pervasive control over both plant and animal physiology. Most cells have circadian clock genes that maintain an endogenous 24-h cycle. In the presence of environmental cues (zeitgebers) the master clock, located in mammals in the pacemaker neurons of the suprachiasmatic nucleus (SCN), is entrained to a diurnal cycle.[38] In humans, the SCN sets the intrinsic 24-h cycle accurately to an average of 24 h and 11 min.[39] Light is the main zeitgeber, it regulates the master clock that synchronizes the light-insensitive peripheral clocks to coordinate a 24-h cycle of waking, sleeping, feeding, *etc*. Before the discovery of artificial light, human lives, like the lives of other animals, were synchronized to the cycles of night and day. Extensive research shows that this rhythmic cycle is essential for normal physiology, health, organ growth and tissue renewal and that disruption of this cycle by the plastic environment is a significant CVD risk factor.

Cardiovascular function, as reflected by heart rate and blood pressure, changes rhythmically in synchrony with the day–night cycle. It is lowest at night and during sleep and it begins to rise before waking up, coinciding with a period of vagal dominance, in anticipation of daytime activities. Cardiovascular genes and proteins undergo similar rhythmic changes. It has been estimated that 13% of cardiac genes are under the rhythmic control of the 24-h diurnal cycle.[40] Diurnal variations in gene cycling have also been reported in vascular tissues such as the aorta.[41] Moreover, the intrinsic clock genes are regulated by changes in cell redox, particularly the changes in NAD^+ levels[42] that accompany fluctuations in cell metabolism Changes in NAD^+ levels are significant, because in addition to regulating energy metabolism NAD^+ is also an essential cofactor for the deacetylase Sirt1 (the molecular target of the lifespan enhancing ingredient of red wine – resveratrol[43]). Although the role of Sirt1 in circadian rhythms is not clear, genome-wide acetylation exhibits time-of-day oscillation.[44] Given the recent findings that a large number of the enzymes involved in glycolysis and the TCA cycle undergo acetylation,[45,46] it is likely that diurnal variations in metabolism may be linked to cycles of protein acetylation–deacteylation reactions.

Metabolic processes such as cell growth and tissue repair also oscillate in phase with the day–night cycle. Myocardial proteins are synthesized at the highest rate late in the sleeping period and cardiovascular growth and renewal occurs during the sleeping hours. In addition, neurohormones that regulate cardiovascular function, such as angiotensin II, rennin, aldosterone, growth hormone and atrial naturetic peptide show diurnal variations.[38] Interestingly, it has been shown that rat hearts isolated during their subjective day (dark phase) contract better than those isolated during their subjective night,[47] indicating that the time of the day may be an important regulator of cardiac performance.

In agreement with diurnal variations in cardiovascular metabolism, function and regulation, the incidence of adverse cardiovascular events varies with the time of the day. Myocardial infarctions are most frequent between 6 AM to 12 PM, with most occurring between 3 to 6 AM.[48,49] These events are three times more likely to occur early in the morning than late at night. The frequency of strokes, arrhythmias, abdominal aortic aneurysm rupture and sudden cardiac death also shows matutinal clustering between 8 to 11 AM.[50–52] The timing of the onset of adverse cardiovascular events has been linked directly to the intrinsic clock mechanism and does not appear to be related to the stress of waking up. When in a new geographic location, the frequency of cardiovascular events in travelers peaks, for a few days, at times that correspond to their time zone of origin.[38]

In view of the tight rhythmic control of cardiovascular function by the day–night cycle and the clock genes, it is not surprising that disruption of this synchrony has devastating effects on cardiovascular health and that a failure to harmonize internal and external rhythms increases CVD risk. A reflection of this failure is the reported increase in cardiovascular morbidity and mortality in shift workers who are subjected to frequent disturbances in their sleep–wake cycle. Many studies show that shift workers,[53,54] transmeridian flight crews, patients with sleep apnea and other sleep disturbances[38] have higher rates of diabetes, obesity and adverse cardiovascular events. Also, a modest increase in the risk of stroke in women after extended periods of rotating night-shift work has been recently reported.[55] Data showing greater mortality in mice subjected to phase advances of the light–dark cycle, simulating chronic jet lag,[56] provide further support to the view that a mismatch between external and internal rhythms adversely affects health and longevity. In humans, short-term circadian misalignment, similar to that which occurs with jet lag or shift work, results in an increase in postprandial levels of blood glucose and insulin and the mean arterial pressure with a systemic decrease in leptin.[57] These changes may be responsible for the increase in the risk of obesity and diabetes,[58,59] and hypertension[60] in shift workers.

In addition to increasing CVD susceptibility, disruption of the day–night cycle also exacerbates cardiovascular disease. Myocardial infarcts that occur in the middle of the night are larger[61] and angioplasties performed at night are less successful.[62] Animal studies show that the day–night rhythm disturbance increases pressure overload-induced myocardial dysfunction.[38] Similarly, mice subjected to myocardial ischemia-reperfusion at the sleep-to-wake transition exhibit a dramatic increase in infarct size compared with those at the wake-to-sleep transition,[63] indicating that an inappropriately synchronized wake–sleep schedule may be an important environmental determinant modulating CVD severity. Collectively, these findings suggest "that maintaining normal diurnal body physiology, treating underlying sleep disorders, and/or restoring the endogenous neuroendocrine hormonal profiles, perhaps by imposing a fixed or regular schedule of zeitgebers such as light/dark, rest/activity, or the timing or meals, may significantly benefit cardiovascular health".[38] Thus, recent work in chronobiology reveals an intricate link between a central feature of our environment and cardiovascular function and disease. It shows us that we are inextricably linked to our natural environment and exquisitely attenuated to its primordial rhythms. The synchrony between our endogenous circadian rhythms and the exogenous day–night cycle is of fundamental importance for the normal cardiovascular growth and function, and even seemingly minor disruptions of this primary link have devastating effects on cardiovascular health.

1.3.2 Four Seasons

An additional invariant feature of the natural environment is the constant changing of seasons. In most geographical locations there are wide variations in temperature and humidity. The changing of seasons also brings with it changes in the length of day. These changes alter human activity, feeding behavior and the duration of exposure to sunlight. As a result, there are profound variations in cardiovascular health and disease susceptibility. By modifying physiological responses and basic metabolism, seasonal variations affect the expression of CVD phenotype and recovery from adverse cardiovascular events. Although the underlying mechanisms remain mostly unknown, a large number of studies demonstrate that cardiovascular risk factors as well as adverse cardiovascular events show pronounced seasonal variations.

Cardiovascular risk factors – hypercholesterolemia, hypertension, thrombosis, and insulin resistance – show consistent seasonal variations. Cyclical seasonal variations in the circulating levels of cholesterol have been known for the last 80 years.[64] Most studies show that cholesterol levels are higher in winter than in summer. Statistically significant sinusoidal seasonal cycles have been observed in many geographic locations, independent of age, gender, ethnicity, and baseline lipid levels. In general, 3–5% increase has been reported in total cholesterol[65] as well as LDL cholesterol.[66,67] Some studies report that HDL cholesterol follows an inverse pattern; with a peak value in late summer and the lowest value in late winter. On average, the overall difference is 16%,[66] although a decrease in HDL levels in summer has been reported also.[67,68] Nevertheless, plasma low- and high-density lipoprotein cholesterol levels, analyzed similarly, showed synchronous seasonal cycles.[67] In addition, more patients with acute coronary syndromes on statin therapy achieve their target cholesterol level in summer than in winter,[69] indicating that cholesterol synthesis varies with season or that the efficacy of drug treatment is under seasonal control. Mechanisms underlying seasonal cholesterol cycles remain obscure, although they appear to be relatively independent of changes in ambient temperature, diet or physical activity.[67,70]

Circulating levels of fibrinogen also display cyclical seasonal variation. Like cholesterol, the fibrinogen levels appear to be the highest during winter,[71–73] although some studies have reported peaks in summer[74] or no association at all.[75] Seasonal variations have also been observed in the plasma levels of tissue plasminogen activator antigen and von Willebrand factor.[76] The average seasonal change is between winter and summer months is 10 to 30% or 0.13 to 0.32 g/L.[64] Some investigators have attributed the increase in fibrinogen to concurrent upper respiratory infection especially in old individuals.[72] Fibrinogen is an acute phase protein, which increases with infection. In agreement with this view, a strong association between fibrinogen and other markers of inflammation was observed.[72] However, changes in fibrinogen have also been observed in younger cohorts[66,76] without signs of concurrent infection[66] and therefore do not appear to be always associated with an acute phase response. Such seasonal variations in components of the blood coagulation pathways indicate that the likelihood of adverse cardiovascular events would be higher in winter than in summer months. Indeed, several studies on the seasonal variation in cardiac events show that the frequency of cardiovascular mortality is much higher in winter than in summer.

A significant increase in cardiovascular mortality in winter has been reported by several investigators from all geographic locations both north[77,78] and south[79,80] of the equator. Most of the excessive deaths in the winter months are due to ischemic heart disease although a marked increase in heart-failure deaths has also been reported.[80,81] The difference between the winter peak and summer trough is large. It has been estimated that in England and Wales the winter peak accounts for 20 000 additional deaths per year.[82] Analysis of the 259 000 cases of acute myocardial infarction in 1474 hospitals across the US showed that 53% more cases are reported in winter than in summer.[83] In the entire year, the month of January was the most lethal. In the Australian MONICA study,[84] both fatal and nonfatal coronary events were 20–40% higher in winter than in summer and a 17% seasonal variation was observed in the German Dessau Registry.[85] Data from Los Angeles show 33% more deaths occur in December–January than June through September.[86] These studies suggest that there is a large increase in CVD deaths during the winter months and that this increase could not be attributed to a higher case fatality rate, but that it reflects an authentic increase in the incidence of acute myocardial infarctions. A similar rhythmic seasonal pattern, with a peak in winter, has also been observed for cases of nontraumatic rupture and dissection of aortic aneurysms[87] and stroke.[88]

Several factors can account for excessive CVD mortality in winter. It may be that much of this mortality could be ascribed to susceptible elderly patients with pre-existing disease. Some studies have reported that the elderly are more susceptible to increased winter mortality.[88] This may be because during winter months they are vulnerable to respiratory infections, which trigger an acute phase response leading to exacerbation of cardiovascular disease. However, excessive mortality has been observed in all ages (<55 to 74 years) at levels comparable to the aged (>75 years),[83] suggesting that the aged are not more vulnerable to excessive winter mortality than the young. Infections, changes in activity levels or diet, however, can only account for part of the excessive mortality associated with winter months, indicating that there may be other explanations for the seasonal pattern of CVD mortality. Although no hard data are available to support any one mechanism, it has been speculated that hemodynamic effects of cold exposure (an increase in sympathetic activity, blood pressure, arterial spasms) could destabilize a vulnerable lesion leading to plaque rupture and occlusive thrombosis.

Cold temperatures, independent of the season, could be an important contributing factor because an excessive number of infarctions has been observed on colder days both in winter and in summer.[89] Exposure to cold temperatures increases vascular resistance and blood pressure, leading to an increase in oxygen demand.[90,91] Cooling of the body is associated with activation of the sympathetic nervous system leading to peripheral vasoconstriction and a decrease in blood flow at rest: a 1 °C decrease in room temperature is associated with a 1.3 mm Hg increase in systolic pressure and 0.6 mm Hg increase in diastolic blood pressure.[92] In the Framingham Offspring Cohort, ambient temperature was found to be a strong determinant of microvascular vasodilation function as measured by hyperemic flow.[93] Low temperature can also increase coronary artery resistance and in some cases induce coronary vasospasm. Changes in temperature can also affect hematologic properties such as blood viscosity and coagulation and even mild surface cooling increases the hematocrit and platelet counts, thereby increasing the likelihood of spontaneous thrombosis.[94] Nevertheless, cold temperatures may not be the only important factor. In some studies, particularly those from the Southern United States suggest that there is an increase in cardiac deaths in summer,[95] and excessive mortality during winter months has been reported even in areas where the temperature is mild throughout the year (*e.g.*, Los Angeles[86]). These studies suggest that changes in the season have a more pervasive effect on cardiovascular function independent of the effects of a change in temperature.

Plasma lipids and fibrinogen levels show seasonal variability and flow-mediated dilation of the brachial artery is the lowest in winter.[93] Together, or by themselves, these changes can trigger plaque rupture. However, the observation that heart-failure deaths are also increased in winter suggests that in addition to plaque rupture, increases in arrhythmia susceptibility, blood pressure or changes in myocardial metabolism *per se* may be important contributors to seasonal clustering of CVD mortality. Direct seasonal variation in cardiac physiology is consistent with the work of Scherlag and coworkers who report that the incidence of sudden cardiac deaths from arrhythmias in dogs subjected to coronary ligation was much higher between November and February than between July and August (42% versus 6%).[96] While these data provide clearer evidence for seasonal variation than the human data (which might be affected by other, social confounders), it remains unclear whether the excessive sudden deaths were due an increase in the sympathetic tone or due to seasonal changes in myocardial metabolism and excitability. Animal studies also show seasonal variations in cholesterol levels. For example, European badgers, under experimental conditions in which diet was held constant and seasonal weight gains were minimal, show large spontaneous changes in blood cholesterol levels.[97] Whether other animals show similar variations in the levels of cholesterol and other plasma constituents remains unknown, but it is tempting to speculate that because cholesterol is needed to repair injured tissue, and lipoproteins decrease endotoxin injury, the seasonal increase in cholesterol may be an evolutionary adaptation in anticipation of an increase in microbial infections.

1.3.3 I'll Follow the Sun

While mechanisms underlying seasonal clustering of CVD mortality remain unclear, a particularly attractive hypothesis is that the increase in cardiovascular mortality during winter may be due to low vitamin D levels. The major source of vitamin D for humans is exposure to sunlight. Diet accounts for a small percentage because only few natural foods contain vitamin D. The photosynthesis of vitamin D involves the conversion of 7 dehydrocholesterol in the epidermis by solar UVB (290–315 nm) radiation to previtamin D3, which then undergoes thermal isomerization to vitamin D3.[98] Vitamin D3 formed in the skin appears in the circulation and it is then transported to the liver where it is converted to 25(OH)D3 – the major index of total vitamin D3 stores. In kidney, 25(OH) D3 undergoes additional hydroxylation to form the biologically active 1,25(OH)2D. Excessive sunlight exposure cannot cause vitamin D toxicity because UVB converts excess vitamin D3 to biologically inert isomers.[99] The efficiency of vitamin D synthesis depends upon the number of photons that penetrate the endothelium. Melanin pigmentation of the skin retards UVB penetration and therefore it decreases sunlight-induced vitamin D synthesis.[100] When exposed to the same amount of sunlight, 20–30% of UVB radiation is transmitted through the epidermis of white skin, whereas in heavily pigmented skin the penetration is $<5\%$. As a results, individuals with darker skin require a much longer time to synthesize the same amount of vitamin D than those with white skin.[101]

The efficiency of vitamin D synthesis depends upon the extent of exposure to UVB radiation. UVB radiation reaching the earth's surface changes with changing zenith angles. When the sun is low in the sky (during winter or during early morning and late evening) incoming radiation has to travel longer and is subject to more scattering and absorption than when the sun is directly overhead. Consequently, the ability of synthesize vitamin D is affected by the time of the day, the season and the latitude. In northern latitudes (*e.g.*, Boston, 42°N), the filtering effect due to an increase in the zenith angle of sun in winter is sufficient to completely prevent vitamin D3 synthesis from November to February and in Edmonton (10° north of Boston) no vitamin D3 could be synthesized from October to April.[102] Thus, residents in northern latitudes are likely to face severe vitamin D deficiency in winter months. Indeed, it has been suggested that skin pigmentation in Northern Europeans was lost due to negative selection upon migration from Africa.[103] Hominids in the tropics were probably deeply pigmented; however, as they moved further north, the more deeply pigmented infants were less likely to survive due to bone malformation caused by vitamin D deficiency. As a result, the northern population lost pigmentation due to natural selection. That skin color is an adaptation to maximize UV penetration in northern latitudes (and minimize UV damage in south) is supported by a significant correlation between skin pigmentation and equatorial latitudes in human populations.[103] Despite this adaptation, residents of northern latitudes face constant vitamin D deficiency. Autopsy studies on 19th century residents of Boston, Leiden and The Netherlands show that there was 80–90% prevalence of rickets in children residing in these areas.[104] Even today, living at higher latitudes and being prone to vitamin D deficiency is associated with an increase in the risk of colon, prostate, breast and ovarian cancer, as well as an increased risk of multiple sclerosis, Crohn disease, type-1 diabetes and hypertension.[105] Remarkably, it has been reported that living above 35° latitude for the first 10 years of life was sufficient to imprint on a child a 100% increase in the risk of developing multiple sclerosis independent of where they lived in later life.[104]

Although residents in northern latitudes are particularly prone to vitamin D deficiency, those living in the south are susceptible as well. A seasonal decrease of vitamin D in winter has been reported both in the Northern and the Southern United States. It has been found that that there is 40% greater prevalence of vitamin D deficiency in fall and winter than in summer and spring and that the deficiency was higher in obese children.[106] In the NHANES III cohort, vitamin D deficiency was fairly frequent in younger individuals, especially in the winter/lower latitude subsample.[107] The serum 25-OHD levels in African-Americans were lower than whites, consistent with the higher efficacy of pigmented skin in preventing UVB absorption. Current estimates indicate that globally, 35–80% of children have vitamin D deficiency.[104] In the US, vitamin D deficiency has been found to be common in all groups of adolescents and adults in the winter/lower latitudes subpopulation[107] and 25(OH)D levels are inversely associated with the winter season.[108]

Seasonal and latitudinal variations in vitamin D levels have been associated with geographic and seasonal variations in blood pressure. With increasing distance from the equator, there is a progressive increase in blood pressure that correlates with a gradual fall in ambient UVB radiation.[109] The prevalence of hypertension shows a similar latitudinal distribution. Moreover, blood pressure is higher in winter,[90,109] when UVB levels are low and decreases in summer with the advent of sunnier days. Although it is not clear whether there is a causal relationship between blood pressure and sunlight, it has been reported that exposure to UVB radiation[110] skin tanning in salons[111] or treatment with high-dose vitamin D2[112] reduces blood pressure. In addition, both experimental and epidemiological studies indicate that vitamin D regulates rennin biosynthesis and blood-pressure homeostasis. Disruption of vitamin D signaling in mice activates the rennin–agiotensin system and induces hypertension and cardiac hypertrophy and in men with the low levels of vitamin D are associated with a 6-fold higher risk of developing incident hypertension.[113] Because vitamin D regulates calcium homeostasis and the secretion of the parathyroid hormone (PTH), chronic vitamin D deficiency causes secondary hyperparathyroidism, which has been linked to an increase in both blood pressure and myocardial contractility.

In addition to blood pressure, vitamin D regulates other cardiovascular functions as well. All cardiovascular tissues express the vitamin D receptor (VDR).[114] This receptor binds to 1,25(OH)2D and the ligand bound receptor, upon association with retinoic acid x-receptor regulates the expression of nearly 200 genes, such as those involved in rennin production, release of insulin by the pancreases, cytokine production by lymphocytes, and the growth of vascular smooth muscle cells and neonatal cardiomyocytes.[114] Overall, 3% of the human genome is regulated directly or indirectly by the vitamin D endocrine system. In mice, the absence of a functional VDR leads not only to a bone and growth plate phenotype, but also high rennin hypertension, cardiac hypertrophy, and increased thrombogenicity.[115] In humans, vitamin D deficiency is associated with an increased risk of hypertension, myocardial infarction, stroke, heart failure, diabetes and peripheral artery disease.[98,114] A correlation between plasma 25-OHD levels and subsequent adverse coronary events has also been observed in the Framingham Offspring Study, which reported that the CVD events were 53–80% higher in people with low vitamin D levels.[116] That these associations may be causal is supported by a recent meta-analysis of 18 randomized controlled trials consisting of 57 000 individuals, which showed that vitamin D intake (> 500 IU/day) decreases all-cause mortality in part by decreasing cardiovascular deaths.[117]

Vitamin D may be a particularly important regulatory factor in obesity and diabetes. Human pancreatic islet cells are capable of calcitriol production[98] and hypovitaminosis D is considered a risk factor for glucose intolerance.[106] Vitamin D status is positively correlated with insulin sensitivity index and individuals with low vitamin D levels display impaired β-cell function and are at a greater risk of developing insulin resistance, type-1 and type-2 diabetes and metabolic syndrome.[98] Supplementation with vitamin D improves insulin resistance, and in one study vitamin D treatment was as effective as troglitazone or metformin in improving insulin sensitivity.[118] Vitamin D is a fat-soluble vitamin and it is readily sequestered in fat, therefore, the bioavailability of vitamin D is obese people is decreased in comparison with nonobese people[108] and in most studies obesity is negatively correlated with plasma 25(OH)D levels and positively correlated with PTH levels.[106] Thus, the obese are likely to be more susceptible to seasonal and latitudinal changes in vitamin D levels. Similarly, vitamin D deficiency in the US is more prevalent in African-Americans than in whites, which may account in part of the high CVD burden in Southern United States, despite lower latitudes and plenty of sunshine.

In Europe, there is a progressive increase in the rates of heart disease from southern to northern Europe. A similar south-to-north gradient is also evident in the distribution of the ancestral ε4 haplotype of apoE (see above). The proportion of apoE ε4 carriers rises from 10–15% in the south to 40–50% in the north. The retention of the apoE ε4 allele in Northern Europeans has been suggested to be related to better intestinal absorption of fats and a better absorption of vitamin D in the kidneys.[119] Both apoE and vitamin D binding proteins share a common receptor in the proximal tubules. Hence, carriers of the ε4 allele may have been less likely to develop vitamin D deficiency and better fit for living in northern latitudes where the sunlight contains low levels of UVB radiation.

1.3.4 In High Places

Altitude is another dimension of the natural environment that affects cardiovascular health and disease. Nearly 150 million people live in areas that are more than 2500 m (8200 feet) above sea levels.[120] Populations living at high altitude have adapted to cold temperatures and low oxygen levels. There are significant anatomical, physiological, metabolic and biochemical differences in the cardiovascular system and functioning of highlanders and sea-level dwellers. For instance, it has been reported that although right ventricular hypertrophy at birth decreases promptly in newborns at sea levels, it persists throughout life in Andean children living at 4540 m.[120] Moreover, the level of altitude has an inverse relationship to arterial oxygen saturation and a direct relationship to pulmonary artery pressure. As a result, some human and animal populations living at high altitudes have right ventricular hypertrophy and thick pulmonary arteries. Alveolar hypoventilation in susceptible highlanders leads to chronic mountain sickness.[121] It is characterized by excessive erythrocytosis, severe hypoxemia, and pulmonary hypertension, which often evolves to cor pulmonale, leading to congestive heart failure. CMS prevalence varies between 5–8% in various populations of highlanders and it increases in association with lung disease, obesity, smoking and environmental pollution.[121]

Of the several highland populations, native Tibetans and Nepalese sherpas appear to be most well-adapted to living at high altitudes. Tibetans have the oldest altitude ancestry in the world and have through successive generations attained a high grade of adaptation to high altitudes; perhaps by natural selection. The prevalence of chronic mountain sickness in Tibet is low (1.2%).[120] In contrast with sea-level dwellers (Chinese Han immigrants, or Caucasians), Tibetans show lower pulmonary pressure response to exercise with less increases in ventilation rates and better preservation of cardiac output. They have greater ventilatory capacity and hypoxic ventrilatory response as well as greater physical performance. Interestingly, Andean natives who are have a shorter history of living at high altitudes than Tibetans are less well adapted. Autopsies of Andeans frequently show greater muscularlization of the distal pulmonary arterial branches and right ventricular hypertrophy.[122] A similar difference is evident in animals. Species native to mountainous areas (yaks, snow pigs, pika) have better cardiopulmonary responses than domestic animals recently transported to high altitudes.[120]

Living in areas of high elevation seems to affect CVD risk as well as CVD mortality. In several small studies it has been found that individuals living at high altitudes (1000 to 5500 m) have significantly lower total cholesterol and/or LDL-cholesterol[123-125] and higher HDL levels.[124,126] Also, serum leptin levels are negatively correlation with altitude.[127] Although in human studies it is difficult to account for cultural, socioeconomic, and physiological (*e.g.*, changes in hematocrit) confounders, a decrease in arterial accumulation of cholesterol in rabbits born and raised at high altitude[128] suggests that high altitude has a favorable effect on cholesterol metabolism. Even brief sojourns for a few days or weeks at high altitudes could lead to favorable changes in blood lipids[129] decrease insulin resistance,[130] and stimulate lipolysis of plasma triglycerides.[131] However, acute exposure to high altitudes can lead to potent illness such as high-altitude pulmonary edema and subacute mountain sickness.[121] This is often seen in nonadapted lowland dwellers such as Han Chinese children who move to high altitude or soldiers engaged in exercise at high altitudes (5800–6700 m).[121] In nonadapted individuals, high altitude hypoxia increases heart rate and cardiac output. Epicardial coronary arteries dilate to compensate for the reduced oxygen content of the blood. There is significant pulmonary vasoconstriction leading to pulmonary hypertension. Subacute mountain sickness results from chronic exposure to high altitudes and is a symptom of right heart failure due to pulmonary hypertension and enlargement of the right ventricles. However, many of the symptoms are resolved upon returning to lowlands.[121] On the other hand, prolonged residence at high altitudes may be beneficial. Tibetan highlanders rarely exhibit systolic hypertension and they have lower levels of serum cholesterol and apoB.[132] Data on CV mortality are scarce and contradictory. Nevertheless, a recent study, which compared ethnically and culturally similar German Swiss residents living in Switzerland, reported an almost continuous decrease in coronary heart disease and stroke mortality with increasing altitude (259 to 1960 m).[133] The mortality risk for CHD was decreased by 22% with an increase of 1000 m in the altitude of the place of residence. Similarly, total and coronary mortality were found to be lower in a mountainous village in Greece than in lowland villages.[134] Remarkably, the hazard ratios for coronary mortality in highland dwelling men and women were found to be 0.39 and 0.46, indicating that residence in mountainous areas protects against heart disease.

Several reasons could account for lower cardiovascular mortality in highlanders. With every 300 m increase in altitude, the UV levels increase by 10%[135] and significantly higher levels of UV radiation have been measured at higher than at lower altitudes in Switzerland.[135] Moreover, vitamin D synthesis is increased at high altitudes[136] In addition, several factors such as changes in diet, exercise, air pollution could account for higher mortality in lowlanders. However, stable differences in the natural environment such as sunlight and climate are likely to be important determinants of CVD mortality because in the Swiss study[133] being born at high altitude had an independent beneficial effect on coronary heart disease. People who were born at a higher residence and moved down in altitude had a lower risk than those who lived at low altitudes their entire life. Thus, the impact of the environment appears to be most prominent during intrauterine development or during early adulthood.

When taken together, the studies showing that the prevalence and the severity of heart disease are affected by geographic and seasonal variables provide strong support to the notion that CVD is significantly affected by the salient features of the natural environment such as sunlight, night and day cycles, changes in seasons, and the geographic location. Human metabolism and physiology are adapted to the constantly changing rhythms and the immutable features of the natural environment. Therefore, a mismatch between the original adaptation and the current environment may be one contributing factor in the development of cardiometabolic diseases. As we have seen, movement to new geographic locations results in the acquisition of new characteristics (*e.g.*, different apoE alleles, changes in skin pigmentation, *etc.*). These adaptations provide temporary harmony with the environment. Nevertheless, with additional migrations they are tested in new environmental contexts. With increasing urbanization the context has shifted dramatically from the natural environment to the plastic urban environment and the community neighborhoods. The influence of this "new" community environment is no less pervasive than the ecological forces of the natural world and this environment, with the personal lifestyle choices it allows, has a major impact on cardiovascular health and disease.

1.4 Cardiovascular Disease and the Plastic Environment

Like other great apes, humans live in discrete communities. It is currently believed that humans evolved in Western Africa and then dispersed to far reaches of the globe. Fragmentation of the ancestral populations and geographical isolation are prerequisites for speciation. Although, humans have not yet diverged into different subspecies, dispersion and isolation of specific human communities have led to the acquisition of new genetic variations and the loss of some ancestral features (*e.g.*, skin pigmentation). However, the impulse to migrate is balanced by a desire to aggregate and settle in small communities. Both tendencies are, however, attempts to avoid being a part of a large amorphous group. In large interbreeding populations, genetic diversity is decreased by gene fixation and genetic variation is diluted in the large gene pool. On the other hand, geographic separation promotes adaptation to new environments, the acquisition of new traits or the loss of old ones. Therefore, the development of discrete communities has evolutionary advantages. Hence, human development and disease can be understood only within the context of a community of shared genes and shared environments.

Settlement into small cohesive communities has additional advantages. It provides a network of social support; it promotes cooperation, collaboration, and commerce; and it helps in creating a cultural and social identity. Therefore, despite globalization and emigration, discrete communities are relatively well preserved. The nature of these communities, however, depends on the natural environment (mountainous communities are different from coastal communities or from people living in the desert) as well as the historical and political heritage of the community. The characteristics of the community and the place of the individual within this community have profound effects on health and disease. These effects are complex and operate at different levels. There is a multilayered network of interactions in which an individual adapts not only to his or her physical and social environment, but also makes up part of the social environment of others. Moreover, these communities do not exist in isolation. They communicate and depend upon other communities for resources and trade. Discrete communities are, therefore, nested in larger, more complex environments. As a result, systematic deconstruction of the mechanisms by which the community structure affects CVD is difficult and is limited at present by the inadequacies of the models available and the lack of interdisciplinary expertise required for simultaneously studying sociology and biology. However, such understanding is urgently needed. Cardiovascular diseases, which have been the leading cause of death in the industrialized countries since 1900, are becoming the leading cause of mortality in the developing world as well. Understanding how community characteristics affect CVD may be helpful in stemming the tide of this looming epidemic by informed urban planning and just environmental laws. In developed countries this understanding may help in preventing and managing the epidemics of diabetes and obesity that are currently unfolding at an ever-increasing pace. It has been estimated that in the US and Western Europe between 2000 and 2005 the number of people with a BMI of $> 50 \, \text{kg/m}^2$ has increased by 75%.[137] Certainly, no genetic mutation or fundamental metabolic or physiological defect could have caused such a profound shift in CVD risk within such a short time. What has changed is not the biology of the individual, but the environment in which individuals live and the modes in which they interact with each other or with their environment.

1.4.1 *It Takes a Village*

There are several ways to examine the influence of the community environment on CVD. The effects of the community can be studied either by looking at CVD risk and mortality in immigrants that migrate to a new community (how does a change in community affect CVD risk?) or when the environmental conditions change within the same community (how do environmental changes within a community affect CVD risk?). We can also delineate the contribution of community by assessing the role of socioeconomic status and social networks to individual health. Finally, we can look at the cardiovascular effects of community characteristics such as its built environment, access to healthy foods and health care and exposure to indoor and outdoor pollutants. All of these issues have been variably examined. But, as we shall see, with few notable examples, much of the evidence is observational. No randomized trials could be designed to put people in different communities or to change their socioeconomic status;

therefore, it is difficult to establish causality or to rigorously rule out confounding and exposure misclassification.[138] Nevertheless, the weight of evidence builds a compelling case. Studies showing that modification of the environment (e.g., the level of pollutants) changes CVD outcomes and that social conditions (e.g., socioeconomic status, social networks) predict CVD risk support a major role of the community environment in cardiovascular health.

Studies on migrants show that changing communities causes substantive changes in CVD risk and mortality. Data collected between the 1960s–1970s indicate that the rates of coronary heart disease deaths in Japanese men increased when they moved from Hiroshima and Nagasaki to Honolulu or San Francisco, whereas the rates of stroke deaths were decreased.[139] These studies show that moving to a new community significantly alters CVD prevalence. Moving to a new community also affects CVD risk factors. Patel et al.,[140] report that Indians living in UK have a higher mean body mass, blood pressure, cholesterol and triglycerides than their counterparts in their native villages in India. Although impaired plasma glucose was more prevalent in India, it was less frequently associated with CVD risk factors than in Indians living in the UK. Changes due to migration to a new community provide additional support to the view that modifications in the environment can significantly modify CVD risk in genetically similar populations. This may be true even in twins who have identical genetic backgrounds. Finnish immigrants in Sweden show a reduced prevalence of coronary heart disease compared with subjects that were always living in Finland[141] and this decrease in CVD risk was evident even in migrant twins, indicating that the environment could modify CVD risk independent of genetics.

Environmental changes with the same community can also alter CVD risk and mortality. Current statistics show that there are large geographic and ethnic variations in the rates of CVD deaths. The rates of CVD deaths in Eastern European countries such as Russia (1555.2 per 100 000 population) and Hungary (709.7) are among the highest in the world, whereas those in Japan (151.6) and France (149.8) are the lowest.[1] At first glance, a 10-fold geographic difference in CVD mortality would suggest that there are strong genetic components and that people of different populations have intrinsically different CVD risk. Alternatively, some of this difference could be related to natural geographic variation and components of the natural environment discussed before. For example, the levels of UVB and therefore vitamin D may be lower in Russia and Northern Europe than in the sun-drenched Mediterranean countries. But independent of these reasons, heart disease rates can change dramatically by changes in the community environment even within the same geographic region. For instance, CVD mortality in Finnish men and women decreased by 75% between 1971 and 1995 largely to the implementation of national nutrition and antismoking programs.[142] Cardiovascular death in Dublin fell by 10.3% due to a decrease in particulate air pollution within 72 months of a ban of coal sales.[143] Similarly, changes in economic and agricultural policies in Poland, even without changes in health policies, led to a 24% drop in coronary mortality from 1990 to 1999, which was related to an increase in the consumption of polyunsaturated fatty acids.[144] Large decreases in CVD mortality have been reported in other countries as well. In England and Wales, for instance, coronary heart disease mortality rate between 1981 and 2000 decreased 62% in men and 45% in women. More than half of this decline was attributed to reduction in environmental risk factors such as smoking.[145] Similarly, a recent study of the decrease in US deaths due to coronary heart disease from 1980 to 2000 suggested that approximately 44% of the decrease could be attributable to change in risk factors in the population due to environmental changes.[146]

While CVD deaths are generally declining in the US and Europe, they are increasing in other parts of the world, particularly in developing countries such as China and India. Indeed, the WHO estimates that 80% of all CVD death in 2004 occurred in developing or underdeveloped countries. Although few comprehensive surveys of CVD deaths in these parts of the world are available, a study from Beijing reported that between 1984 and 1999 age-adjusted coronary heart disease mortality rates increased by 50% in men and 27% in women. Much of this increase in mortality was attributed to an increase in cholesterol levels reflecting an increasingly Western diet.[147] Thus, cultural changes within a few decades could alter CVD risk factors and therefore dramatically affect cardiovascular health and the prevalence of cardiovascular disease.

A specific example of how civic policies affect heart disease is the implementation of a smoking ban in several parts of the United States and Europe. Data from 11 independent studies show that institution of the smoking ban was associated with a 6 to 47% reduction in acute coronary disease mortality in that community (Second Hand Smoke Exposure and Cardiovascular Effects: Making Sense of the Evidence, Institute of Medicine, 2009). This decrease in CVD mortality, and hospital admissions was apparent within a few (1 to 3) years. On average, the incidence rate ratio (IRR) decreased incrementally by 26% for each year after the implementation of the ban and the risk of acute myocardial infarction was decreased by 17%.[148] Significantly, the greatest effect was seen among younger individuals[148] indicating that reduction in second-hand exposure to tobacco smoke decreases the risk of coronary mortality within the short span of a few years. Thus, changes in the environment could significantly and rapidly change CVD risk and outcomes.

1.4.2 Wealth is Health

Studies showing that socioeconomic status within a community affects CVD also support the role of the environment. Poor health has always been known to be associated with poverty, but perhaps not in the same context as in modern society. Cardiovascular disease in the early part of this century was considered to be a disease of affluence. Studies from 1930s and 1940s from the US and UK reported that the rates of coronary heart disease were higher in men with higher socioeconomic status (SES).[149] Despite much evidence to the contrary, CVD is still considered to be a disease of affluent countries – a claim that is re-enforced by the increase in CVD rates in developing countries as they become more affluent and more Westernized. Ironically, within industrialized countries the relationship between SES and CVD has reversed from its original association with higher SES seen in the 1940s to the 1960s. Since 1961, all-cause mortality as well as CVD mortality is robustly and negatively associated with SES.[138]

The most commonly used indices of SES are education, income, occupation and employment status, although indices of social class, measures of living conditions and area-based measures are also used frequently. It is unclear whether these measures accurately report social stratification, which could be either based on class, status, and power (American definition) or on power, control, and ownership (Marxist criteria) or whether there are other more useful matrices to include. Certainly, the limitations in understanding the impact of SES on CVD are, in part, due to the lack of understanding of what exactly is being measured. Indeed, as discussed later, area-based measures and assessment of the built environment improve the strength of the relationship between CVD risk and the community environment. Nevertheless, all existing criteria of SES show a consistent association with CVD mortality and the extant evidence taken together strongly supports the notion that SES is an important contributing factor in the etiology and the progression of CVD.

There is consistent evidence that low education is a significant predictor of myocardial infarction and sudden death. In a remarkable study on 270 000 Bell employees in the US, Hinkle et al.,[150] found that men who entered the organization with a college degree had a lower incidence and death rate from coronary disease in every part of the country and in all departments. Several other studies report similar association between low education and low income and excessive death from ischemic heart disease.[138] The contribution of a number of psychosocial factors such as social support, coping styles, behavior, job strain or stress, and anger or hostility has also been studied, but their direct role in contributing to CVD risk remains uncertain.

Socioeconomic status also bears a strong inverse relationship with several CVD risk factors such as hypertension, smoking, obesity and diabetes. Its relation to cholesterol, if any, is, however, unclear. Because of low income and low education, individuals with a lower SES are more likely to be smokers, overweight, physically inactive and have unhealthy eating habits. In most industrialized countries smoking prevalence is directly related to SES, particularly the level of education. The importance of education as a measure of SES is also underscored by studies showing that the level of education is inversely related to hypertension in all of the 14 communities followed by the Hypertension and Detection Follow-up Program in the US. Interestingly, even at higher education levels, the adjusted prevalence of hypertension remained nearly twice as high in blacks as in whites.[151] Reasons for this disparity may relate to other environmental and genetic factors such as differences in fluid regulation and/or vitamin D synthesis.

Although SES correlates with CVD risk factors, it is not clear whether it is an independent risk factor or independent predictor of risk, or whether it is a preceding causal variable that is important in the adoption and maintenance of risk, that is, whether it is important in determining the levels of other risk factors. Despite this uncertainty, differences in CVD risk factors do not entirely account for the strong effect of SES. For instance, in the Whitehall study men in the lowest socioeconomic grade have 2.7 times the 10-year CHD death risk than those in the highest grade. After adjusting for smoking, systolic blood pressure, plasma cholesterol and glucose and height the relative risk was reduced to only 2.1.[152] Other studies have also shown that the risk imparted by low SES is relatively independent of the major CVD risk factors. At best, conventional risk factors seem to account for 15–30% of the CVD risk imposed by SES. Clearly, other unappreciated environmental factors are at work. Indeed, recent work suggests that the some of the effects of SES may be due to the unique structuring, civic architecture and characteristics of disadvantaged neighborhoods, particularly in the US and other industrialized countries.

1.4.3 People or Places?

In addition to looking at SES as a global index of social influence, some investigators have examined the role of physical neighborhoods. Results of many ecological studies indicate that rates of CVD mortality vary across communities with different socioenvironmental characteristics, indicating that area characteristics contribute to CVD mortality risk.[153] Some studies have also found that social inequalities are related to mortality and other health outcomes independent of individual level indicators. After adjustment of individual-level variables, Diez-Roux and coworkers[154] reported that living in deprived neighborhoods was associated with increased prevalence of coronary heart disease and increased levels of risk factors. Analysis of data from the Atherosclerosis Risk in Communities (ARIC) Study cohort showed that residents of disadvantaged neighborhoods had a higher incidence of coronary heart disease ($RR = 3.1$ for whites and 2.5 for blacks).[155] The relationship persisted even after controlling for personal income, education and occupation or after adjusting for established CVD risk factors. In aggregate, over 40 published studies report that living in socially deprived areas increases CVD risk that could not be accounted for by individual socioeconomic characteristics.[153] Although measures of social environment and outcomes vary between studies, the fact that in most studies the association between area factors and CVD remain unchanged or are only slightly affected after the introduction of risk factors indicate that there is strong neighborhood effect on CVD risk and mortality.

Several mechanisms could account for the observed association between CVD and the neighborhood environment. Neighborhood characteristics such as the availability and costs of various types of foods, publicity and availability of cigarettes, the distribution of recreational spaces and differences in the built environment could all contribute to excessive CVD mortality in disadvantaged neighborhoods. Services such as transportation, healthcare resources and social interactions such as networks and social ties and neighborhood identity may be important as well. In addition, it has been argued[153] that experiential factors such as affective experience (attachment, sense of community), cognitive experience (satisfaction with the neighborhood) and relational experience (social integration, social support and stressful interactions) should also be taken into account, but no such comprehensive environmental assessment has been attempted. Communities and neighborhoods are complex entities that are a product of economics, historical events, social structure, public policy and cultural practices and, therefore, it is difficult to capture all their dimensions into a simple matrix that could be readily correlated with CVD risk.

Social characteristics of neighborhoods and the role of the built environment may be particularly important in understanding the currently unfolding epidemic of obesity. While obesity is not necessarily related to heart disease, it increases the risk of developing diabetes, which in turn is a major CVD risk factor. Diabetes markedly increases CVD risk and most diabetics (70–75%) die of heart disease. Several investigators have suggested that the recent increase in obesity is due to constant exposure to an "obesogenic" environment. While several environmental factors can promote overeating and physical inactivity, the built environment may be particularly obesogenic for disadvantaged communities. Meta-analysis of over 60 studies shows that aspects of the built environment are positively correlated with obesity,[156] particularly in disadvantaged groups.[157] The strongest evidential support was found for food stores (supermarkets instead of smaller grocery stores), places to exercise, and safety. Each of these neighborhood characteristics was found to be correlated with body mass index.[157] Greater neighborhood physical activity resources were associated with lower insulin resistance[158] and high walkability neighborhoods were associated with decreases in weight and waist circumference.[159] These measures of obesity were also associated with a high density of fast-food restaurants.[159] For neighborhoods with a high-density of fast-food restaurants an odds ratio of 1.8 has been reported.[160] Obesity has also been linked to land-use mix and it has been reported that each quartile increase in land-use mix is associated with a 12% reduction in the likelihood of obesity.[161] Moreover, each additional hour spent in a car per day was associated with a 6% increase in obesity risk and each kilometer walked per day was found to be associated with a 55% reduction in the likelihood of obesity. In a recent study it was found that 52% of the inverse association between neighborhood education and blood pressure was mediated by body mass index/waist circumference,[162] indicating that obesity mediates some of the CVD risk associated with poor neighborhoods.

When taken together, the work done in this area points to the important role of the neighborhood environment in CVD prevalence and risk. Neighborhoods provide order and structure to the lives of their residents and are important determinants of health and well being. Historically, because of the close clustering of buildings, highly populated cities were more prone to the accumulation of pollutants and waste and the spread of contagious disease. With improvements in urban planning, neighborhoods in most industrialized countries have succeeded, albeit partially, in providing better civic amenities and in minimizing the spread of communicable disease. However, residential structuring continues to be an important mechanism of class relation and social differentiation. Neighborhoods provide social support and foster social interactions from which individuals, to a large extent, derive their values, expectations, consumption habits and market capabilities. Therefore, it is not surprising that many components of the environment, structured within a neighborhood, (food stores, recreational spaces, walkability and transportation) or associated with its residents (attachment, social integration, sense of security) are important determinants of CVD risk.

1.4.4 With the Help of My Friends

Although neighborhood characteristics and the built environment have a strong impact on CVD, the most important components of our environment are not its physical attributes but other people. We live in families and communities, embedded to a varying degree, in complex social networks. Relationships within these networks, more than any other dimension of our environment, provide us meaning and purpose and, perhaps more importantly, a context within which we define our identity. It follows then that the activities of other people in our environment profoundly affect our health and well-being. These activities include direct social interactions or the indirect effects of the community such as health-care access, civic policies, economic activity, the structuring of neighborhoods and cities or the generation of environmental pollutants. Such relationships are mutual. We are a part of the environment of others as they are a part of our environment and therefore to delineate the environmental basis of CVD, it is essential to understand how the differentiated and the combined activities of other people affect an individual's risk for developing cardiovascular disease.

Consistent with the high impact of social interactions on a person's well being, CVD risk factors such as obesity[163] and smoking[164] form discernible clusters within a social network. Analysis of the obesity within a network of 12 067 participants of the Framingham Heart Study over 32 years shows that obesity spreads through social ties. A person's chances of becoming obese in a given period were found to increase by 57% if he or she had a friend who became obese within the same period. If one sibling became obese the other was at an increased risk as well. Spouses showed a similar codependence. Smoking cessation by a spouse decreased a person's chance of smoking by 67%, indicating that like obesity, smoking behavior spreads through close and distant social ties. Groups of interconnected people stop smoking or gain weight together. Although a decrease in smoking is associated with weight gain, smoking behavior did not appear to be instrumental to the spread of obesity.[164] Remarkably, geography and SES did not seem to make a difference and the effect of geographic distance did not modify the effect of one person on the other. These observations, contrary to the work on neighborhoods, suggest that exposure to local environmental factors does not account for the clustering of obesity. Even for smoking cessation, interpersonal effects were observed in geographically separated contacts, but not by immediate neighbors, indicating that joint exposure to local environmental factors (tobacco marketing, taxes, cigarette availability) may not be an explanation of clustering in this study.

Clustering of CVD risk factors within a social network underscores the importance of social interactions in modifying heart-disease risk. The estimated strength of this influence exceeds that of the major known genetic influences. As pointed out by Barabasi,[165] the strongest genotype–phenotype association discovered by genome screening has been demonstrated for the FTO gene. Persons who carry this gene have a 67% increased risk of obesity, however, risk of obesity increases by 171% by having just one friend who is obese, indicating that, at least in this analysis, friends have a stronger influence on CVD risk than genetics. The influence of other people, however, diminishes with the degree of separation and no influence could be detected after three degrees of separation both for smoking cessation and for obesity.[163,164] Thus, the behavior of those nearest to a person has the largest impact on the person's behavior. If one spouse became obese, the chances that the other will become obese are increased by 37%. Clearly, no genetic factors could be implicated in this case and therefore, it seems that social interactions, independent of genetic relationships, could have a significant effect on a person's CVD risk

Heart disease has long been known to segregate in families. In clinical practice, family history is considered to be of high prognostic value. In the National Cholesterol Education Program (NCEP) family history of premature heart disease is considered to be a CVD risk factor and is used to define risk status.[166] Framingham data also indicate that family history is a risk correlate.[167] Family history, however, is not included in Framingham risk charts because it is not clear whether the effects of family history are independent of the major CVD risk factors.[168] In other words, the genetic and the environmental risks that segregate in families are not independent of smoking, hypertension, dyslipidemia or diabetes. Because these risk factors are emergent properties of complex integrated systems, they could not be ascribed to unitary causes or linked to a single gene or even a few specific genes, even though the constituents of these systems may have direct genetic correspondence. Thus, even though blood pressure is heritable, essential hypertension does not show a clear pattern of inheritance. Similar difficulties arise in understanding the genetic "causes" of dyslipidemia or insulin resistance.

Evidence from twin studies has often been marshaled to support a genetic basis of complex CVD risk factors and coronary heart disease itself. Data from these studies clearly show that an individual's risk of diabetes, obesity and CVD is increased when he or she has a relative with these conditions. From a cohort of 20 966 Swedish twins it was estimated that the heritability of fatal coronary artery disease is 57% for men and 38% for women.[169] Similar heritability was reported for coronary heart disease in the Danish twin registry (53% for men and 58% for women).[170] Based on these data it has been argued that much of CVD risk is heritable. However, the obverse evidence from network analysis and from studies on the effect of neighborhoods and SES (*vide supra*) suggests that the effects of shared genes cannot be readily separated from the effects of shared environments. Families are genetically related, but they also share the same environment and usually the same SES and often the same neighborhood. They have similar diets and they make similar lifestyle choices. In addition, they dynamically affect each other's behavior such that if one member of the family changes (for better or for worse) the likelihood of the other changing increases as well. For instance, social network studies report that if one sibling became obese the chances that the other would become obese increases by 40%.[163] Similarly, smoking cessation by sibling decreases the chances of an individual being a smoker by 36%.[164] Hence, it appears likely that siblings, twins and families have similar CVD risk because they might be sharing the same environment – they show similar behavior, they imitate each other, and they share similar cultural imprinting. Even in studies in which twins are separated at birth the change in environment is often not large enough to allow for a clear distinction between genes and environment. Nevertheless, arguments in which genes are pitted against the environment are counter-productive because chronic diseases such as CVD cannot be attributed to static, individual components of the environment or the genes but only to their interaction. Genetic information is expressed only in the right environment and whether a genetic variance promotes health or disease depends upon the environmental context and conditions. This is true for all life forms, however, the role of the environment in human development and disease is even more complex because humans constantly shape their own environment and transmit these self-created environments to their progeny.

1.5 Heritability of the Environment

All animals interact with and alter their environments to varying degrees. Humans do so too, but their interactions with the environment are more complex in part because, in comparison with other animals, humans can more profoundly and radically shape their immediate surroundings. Moreover, animals also inherit a relatively stable environment, however, the inheritable human is more complex, and more importantly it changes far more rapidly than the environment of other animals. As has become abundantly clear, our ability to modify our own environment affects the environment of other animals as well as the natural environment, often not for the better. Our capacity to modify and mold our environment stems from our unique ability to transmit our learning and, in a general sense, our environment across generations. We pass along not only our genes and our instinctive behaviors but also bequeath to the next generation our environmental legacy. This inheritable environment is the sum total of human culture and knowledge that is passed from one generation to the next. We fiercely guard and zealously protect this legacy as animals do their genes. The major biological consequence of this is that our genes change in unison with the environment we create and both are transmitted across generations. Several examples from evolutionary biology demonstrate such transgenerational environmental effects. For instance, in most mammals, the ability to digest lactose is lost after weaning; however, populations with a long history of cattle domestication and milk drinking have acquired a "persistence of lactase" trait.[171] It is believed that this allele arose and was selected for among dairy farmers in the central Balkans and central Europe. Other human populations who have no history of dairy farming (e.g., Najavo Indians) are mostly (> 95%) lactose intolerant, indicating that due to positive selection, the persistence of lactase trait was enriched in communities of dairy farmers. The lactase trait may have conferred survival advantage in these populations by facilitating the digestion of milk and preventing lactose intolerance in communities where milk and milk products are important food sources. Thus, environments (dairy farming) transmitted (inherited) over several generations can become important agents of natural selection.

Changes in the apoE gene are another example of such adaptation. As discussed before, the ε4 variant of the apoE gene is the ancestral allele and ε3 and ε2 alleles arose with the development of modern man. It has been suggested that ε3 was selected in some human populations because it increased resistance to infectious diseases that originate from domestic animals or those that become endemic in large communities (smallpox, tuberculosis, etc.).[119] It is believed that ε3 carriers may have been better suited to survive in densely populated agricultural communities in close contact with domestic animals. In this scenario as well, the selection of the gene variant is possible only if one assumes that the environment (an agrarian community) is coinherited with the gene over several successive generations.

Another explanation for the selection of ε3 is that it is related to the evolution of "grandmothering" in humans.[119] Some investigators have suggested that humans show a prolonged dependence on maternal care for learning skills. Often, this requires the participation of grandmothers in rearing their grandchildren. Because ε3 is associated with better preservation of cognitive function during aging, it is suggested that grandmothers with the ε3 allele were better able to ensure the safety and survival of their grandchildren. This could have provided a positive selection pressure for ε3 and may have led to the enrichment of this variant in the modern human population. Again, the continuity of environmental conditions over several generations (the need for prolonged care and teaching and transmission of inherited culture) provides a constant, relatively unchanging driving force for natural selection.

Continuity of human environmental conditions over several successive generations has also been linked to the evolution of obesity. The thrifty-gene hypothesis put forth by Neel suggests that genes for obesity were favored by natural selection because these genes conferred a phenotype that was exceptionally efficient in the intake and the utilization of food. Carriers of these genes could efficiently store food as fat, which allowed them to survive during periods of famine. In modern society food is plentiful and therefore, the thrifty gene has become a liability and has led to the recent explosion of diabetes and obesity. The hypothesis represents an interesting example of how recurrent changes in the food environment occurring over several human generations could lead to the selection of a specific (obese) phenotype. However, the hypothesis has been recently challenged[172] on the grounds that famines are a recent phenomenon (since the spread of agriculture) and that they are rare events. Even when they do happen, they usually kill <5% of the population mostly young children and old postreproductive adults. Therefore, it is argued that recurrent famines cannot contribute to natural selection. An alternative theory, suggested by Speakman[172] is that like other animals, early hominoids maintained a constant body weight. The lower limit was controlled by the threat of starvation and excessive disease, while the upper limit was set by the threat of predation. Obese individuals may have had a greater survival advantage during episodes of disease outbreak or food shortage, but they were less able to avoid predators. With the advent of civilization, particularly the discovery of fire and the use of weapons, the predatory threat was gradually removed. This new man-made environment was transmitted over several generations and gradually, a decrease in the threat of predation removed the selective pressure for maintaining the upper limit for body weight. The hypothesis posits that obesity arises not because of positive selection (as suggested by the "thrifty-gene hypothesis") but by random genetic drift due to loss of selection pressure for the upper body weight limit.

Mathematical modeling supports the view that the increase in BMI is due to a nonadaptive drift but the idea that this is due to relaxation in predation risk remains conjectural. Yet, despite their different mechanisms (positive selection versus random drift) both hypotheses assume that a constant, humanly constructed, environment was maintained over several successive generations and that a change in the environment (again over several generations) changed natural selection. It is also likely that there is no significant selection of specific genotypes and that the exponential increase in obesity in some human populations is due to a relatively new environment interacting in new ways with genetic predisposition. Certainly, the relatively sudden increase in obesity (within two generations) in restricted geographic areas (*e.g.*, South United States) populated by genetically dissimilar populations (African-Americans, Hispanic-Americans and Americans of European descent) does not help in building a convincing case for a genetic disposition selected by evolution. Therefore, a more complete theory must integrate physiological and genetic frameworks with the socioeconomic and the environmental perspectives.[172]

1.6 Pollution and Heart Disease

Of all the components of our self-made plastic environment, pollution has the most insidious effect on CVD. The problem of pollution is not new. Ceilings of prehistoric caves show signs of soot indicating that early hominoids were exposed to high levels of pollution created by burning wood without adequate ventilation. Analysis of the Greenland ice core[173] covering a period of 3000 to 500 years ago shows that lead was present at concentrations 4 times higher than natural levels. Values from 2500 to 1700 years ago suggest that Greek, Roman and Chinese lead and silver mining and smelting activities polluted the middle troposphere of the Northern Hemisphere. Pronounced lead pollution was also observed during the Middle Ages and the Renaissance. During the Middle Ages in Europe there were frequent outbreaks of infectious diseases due to unsanitary living conditions. With better understanding, urban living conditions improved in the 1700s–1800s, however, the advent of the industrial revolution and the emergence of industries and factories led to severe air and water pollution. The problem was further exacerbated by the increasing use of synthetic chemicals, pesticides, and metals, all of which have had a significant effect on the natural environment as well as human health.

1.6.1 Brave New World

The modern environment is awash with synthetic chemicals and pollutants. By some estimates, more than 30 000 synthetic chemicals are in current use, of which at least 5500 are produced at more than 100 tonnes per year.[174] Most of the major rivers and lakes show significant contamination by synthetic chemicals, pesticides or metals. High levels of pollutants are also released in the air. Although the level of air pollution in the developed world today is much lower than during its peak in the 1950s–1970s, pollutant levels in the developing world remain extraordinarily high. By some estimates, 656 000 people in China and 527 000 people in India die prematurely every year because of air pollution (National Geographic New, July 9, 2007). The WHO ranks air pollution as the 13th leading cause of world-wide mortality (World Health Report, 2002).

Human activity has caused significant changes in the natural environment. While humans have always altered their environment, human activity in the past 50 years has had a dominant, detectable influence on global climate. The burning of fossil fuels and other industrial activities has changed the composition of the atmosphere by generating greenhouse gases. For instance, carbon dioxide levels have increased 31% since the preindustrial times, and half the increase has been since 1965.[175] European pollution of the Mediterranean lower troposphere has increased the levels of ozone and carbon monoxide 2–4-fold,[176] and it has been estimated that the generation of black carbon and soot in the Indian subcontinent has reduced solar heating of the Northern Indian Ocean by 15%.[177] Human activity also causes the emission of large amounts of aerosol particles and their gaseous precursors and estimates of prehuman aerosol burden suggest that anthropogenic emissions have caused a 50–300% increase in the total aerosol load even over remote regions of Asia and North and South America.[178] Because greenhouse gases have long lifetimes they accumulate in the atmosphere for a long time and are transported across vast expanses of sea and land. Other airborne toxicants also travel well. Pesticides such as lindane, chlordane and DTT from Asia have been detected in the Canadian Rockies and mercury generated by human activity has been detected in Arctic wild-life.[179] As a result, there are no pristine, unpolluted places left, especially in the Northern Hemisphere. Perhaps as a result, every year, for the last 30 years, the Earth's climate has consistently exceeded the bounds of natural variability.[175]

Changes in global climate are likely to have a profound effect on human health. The earth's average surface temperature is now higher than it has been in the last 100 000 years. This increase in temperature is likely to affect global climate patterns, food production as well as social and economic conditions particularly for the resource-scarce vulnerable poor. Changes in global climate could also affect patterns of infectious disease and the severity and the outcomes of chronic diseases such as diabetes, cancer and heart disease. In addition, local anthropogenic pollution is likely to have a substantive effect on human health. Changes in land use due to urbanization and changes in agricultural practices change the global climate but they affect local populations as well. Moreover, congregation of large human populations in megacities has created unique local environments that differ significantly from the surrounding green rural areas. These urban microenvironments have high levels of pollutants and wide socioeconomic diversity. Large populations are constrained into small geographic areas with restricted access to fresh food, limited opportunity for physical activity, and high exposure to local sources of pollution such as traffic, industry, ship yards, and airports. These environments create overstressed, overfed, overweight but underactive citizens.

1.6.2 Weaknesses of the Heart

How anthropogenic pollution affects cardiovascular health is not well known. Pollution could impair cardiovascular health by affecting cardiovascular function directly or indirectly by changes in global climate, temperature, sunlight availability, food sources or socioeconomic status. Therefore, the effects of pollution on CVD are likely to be complex and multifactorial. The effects of several pollutants have never been studied and there are no toxicological data on 86% of the 30 000 chemicals in the environment.[174] Even when available, most toxicity data relate to the carcinogenic, pulmonary, reproductive or neurological effects, with little information on cardiovascular toxicity. In part, this is because of the unique complexity associated with studying cardiovascular disease. Because of its large background rate, it is often difficult to discern the effects of pollutant exposure on CVD. Moreover, cardiovascular tissue is almost never directly exposed to pollutants; therefore, cardiovascular effects are likely to be modified by metabolism or transformation at the primary site of exposure or during systemic transit to cardiovascular sites. Pollutants can impact CVD even without reaching or affecting cardiovascular tissue, for instance by inducing pulmonary inflammation or autonomic dysfunction. Similarly, changes in blood pressure could arise from renal toxicity and hepatotoxicity could lead to changes in blood lipids, resulting in cardiovascular disease secondary to liver or kidney toxicity. Finally, CVD risk correlates such as hypertension, insulin resistance and dyslipidemia are emergent properties of complex systems and hence overall CVD toxicity is difficult to predict from *in vitro* cellular experiments or even by studying individual cardiovascular tissues.

Despite being a secondary target, the cardiovascular system appears to be highly vulnerable to environmental toxins.[180–182] This vulnerability is revealed and underscored by the high cardiovascular toxicity of exposure to cigarette smoke, traffic pollutants and ambient particulate matter (PM). These pollutants gain systemic ingress *via* lungs; however, their cardiovascular toxicity often far outweighs their pulmonary effects. For instance, CVD accounts for 70% of the excessive deaths attributable to secondhand tobacco smoke (SHS) exposure.[183] According to a recent analysis of the California EPA, exposure to SHS was related to 7000 excessive deaths due to lung cancer, whereas CVD was implicated in nearly 40 000 deaths. Similarly, 69% of premature deaths due to exposure to particulate air pollution are due to cardiovascular causes and 28% to respiratory causes.[181] In addition, 32.7% of 443 000 premature deaths due to direct smoking are related to CVD and smokers are 2- to 4-times more likely to develop coronary heart disease and stroke.[1] That the cardiovascular system suffers a disproportionate disease burden induced by inhaled pollutants and toxicant highlights the unique vulnerability of the blood vessels and the heart to environmental toxins. This vulnerability is abundantly demonstrated in specific chapters of this monograph discussing the profound and pervasive cardiovascular effects of particulate air pollution (Chapters 2–8), traffic-generated pollutants (Chapters 9 and 10), manufactured nanoparticles (Chapter 11), metals (Chapter 12) and aldehydes (Chapter 13).

1.7 Personal Environment and Lifestyle Choices

The substantive cardiovascular effects of the natural and community environment are significantly modified by our personal environment and the lifestyle choices that we make. In principle, we can choose not to live in communities of high pollution or make choices that alter our socioeconomic status or otherwise mitigate or amplify the influences of the natural and the community environment. But in practice, regardless of our intentions, some of our choices are likely to be inevitably constrained by the community environment as well as the natural boundaries of the human condition. Yet, conscious, unconscious or predetermined, the choices that populate our personal environment can significantly alter CVD susceptibility as well as the course and severity of heart disease. CVD is most often the clinical manifestation of atherosclerosis. The process of plaque formation begins early in life and in its inexorable progression registers the cumulative effects of environmental influences through all life stages. Like the hidden portrait of Dorian Gray, the heart and the vessels accumulate the disfiguring scars of our lifestyle choices and environmental exposures, with little outward manifestation. And then one day, under the influence of the same environmental factors, the plaque ruptures, unmasking the gruel that we have accrued to disastrous consequence. Thus, throughout the life of an individual the environment exerts a continuous, though variable influence on CVD and the arterial plaques bear the sum of a lifetime of environmental influences. It follows that to understand the effects of the environment on CVD, it is necessary to delineate life-stage specific effects of the environment and how the environment affects CVD risk at different stages of plaque formation and rupture.

The influence of the environment begins with the earliest stages of CVD development. Several studies show that the *in utero* environment determines CVD susceptibility later in life. An unfavorable *in utero* environment could induce the formation of atherosclerotic lesions during fetal development. The aorta of premature human fetuses shows fatty streaks, inflammation and the accumulation of oxidized lipids and the coronary arteries show intimal thickening.[184] Low birth weight, which is an indicator of an unfavorable uterine environment, shows a strong negative correlation with ischemic heart disease. This association persists even when adjusted for gestational duration, indicating that high CVD risk could be attributed to fetal growth restriction rather than to premature birth.[185]

In the face of adversity the fetus attempts to adapt to the unfavorable environment. It undergoes predictive adaptive programming to diminish the impact of an adverse environment. These adaptations persist and could be beneficial if the *in utero* conditions continue upon birth, however, very often the postnatal environment is different and therefore the individual, already set off on a different course by fetal reprogramming, fails as an adult to adequately adapt to the new postnatal environment. An adverse fetal environment may be created by a variety of conditions such as the obstruction of the uterine artery, maternal undernutrition, smoking, alcohol consumption, diabetes or drugs and pharmaceuticals. While our understanding of the contribution of each of these factors to CVD risk is still in its infancy, both human and animal studies show that maternal hypercholesterolemia is associated with an increase in fatty streak formation in fetal arteries. The Fate of Early Lesions in Children (FELIC) study showed that the progression of atherosclerosis was markedly faster in offspring of hypercholesterolemic than normocholesterolemic mothers.[184] Thus, the disease risk burden of the mother is at least partially transmitted to the fetus. This view is reinforced by the observation that immunization of the mother protects against postnatal atherogenesis, even in the absence of gestational diabetes or maternal hypercholesterolemia.[185] Similarly, we have found that in mice prenatal exposure to low levels of environmental tobacco smoke affects offspring weight gain and induces a lipid profile that could alter the offspring's CVD risk in later life.[186] These findings are consistent with human data showing that infants of mothers who smoke during pregnancy have increased risk of subsequent obesity.[187] Thus, exposure to environmental pollutants or other adverse environmental factors that contribute to CVD in the mother could induce fetal reprogramming that could alter the child's future CVD risk.

The environment continues to exert its influence after birth. Breast-fed infants show a dose-dependent reduction in obesity and formula-fed infants have a higher fat mass, perhaps because breast milk contains leptin, which suppresses appetite and increases fat consumption.[187] Environmental factors may be related also to the epidemic of childhood obesity. In the last 30 years, the prevalence of obesity has increased 3-fold in children 2- to 5-years of age and 4-fold in children 6 to 11 years. Overweight children tend to grow into obese adults. In one study it was found that 80% of the overweight children were obese adults. The increase in childhood obesity has resulted in the higher prevalence of CVD risk factors such as high blood pressure and hypercholesterolemia leading to premature atherosclerosis and type-2 diabetes. These diseases, particularly type-2 diabetes, once thought to be occur only in late adulthood have now been shown to occur in children and adolescents, and it has been predicted that the current generation of children may have a shorter lifespan than their parents, in part because of a higher incidence of diabetes and cardiovascular disease.[187] Such sudden changes across the entire population are unlikely to be genetic, but may be related to technological advances, urbanization and economic issues that impact a child's eating and physical activity behavior.[187] These behaviors are heavily influenced by the social and physical environments, which to some extent dictate specific lifestyle choices relating to diet, exercise and smoking. Indeed, extensive research suggests that individual lifestyle choices are key determinants of CVD risk.

1.7.1 *Sum of Our Choices*

The most persuasive evidence supporting a preeminent role of personal choices comes from studies reporting a direct and robust association between lifestyle and CVD risk. These studies show that people with a healthy lifestyle have a relatively low burden of CVD. In the landmark Nurses' Health Study, Stampler and colleagues[188] reported that women who abstained from smoking, maintained a healthy weight and diet, engaged in moderate physical exercise and consumed moderate levels of alcohol had a relative CHD risk of 0.17, with the population-attributable risk of 82%, indicating that 82% of the coronary events in this cohort might have been prevented if all the women in group had adhered to a low-risk lifestyle. The lowest risk was associated with 5 healthy lifestyle and dietary factors, but the response was graded. Women with only 3 of the 5 healthy lifestyle factors had an incidence of coronary events that was about 57% lower than in the rest of the population. Moreover, a combination of healthy lifestyle choices had a greater impact than any single lifestyle factor.

Similar reductions in CVD risk have been reported by other investigators. In the Health Professionals Follow-up Study, it was found that 62% of all coronary events may have been avoided if all the men in the group have adhered to a low-risk lifestyle. A healthy lifestyle was inversely associated with risk even among men taking medication for hypertension or hypercholesterolemia.[189] Data combined from both the Health Professionals Follow-up and the Nurses' Health Study also show that 47% of stroke in women and 35% in men could be attributed to the lack of adherence to low-risk lifestyle choices.[190] In a cohort of Swedish women, the association between low-risk lifestyle and CVD was even stronger. In this group, Akesson and colleagues found that low-risk behavior was associated with a 92% decrease in risk of myocardial infarction.[191] Even in a cohort of 70–90-year-old individuals, adherence to a healthy lifestyle was found to be associated with more than 60% lower rates of cardiovascular death.[192] Thus, simple lifestyle choices – healthy food, moderate levels of exercise and alcohol consumption, and abstinence from smoking, more than any other environmental factors could profoundly and significantly affect CVD risk. Significantly, in most studies, socioeconomic status did not seem to modify the strength of the relationship between lifestyle choices and CVD risk, but the cohort characteristics were relatively uniform, consisting mostly of white educated professionals. Even in this group, however, less than 5% of individuals adopted all healthy practices, suggesting that current CVD prevalence is in most part (80–90%) due to lifestyle choices determined by the personal or the community environment.

The mechanism by which lifestyle choices influence CVD risk appears to involve clinical risk factors. Healthy lifestyle choices improve lipid levels, blood pressure and insulin resistance and therefore their ability to reduce overall CVD risk may be related to beneficial changes in the classical risk-factor profile. Environmental factors, by modifying classical risk factors, could alter the risk of developing CVD and modify the influence of CVD risk factors on preexisting or incipient disease. Indeed, low-risk lifestyle in nurses has been reported to be associated with a 90% lower incidence of type-2 diabetes and even in patients with impaired glucose tolerance lifestyle modification reduces the incidence of type-2 diabetes by 58%.[193] Healthy lifestyle choices, individually and jointly, have also been found to be associated with a lower lifetime risk of heart failure.[194] Thus, modification of lifestyle choices may be important for both primary and secondary risk prevention and adherence to healthy lifestyle could have a large and significant effect on CVD prevalence. The extent to which lifestyle factors affect CVD is truly astonishing, but the conclusion that 80–90% cases of CVD and diabetes could be avoided by a low-risk lifestyle strongly reinforces the predominance of environmental factors as primary determinants of CVD risk and mortality.

1.7.2 Food for Thought

Of the several lifestyle choices that affect CVD, few are as influential as diet. Dietary patterns are defining characteristics of human cultures. Like language, they provide cultural identity. How food is prepared and eaten are characteristic social features unique to each culture and therefore part of the inherited environment that is transmitted across generations. Different eating patterns reflect different dietary traditions and may be related to geographic and ethnic differences in the incidence and prevalence of CVD. Hence, the segregation of heart disease among different communities or families may be related to culturally transmitted and geographically preserved dietary patterns. Even within a community, personal dietary choices are important environmental determinants of CVD risk. Although intuitively it has been understood for a long time that diet is a major determinant of health, the effects of diet of CVD risk, progression and incidence have been difficult to study. Large populations are studied to maximize power. These populations, however, are often heterogeneous in their health characteristics and genetic backgrounds, making it hard to draw firm conclusions. Moreover, it is difficult to vary single dietary components in one group and not the other or to maintain people on specified diets for the long time required to discern cardiovascular effects. Nevertheless, the effects of several dietary components on CVD have been demonstrated clearly in several well-designed studies. These studies show that CVD risk could be substantially altered by making changes in the diet alone. In the Nurses' Study, replacement of just 5% energy from saturated fat with unsaturated fat was found to be associated with a 42% reduction in CVD risk,[195] indicating that the composition of the diet, independent of its energy content is a powerful environmental determinant of CVD. This view is reinforced by data on trans fatty acids. On a per-calorie basis, these fats increase CHD risk more than any other micronutrient. In a meta-analysis of 4 studies it was found that a 2% increase in energy consumption from transfatty acids was associated with a 23% increase in cardiovascular events. High trans fatty acid levels have also been reported to be associated with 3-fold increase in sudden cardiac death.[196] Thus, in addition to total energy intake the composition of the diet seems important. Men who adhered to a "prudent" diet (characterized by high intake of vegetables, fruit, legumes, whole grains, fish and poultry) have half the CVD risk of men on a "Western" diet (red meat, processed meat, refined grains, sweets and deserts, French fries and high-fat dairy products).[197] Taken together, these studies suggest that independent of other lifestyle variables; dietary choices modify CVD risk.

Systemic mechanisms by which dietary choices affect CVD risk are not well understood. However, diet affects all the major CVD risk factors – hypercholesterolemia, hypertension, diabetes and obesity and therefore, the effects of diet on CVD may be related to changes in the major determinants of heart disease. For instance, consumption of saturated fat increases cholesterol levels and a high salt intake increase CVD risk, in part, by increasing the prevalence of hypertension and the consumption of trans fatty acids raised LDL cholesterol, reduces HDL levels and it reduces the size of the LDL particle, making LDL more atherogenic.[196] Culturally-segregated communities such as the Tarahumara Indians of Mexico that consume relatively low levels of cholesterol (<100 mg/day) and low-fat diets have low levels of plasma LDL-cholesterol and little cardiovascular disease.[198] Similarly, in societies in which the consumption of sodium is <1 g/day, the prevalence of hypertension is only 1% of that in industrialized communities. Dietary patterns are also the major environmental factors in the development of diabetes and obesity. It has been shown that a "Western" dietary pattern is associated with a significantly higher ($RR = 1.59$) risk of type-2 diabetes when compared with a "prudent" dietary pattern.[199] Also, the risk of type-2 diabetes is positively associated with total and saturated fat intake as well as frequent consumption of processed meat.[200] In contrast, moderate consumption of alcohol decreases CVD risk in diabetic women.[201] Consumption of alcohol at moderate levels has been found to be inversely associated with the risk of myocardial infarction in nondiabetic men as well, independent of the type of beverage. The robust association between moderate alcohol consumption and CVD offers another example of the powerful effect of dietary environmental factors on heart disease.[202]

Although multiple studies support a strong link between diet and CVD risk, most of this evidence is derived from epidemiological studies. Correlative epidemiological studies, by their very nature, cannot establish causality. Prospective randomized studies with interventions and experimental models are required to establish the veracity of the relationship and to identify cellular and molecular mechanisms. Indeed, emerging data show that dietary interventions can modify CVD risk. This risk could be modified either by interventions that change the food environment within communities or by altering individual nutrition. To reiterate previous examples, it has been reported that in Finland, implementation of nutritional programs decreased CVD risk by 75% within 20 years,[142] and within 9 years, changes in agricultural policies, which increased the consumption of polyunsaturated fatty acids in Poland, led to a 24% drop in coronary mortality.[144] These data show that changes in the community food environment have a dramatic impact on CVD risk. This is important because even though what we eat is a personal lifestyle choice, it is, in substantial measure, determined by our cultural inheritance and food sources within our neighborhoods. Hence, community-wide changes in food sources significantly affect personal choices. Often, these choices are not deliberate. We chose foods without knowing their health impact or whether they contain unhealthy constituents (*e.g.*, transfat, high salt, *etc.*). Therefore, CVD risk is distributed diffusely throughout the population and cardiovascular health is an issue for the entire community and not just the individual. For instance, approximately 80% of the dietary salt is consumed from processed foods,[203] therefore a decrease in salt consumption can be achieved more readily by community-wide efforts rather than by individual attempts at abstinence.

Dietary interventions at the level of an individual also modify CVD risk. Modifications in diet and lifestyle are particularly effective in reducing the incidence of type-2 diabetes and the rate of progression from impaired glucose tolerance to diabetes.[193] Moreover, appropriate changes in individual nutrition can even reverse CVD. The results of recently completed a 2-year Dietary Intervention Randomized Controlled Trial-Carotid (DIRECT-Carotid) trail[204] showed on average a 5% regression in carotid-wall volume and 1.1% decrease in carotid artery thickness in men on a low-fat Mediterranean diet. These data provide credence to the notion that there is a causative link between diet and CVD.

This link is further supported by experimental evidence; however, we do not fully understand the molecular and cellular mechanisms by which dietary components affect CVD risk factors. Although consumption of unsaturated fatty acids is strongly associated with low levels of plasma cholesterol, it is not clear why unsaturated fatty acids decrease cholesterol levels.[205] It has been suggested that unsaturated fatty acids decrease hepatic production of cholesterol by reducing the transcription of SREBP genes or that they increase the breakdown of lipoproteins. Unsaturated fatty acids are more susceptible to oxidation and the accumulation of fatty-acid products in the liver has been linked to a decrease lipoprotein secretion from the liver. Paradoxically, lipid oxidation products are toxic and their accumulation in the aorta is believed to be a key step in the formation of the development of atherosclerotic plaques, and so, it is unclear how toxic products generated by unregulated oxidative damage exert beneficial effects on fatty-acid metabolism. Mechanisms by which trans fats increase CVD risk or those that impart cardioprotective properties to alcohol are also not clear. Such understanding is, however, critically needed not only to establish causative links between the food environment and CVD but also to determine which specific foods and food constituent affect cardiovascular physiology and disease so that better prevention and intervention strategies can be devised.

1.7.3 Rolling Stone Gathers no Moss

Another lifestyle choice that modifies CVD risk is physical activity. Physical activity is a central feature of healthy living and throughout evolution, it must have been important for obtaining means for survival (food, material goods, protection, mates, *etc.*) and for avoiding natural or predatory threats. As a result, it may have exerted a strong selection pressure on human development. Less physically fit individuals who were unable to acquire the needed elements for survival or escape predators may not have lived long enough to reproduce or may have been otherwise selected against. Therefore, these individuals may not have had an evolutionary impact as strong as those who were more physically fit. Indeed, as discussed earlier, one of the leading hypotheses for explaining the current increase in obesity is that during human development, social behavior, weapons and fire removed the threat of predation. This led to a population distribution of body fatness due to random mutations and drift.[172] Thus, some, but not all, humans became obese. It is likely that this trend was continued by cultural development that created environments in which the demand for physical activity was progressively decreased, resulting in random drift toward higher body weights.

Acculturation to industrialization could have also progressively lifted ancient selective pressures of survival on the genome.[206] While in the past, physically unfit individuals who were unable to obtain food (or escape predators) could not contribute to the gene pool, the current society, with its cheap food and abundant shelter, labor-saving devices, and advanced medical care, imposes no such constraints on reproductive success. Members of modern, high-industrialized societies no longer need to be physically active to survive, and there is no selection pressure against physical inactivity. Nonetheless, physical activity has been an important evolutionary force during most (99.99%) of human evolution and therefore it continues to regulate human physiology. Physical inactivity increases the relative risk of coronary heart disease by 45%; stroke by 60%; hypertension by 30%; and type-2 diabetes by 50%. An inactive physical lifestyle is considered to be an important CVD risk factor and it has been estimated that 13% of all premature deaths in US could be attributed to physical inactivity.[206]

Continuous activity and a constant gravitation load are important for normal cardiovascular function. Without adequate physical activity cardiovascular function deteriorates rapidly. When constrained, due to prolonged bed rest, for example, the whole body insulin sensitivity decreases within the first 3 days of inactivity.[206] In experimental studies, bed rest has been shown to result in an 8% decrease in left ventricular (LV) mass in 12 weeks, which was associated with a decrease in mean wall thickness.[207] A similar decrease in LV mass was observed in astronauts after spaceflight.[207] Microgravity in space induces cardiovascular deconditioning characterized by a decrease in circulating blood volume, diastolic pressure, ventricular stroke volume and a resetting of the carotid baroreceptors, and orthostatic intolerance.[208] Thus, cardiovascular function appears to be dynamically controlled by a constant gravitational load and physical activity. Inactivity leads to cardiac atrophy as a physiological adaptation to a decrease in myocardial load. Conversely, exercise training induces physiological hypertrophy characterized by normal myocardial growth.

Extensive epidemiological data suggest that physical activity decreases CVD risk and that it is a powerful environmental regulator of disease susceptibility. Physically inactive individuals have almost twice the risk of coronary heart disease and they bear a higher CVD risk burden. Several studies show that even moderate levels of exercise impart significant health benefits. In the Health Professional's Follow-up Study, physical activity was associated with a reduction in CVD risk.[209] The effects were dose-dependent. Total physical activity was associated with a 26% reduction, whereas high intensity activities such as running, weight training, and rowing were associated with 42% risk reduction. Walking pace was also strongly related to a decrease in risk, independent of walking hours. Other studies have reported similar effects and there is substantive agreement among studies. By one estimate, moderate to high intensity exercise increases life expectancy by 1.3 to 3.7 years and active individuals remain free of CVD 1–3 years longer than their sedentary peers.[210]

Physical activity also delays CVD in obese and diabetic individuals. Conversely, physical inactivity increases the risk of diabetes independent of obesity. Although both obesity and physical inactivity contribute to CVD, CVD risk is increased monotonically with increasing BMI (for each unit of BMI increment, CVD risk increases by 8%).[211] At least for women, the effects of obesity and physical inactivity appear to be additive. In the NHS cohort of female nurses, obese and sedentary women had the highest relative risk ($RR = 3.44$), followed by women who were active but obese ($RR = 2.48$) and women who had a healthy weight and were sedentary ($RR = 1.48$), compared with women who were active and who maintained a healthy weight.[211] Being physically active moderately attenuated, but did not eliminate, the adverse effects of obesity on CVD and being lean did not overcome the effect of physical inactivity, indicating that physical activity affects CVD risk independent of body weight. Similar data have been reported for men with type-2 diabetes in whom physical activity was found to be strong predictor of fatal CVD events. The highest quintile of total physical activity has 33% less CVD and 42% fewer deaths than the lowest quintile.[212] Overall, these studies provide strong associative data supporting the notion that physical activity lowers CVD risk.

The beneficial role of physical activity is further supported by randomized intervention studies and experimental data. In a small study from Denmark it was shown that diabetic patients assigned to moderate exercise experienced a 50% reduction in adverse cardiovascular events over an 8-year follow-up than those who were inactive.[213] In addition, a meta-analysis of clinical trials on the effects of exercise on glucose control found that interventions reduce HbA1c levels (on average from 8.31% to 7.65%). This magnitude of decrease was thought to be clinically important and enough to significantly decrease secondary diabetic complications.[214]

Several mechanisms can account for the salubrious effects of exercise.[209] Vigorous physical activity increases myocardial oxygen supply and improves myocardial contraction and electrical stability. In addition, exercise increases HDL levels, while decreasing LDL-C, blood pressure, blood coagulation, systemic inflammation and insulin resistance. Even moderate levels of activity improve lipoprotein profiles and glucose homeostasis. At least some of these effects may be related to an improvement in NO production. Nevertheless, it is unclear how physical activity impacts CVD risk correlates such as cholesterol, blood pressure and insulin sensitivity – which specific metabolic, cellular and metabolic processes are affected and how physical activity modifies the nature and the interrelationships between the emergent properties of complex systems that regulate cardiovascular homeostasis and health. A better understanding of the effects of physical activity and exercise is required not only to devise new strategies to promote cardiovascular health and fitness or improve athletic performance, but also to intervene in conditions that promote CVD by preventing physical activity. Inactivity imposed by physical or mental disabilities or extended bed rest and physical restrain compromises cardiovascular health. Indeed, it has been reported that the association between depression and CVD could be largely explained by physical inactivity.[215] Hence, a better understanding of the metabolic processes that could be modified to simulate the effects of physical activity could provide new therapeutic approaches to minimize the impact of physical inactivity. For instance, protein and branched-chain amino acid supplements can partially prevent cardiac atrophy in healthy women following bed rest for 60 days.[216] Other similar interventions could be developed from a better understanding of the molecular and cellular mechanisms behind the benefits of physical activity.

Physical activity is a defining feature of animal life and a primordial response to the non-negotiable, albeit intermittent, demands of the natural environment. Accordingly, most if not all, physiological processes have evolved under environments that needed constant physical activity. However, a gradual decrease in this demand due to industrialization, the development of labor-saving devices, efficient transportation, and entertainment modalities that do not require physical activity has resulted in an environment in which physical activity is not a necessary condition but a personal choice. The community environment in its current urban incarnation does not demand, support, or even encourage physical activity. Thus, like diet, inactivity is an issue not just for individuals, but for the entire community. The risk is mostly derived from the environment because, in large measure, it has been imposed by societal changes. Although few individuals chose to engineer physical activity into daily life, 50% of adults and 65% of school children in the US do not participate in physical activity required to derive cardiovascular benefits.[206] While personal education and individual motivation remain the bedrock of prevention, community-wide changes in neighborhood characteristics, the built environment, recreational opportunities, and academic curricula are required to create an environment conducive to physical activity.

1.7.4 Smoke and Mirrors

No other personal choice has a more negative impact on cardiovascular health than smoking. The CVD risk burden is enormous. Cigarette smoking results in a 2- to 3-fold increased risk of dying of coronary heart disease and stroke and on average adults who smoke die 13 to 14 years earlier than nonsmokers. In the US, smoking results in 443 000 premature deaths leading to a loss of over 5 million potential life years and $193 billion in direct medical costs and lost productivity every year.[1] Yet, nearly 20% Americans continue to smoke and 6100 new smokers are added every day (2007–2008 data). World-wide estimates suggest that 35% of men in developed countries and 50% in the developing countries smoke (WHO report) – over 300 million in China alone. About 5 trillion cigarettes are manufactured each year – nearly 1000 cigarettes for every person on the planet and 15 billion cigarettes are smoked world-wide every day. Although from its peak in 1960s, smoking has declined by 40–50% in the developed countries,[1] it continues to be a personal choice for nearly one billion people in the world.

The high CVD risk associated with cigarette smoking is indicated by overwhelming epidemiological and experimental data.[182,217] These data provide incontrovertible evidence that smoking affects nearly all types of CVD. Meta-analysis of several large studies suggests that smoking is associated with a relative risk of coronary artery disease risk of 1.8–2.0. Even in light smokers (1 to 9 cigarettes per day) the relative risk is 1.66 and the risk of ischemic heart disease in light smokers (1 to 4 cigarettes per day) is nearly 3 times that of a nonsmoker. Men who smoke 6 to 9 cigarettes per day have a relative risk of 2.1 for myocardial infarction compared with nonsmokers.[218] Recent analysis suggests that the exposure–response relationship is nonlinear.[219] In this analysis, substantial excess risk was found to be associated with even very light smoking. The curve rises steeply at low exposure levels and flattens out at higher levels, such that 80% of the risk of smoking > 20 cigarettes per day is associated with smoking < 3 cigarettes a day. This is significant because it explains why low level of exposure to $PM_{2.5}$ in ambient air or second-hand smoke is associated with much higher CVD risk and mortality than would be expected from a linear relationship.

The nonlinear exposure–response relationship between ischemic heart disease and smoking illustrates and underscores the exquisite sensitivity of the cardiovascular system to environmental pollutants particularly combustion products. In comparison, the dose–response relationship of cigarette smoking with lung cancer shows no threshold level and the risk is more monotonically distributed. For those smoking > 20 cigarettes per day the risk of dying from lung cancer is >23 times higher, while those who smoke 1–4 cigarettes per day have a 3-fold higher risk than nonsmokers.[218] Clearly, the mechanisms by which cigarette smoke causes heart disease and cancer differ; perhaps because the chemical constituents of tobacco that cause heart disease are different from those that cause cancer or because the intrinsic sensitivity of heart disease to tobacco smoke is higher than that of lung cancer. Moreover, even though chronic smoking exacerbates both heart disease and cancer, acute exposure to tobacco smoke triggers adverse cardiovascular changes, whereas chronic exposures are required for carcinogenesis, indicating that smoking has both chronic and acute effects on CVD, while the effects on lung cancer are primarily chronic.

Even though the most prevalent and pronounced effects of tobacco smoke are on acute myocardial ischemia, unstable angina and sudden cardiac death, smoking also increases the risk of other cardiac diseases.[182] Several studies show that smoking increases the risk of peripheral artery disease as well as stroke. Smokers are also more susceptible to arrhythmias and smoke cessation has been reported to reduce arrhythmic death in patients after acute myocardial infarction. Smoking decreases regional left ventricular function even in asymptomatic individuals and significantly (45–80%) increases the risk of heart failure. Because most CVD manifestations are affected by smoking, it appears that exposure to tobacco smoke affects fundamental mechanisms that are vital for cardiovascular health and function.

The cardiovascular pathology induced by smoking is unique. Extensive epidemiological research supports the view that smoking is an independent CVD risk factor.[168] This means that the effects of smoking cannot be entirely explained by changes in other risk factors such as hypercholesterolemia, hypertension and diabetes. In other words, the CVD risk imposed by smoking may be related to mechanisms that do not significantly affect other risk factors. A meta-analysis of 54 different studies suggests that although smoking increases LDL-C and decreases HDL, changes in lipids account for < 10% of the excessive CVD risk in smokers.[182] Similarly, even though smoking acutely affects blood pressure, smokers tend to maintain a lower blood pressure and antihypertensive therapy does not completely mitigate the CVD risk of smoking. The contribution of insulin resistance is also uncertain. Although some studies show that smoking increases insulin resistance and smokers have higher insulin levels, others have found no difference in insulin sensitivity or insulin resistance between smokers and nonsmokers.[220] Additional prospective studies are required to determine whether treatment with insulin-sensitizers could attenuate the CVD risk of smoking or whether the effects of smoking persist despite normalization of insulin sensitivity.

The cardiovascular effects of tobacco smoke have been extensively studied on experimental animals. The laboratory data are consistent with the epidemiological evidence. Major epidemiological findings have been validated in the laboratory setting.[182] These studies show that exposure to tobacco smoke increases atherosclerosis and the severity of myocardial infarction, although the effects of smoke exposure on heart failure, arrhythmogensis and plaque rupture have not been studied. Exposure to tobacco smoke has also been shown to induce dyslipidemia, endothelial dysfunction and increased thrombosis, but whether any or all of these processes contribute to smoking-induced CVD remains unclear. Part of the problem arises from the complexity of the tobacco smoke. More than 4000 different chemicals have been detected in tobacco smoke. Several of these could injure cardiovascular tissues by a variety of mechanisms: their effects could also be additive or synergistic. In addition, tobacco smoke constituents could potentially interact with each other or they could become bioactivated *in vivo* and thereby acquire new toxicological properties distinct from the parent compounds. Hence, extensive new investigations are required to develop a rigorous understanding of the molecular mechanisms by which the major constituents of tobacco smoke affect cardiovascular function and to assess their individual efficacy. A clear-eyed mechanistic evaluation of tobacco-smoke toxicity, however, is obfuscated by a haze of social and ethical concerns fanned by good-intentioned preventive efforts and righteous indignation. That smoking is a bad personal choice has been abundantly demonstrated and widely accepted. Nevertheless, scientific elucidation of the mechanisms of smoking-induced CVD is still necessary. Smoking is a paradigmatic environmental exposure and therefore its understanding can shed new light on how primarily pulmonary insults affect CVD incidence, severity and mortality and whether similar mechanisms can account for the toxicity of other similar combustion-derived pollutants such as PM or traffic-generated pollutants. Moreover, and perhaps more importantly, it continues to affect the lives and the cardiovascular health of over 1 billion people.

Although smoking is a personal choice for many, for others, secondhand smoke is a part of their personal and community environment. Despite a nearly 50% decrease in exposure from the 1980s, 126 million nonsmokers in the US are still exposed to secondhand smoke (1999–2002).[1] According to the 2006 report of the US Surgeon General, 60% or 22 million US children (3–11 years) are exposed to secondhand smoke. The effects of secondhand smoke on CVD are nearly as large as smoking.[221] As is evident from the nonlinear nature of the exposure–response curve for tobacco-smoke exposure (*vide supra*), even low levels of exposure to tobacco smoke can significantly increase CVD mortality. In agreement with this view, the risk associated with passive smoke has been estimated to be between 1.45 and 1.57, or in other words, passive smoking is associated with 68–86% of the risk of light smoking.

The high CVD risk of exposure to SHS may be related in part to the high chemical toxicity of sidestream tobacco smoke. In experimental studies it has been found that at equal doses extracts of sidestream are more potent in activating platelets than those of mainstream smoke. The high toxicity of sidestream smoke may be related to differences in the combustion chemistry of cigarettes during puffing or smoldering, which results in greater formation of reactive combustion products such as acrolein in sidestream than in mainstream smoke. Indeed, the toxicological effects of acrolein and related aldehydes are similar to those observed with smoking or secondhand smoke; bolstering the view that these aldehydes may be important mediators of cardiovascular toxicity of tobacco smoke (see Chapter 13 for further discussion on aldehyde toxicity).

Several mechanisms can account for the high cardiovascular toxicity of secondhand smoke. Significantly, the mechanisms affected by exposure to secondhand smoke and those that mediate its toxicity are similar to those involved in the toxicity of smoking as well as exposure to PM (Chapters 2 and 3), traffic-generated pollutants (Chapters 9 and 10), manufactured nanoparticles (Chapter 11), metals (Chapter 12) and aldehydes (Chapter 13). This similarity is not surprising, given the fact that tobacco smoke is a multi-component pollutant that contains high levels of particulate matter, metals and aldehydes, and therefore, cardiovascular responses to tobacco-smoke exposure are likely to be a summation of the responses to all its constituents. Significantly, nicotine, which is the main addictive component of tobacco smoke, seems to be rather innocuous, because the atherogenic effects of secondhand smoke in animal models persist despite removal of nicotine from the cigarettes.[221] As with other pollutants, the mechanisms of the cardiovascular toxicity of secondhand smoke have been linked to endothelial dysfunction, increased blood coagulation and dyslipidemia leading to an increase in atherosclerosis and vascular function induced by oxidative stress and low-grade chronic inflammation.[221] These mechanisms are discussed in detail in several chapters of this monograph.

1.8 Mechanisms of Environmental CVD

1.8.1 *Risky Business*

Current strategies to prevent or treat CVD are based on the concept of risk factors. The major risk factors identified by the Framingham Heart Study are – cigarette smoking, diabetes, hypertension and high serum cholesterol and low HDL levels. Many other secondary risk factors have also been associated with CVD risk. These include increased levels of triglycerides, C-reactive protein, lipoprotein (a), and homocysteine, and obesity, physical inactivity, small low-density lipoprotein particles, family history of premature CVD and coagulation factor abnormalities. The secondary risk factors are not included in current Framingham risk assessment because they impart CVD risk in part by affecting the major risk factors. For instance, serum triglycerides levels are inversely related to HDL and therefore, the risk from elevated triglycerides is partly reflected by low HDL levels. Similarly, even though family history is a positive risk correlate and a useful clinical measure of CVD risk, it is difficult to evaluate the effect of a positive family history independent of other major risk factors. More than 85% of the excess risk for coronary heart disease in the US population could be accounted for by the sum of the 5 major risk factors and the addition of other risk factors does not increase the predictive efficacy of risk estimates.[168] Analysis of data from 14 randomized trials and 2 observational studies indicates that 80 to 90% of the patients who develop coronary heart disease and 95% of those who experience a fatal coronary event have at least 1 of the 5 major risk factors.[222]

The major risk factors have a continuous dose-dependent effect on CVD risk. Individuals with higher levels of risk factors have higher risk and the total weight of risk-factor burden bears a continuous relationship to CVD risk. Although cut points of risk are used to design interventions, in terms of mechanisms, there appear to be no threshold levels that could be considered risk-free. Nevertheless, some people with low risk-factor burden may have CVD, while others with a high risk profile may remain disease free, suggesting that the presence of a risk factor may be necessary, but not sufficient, for developing CVD. In addition, the impact of one risk factor could be modified by other risk factors. For example, high HDL levels could mitigate the effects of high LDL. Alternatively, environmental factors could also modify the contribution of one risk factor to overall CVD risk. For instance, diabetic individuals could reduce their CVD risk by being physically active or people prone to hypertension could decrease CVD risk by avoiding salt. Thus, the effects of risk factors could be significantly modified by the environment and the environmental factors that affect CVD development.

Major risk factors have a high clinical utility because they are related to the development of atherosclerosis and therefore risk-factor treatment decreases CVD. High LDL and low HDL levels promote the formation of large, fatty unstable plaques that are more prone to rupture;[223] therefore anti-cholesterol therapy decreases the risk of myocardial infarction and unstable angina. Hypertension appears to promote plaque development rather than plaque rupture. It is associated with the development of stable plaques[223] and lowering blood pressure does not seem to reduce the risk of acute events even though it decreases disease progression. The role of diabetes is less clear. Although hyperglycemia has been suggested to promote plaque development, strict glycemic control does not decrease CVD risk.[224] Nevertheless, diabetes leads to a more diffuse and inflammatory atherosclerosis, suggesting that it may be affecting both plaque progression and rupture. Cigarette smoking also accelerates atherogenesis. However, autopsy studies show that smoking is associated with eroded plaques with acute thrombus.[223] In comparison with nonsmokers, the plaques of smokers are richer in smooth muscle cell and proteoglycans, have infrequent calcification, but more pronounced intimal thickening. These plaques tend to embolize more frequently and may be associated with coronary vasospasm. The distinct features of plaques in smokers suggest that the mechanisms by which smoking affects atherogenesis may be different from those that mediate the effects of high cholesterol or diabetes.

The mechanistic dissimilarity between the major risk factors is consistent with their statistical independence. This independence implies that the effects of one risk factor are not related to those of the other. The observation that diabetes is an independent CVD risk factor suggests that the effects of diabetes cannot be alleviated completely by treating hypertension or hypercholesterolemia. Lowering blood pressure or cholesterol levels will decrease CVD in diabetics by decreasing the total disease risk, but it may not completely alleviate the effects of diabetes. Similarly, reductions in cholesterol, blood pressure or insulin resistance are unlikely to mitigate completely the effects of smoking (leaving cessation as the only completely affective intervention). Given that the effects of other pollutants – PM, traffic-generated particles, *etc.*, are similar to smoking, it is likely that their risks are also not affected by current risk-reduction strategies targeted at reducing the burden of other risk factors.

Currently, risk-factor reduction is the main strategy for treating CVD. Risk factors contribute to disease mechanisms and they show a dose-dependent association with risk. Nevertheless, the risk factors themselves are a summation of several physiological and metabolic processes. The levels of LDL and HDL, for instance, are regulated by hepatic metabolism as well as a variety of other systemic processes. Similarly, blood pressure is regulated by renal physiology, the rennin–angiotensin system and responses localized to the vascular endothelium. Likewise, diabetes is the summation of insulin resistance in the liver, skeletal muscle, adipocytes and the pancreas. Because risk factors are emergent properties of complex systems, they are highly sensitive to changes in several internal mechanisms as well as a range of environmental factors. Therefore, risk-factor therapy only treats the outcomes or the symptoms of the disease but not the disease itself. For example, cholesterol-lowering drugs do not treat the causes of hypercholesterolemia nor do they correct the underlying metabolic defect. The question why cholesterol is up in the first place is not addressed. This is important because cholesterol may be elevated as a protective response to repair injured tissue; therefore decreasing cholesterol may decrease plaque formation, but in the long run the therapy does not address the demand that was driving the need for high cholesterol in the first place. Similar problems are also associated with treatments of hypertension and diabetes. Because there is no consensus regarding the mechanisms that cause hypertension or diabetes current therapy is symptomatic and largely focused on decreasing blood pressure or insulin resistance, without addressing underlying mechanisms. The failure of symptomatic risk-factor management is reflected by the fact that patients with drug-treated hypertension, high cholesterol, or type-2 diabetes have a higher CVD risk than individuals who do not have these conditions.[225] Clearly risk-factor reduction does not cure heart disease. Hence, it may be necessary to understand the molecular and cellular mechanisms that contribute to risk-factor burden. Because environment factors are important regulators of the processes (*e.g.*, high-fat diet, inactivity) that contribute to risk (hypercholesterolemia, diabetes) it may be more useful to remove environmental causes or correct metabolic defects (bottom-up approach) rather than to treat symptomatic risk factors (top-down approach).

1.9 Implications of an Environmental Perspective

The forgoing discussion reveals the many ways in which the environment affects heart disease. As we have seen, several aspects of the natural, community and personal environment significantly affect CVD. Cardiovascular health and function are intricately tied to the natural rhythms of the night and day, sunlight and altitude. The social structures we create, the urban environments we live in, and the personal choices we make, significantly impact CVD risk. We have seen that elements of the human environment, like genes, are transmitted across generations and our current genetics is a product of our past environments. We have found extensive evidence that the effects of the heritable, but constantly changing, environment are as large, if not larger, than the genes we inherit. The evidence supporting the environmental basis of CVD is overwhelming to the extent that it appears that all aspects of our lives and all things that we do, directly or indirectly, affect CVD risk. This is because cardiovascular health is inextricably linked to all aspects of life and is the summation of our genetic and metabolic responses to the environments that we inherit, chose or create.

Much of what is discussed here – the cardiovascular effects of diet, smoking, physical inactivity, socioeconomic status, pollution, *etc.*, is widely known. Yet, when examined in aggregate, the totality of this evidence builds a more compelling case for the environmental basis of heart disease. Moreover, this redemptive appropriation of extant facts provides a new perspective. It shows us that the environment is the primary determinant of heart disease and that the relationships between individual elements of our environment are complex and interdependent. Together, these environmental factors influence and modify equally complex physiological responses that regulate the adoption and the development of the major risk factors – dyslipidemia, hypertension and insulin resistance. In turn, these risk factors are themselves properties of complex systems that emerge from the working of multiple cellular and molecular networks. These networks form the proximal interface between the individual and his or her environment. But, despite this vision of staggering complexity, the new perspective provides a firm view of the major outlines of CVD etiology, and has important implications for prevention, treatment and understanding of heart disease.

The clearest direction provided by the environmental perspective points toward prevention. Extensive evidence shows that most CVD is preventable. Simple healthy lifestyle choices can alleviate CVD risk burden in a large percentage of the population. Lifestyle modifications are also effective in secondary prevention. However, as we have seen, lifestyle choices are not entirely independent of the community environment and are indeed biased by social networks, cultural and familial backgrounds and socioeconomic factors such as education. Hence, a narrow focus on lifestyle choices alone overlooks the impact of other factors that determine these choices. It is akin to addressing risk factors without addressing the underlying mechanisms. Moreover, CVD risk is evenly distributed between the individual and the community. Hence, population-based prevention strategies are required to create healthy environments that encourage, or at least permit, physical activity; provide access to fresh produce; and afford living in clean, verdant neighborhoods. Intensive, community-wide efforts are also required to decrease the levels of pollution, toxicant and chemicals in our environment and to address the problem of global climate change.

Because CVD often starts *in utero* and progresses throughout life, prevention strategies cannot be targeted only to those who have adverse risk factors, but that they must be extended universally to prevent the development of risk factors in the first place. Such primordial prevention strategies are likely to be more effective in preventing heart disease because CVD risk is widely diffused throughout the whole population.[226] They are based on promoting ideal cardiovascular health assessed by the simultaneous presence of favorable health behaviors such as abstinence from smoking, maintenance of an ideal body weight and healthy diet. They advocate that cardiovascular health should be assessed by the presence of health factors rather than the absence of risk factors so they could be prevented before the manifestation of clinical disease or risk. But the specifics of such prevention efforts have to take into account the genetic and cultural differences between human populations. Such efforts should be based on minimizing the mismatch between our current environment and our ancestral genes with a clear understanding of our past evolutionary history and of the gene–environment interactions unique to each individual.

The current concept of personalized medicine recognizes that each individual is unique. Each person has a distinctive genetic make-up and, therefore, this uniqueness must be taken into account when designing any preventive or therapeutic strategy. However, it should also be recognized that each person lives in a unique environment, which has strong, pervasive and dominant effects on CVD risk. Therefore, in addition to unique genes, we must also take into account the personal environmental history of an individual. But perhaps, by themselves, neither genetic nor environmental histories are sufficient. What is more important is the understanding of the interactions between the genes and the environment and how they result in the creation of a diseased phenotype. Understanding of these interactions is important because it appears that cardiovascular health is maintained by a harmony between genetic disposition and environmental circumstance.

Current medical therapies and preventive measures are mostly targeted at the individual. This approach is based on the assumption that disease arises solely from physiological disorders and that the cause of the disease resides within that person. It is thought that differences in CVD susceptibility are related primarily to individual genetic differences. This view fosters the development of pharmacological strategies, and potentially gene therapy, to attenuate disease in "at risk" individuals. On the other hand, epidemiological studies find causes of CVD within the population and point towards social and environmental factors from which common-source epidemics such as CVD arise. However, a more integrated approach will be to target the interactions between susceptibility genes and environmental factors that contribute to CVD. The risk imparted by these interactions could be minimized to a large extent by modifying the personal environment and lifestyle choices but also by changing the community environment from which a sizeable portion of the risk is derived. This would require a better understanding of how various aspects of the environment interact with specific metabolic, physiologic and genetic processes.

Perhaps the most important conclusion from this survey is the realization that further studies are urgently needed. In physiological, biochemical and genetic experiments the environment is carefully controlled and the impact of environmental variations is minimized. By their very design, these experiments preclude any understanding of environmental factors. Although such experimental designs are important for studying intrinsic physiological function, they fail to take into account the effects of the environment. Hence, to understand a predominantly environmental disease such as CVD, we need to examine the environment itself and how changes in the environment affect specific physiological processes. Examples of such designs are documented in several chapters of this monograph. These designs show how well-defined components of the environments such as particulate matter affect the development and the manifestations of CVD. These studies serve as a paradigm for developing other experimental strategies to elucidate the influence of other environmental factors.

Even studying one component of the environment may not be enough. As we have seen, specific aspects of the environment are interrelated and therefore additional work is required to understand how individual component of the environment interact with each other and how these interactions modify their ability to impart CVD risk. For instance, even though episodic increases in PM are associated with increased CVD mortality, it is not known how variations in season, the time of the day, the altitude, geography or the sunlight modify this risk. Similarly, chronic exposure to pollutants is associated with an increase in CVD prevalence, but it is unclear whether this relationship is independent of the classic risk factors and how it is modified by components of the community and personal environment, such as neighborhood characteristics, diet, nutrition or physical activity.

Finally, the complex relationships between the environment and CVD can only be understood by elucidating causative mechanisms. Most of the evidence supporting the link between the environment and CVD is derived from epidemiological studies, which, as pointed out before, cannot establish cause-and-effect relationships. But, often, these relationships are considered scientifically valid even when they lack specific hypothesis and a causative explanation. Results of logistic regression equations are interpreted as the cause of disease without biological interpretation of the data.[227] It is believed that we do not need to know the specific molecular biology of a relationship to implement prevention and therapy. Without knowing the mechanisms, prevention could be successful, but perhaps only when the relationship is overwhelmingly strong. For example, the strength of the epidemiological relationship between smoking and CVD justifies prevention without understanding the molecular and cellular effects of smoking. However, in most other cases, the relationship is not so strong or clear-cut; therefore, approaches based on a descriptive understanding of phenomena are unlikely to be successful and may even prove to be harmful. Recommendations to decrease total dietary fat led to a reduction in US fat consumption from 36.4% of total calories in 1977 to 32.8% in 2009,[225] however, at the same time the rates of diabetes and obesity have exploded. Carbohydrate consumption increased from 44.3% to 50.3% and the total caloric intake from 2000 to 2250 kcal/d. Although it is unclear whether these changes are interconnected, we now know that some fats (omega-3 and polyunsaturated fatty acids) could prevent CVD. Therefore, in the future preventive strategies must be based on more thorough understanding of the basic mechanisms that contribute to epidemiologic relationships and the biological causes from which these relationships are derived. After all, scientific understanding is occasioned by causal explanation. It is brought about and affected by describing natural phenomenon in such a way that we understand what is responsible for the feature in the natural world that we are interested in. In our current understanding of the environmental basis of CVD we have succeeded in identifying some of the features of natural world and our self-created environment that affect CVD, and in some cases we have even found causal links. But most of the evidence remains observational and additional work is required to test its scientific validity.

To develop a better understanding of the effects of the environment on CVD we must seek additional mechanistic explanations. For this, we will need to understand the proximal causes that could be attributed to the personal environment as well as distal causes related to the population risk and evolutionary adaptation. We also have to understand more completely the interrelationships between different components of our environment and how they impact specific molecular and cellular events and their complex interactions that give rise to emergent properties particularly those that regulate blood pressure, cholesterol levels and insulin sensitivity. In each case we will need to establish causality. We have to understand the processes by which environmental influences contribute to clinical and subclinical CVD, with the overall aim of not just preventing disease but promoting health.

The environment and heart disease are the two most important problems that confront us today. Cardiovascular diseases, with their risk correlates of diabetes and obesity are the leading causes of death in the world (Figure 1.1). The environment remains a major global issue as well. Not only because of concerns regarding world-wide climate change, but also because we are beginning to realize that the environments we have created around ourselves are unhealthy. Millions of citizens live in large urban centers or megacities that provide limited access to fresh food, few opportunities for physical activity, but are replete with pollutants, chemicals and toxicants. Although it is widely accepted that our personal environments, populated by our lifestyle choices, have a significant impact on heart disease, the effects of the natural and community environment are less well appreciated. But, as we have seen, the influence of these environments permeates through all aspects and stages of heart disease. We are irrevocably tethered to the primordial rhythms of our natural environment and we are exquisitely sensitive to the contours of our social identity. Thus, changes in our natural and social environments have an indelible effect on our health. Disruption of the sleep–wake cycle, lack of exposure to natural sunlight, low socioeconomic status, unhealthy social networks, constrained urban planning and architecture and exposure to pollutants impair cardiovascular health. These unhealthy influences of the modern environment contribute to the growing epidemic of cardiometabolic diseases and may even be changing our evolutionary trajectory. To thrive, therefore, we must learn healthful individual behavior and commit ourselves to promoting the development of local and global environments that are conducive to cardiovascular health.

References

1. D. Lloyd-Jones, R. J. Adams, T. M. Brown, M. Carnethon, S. Dai, G. De Simone, T. B. Ferguson, E. Ford, K. Furie, C. Gillespie, A. Go, K. Greenlund, N. Haase, S. Hailpern, P. M. Ho, V. Howard, B. Kissela, S. Kittner, D. Lackland, L. Lisabeth, A. Marelli, M. M. McDermott, J. Meigs, D. Mozaffarian, M. Mussolino, G. Nichol, V. L. Roger, W. Rosamond, R. Sacco, P. Sorlie, R. Stafford, T. Thom, S. Wasserthiel-Smoller, N. D. Wong and J. Wylie-Rosett, Heart disease and stroke statistics – 2010 update: a report from the american heart association, *Circulation*, 2010, **121**, e46–e215.
2. S. T. Stewart, D. M. Cutler and A. B. Rosen, Forecasting the effects of obesity and smoking on US life expectancy, *N. Engl. J. Med.*, 2009, **361**, 2252–2260.
3. H. Watkins and M. Farrall, Genetic susceptibility to coronary artery disease: from promise to progress, *Nature Rev. Genet.*, 2006, **7**, 163–173.
4. H. C. Harpending, M. A. Batzer, M. Gurven, L. B. Jorde, A. R. Rogers and S. T. Sherry, Genetic traces of ancient demography, *Proc. Natl. Acad. Sci. USA*, 1998, **95**, 1961–1967.
5. M. V. Olson and A. Varki, Sequencing the chimpanzee genome: insights into human evolution and disease, *Nature Rev. Genet.*, 2003, **4**, 20–28.
6. A. Varki, A chimpanzee genome project is a biomedical imperative, *Genome Res.*, 2000, **10**, 1065–1070.
7. K. Hill, C. Boesch, J. Goodall, A. Pusey, J. Williams and R. Wrangham, Mortality rates among wild chimpanzees, *J. Hum. Evol.*, 2001, **40**, 437–450.

8. J. M. Williams, E. V. Lonsdorf, M. L. Wilson, J. Schumacher-Stankey, J. Goodall and A. E. Pusey, Causes of death in the Kasekela chimpanzees of Gombe National Park, Tanzania, *Am. J. Primatol.*, 2008, **70**, 766–777.
9. K. Benirschke and F. D. Adams, Gorilla diseases and causes of death, *J. Reprod. Fertil. Suppl*, 1980, **Suppl 28**, 139–148.
10. R. E. Junge, L. E. Mezei, M. C. Muhlbauer and M. Weber, Cardiovascular evaluation of lowland gorillas, *J. Am. Vet. Med. Assoc.*, 1998, **212**, 413–415.
11. C. J. Doane, D. R. Lee and M. M. Sleeper, Electrocardiogram abnormalities in captive chimpanzees (Pan troglodytes), *Comp. Med.*, 2006, **56**, 512–518.
12. B. M. Seiler, E. J. Dick Jr, R. Guardado-Mendoza, J. L. VandeBerg, J. T. Williams, J. N. Mubiru and G. B. Hubbard, Spontaneous heart disease in the adult chimpanzee (Pan troglodytes), *J. Med. Primatol.*, 2009, **38**, 51–58.
13. M. L. Lammey, D. R. Lee, J. J. Ely and M. M. Sleeper, Sudden cardiac death in 13 captive chimpanzees (Pan troglodytes), *J. Med. Primatol.*, 2008, **37**(Suppl 1), 39–43.
14. M. M. Sleeper, C. J. Doane, P. H. Langner, S. Curtis, K. Avila and D. R. Lee, Successful treatment of idiopathic dilated cardiomyopathy in an adult chimpanzee (Pan troglodytes), *Comp. Med.*, 2005, **55**, 80–84.
15. J. F. Hansen, P. L. Alford and M. E. Keeling, Diffuse myocardial fibrosis and congestive heart failure in an adult male chimpanzee, *Vet. Pathol.*, 1984, **21**, 529–531.
16. C. L. Miller, A. M. Schwartz, J. S. Barnhart Jr and M. D. Bell, Chronic hypertension with subsequent congestive heart failure in a western lowland gorilla (Gorilla gorilla gorilla), *J. Zoo. Wildl. Med.*, 1999, **30**, 262–267.
17. E. J. Baitchman, P. P. Calle, T. L. Clippinger, S. L. Deem, S. B. James, B. L. Raphael and R. A. Cook, Preliminary evaluation of blood lipid profiles in captive western lowland gorillas (Gorilla gorilla gorilla), *J. Zoo. Wildl. Med.*, 2006, **37**, 126–129.
18. M. Caceres, J. Lachuer, M. A. Zapala, J. C. Redmond, L. Kudo, D. H. Geschwind, D. J. Lockhart, T. M. Preuss and C. Barlow, Elevated gene expression levels distinguish human from non-human primate brains, *Proc. Natl. Acad. Sci. USA*, 2003, **100**, 13030–13035.
19. M. V. Olson, When less is more: gene loss as an engine of evolutionary change, *Am. J. Hum. Genet.*, 1999, **64**, 18–23.
20. The Chimpanzee Sequencing and Analysis Consortium, Initial sequence of the chimpanzee genome and comparison with the human genome. *Nature*, 2005, **437**, 69–87.
21. K. Ding and I. J. Kullo, Evolutionary genetics of coronary heart disease, *Circulation*, 2009, **119**, 459–467.
22. D. Y. Hui, T. L. Innerarity and R. W. Mahley, Defective hepatic lipoprotein receptor binding of beta-very low density lipoproteins from type III hyperlipoproteinemic patients. Importance of apolipoprotein, *E. J. Biol. Chem.*, 1984, **259**, 860–869.

23. C. S. Hanlon and D. C. Rubinsztein, Arginine residues at codons 112 and 158 in the apolipoprotein E gene correspond to the ancestral state in humans, *Atherosclerosis*, 1995, **112**, 85–90.
24. P. de Knijff, A. M. van den Maagdenberg, R. R. Frants and L. M. Havekes, Genetic heterogeneity of apolipoprotein E and its influence on plasma lipid and lipoprotein levels, *Hum. Mutat.*, 1994, **4**, 178–194.
25. J. H. Stengard, K. E. Zerba, J. Pekkanen, C. Ehnholm, A. Nissinen and C. F. Sing, Apolipoprotein E polymorphism predicts death from coronary heart disease in a longitudinal study of elderly Finnish men, *Circulation*, 1995, **91**, 265–269.
26. R. M. Corbo and R. Scacchi, Apolipoprotein E (APOE) allele distribution in the world. Is APOE*4 a 'thrifty' allele?, *Ann. Hum. Genet.*, 1999, **63**, 301–310.
27. L. U. Gerdes, C. Gerdes, P. S. Hansen, I. C. Klausen and O. Faergeman, Are men carrying the apolipoprotein epsilon 4- or epsilon 2 allele less fertile than epsilon 3 epsilon 3 genotypes?, *Hum. Genet.*, 1996, **98**, 239–242.
28. J. Davignon, R. E. Gregg and C. F. Sing, Apolipoprotein E polymorphism and atherosclerosis, *Arteriosclerosis*, 1988, **8**, 1–21.
29. A. B. Weder, Evolution and hypertension, *Hypertension*, 2007, **49**, 260–265.
30. N. E. Morton, C. L. Gulbrandsen, D. C. Rao, G. G. Rhoads and A. Kagan, Determinants of blood pressure in Japanese-American Families, *Hum. Genet.*, 1980, **53**, 261–266.
31. J. L. Annest, C. F. Sing, P. Biron and J. G. Mongeau, Familial aggregation of blood pressure and weight in adoptive families. II. Estimation of the relative contributions of genetic and common environmental factors to blood pressure correlations between family members, *Am. J. Epidemiol.*, 1979, **110**, 492–503.
32. R. S. Danziger, Hypertension in an anthropological and evolutionary paradigm, *Hypertension*, 2001, **38**, 19–22.
33. V. L. Burt, P. Whelton, E. J. Roccella, C. Brown, J. A. Cutler, M. Higgins, M. J. Horan and D. Labarthe, Prevalence of hypertension in the US adult population. Results from the Third National Health and Nutrition Examination Survey, 1988–1991, *Hypertension*, 1995, **25**, 305–313.
34. S. M. Fullerton, A. G. Clark, K. M. Weiss, D. A. Nickerson, S. L. Taylor, J. H. Stengard, V. Salomaa, E. Vartiainen, M. Perola, E. Boerwinkle and C. F. Sing, Apolipoprotein E variation at the sequence haplotype level: implications for the origin and maintenance of a major human polymorphism, *Am. J. Hum. Genet.*, 2000, **67**, 881–900.
35. E. M. Crimmins and C. E. Finch, Infection, inflammation, height, and longevity, *Proc. Natl. Acad. Sci. USA*, 2006, **103**, 498–503.
36. C. E. Finch and E. M. Crimmins, Inflammatory exposure and historical changes in human life-spans, *Science*, 2004, **305**, 1736–1739.
37. W. C. Willett, Balancing life-style and genomics research for disease prevention, *Science*, 2002, **296**, 695–698.

38. T. A. Martino and M. J. Sole, Molecular time: an often overlooked dimension to cardiovascular disease, *Circ. Res.*, 2009, **105**, 1047–1061.
39. C. A. Czeisler, J. F. Duffy, T. L. Shanahan, E. N. Brown, J. F. Mitchell, D. W. Rimmer, J. M. Ronda, E. J. Silva, J. S. Allan, J. S. Emens, D. J. Dijk and R. E. Kronauer, Stability, precision, and near-24-hour period of the human circadian pacemaker, *Science*, 1999, **284**, 2177–2181.
40. T. Martino, S. Arab, M. Straume, D. D. Belsham, N. Tata, F. Cai, P. Liu, M. Trivieri, M. Ralph and M. J. Sole, Day–night rhythms in gene expression of the normal murine heart, *J. Mol. Med.*, 2004, **82**, 256–264.
41. P. McNamara, S. B. Seo, R. D. Rudic, A. Sehgal, D. Chakravarti and G. A. FitzGerald, Regulation of CLOCK and MOP4 by nuclear hormone receptors in the vasculature: a humoral mechanism to reset a peripheral clock, *Cell*, 2001, **105**, 877–889.
42. J. Rutter, M. Reick, L. C. Wu and S. L. McKnight, Regulation of clock and NPAS2 DNA binding by the redox state of NAD cofactors, *Science*, 2001, **293**, 510–514.
43. S. Imai, C. M. Armstrong, M. Kaeberlein and L. Guarente, Transcriptional silencing and longevity protein Sir2 is an NAD-dependent histone deacetylase, *Nature*, 2000, **403**, 795–800.
44. B. Grimaldi, Y. Nakahata, M. Kaluzova, S. Masubuchi and P. Sassone-Corsi, Chromatin remodeling, metabolism and circadian clocks: the interplay of CLOCK and SIRT1, *Int. J. Biochem. Cell Biol.*, 2009, **41**, 81–86.
45. C. Choudhary, C. Kumar, F. Gnad, M. L. Nielsen, M. Rehman, T. C. Walther, J. V. Olsen and M. Mann, Lysine acetylation targets protein complexes and co-regulates major cellular functions, *Science*, 2009, **325**, 834–840.
46. S. Zhao, W. Xu, W. Jiang, W. Yu, Y. Lin, T. Zhang, J. Yao, L. Zhou, Y. Zeng, H. Li, Y. Li, J. Shi, W. An, S. M. Hancock, F. He, L. Qin, J. Chin, P. Yang, X. Chen, Q. Lei, Y. Xiong and K. L. Guan, Regulation of cellular metabolism by protein lysine acetylation, *Science*, 2010, **327**, 1000–1004.
47. M. E. Young, P. Razeghi, A. M. Cedars, P. H. Guthrie and H. Taegtmeyer, Intrinsic diurnal variations in cardiac metabolism and contractile function, *Circ. Res.*, 2001, **89**, 1199–1208.
48. J. E. Muller, P. H. Stone, Z. G. Turi, J. D. Rutherford, C. A. Czeisler, C. Parker, W. K. Poole, E. Passamani, R. Roberts, T. Robertson, B. E. Sobel, J. T. Wilterson and E. Braunwald and MILIS Study Group, Circadian variation in the frequency of onset of acute myocardial infarction, *N. Engl. J. Med.*, 1985, **313**, 1315–1322.
49. M. C. Cohen, K. M. Rohtla, C. E. Lavery, J. E. Muller and M. A. Mittleman, Meta-analysis of the morning excess of acute myocardial infarction and sudden cardiac death, *Am. J. Cardiol.*, 1997, **79**, 1512–1516.
50. J. E. Muller, G. H. Tofler and P. H. Stone, Circadian variation and triggers of onset of acute cardiovascular disease, *Circulation*, 1989, **79**, 733–743.

51. R. H. Mehta, R. Manfredini, F. Hassan, U. Sechtem, E. Bossone, J. K. Oh, J. V. Cooper, D. E. Smith, F. Portaluppi, M. Penn, S. Hutchison, C. A. Nienaber, E. M. Isselbacher and K. A. Eagle, Chronobiological patterns of acute aortic dissection, *Circulation*, 2002, **106**, 1110–1115.
52. M. Sumiyoshi, S. Kojima, M. Arima, S. Suwa, Y. Nakazato, H. Sakurai, T. Kanoh, Y. Nakata and H. Daida, Circadian, weekly, and seasonal variation at the onset of acute aortic dissection, *Am. J. Cardiol.*, 2002, **89**, 619–623.
53. I. Kawachi, G. A. Colditz, M. J. Stampfer, W. C. Willett, J. E. Manson, F. E. Speizer and C. H. Hennekens, Prospective study of shift work and risk of coronary heart disease in women, *Circulation*, 1995, **92**, 3178–3182.
54. A. Knutsson and H. Boggild, Shiftwork and cardiovascular disease: review of disease mechanisms, *Rev. Environ. Health*, 2000, **15**, 359–372.
55. D. L. Brown, D. Feskanich, B. N. Sanchez, K. M. Rexrode, E. S. Schernhammer and L. D. Lisabeth, Rotating night shift work and the risk of ischemic stroke, *Am. J. Epidemiol.*, 2009, **169**, 1370–1377.
56. A. J. Davidson, M. T. Sellix, J. Daniel, S. Yamazaki, M. Menaker and G. D. Block, Chronic jet-lag increases mortality in aged mice, *Curr. Biol.*, 2006, **16**, R914–916.
57. F. A. Scheer, M. F. Hilton, C. S. Mantzoros and S. A. Shea, Adverse metabolic and cardiovascular consequences of circadian misalignment, *Proc. Natl. Acad. Sci. USA*, 2009, **106**, 4453–4458.
58. K. L. Knutson, A. M. Ryden, B. A. Mander and E. Van Cauter, Role of sleep duration and quality in the risk and severity of type-2 diabetes mellitus, *Arch. Intern. Med.*, 2006, **166**, 1768–1774.
59. N. D. Kohatsu, R. Tsai, T. Young, R. Vangilder, L. F. Burmeister, A. M. Stromquist and J. A. Merchant, Sleep duration and body mass index in a rural population, *Arch. Intern. Med.*, 2006, **166**, 1701–1705.
60. J. E. Gangwisch, S. B. Heymsfield, B. Boden-Albala, R. M. Buijs, F. Kreier, T. G. Pickering, A. G. Rundle, G. K. Zammit and D. Malaspina, Short sleep duration as a risk factor for hypertension: analyses of the first National Health and Nutrition Examination Survey, *Hypertension*, 2006, **47**, 833–839.
61. K. J. Mukamal, J. E. Muller, M. Maclure, J. B. Sherwood and M. A. Mittleman, Increased risk of congestive heart failure among infarctions with nighttime onset, *Am. Heart J.*, 2000, **140**, 438–442.
62. J. P. Henriques, A. P. Haasdijk and F. Zijlstra, Outcome of primary angioplasty for acute myocardial infarction during routine duty hours versus during off-hours, *J. Am. Coll. Cardiol.*, 2003, **41**, 2138–2142.
63. D. J. Durgan, T. Pulinilkunnil, C. Villegas-Montoya, M. E. Garvey, N. G. Frangogiannis, L. H. Michael, C. W. Chow, J. R. Dyck and M. E. Young, Short communication: ischemia/reperfusion tolerance is time-of-day-dependent: mediation by the cardiomyocyte circadian clock, *Circ. Res.*, 2010, **106**, 546–550.
64. G. S. Kelly, Seasonal variations of selected cardiovascular risk factors, *Altern. Med. Rev.*, 2005, **10**, 307–320.

65. D. Robinson, E. A. Bevan, S. Hinohara and T. Takahashi, Seasonal variation in serum cholesterol levels – evidence from the UK and Japan, *Atherosclerosis*, 1992, **95**, 15–24.
66. M. Frohlich, M. Sund, S. Russ, A. Hoffmeister, H. G. Fischer, V. Hombach and W. Koenig, Seasonal variations of rheological and hemostatic parameters and acute-phase reactants in young, healthy subjects, *Arterioscler. Thromb. Vasc. Biol.*, 1997, **17**, 2692–2697.
67. D. J. Gordon, D. C. Trost, J. Hyde, F. S. Whaley, P. J. Hannan, D. R. Jacobs Jr and L. G. Ekelund, Seasonal cholesterol cycles: the Lipid Research Clinics Coronary Primary Prevention Trial placebo group, *Circulation*, 1987, **76**, 1224–1231.
68. J. Sasaki, G. Kumagae, T. Sata, M. Ikeda, S. Tsutsumi and K. Arakawa, Seasonal variation of serum high density lipoprotein cholesterol levels in men, *Atherosclerosis*, 1983, **48**, 167–172.
69. P. Tung, S. D. Wiviott, C. P. Cannon, S. A. Murphy, C. H. McCabe and C. M. Gibson, Seasonal variation in lipids in patients following acute coronary syndrome on fixed doses of Pravastatin (40 mg) or Atorvastatin (80 mg) (from the Pravastatin or Atorvastatin Evaluation and Infection Therapy-Thrombolysis In Myocardial Infarction 22 [PROVE IT-TIMI 22] Study), *Am. J. Cardiol.*, 2009, **103**, 1056–1060.
70. M. Bluher, B. Hentschel, F. Rassoul and V. Richter, Influence of dietary intake and physical activity on annual rhythm of human blood cholesterol concentrations, *Chronobiol. Int.*, 2001, **18**, 541–557.
71. R. W. Stout and V. Crawford, Seasonal variations in fibrinogen concentrations among elderly people, *Lancet*, 1991, **338**, 9–13.
72. P. R. Woodhouse, K. T. Khaw, M. Plummer, A. Foley and T. W. Meade, Seasonal variations of plasma fibrinogen and factor VII activity in the elderly: winter infections and death from cardiovascular disease, *Lancet*, 1994, **343**, 435–439.
73. R. C. Hermida, C. Calvo, D. E. Ayala, J. E. Lopez, J. R. Fernandez, A. Mojon, M. J. Dominguez and M. Covelo, Seasonal variation of fibrinogen in dipper and nondipper hypertensive patients, *Circulation*, 2003, **108**, 1101–1106.
74. V. L. Crawford, S. E. McNerlan and R. W. Stout, Seasonal changes in platelets, fibrinogen and factor VII in elderly people, *Age Ageing*, 2003, **32**, 661–665.
75. C. Otto, M. G. Donner, P. Schwandt and W. O. Richter, Seasonal variations of hemorheological and lipid parameters in middle-aged healthy subjects, *Clin. Chim. Acta*, 1996, **256**, 87–94.
76. A. R. Rudnicka, A. Rumley, G. D. Lowe and D. P. Strachan, Diurnal, seasonal, and blood-processing patterns in levels of circulating fibrinogen, fibrin D-dimer, C-reactive protein, tissue plasminogen activator and von Willebrand factor in a 45-year-old population, *Circulation*, 2007, **115**, 996–1003.

77. R. R. West and C. R. Lowe, Mortality from ischaemic heart disease – inter-town variation and its association with climate in England and Wales, *Int. J. Epidemiol.*, 1976, **5**, 195–201.
78. H. Tanaka, M. Shinjo, H. Tsukuma, Y. Kawazuma, S. Shimoji, N. Kinoshita and T. Morita, Seasonal variation in mortality from ischemic heart disease and cerebrovascular disease in Okinawa and Osaka: the possible role of air temperature, *J. Epidemiol.*, 2000, **10**, 392–398.
79. A. G. Barnett, M. de Looper and J. F. Fraser, The seasonality in heart-failure deaths and total cardiovascular deaths, *Aust. N. Z. J. Public Health*, 2008, **32**, 408–413.
80. S. C. Inglis, R. A. Clark, S. Shakib, D. T. Wong, P. Molaee, D. Wilkinson and S. Stewart, Hot summers and heart failure: seasonal variations in morbidity and mortality in Australian heart failure patients (1994–2005), *Eur. J. Heart Fail.*, 2008, **10**, 540–549.
81. S. Stewart, K. McIntyre, S. Capewell and J. J. McMurray, Heart failure in a cold climate. Seasonal variation in heart failure-related morbidity and mortality, *J. Am. Coll. Cardiol.*, 2002, **39**, 760–766.
82. J. P. Pell and S. M. Cobbe, Seasonal variations in coronary heart disease, *QJM*, 1999, **92**, 689–696.
83. F. A. Spencer, R. J. Goldberg, R. C. Becker and J. M. Gore, Seasonal distribution of acute myocardial infarction in the second National Registry of Myocardial Infarction, *J. Am. Coll. Cardiol.*, 1998, **31**, 1226–1233.
84. F. Enquselassie, A. J. Dobson, H. M. Alexander and P. L. Steele, Seasons, temperature and coronary disease, *Int. J. Epidemiol.*, 1993, **22**, 632–636.
85. C. Spielberg, D. Falkenhahn, S. N. Willich, K. Wegscheider and H. Voller, Circadian, day-of-week, and seasonal variability in myocardial infarction: comparison between working and retired patients, *Am. Heart J.*, 1996, **132**, 579–585.
86. R. A. Kloner, W. K. Poole and R. L. Perritt, When throughout the year is coronary death most likely to occur? A 12-year population-based analysis of more than 220 000 cases, *Circulation*, 1999, **100**, 1630–1634.
87. R. Manfredini, B. Boari, M. Gallerani, R. Salmi, E. Bossone, A. Distante, K. A. Eagle and R. H. Mehta, Chronobiology of rupture and dissection of aortic aneurysms, *J. Vasc. Surg.*, 2004, **40**, 382–388.
88. T. Sheth, C. Nair, J. Muller and S. Yusuf, Increased winter mortality from acute myocardial infarction and stroke: the effect of age, *J. Am. Coll. Cardiol.*, 1999, **33**, 1916–1919.
89. B. Marchant, K. Ranjadayalan, R. Stevenson, P. Wilkinson and A. D. Timmis, Circadian and seasonal factors in the pathogenesis of acute myocardial infarction: the influence of environmental temperature, *Br. Heart. J.*, 1993, **69**, 385–387.
90. A. Argiles, G. Mourad and C. Mion, Seasonal changes in blood pressure in patients with end-stage renal disease treated with hemodialysis, *N. Engl. J. Med.*, 1998, **339**, 1364–1370.

91. J. M. Hayward, W. F. Holmes and B. A. Gooden, Cardiovascular responses in man to a stream of cold air, *Cardiovasc. Res.*, 1976, **10**, 691–696.
92. P. R. Woodhouse, K. T. Khaw and M. Plummer, Seasonal variation of blood pressure and its relationship to ambient temperature in an elderly population, *J. Hypertens.*, 1993, **11**, 1267–1274.
93. M. E. Widlansky, J. A. Vita, M. J. Keyes, M. G. Larson, N. M. Hamburg, D. Levy, G. F. Mitchell, E. W. Osypiuk, R. S. Vasan and E. J. Benjamin, Relation of season and temperature to endothelium-dependent flow-mediated vasodilation in subjects without clinical evidence of cardiovascular disease (from the Framingham Heart Study), *Am. J. Cardiol.*, 2007, **100**, 518–523.
94. G. M. Bull, M. Brozovic, R. Chakrabarti, T. W. Meade, J. Morton, W. R. North and Y. Stirling, Relationship of air temperature to various chemical, haematological, and haemostatic variables, *J. Clin. Pathol.*, 1979, **32**, 16–20.
95. R. A. Kloner, Natural and unnatural triggers of myocardial infarction, *Prog. Cardiovasc. Dis.*, 2006, **48**, 285–300.
96. B. J. Scherlag, E. Patterson and R. Lazzara, Seasonal variation in sudden cardiac death after experimental myocardial infarction, *J. Electrocardiol.*, 1990, **23**, 223–230.
97. P. M. Laplaud, L. Beaubatie and D. Maurel, A spontaneously seasonal hypercholesterolemic animal: plasma lipids and lipoproteins in the European badger (Meles meles L.), *J. Lipid Res.*, 1980, **21**, 724–738.
98. D. E. Wallis, S. Penckofer and G. W. Sizemore, The "sunshine deficit" and cardiovascular disease, *Circulation*, 2008, **118**, 1476–1485.
99. M. F. Holick, J. A. MacLaughlin and S. H. Doppelt, Regulation of cutaneous previtamin D3 photosynthesis in man: skin pigment is not an essential regulator, *Science*, 1981, **211**, 590–593.
100. T. L. Clemens, J. S. Adams, S. L. Henderson and M. F. Holick, Increased skin pigment reduces the capacity of skin to synthesise vitamin D3, *Lancet*, 1982, **1**, 74–76.
101. M. F. Holick, Photosynthesis of vitamin D in the skin: effect of environmental and life-style variables, *Fed. Proc.*, 1987, **46**, 1876–1882.
102. A. R. Webb, L. Kline and M. F. Holick, Influence of season and latitude on the cutaneous synthesis of vitamin D3: exposure to winter sunlight in Boston and Edmonton will not promote vitamin D3 synthesis in human skin, *J. Clin. Endocrinol. Metab.*, 1988, **67**, 373–378.
103. W. F. Loomis, Skin-pigment regulation of vitamin-D biosynthesis in man, *Science*, 1967, **157**, 501–506.
104. M. F. Holick, Resurrection of vitamin D deficiency and rickets, *J. Clin. Invest.*, 2006, **116**, 2062–2072.
105. M. F. Holick and T. C. Chen, Vitamin D deficiency: a worldwide problem with health consequences, *Am. J. Clin. Nutr.*, 2008, **87**, 1080S–1086S.
106. R. Alemzadeh, J. Kichler, G. Babar and M. Calhoun, Hypovitaminosis D in obese children and adolescents: relationship with adiposity, insulin sensitivity, ethnicity, and season, *Metabolism*, 2008, **57**, 183–191.

107. A. C. Looker, B. Dawson-Hughes, M. S. Calvo, E. W. Gunter and N. R. Sahyoun, Serum 25-hydroxyvitamin D status of adolescents and adults in two seasonal subpopulations from NHANES III, *Bone*, 2002, **30**, 771–777.
108. S. Cheng, J. M. Massaro, C. S. Fox, M. G. Larson, M. J. Keyes, E. L. McCabe, S. J. Robins, C. J. O'Donnell, U. Hoffmann, P. F. Jacques, S. L. Booth, R. S. Vasan, M. Wolf and T. J. Wang, Adiposity, cardiometabolic risk, and vitamin D status: the Framingham Heart Study, *Diabetes*, 2010, **59**, 242–248.
109. S. G. Rostand, Ultraviolet light may contribute to geographic and racial blood pressure differences, *Hypertension*, 1997, **30**, 150–156.
110. F. Kokot, H. Schmidt-Gayk, A. Wiecek, Z. Mleczko and B. Bracel, Influence of ultraviolet irradiation on plasma vitamin D and calcitonin levels in humans, *Kidney Int. Suppl.*, 1989, **27**, S143–146.
111. R. Krause, M. Buhring, W. Hopfenmuller, M. F. Holick and A. M. Sharma, Ultraviolet B and blood pressure, *Lancet*, 1998, **352**, 709–710.
112. A. Sugden, J. Smith and E. Pennisi, The future of forests, *Science*, 2008, **320**, 1435.
113. J. P. Forman, E. Giovannucci, M. D. Holmes, H. A. Bischoff-Ferrari, S. S. Tworoger, W. C. Willett and G. C. Curhan, Plasma 25-hydroxyvitamin D levels and risk of incident hypertension, *Hypertension*, 2007, **49**, 1063–1069.
114. J. H. Lee, J. H. O'Keefe, D. Bell, D. D. Hensrud and M. F. Holick, Vitamin D deficiency an important, common, and easily treatable cardiovascular risk factor?, *J. Am. Coll. Cardiol.*, 2008, **52**, 1949–1956.
115. R. Bouillon, G. Carmeliet, L. Verlinden, E. van Etten, A. Verstuyf, H. F. Luderer, L. Lieben, C. Mathieu and M. Demay, Vitamin D and human health: lessons from vitamin D receptor null mice, *Endocr. Rev.*, 2008, **29**, 726–776.
116. T. J. Wang, M. J. Pencina, S. L. Booth, P. F. Jacques, E. Ingelsson, K. Lanier, E. J. Benjamin, R. B. D'Agostino, M. Wolf and R. S. Vasan, Vitamin D deficiency and risk of cardiovascular disease, *Circulation*, 2008, **117**, 503–511.
117. P. Autier and S. Gandini, Vitamin D supplementation and total mortality: a meta-analysis of randomized controlled trials, *Arch. Intern. Med.*, 2007, **167**, 1730–1737.
118. K. C. Chiu, A. Chu, V. L. Go and M. F. Saad, Hypovitaminosis D is associated with insulin resistance and beta cell dysfunction, *Am. J. Clin. Nutr.*, 2004, **79**, 820–825.
119. L. U. Gerdes, The common polymorphism of apolipoprotein E: geographical aspects and new pathophysiological relations, *Clin. Chem. Lab. Med.*, 2003, **41**, 628–631.
120. D. Penaloza and J. Arias-Stella, The heart and pulmonary circulation at high altitudes: healthy highlanders and chronic mountain sickness, *Circulation*, 2007, **115**, 1132–1146.
121. P. Bartsch and J. S. Gibbs, Effect of altitude on the heart and the lungs, *Circulation*, 2007, **116**, 2191–2202.

122. J. Arias-Stella and M. Saldana, The Terminal Portion of the Pulmonary Arterial Tree in People Native to High Altitudes, *Circulation*, 1963, **28**, 915–925.
123. S. de Mendoza, H. Nucete, E. Ineichen, E. Salazar, A. Zerpa and C. J. Glueck, Lipids and lipoproteins in subjects at 1,000 and 3,500 meter altitudes, *Arch. Environ. Health*, 1979, **34**, 308–311.
124. S. Sharma, Clinical, biochemical, electrocardiographic and noninvasive hemodynamic assessment of cardiovascular status in natives at high to extreme altitudes (3000–5500 m) of the Himalayan region, *Ind. Heart J.*, 1990, **42**, 375–379.
125. S. Mohanna, R. Baracco and S. Seclen, Lipid profile, waist circumference, and body mass index in a high altitude population, *High Alt. Med. Biol.*, 2006, **7**, 245–255.
126. S. Dominguez Coello, A. Cabrera De Leon, F. Bosa Ojeda, L. I. Perez Mendez, L. Diaz Gonzalez and A. J. Aguirre-Jaime, High density lipoprotein cholesterol increases with living altitude, *Int. J. Epidemiol.*, 2000, **29**, 65–70.
127. A. Cabrera de Leon, D. A. Gonzalez, L. I. Mendez, A. Aguirre-Jaime, M. del Cristo Rodriguez Perez, S. D. Coello and I. C. Trujillo, Leptin and altitude in the cardiovascular diseases, *Obes. Res.*, 2004, **12**, 1492–1498.
128. K. Fronek and N. Alexander, Sympathetic activity, lipids accumulation and arterial wall morphology in rabbits at high altitude, *Am. J. Physiol.*, 1986, **250**, R485–492.
129. J. Ferezou, J. P. Richalet, T. Coste and C. Rathat, Changes in plasma lipids and lipoprotein cholesterol during a high altitude mountaineering expedition (4800 m), *Eur. J. Appl. Physiol. Occup. Physiol.*, 1988, **57**, 740–745.
130. W. Schobersberger, P. Schmid, M. Lechleitner, S. P. von Duvillard, H. Hortnagl, H. C. Gunga, A. Klingler, D. Fries, K. Kirsch, R. Spiesberger, R. Pokan, P. Hofmann, F. Hoppichler, G. Riedmann, H. Baumgartner and E. Humpeler, Austrian Moderate Altitude Study 2000 (AMAS 2000). The effects of moderate altitude (1,700 m) on cardiovascular and metabolic variables in patients with metabolic syndrome, *Eur. J. Appl. Physiol.*, 2003, **88**, 506–514.
131. J. Ferezou, J. P. Richalet, C. Serougne, T. Coste, E. Wirquin and D. Mathe, Reduction of postprandial lipemia after acute exposure to high altitude hypoxia, *Int. J. Sports Med.*, 1993, **14**, 78–85.
132. N. Fujimoto, K. Matsubayashi, T. Miyahara, A. Murai, M. Matsuda, H. Shio, H. Suzuki, M. Kameyama, A. Saito and L. Shuping, The risk factors for ischemic heart disease in Tibetan highlanders, *Jpn. Heart J.*, 1989, **30**, 27–34.
133. D. Faeh, F. Gutzwiller and M. Bopp, Lower mortality from coronary heart disease and stroke at higher altitudes in Switzerland, *Circulation*, 2009, **120**, 495–501.
134. N. Baibas, A. Trichopoulou, E. Voridis and D. Trichopoulos, Residence in mountainous compared with lowland areas in relation to total and

coronary mortality. A study in rural Greece, *J. Epidemiol. Community Health*, 2005, **59**, 274–278.
135. A. Zittermann, S. S. Schleithoff and R. Koerfer, Putting cardiovascular disease and vitamin D insufficiency into perspective, *Br. J. Nutr.*, 2005, **94**, 483–492.
136. M. F. Holick, T. C. Chen, Z. Lu and E. Sauter, Vitamin D and skin physiology: a D-lightful story, *J. Bone Miner Res*, 2007, **22**(Suppl 2), V28–33.
137. R. Sturm, Increases in morbid obesity in the USA: 2000–2005, *Public Health*, 2007, **121**, 492–496.
138. G. A. Kaplan and J. E. Keil, Socioeconomic factors and cardiovascular disease: a review of the literature, *Circulation*, 1993, **88**, 1973–1998.
139. R. M. Worth, H. Kato, G. G. Rhoads, K. Kagan and S. L. Syme, Epidemiologic studies of coronary heart disease and stroke in Japanese men living in Japan, Hawaii and California: mortality, *Am. J. Epidemiol.*, 1975, **102**, 481–490.
140. J. V. Patel, A. Vyas, J. K. Cruickshank, D. Prabhakaran, E. Hughes, K. S. Reddy, M. I. Mackness, D. Bhatnagar and P. N. Durrington, Impact of migration on coronary heart-disease risk factors: comparison of Gujaratis in Britain and their contemporaries in villages of origin in India, *Atherosclerosis*, 2006, **185**, 297–306.
141. E. Hedlund, J. Kaprio, A. Lange, M. Koskenvuo, L. Jartti, T. Ronnemaa and N. Hammar, Migration and coronary heart disease: A study of Finnish twins living in Sweden and their co-twins residing in Finland, *Scand. J. Public Health*, 2007, **35**, 468–474.
142. P. Pekka, P. Pirjo and U. Ulla, Influencing public nutrition for non-communicable disease prevention: from community intervention to national programme – experiences from Finland, *Public Health Nutr.*, 2002, **5**, 245–251.
143. L. Clancy, P. Goodman, H. Sinclair and D. W. Dockery, Effect of air-pollution control on death rates in Dublin, Ireland: an intervention study, *Lancet*, 2002, **360**, 1210–1214.
144. W. A. Zatonski and W. Willett, Changes in dietary fat and declining coronary heart disease in Poland: population based study, *BMJ*, 2005, **331**, 187–188.
145. B. Unal, J. A. Critchley and S. Capewell, Explaining the decline in coronary heart disease mortality in England and Wales between 1981 and 2000, *Circulation*, 2004, **109**, 1101–1107.
146. E. S. Ford, U. A. Ajani, J. B. Croft, J. A. Critchley, D. R. Labarthe, T. E. Kottke, W. H. Giles and S. Capewell, Explaining the decrease in US deaths from coronary disease, 1980–2000, *N. Engl. J. Med.*, 2007, **356**, 2388–2398.
147. J. Critchley, J. Liu, D. Zhao, W. Wei and S. Capewell, Explaining the increase in coronary heart disease mortality in Beijing between 1984 and 1999, *Circulation*, 2004, **110**, 1236–1244.
148. D. G. Meyers, J. S. Neuberger and J. He, Cardiovascular effect of bans on smoking in public places: a systematic review and meta-analysis, *J. Am. Coll. Cardiol.*, 2009, **54**, 1249–1255.

149. A. Antonovsky, Social class and the major cardiovascular diseases, *J. Chronic Dis.*, 1968, **21**, 65–106.
150. L. E. Hinkle Jr, L. H. Whitney, E. W. Lehman, J. Dunn, B. Benjamin, R. King, A. Plakun and B. Flehinger, Occupation, education, and coronary heart disease. Risk is influenced more by education and background than by occupational experiences, in the Bell System, *Science*, 1968, **161**, 238–246.
151. Hypertension Detection and Follow-up Program Cooperative Group, Race, education and prevalence of hypertension. *Am. J. Epidemiol.*, 1977, **106**, 351–361.
152. M. G. Marmot, M. J. Shipley and G. Rose, Inequalities in death – specific explanations of a general pattern?, *Lancet*, 1984, **1**, 1003–1006.
153. B. Chaix, Geographic life environments and coronary heart disease: a literature review, theoretical contributions, methodological updates, and a research agenda, *Annu. Rev. Public Health*, 2009, **30**, 81–105.
154. A. V. Diez-Roux, F. J. Nieto, C. Muntaner, H. A. Tyroler, G. W. Comstock, E. Shahar, L. S. Cooper, R. L. Watson and M. Szklo, Neighborhood environments and coronary heart disease: a multilevel analysis, *Am. J. Epidemiol.*, 1997, **146**, 48–63.
155. A. V. Diez Roux, S. S. Merkin, D. Arnett, L. Chambless, M. Massing, F. J. Nieto, P. Sorlie, M. Szklo, H. A. Tyroler and R. L. Watson, Neighborhood of residence and incidence of coronary heart disease, *N. Engl. J. Med.*, 2001, **345**, 99–106.
156. M. A. Papas, A. J. Alberg, R. Ewing, K. J. Helzlsouer, T. L. Gary and A. C. Klassen, The built environment and obesity, *Epidemiol. Rev.*, 2007, **29**, 129–143.
157. G. S. Lovasi, M. A. Hutson, M. Guerra and K. M. Neckerman, Built environments and obesity in disadvantaged populations, *Epidemiol. Rev.*, 2009, **31**, 7–20.
158. A. H. Auchincloss, A. V. Diez Roux, D. G. Brown, C. A. Erdmann and A. G. Bertoni, Neighborhood resources for physical activity and healthy foods and their association with insulin resistance, *Epidemiology*, 2008, **19**, 146–157.
159. F. Li, P. Harmer, B. J. Cardinal, M. Bosworth, D. Johnson-Shelton, J. M. Moore, A. Acock and N. Vongjaturapat, Built environment and 1-year change in weight and waist circumference in middle-aged and older adults: Portland Neighborhood Environment and Health Study, *Am. J. Epidemiol.*, 2009, **169**, 401–408.
160. F. Li, P. Harmer, B. J. Cardinal, M. Bosworth and D. Johnson-Shelton, Obesity and the built environment: does the density of neighborhood fast-food outlets matter?, *Am. J. Health Promot.*, 2009, **23**, 203–209.
161. L. D. Frank, M. A. Andresen and T. L. Schmid, Obesity relationships with community design, physical activity, and time spent in cars, *Am. J. Prev. Med.*, 2004, **27**, 87–96.
162. B. Chaix, K. Bean, C. Leal, F. Thomas, S. Havard, D. Evans, B. Jego and B. Pannier, Individual/neighborhood social factors and blood pressure in

the RECORD Cohort Study: which risk factors explain the associations?, *Hypertension*, 2010, **55**, 769–775.
163. N. A. Christakis and J. H. Fowler, The spread of obesity in a large social network over 32 years, *N. Engl. J. Med.*, 2007, **357**, 370–379.
164. N. A. Christakis and J. H. Fowler, The collective dynamics of smoking in a large social network, *N. Engl. J. Med.*, 2008, **358**, 2249–2258.
165. A. L. Barabasi, Network medicine – from obesity to the "diseasome", *N. Engl. J. Med.*, 2007, **357**, 404–407.
166. Summary of the second report of the National Cholesterol Education Program (NCEP) Expert Panel on Detection, Evaluation, and Treatment of High Blood Cholesterol in Adults (Adult Treatment Panel II), *JAMA*, 1993, **269**, 3015–3023.
167. R. H. Myers, D. K. Kiely, L. A. Cupples and W. B. Kannel, Parental history is an independent risk factor for coronary artery disease: the Framingham Study, *Am. Heart J.*, 1990, **120**, 963–969.
168. S. M. Grundy, G. J. Balady, M. H. Criqui, G. Fletcher, P. Greenland, L. F. Hiratzka, N. Houston-Miller, P. Kris-Etherton, H. M. Krumholz, J. LaRosa, I. S. Ockene, T. A. Pearson, J. Reed, R. Washington and S. C. Smith Jr, Primary prevention of coronary heart disease: guidance from Framingham: a statement for healthcare professionals from the AHA Task Force on Risk Reduction, American Heart Association, *Circulation*, 1998, **97**, 1876–1887.
169. S. Zdravkovic, A. Wienke, N. L. Pedersen, M. E. Marenberg, A. I. Yashin and U. De Faire, Heritability of death from coronary heart disease: a 36-year follow-up of 20 966 Swedish twins, *J. Intern. Med.*, 2002, **252**, 247–254.
170. A. Wienke, N. V. Holm, A. Skytthe and A. I. Yashin, The heritability of mortality due to heart diseases: a correlated frailty model applied to Danish twins, *Twin Res.*, 2001, **4**, 266–274.
171. G. S. Omenn, Evolution in health and medicine Sackler colloquium: Evolution and public health, *Proc. Natl. Acad. Sci. USA*, 2010, **107**(Suppl 1), 1702–1709.
172. J. R. Speakman, A nonadaptive scenario explaining the genetic predisposition to obesity: the "predation release" hypothesis, *Cell Metab.*, 2007, **6**, 5–12.
173. S. Hong, J. P. Candelone, C. C. Patterson and C. F. Boutron, Greenland ice evidence of hemispheric lead pollution two millennia ago by Greek and Roman civilizations, *Science*, 1994, **265**, 1841–1843.
174. T. Hartung, Toxicology for the twenty-first century, *Nature*, 2009, **460**, 208–212.
175. T. R. Karl and K. E. Trenberth, Modern global climate change, *Science*, 2003, **302**, 1719–1723.
176. J. Lelieveld, H. Berresheim, S. Borrmann, P. J. Crutzen, F. J. Dentener, H. Fischer, J. Feichter, P. J. Flatau, J. Heland, R. Holzinger, R. Korrmann, M. G. Lawrence, Z. Levin, K. M. Markowicz, N. Mihalopoulos, A. Minikin, V. Ramanathan, M. De Reus, G. J. Roelofs, H. A. Scheeren, J. Sciare, H. Schlager, M. Schultz, P. Siegmund, B. Steil, E. G. Stephanou,

P. Stier, M. Traub, C. Warneke, J. Williams and H. Ziereis, Global air pollution crossroads over the Mediterranean, *Science*, 2002, **298**, 794–799.
177. J. Lelieveld, P. J. Crutzen, V. Ramanathan, M. O. Andreae, C. M. Brenninkmeijer, T. Campos, G. R. Cass, R. R. Dickerson, H. Fischer, J. A. de Gouw, A. Hansel, A. Jefferson, D. Kley, A. T. de Laat, S. Lal, M. G. Lawrence, J. M. Lobert, O. L. Mayol-Bracero, A. P. Mitra, T. Novakov, S. J. Oltmans, K. A. Prather, T. Reiner, H. Rodhe, H. A. Scheeren, D. Sikka and J. Williams, The Indian Ocean experiment: widespread air pollution from South and Southeast Asia, *Science*, 2001, **291**, 1031–1036.
178. M. O. Andreae, Atmosphere. Aerosols before pollution, *Science*, 2007, **315**, 50–51.
179. K. E. Wilkening, L. A. Barrie and M. Engle, Atmospheric science. trans-Pacific air pollution, *Science*, 2000, **290**, 65–67.
180. A. Bhatnagar, Cardiovascular pathophysiology of environmental pollutants, *Am. J. Physiol. Heart Circ. Physiol.*, 2004, **286**, H479–485.
181. A. Bhatnagar, Environmental cardiology: studying mechanistic links between pollution and heart disease, *Circ. Res.*, 2006, **99**, 692–705.
182. T. E. O'Toole, D. J. Conklin and A. Bhatnagar, Environmental risk factors for heart disease, *Rev. Environ. Health*, 2008, **23**, 167–202.
183. S. A. Glantz and W. W. Parmley, Passive smoking and heart disease. Mechanisms and risk, *JAMA*, 1995, **273**, 1047–1053.
184. W. Palinski and C. Napoli, The fetal origins of atherosclerosis: maternal hypercholesterolemia, and cholesterol-lowering or antioxidant treatment during pregnancy influence in utero programming and postnatal susceptibility to atherogenesis, *FASEB J*, 2002, **16**, 1348–1360.
185. W. Palinski and C. Napoli, Impaired fetal growth, cardiovascular disease, and the need to move on, *Circulation*, 2008, **117**, 341–343.
186. S. P. Ng, D. J. Conklin, A. Bhatnagar, D. D. Bolanowski, J. Lyon and J. T. Zelikoff, Prenatal exposure to cigarette smoke induces diet- and sex-dependent dyslipidemia and weight gain in adult murine offspring, *Environ. Health Perspect.*, 2009, **117**, 1042–1048.
187. S. R. Daniels, M. S. Jacobson, B. W. McCrindle, R. H. Eckel and B. M. Sanner, American Heart Association Childhood Obesity Research Summit Report, *Circulation*, 2009, **119**, e489–517.
188. M. J. Stampfer, F. B. Hu, J. E. Manson, E. B. Rimm and W. C. Willett, Primary prevention of coronary heart disease in women through diet and lifestyle, *N. Engl. J. Med.*, 2000, **343**, 16–22.
189. S. E. Chiuve, M. L. McCullough, F. M. Sacks and E. B. Rimm, Healthy lifestyle factors in the primary prevention of coronary heart disease among men: benefits among users and nonusers of lipid-lowering and antihypertensive medications, *Circulation*, 2006, **114**, 160–167.
190. S. E. Chiuve, K. M. Rexrode, D. Spiegelman, G. Logroscino, J. E. Manson and E. B. Rimm, Primary prevention of stroke by healthy lifestyle, *Circulation*, 2008, **118**, 947–954.

191. A. Akesson, C. Weismayer, P. K. Newby and A. Wolk, Combined effect of low-risk dietary and lifestyle behaviors in primary prevention of myocardial infarction in women, *Arch. Intern. Med.*, 2007, **167**, 2122–2127.
192. K. T. Knoops, L. C. de Groot, D. Kromhout, A. E. Perrin, O. Moreiras-Varela, A. Menotti and W. A. van Staveren, Mediterranean diet, lifestyle factors, and 10-year mortality in elderly European men and women: the HALE project, *JAMA*, 2004, **292**, 1433–1439.
193. F. B. Hu, J. E. Manson, M. J. Stampfer, G. Colditz, S. Liu, C. G. Solomon and W. C. Willett, Diet, lifestyle, and the risk of type-2 diabetes mellitus in women, *N. Engl. J. Med.*, 2001, **345**, 790–797.
194. L. Djousse, J. A. Driver and J. M. Gaziano, Relation between modifiable lifestyle factors and lifetime risk of heart failure, *JAMA*, 2009, **302**, 394–400.
195. F. B. Hu, M. J. Stampfer, J. E. Manson, E. Rimm, G. A. Colditz, B. A. Rosner, C. H. Hennekens and W. C. Willett, Dietary fat intake and the risk of coronary heart disease in women, *N. Engl. J. Med.*, 1997, **337**, 1491–1499.
196. D. Mozaffarian, M. B. Katan, A. Ascherio, M. J. Stampfer and W. C. Willett, Trans fatty acids and cardiovascular disease, *N. Engl. J. Med.*, 2006, **354**, 1601–1613.
197. F. B. Hu, E. B. Rimm, M. J. Stampfer, A. Ascherio, D. Spiegelman and W. C. Willett, Prospective study of major dietary patterns and risk of coronary heart disease in men, *Am. J. Clin. Nutr.*, 2000, **72**, 912–921.
198. W. E. Connor, M. T. Cerqueira, R. W. Connor, R. B. Wallace, M. R. Malinow and H. R. Casdorph, The plasma lipids, lipoproteins, and diet of the Tarahumara indians of Mexico, *Am. J. Clin. Nutr.*, 1978, **31**, 1131–1142.
199. R. M. van Dam, E. B. Rimm, W. C. Willett, M. J. Stampfer and F. B. Hu, Dietary patterns and risk for type-2 diabetes mellitus in US men, *Ann. Intern. Med.*, 2002, **136**, 201–209.
200. R. M. van Dam, W. C. Willett, E. B. Rimm, M. J. Stampfer and F. B. Hu, Dietary fat and meat intake in relation to risk of type-2 diabetes in men, *Diabetes Care*, 2002, **25**, 417–424.
201. C. G. Solomon, F. B. Hu, M. J. Stampfer, G. A. Colditz, F. E. Speizer, E. B. Rimm, W. C. Willett and J. E. Manson, Moderate alcohol consumption and risk of coronary heart disease among women with type-2 diabetes mellitus, *Circulation*, 2000, **102**, 494–499.
202. K. J. Mukamal, K. M. Conigrave, M. A. Mittleman, C. A. Camargo Jr., M. J. Stampfer, W. C. Willett and E. B. Rimm, Roles of drinking pattern and type of alcohol consumed in coronary heart disease in men, *N. Engl. J. Med.*, 2003, **348**, 109–118.
203. L. J. Appel, M. W. Brands, S. R. Daniels, N. Karanja, P. J. Elmer and F. M. Sacks, Dietary approaches to prevent and treat hypertension: a scientific statement from the American Heart Association, *Hypertension*, 2006, **47**, 296–308.

204. I. Shai, J. D. Spence, D. Schwarzfuchs, Y. Henkin, G. Parraga, A. Rudich, A. Fenster, C. Mallett, N. Liel-Cohen, A. Tirosh, A. Bolotin, J. Thiery, G. M. Fiedler, M. Bluher, M. Stumvoll and M. J. Stampfer, Dietary intervention to reverse carotid atherosclerosis, *Circulation*, 2010, **121**, 1200–1208.
205. G. S. Getz and C. A. Reardon, Nutrition and cardiovascular disease, *Arterioscler. Thromb. Vasc. Biol.*, 2007, **27**, 2499–2506.
206. F. W. Booth and S. J. Lees, Fundamental questions about genes, inactivity, and chronic diseases, *Physiol. Genomics*, 2007, **28**, 146–157.
207. M. A. Perhonen, F. Franco, L. D. Lane, J. C. Buckey, C. G. Blomqvist, J. E. Zerwekh, R. M. Peshock, P. T. Weatherall and B. D. Levine, Cardiac atrophy after bed rest and spaceflight, *J. Appl. Physiol.*, 2001, **91**, 645–653.
208. G. Antonutto and P. E. di Prampero, Cardiovascular deconditioning in microgravity: some possible countermeasures, *Eur. J. Appl. Physiol.*, 2003, **90**, 283–291.
209. M. Tanasescu, M. F. Leitzmann, E. B. Rimm, W. C. Willett, M. J. Stampfer and F. B. Hu, Exercise type and intensity in relation to coronary heart disease in men, *JAMA*, 2002, **288**, 1994–2000.
210. O. H. Franco, C. de Laet, A. Peeters, J. Jonker, J. Mackenbach and W. Nusselder, Effects of physical activity on life expectancy with cardiovascular disease, *Arch. Intern. Med.*, 2005, **165**, 2355–2360.
211. T. Y. Li, J. S. Rana, J. E. Manson, W. C. Willett, M. J. Stampfer, G. A. Colditz, K. M. Rexrode and F. B. Hu, Obesity as compared with physical activity in predicting risk of coronary heart disease in women, *Circulation*, 2006, **113**, 499–506.
212. M. Tanasescu, M. F. Leitzmann, E. B. Rimm and F. B. Hu, Physical activity in relation to cardiovascular disease and total mortality among men with type-2 diabetes, *Circulation*, 2003, **107**, 2435–2439.
213. P. Gaede, P. Vedel, N. Larsen, G. V. Jensen, H. H. Parving and O. Pedersen, Multifactorial intervention and cardiovascular disease in patients with type-2 diabetes, *N. Engl. J. Med.*, 2003, **348**, 383–393.
214. N. G. Boule, E. Haddad, G. P. Kenny, G. A. Wells and R. J. Sigal, Effects of exercise on glycemic control and body mass in type-2 diabetes mellitus: a meta-analysis of controlled clinical trials, *JAMA*, 2001, **286**, 1218–1227.
215. M. A. Whooley, P. de Jonge, E. Vittinghoff, C. Otte, R. Moos, R. M. Carney, S. Ali, S. Dowray, B. Na, M. D. Feldman, N. B. Schiller and W. S. Browner, Depressive symptoms, health behaviors, and risk of cardiovascular events in patients with coronary heart disease, *JAMA*, 2008, **300**, 2379–2388.
216. T. A. Dorfman, B. D. Levine, T. Tillery, R. M. Peshock, J. L. Hastings, S. M. Schneider, B. R. Macias, G. Biolo and A. R. Hargens, Cardiac atrophy in women following bed rest, *J. Appl. Physiol.*, 2007, **103**, 8–16.
217. J. A. Ambrose and R. S. Barua, The pathophysiology of cigarette smoking and cardiovascular disease: an update, *J. Am. Coll. Cardiol.*, 2004, **43**, 1731–1737.

218. R. E. Schane, P. M. Ling and S. A. Glantz, Health effects of light and intermittent smoking: a review, *Circulation*, 2010, **121**, 1518–1522.
219. C. A. Pope 3rd, R. T. Burnett, D. Krewski, M. Jerrett, Y. Shi, E. E. Calle and M. J. Thun, Cardiovascular mortality and exposure to airborne fine particulate matter and cigarette smoke: shape of the exposure–response relationship, *Circulation*, 2009, **120**, 941–948.
220. G. Reaven and P. S. Tsao, Insulin resistance and compensatory hyperinsulinemia: the key player between cigarette smoking and cardiovascular disease?, *J. Am. Coll. Cardiol.*, 2003, **41**, 1044–1047.
221. J. Barnoya and S. A. Glantz, Cardiovascular effects of secondhand smoke: nearly as large as smoking, *Circulation*, 2005, **111**, 2684–2698.
222. J. G. Canto and A. E. Iskandrian, Major risk factors for cardiovascular disease: debunking the "only 50%" myth, *JAMA*, 2003, **290**, 947–949.
223. A. P. Burke, A. Farb, G. T. Malcom, Y. Liang, J. Smialek and R. Virmani, Effect of risk factors on the mechanism of acute thrombosis and sudden coronary death in women, *Circulation*, 1998, **97**, 2110–2116.
224. Intensive blood-glucose control with sulphonylureas or insulin compared with conventional treatment and risk of complications in patients with type-2 diabetes (UKPDS 33). UK Prospective Diabetes Study (UKPDS) Group. *Lancet*, 1998, **352**, 837–853.
225. D. Mozaffarian, P. W. Wilson and W. B. Kannel, Beyond established and novel risk factors: lifestyle risk factors for cardiovascular disease, *Circulation*, 2008, **117**, 3031–3038.
226. D. M. Lloyd-Jones, Y. Hong, D. Labarthe, D. Mozaffarian, L. J. Appel, L. Van Horn, K. Greenlund, S. Daniels, G. Nichol, G. F. Tomaselli, D. K. Arnett, G. C. Fonarow, P. M. Ho, M. S. Lauer, F. A. Masoudi, R. M. Robertson, V. Roger, L. H. Schwamm, P. Sorlie, C. W. Yancy and W. D. Rosamond, Defining and setting national goals for cardiovascular health promotion and disease reduction: the American Heart Association's strategic Impact Goal through 2020 and beyond, *Circulation*, 2010, **121**, 586–613.
227. L. Kuller, Is phenomenology the best approach to health research?, *Am. J. Epidemiol.*, 2007, **166**, 1109–1115.

CHAPTER 2
Cardiovascular Effects of Particulate-Matter Air Pollution: An Overview and Perspectives

J. A. ARAUJO[1] AND R. D. BROOK[2]

[1] Division of Cardiology, David Geffen School of Medicine at UCLA, Los Angeles, CA, USA; [2] Division of Cardiovascular Medicine, University of Michigan, Ann Arbor, MI, USA

2.1 Introduction

Extensive epidemiological evidence supports the association of air pollution with adverse health effects leading to increased morbidity and mortality.[1–3] In fact, the World Health Organization ranks air pollution as the 13th leading cause of world-wide mortality.[4] Before the 1990s, these adverse health effects were generally ascribed to pulmonary diseases. However, evidence accumulated during the last decade supports the notion that the largest portion of air pollution-related mortality is in fact due to cardiovascular diseases.[5] Pope has estimated that assuming similar relative risks for mortality, 68% of the total deaths attributed to daily changes in particulate air-pollution levels are due to cardiovascular causes, while only 12% are due to respiratory diseases (C. A. Pope, personal communication). Extensive reviews have implicated air pollution as a novel, potentially "modifiable" cardiovascular risk factor of high importance. Whilst the relative risk of exposure for one individual is small (*e.g.* approximately 1 excess death per 5 million people per day for a $10\,\mu\text{g/m}^3$ increase in PM_{10}) the overall adverse impact on the global public health is enormous (~800 000 deaths/year) because air pollution continuously affects a large number of people worldwide.[6–9] Indeed, extant data also demonstrates the public-health importance and benefits that result from improvements in air quality.[10] The purpose of this chapter is to provide a general summary and perspective overview of the epidemiological, clinical and experimental evidence that supports the association of air pollution with increased cardiovascular disease risk. More detailed discussions on specific topics are the subjects of other chapters.

2.2 Air-Pollution Components and Characterization

Air pollution is a complex mixture of compounds in gaseous and particle phases.[8] Examples of gaseous pollutants include ozone, carbon monoxide (CO), along with sulfur and nitrogen oxides. In addition, the particulate matter (PM) phase itself is comprised of heterogenous solid or liquid compounds varying in size, number, chemical composition, surface area, concentration and source.[8,11] PM includes primary particles that are emitted directly from sources such as fossil-fuel combustion (*e.g.* diesel-exhaust particles) and secondary particles, that are generated from gases through chemical reactions involving atmospheric oxygen (O_2), water vapor (H_2O), reactive species such as ozone (O_3), free radicals such as hydroxyl ($\cdot$OH) and nitrate ($\cdot NO_3$) radicals, pollutants such as sulfur dioxide (SO_2), nitrogen oxides (NO_x), and organic gases from natural and anthropogenic sources.[11] While both gases and particles have been linked to detrimental health effects, current evidence suggests that components of PM are most principally responsible for a major portion of the cardiovascular effects of air pollution.[8,11–16] Therefore, this chapter will focus upon studies demonstrating adverse cardiovascular effects of PM.

Particles are classified according to their aerodynamic diameter into size fractions such as PM_{10} ("thoracic" particles, <10 µm), $PM_{2.5-10}$ ("coarse" particles, 2.5 to 10 µm), $PM_{2.5}$ (fine particles, <2.5 µm) and UFP (ultrafine particles, <0.1 µm). These particles are derived from various sources and by various mechanisms as shown in Table 2.1, producing distinct lognormal modes in the particle-size distributions by number and volume (nucleation, Aitken, accumulation and coarse modes). Particles in the nucleation and Aitken modes are typically derived from combustion of fossil fuels and are the result of nucleation of gas-phase compounds to form condensed-phase species in newly formed particles that have had little chance to grow (nucleation mode) or in recently formed particles that are actively undergoing coagulation (Aitken mode). Particles in the accumulation mode form by growing in size and "accumulate" by coagulation (two particles combining to form one) or by condensation (gas molecules condensing on a particle). Particles in the coarse mode are mainly formed by the mechanical breakdown (crushing, grinding, abrasion of surfaces) of crustal material, minerals and organic debris.[11] In brief, it appears that particles that have significant redox potential (*e.g.* metals, organic carbon compounds, semiquinones, polycyclic aromatic hydrocarbons)

Table 2.1 Classification of particles based on size.

Particle	Aerodynamic diameter (μm)	Sources	Mode of generation	Atmospheric half-life
Thoracic particles (PM_{10})	<10	–	–	–
Coarse particles ($PM_{2.5-10}$)	2.5–10	Suspension from disturbed soil (farming, mining, unpaved roads), construction, plant and animal fragments	Mechanical disruption (crushing, grinding, abrasion of surfaces), evaporation of sprays, suspension of dusts.	Minutes to hours
Fine particles ($PM_{2.5}$)	<2.5	Power plants, oil refineries, wildfires, residential fuel combustion, tailpipe and brake emissions	Gas-to-particle conversion by condensation, coagulation (accumulation mode)	Days to weeks
Ultrafine particles (UFP)	<0.1	Fuel combustion (diesel, gasoline) and tailpipe emissions from mobile sources (motor vehicles, aircrafts, ships)	Fresh emissions, secondary photochemical reactions (nucleation mode)	Minutes to hours

PM_{10} includes all other PM fractions.
Source: Araujo and Nel.[9]

that can elicit oxidative stress in biological tissues (lung, heart, vasculature) and thereby activate inflammatory cellular pathways are most likely principally responsible for the health effects.[7,8]

2.3 PM Exposure and Cardiovascular Morbidity and Mortality

Numerous studies conducted throughout the world have shown that both short- and long-term exposures to PM increase the risk for cardiovascular morbidity and mortality. A concise overview of the findings from selected studies of major importance is provided in Table 2.2.

2.3.1 Short-Term Exposures

Short-term exposure (1 day to several weeks' duration) to PM_{10} results in small increases in cardiovascular mortality over the ensuing few days, as evidenced by time-series and case crossover studies[7,9] (Table 2.2). These studies typically

Table 2.2 Selected studies that support an association of exposure to air pollution with increased CV morbidity and/or mortality.

Type of Study	Study	No. of subjects	Exposure variable	Outcome variable	Major Findings	Ref
Time Series	NMMAPS	~50 million	PM_{10}	Daily CP mortality	Daily CP mortality increased by 0.6% per 20 µg/m³ increase in PM_{10} in adults from 20 to 100 US cities and hundreds of counties	83–86
	APHEA2	~43 million	PM_{10}	Daily CV mortality	Daily CV mortality increased by 1.5% per 20 µg/m³ increase in PM_{10} in adults from 29 European cities	18,87
	APHENA	NMMAPS, APHEA2 & Canadian studies	PM_{10}	Daily CV mortality	Daily all-cause mortality increased from 0.2 to 0.6% per 10 µg/m³ increase in PM_{10} in individuals from the US (90 cities), Europe (22 cities) and Canada (12 cities), mostly in individuals >75-year-old.	88
	NCHS	NCHS data on 112 US cities	$PM_{2.5}$	Daily CV mortality	Daily all-cause and CV mortality increase of 0.98% and 0.85% per 10 µg/m³ increase in 2-day averaged $PM_{2.5}$	20

Table 2.2 (continued)

Type of Study	Study	No. of subjects	Exposure variable	Outcome variable	Major Findings	Ref
Cohort Survival Analysis	American Cancer Society II (extended analysis)	~500 000	$PM_{2.5}$	CV mortality	Risk for CV death increased by 12% per 10 μg/m³ increase in long-term $PM_{2.5}$ exposure in an extended 16-year follow-up. Ischemic heart disease was the single largest cause of mortality. Smaller absolute numbers of people (although with similar relative risk elevations) died from arrhythmias and heart failure.	5
	Harvard Six Cities (extended analysis)	8096	$PM_{2.5}$	CV mortality	Relative risk for CV death increased by 1.28 per 10 μg/m³ increase in long-term $PM_{2.5}$ in an extended 28-year follow-up of people leaving in six US cities. Decrease in $PM_{2.5}$ was observed in some cities during the study which correlated with significant reduction in CV mortality.	10

Study	N	Exposure	Outcome	Results	Ref
Woman's Health Initiative	65 893	$PM_{2.5}$	CV acute events and CV mortality	CV acute events and CV mortality increased by 24% and 76% per 10 μg/m³ increase in long-term $PM_{2.5}$ exposure in healthy postmenopausal women in 36 US cities.	23
NHS	66 250	PM_{10}	CV acute events and CV mortality	Fatal CAD disease increased by 43% per 10 μg/m³ increase in long-term PM_{10} exposure among nurses from northeastern US	29
IHCS	~12 865	$PM_{2.5}$	Ischemic coronary events	Daily acute ischemic coronary events increased by 4.5% per 10 μg/m³ increase in $PM_{2.5}$ in patients in UTAH. Increased frequency was only significant among patients with pre-existent CAD.	19
MI Registry in Augsburg (data from KORA)	691	Traffic	MI	Relative risk for an MI increased by 2.92 after exposure to traffic 1 h before the event.	39

CV = cardiovascular, CP = cardiopulmonary, MI = myocardial infarction, APHEA = Air Pollution and Health: A Combined European and North American Approach study, APHENA = Air Pollution and Health: A European Approach study, IHCS = Intermountain Heart Collaborative Study, KORA = Cooperative Health Research in the Region of Augsburg, NCHS = National Center for Health Statistics, NMMAPS = National Morbidity, Mortality, and Air Pollution Study.

capture the effects of temporal variations – meaning the effect of acute changes in PM levels within 1 or more locations on health outcomes. *On average, these studies demonstrate an approximate 1.0% increase in daily cardiovascular mortality (or hospitalization) per 10 µg/m³ increase in PM levels.*

One of the largest studies, of approximately 50 million people from National Morbidity, Mortality, and Air Pollution Study (NMMAPS), found a 0.6% increase in daily cardiopulmonary mortality per 20 µg/m³ increase in PM_{10}.[17] A similar association was confirmed in Europe, where ~43 million adults in 29 European cities from the Air Pollution and Health: A European Approach (APHEA2) study exhibited a 1.5% increase in daily cardiovascular mortality per 20 µg/m³ increase in PM_{10}.[18] Other studies have shown that the magnitude of these associations may be similar to or perhaps even larger in association with the smaller $PM_{2.5}$ fraction *per se*. In 12 865 patients enrolled in the Intermountain Heart Collaborative Study (IHCS) a daily increase of 4.5% in acute ischemic coronary events was observed per 10 µg/m³ increase in ambient $PM_{2.5}$.[19] Recently, Zanobetti and Schwartz specifically evaluated the health effects of short-term $PM_{2.5}$ exposure, as compared with coarse $PM_{10-2.5}$, in 112 US cities. They reported that a 2-day average increase in $PM_{2.5}$ of 10 µg/m³ resulted in 0.85%, 1.18%, and 1.78% increases in cardiovascular, myocardial infarction, and stroke deaths, respectively. Mortality attributed to the coarse PM fraction alone, after accounting for the effects of $PM_{2.5}$, was only significant for stroke. Thus, the risk of cardiovascular mortality posed by short-term PM exposure appears to be largely attributed to particles <2.5 µm in diameter.[20]

Numerous studies have also evaluated the risk of hospitalization for cardiovascular diseases associated with short-term changes in PM levels. In one of the largest studies, Dominici *et al.* that among 11.5 million US medicare enrollees >65 years old that admission rates for all types of cardiovascular events increased in association with a 10 µg/m³ increase in $PM_{2.5}$. The increase in risk was largest for heart failure (1.28%), but was also significant for ischemic heart (0.44%) and cerebrovascular disease (0.81%).[21] Similarly, admissions for chronic obstructive pulmonary diseases were also increased (0.91%). On the other hand, similar to the mortality study results, the investigators did not find significant independent risks for cardiovascular hospitalizations due to coarse particles exposure alone.[22]

2.3.2 Longer-Term Exposures

Associations of PM air-pollution levels with cardiovascular morbidity and mortality have also been demonstrated in exposure studies, as evidenced by cohort survival analysis (Table 2.2). These studies assess the differences in health outcomes across differing geographic locations that vary in long-term PM levels, while controlling for covariates and typically demonstrate larger cardiovascular health risks than do the short-term time-series studies. The risks for cardiovascular-related mortality also appear to be larger (or more statistically robust) than for nonfatal events. These associations may reflect the cumulative health effects of longer-term exposures over months-to-years duration. *On average, these studies demonstrate an approximately 10% increase in cardiovascular mortality per 10 μg/m³ increase in PM levels.*

The American Cancer Society II study demonstrated in ~500 000 adults, a 12% increase in cardiovacular deaths per 10 μg/m³ increase in long-term (annual average) $PM_{2.5}$ exposure.[5] Importantly, this study showed that deaths from ischemic heart disease was the single largest cause of mortality (18% increase). The association was found to be even larger in the extended analysis of the Harvard Six Cities Study that showed a 28% increase in cardiovascular (CV) deaths per 10 μg/m³ increase in long-term $PM_{2.5}$ exposure,[10] and much larger in the Women Health Initiative Study,[23] which showed in 65 893 postmenopasual women, a 24% and 76% increase in the incidence of CV events and CV mortality, respectively. It appears that as the studies have become better tailored to assess CV morbidity and mortality and have been more capable of controlling for personal-level and ecological confounders, the strength and magnitude of adverse health associations with PM have become stronger (not weaker). This supports the overall veracity of these cohort association studies.

The data also support the view that sustained reductions in PM exposures over months-to-years result in a decrease in mortality. The extended analysis of the Harvard Six Cities Study showed that in 8096 participants from six US cities followed up over two time periods (1974 through 1989 and 1990–1998) that a decrease in CV mortality paralleled a decline in annual mean $PM_{2.5}$ concentrations. CV mortality exhibited a 31% reduction per each 10 μg/m³ improvement in city-specific mean $PM_{2.5}$ in between the two periods.[10] More importantly, the drop in the adjusted mortality rate was largest in the cities with the largest reductions in $PM_{2.5}$, underscoring the importance of PM monitoring, regulation and effective reduction, that could potentially yield substantial benefits in public-health outcomes. It has been suggested that reduction in the level of air PM may even account for some of the reduction observed in US mortality rates over the last 2 decades.[24] Indeed, Pope *et al.* reported that a decrease in $PM_{2.5}$ of 10 μg/m³ during the past 20 years has been associated with an estimated increase in life expectancy of 0.61 years.[25]

2.3.3 Additional Epidemiological Findings

Although the epidemiological studies with PM_{10} and $PM_{2.5}$ support the notion that the smaller particle sizes are associated with greater adverse cardiovascular effects, only a few studies have evaluated the effects of the smallest size fraction, ultrafine particles (UFP), with cardiorespiratory morbidity and mortality.[26–28] This paucity of available studies is likely largely due to the difficulty in accurately quantifying UFP exposure (as the levels are much more heterogeneous within a city or region than $PM_{2.5}$) and due to the lack of routine air-pollution monitoring equipment to assess UFP levels.[9] Future studies may help to clarify the public-health effects and to estimate the additional risk of cardiovascular morbidity imposed specifically by UFP levels.

It is important to note that the overall epidemiological evidence also supports the notion that PM exposure principally enhances the risk for cardiovascular morbidity and mortality among subsets of susceptible individuals. For example, pre-existent coronary atherosclerosis constitutes a predisposing substrate. In the only study in which participating subjects had information regarding coronary angiogram results, ischemic events in relation to recent $PM_{2.5}$ levels only occurred among individuals with obstructive disease in at least 1 vessel.[19] It has also been suggested in reviews of the literature that people with diabetes, of elderly age, or with underlying heart/lung diseases are at greater risk.[7,8] Finally, recent cohort studies also suggest the possibility that women and obese individuals are more susceptible to PM-induced cardiovascular mortality.[23,29]

While these positive statistical associations in many large epidemiological studies between PM and cardiovascular morbidity and mortality have continued to be reported from regions across the world over the past 20 years, the greatest insights into the potential mechanisms underlying these relationships (Figure 2.1) come from studies published in the past 5–10 years. The following sections briefly outline some of the major findings.

2.4 PM Exposure and Clinical and Subclinical Cardiovascular Outcomes

Exposure of particulate-matter air pollution is associated with several cardiovascular diseases including hard outcomes, as well as subclinical measures of cardiovascular damage. Table 2.3 provides an overall summary of many studies demonstrating associations between particle exposures of varying types and duration with a variety cardiovascular outcomes. Numerous subclinical measures that have been reported reflect the involvement of various pathophysiological aspects such as the promotion of systemic oxidative stress, systemic inflammation, atherosclerosis progression, vascular endothelial dysfunction, hypertension, vasoconstriction, enhanced coagulation or thrombotic status, altered heart-rate variability, increased arrhythmia potential, and even overt myocardial ischemia. Multiple studies report associations between air pollution and specific clinical manifestations of ischemic heart disease such as ischemic deaths,[5,23,30,31] increased incidence of fatal and nonfatal myocardial infarctions (MI),[19,23,32–39] various post-MI adverse outcomes (deaths, recurrent MI, CHF)[35,40] that lead to increased visits to the emergency rooms and hospital admissions for CV ischemic events[21,41,42] (Table 2.3). These associations have been observed with both short- and long-term PM exposure. Whilst most studies show associations between PM exposure and increased rates of acute cardiovascular events (or subclinical measures such as systemic inflammation) a few days later, even brief exposures can trigger events as rapidly as a few hours. Longer-term exposures on the other hand, have also been associated with effects, such as enhanced atherosclerosis or thrombosis. However, for ischemic

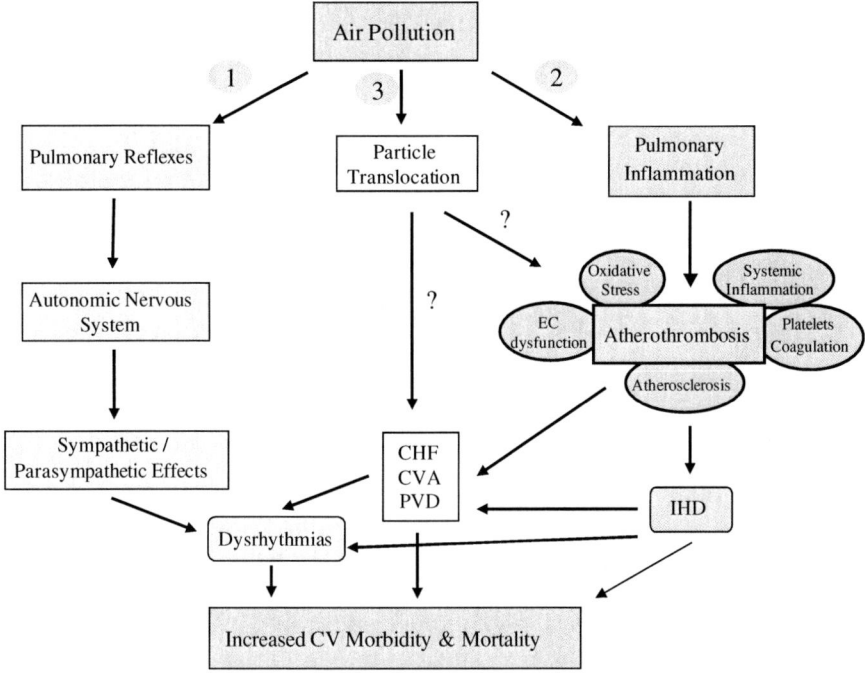

Figure 2.1 Potential mechanisms of how exposure to PM leads to CV disease. Three main pathways are proposed. (1) Induction of autonomic nervous system (ANS) imbalance, (2) Development of pulmonary oxidative stress and inflammation with systemic "spill-over" of inflammatory mediators (*e.g.* cytokines, activated cells), (3) Translocation of particles and/or chemical constituents to the systemic circulation. Various factors such as the course and length of exposures, particle size, chemical composition, physical properties, redox potential, interaction with vapor and other copollutants, among others, may determine the specific pathway(s) that is/are to be activated, degree of overlap and timing (hyperacute *vs.* acute *vs.* chronic effects). CHF = Congestive heart failure, CV = cardiovascular, CVA = Cerebrovascular accident, EC = endothelial cells, IHD = ischemic heart disease, PVD = Peripheral vascular disease. Modified from Araujo and Nel.[9]

heart disease, the associations with mortality are typically stronger than for nonfatal events. This suggests that PM exposure, rather than just serving as an acute triggering of events, may act to augment the severity of an ischemic event precipitated by unrelated causes.[23]

Recent studies demonstrate that congestive heart failure (CHF), dysrhythmias, cerebrovascular and peripheral vascular disease are clinical CV entities that are also adversely affected by particulate air pollution (Table 2.3). Exposure to air pollutants has been shown to be associated with increased hospital admissions[21,43,44] and deaths due to CHF,[45,46] ventricular and supraventricular arrhythmias[47–54] and cardiac arrest.[55,56] These effects appear after a few hours

Table 2.3 Association of Air Pollution with subclinical and clinical CV measures.

CV Measures	Indicators	Refs
Clinical Measure		
Ischemic heart disease	Increased mortality	5,23,30,31
	Increased post-MI adverse outcomes (recurrent MI, CHF, deaths)	35,40
	Increased hospitalizations	21,41,42
	Increased MI	19,23,32–39
Congestive heart failure	Increased mortality	45,46
	Increased hospital admissions	21,43,44
Dysrhythmias	Ventricular and/or supraventricular arrhythmias	47–54
	Cardiac arrest	55,56
Cerebrovascular disease	Increased stroke mortality	23,57–59
	Increased hospital admissions	41,42,84,89–92
Peripheral vascular disease	Increased hospital admissions	21,22
	Increased risk for deep vein thrombosis	93
Subclinical measures		
Systemic inflammation	Elevated circulating pro-inflammatory biomarkers/mediators (fibrinogen, CRP, TNF-α, IL 1-β, IL-7, ICAM-1)	45,94–102
Systemic oxidative stress	Elevated markers of lipid, protein or DNA oxidation	103–105
	Elevated plasma homocysteine	106,107
	Inhibition of oxidative response by w-3 PUFA	108
Atherosclerosis	Increased carotid-media thickness	80,81,109
	Increased coronary or aortic calcium scores	110,111
Vascular function	Impaired endothelial-dependent vasodilation, especially in diabetic patients	112–116
	Impaired endothelial-independent vasodilation	112
Systemic and pulmonary blood pressure	Elevated systemic diastolic and/or systolic BP	15,40,117–121
	Elevated pulmonary artery pressure	122,123
	Elevated Endothelin-1 levels	123
Thrombosis and coagulation	Elevated plasma fibrinogen, VWF, sCD40-L, PAI-1	95–97,100,101,124,125
	Decreased plasma protein C	126
Cardiac ischemia and repolarization abnormalities	Promotion of ST depression	127–130
Cardiac autonomic function	Increased heart rate and/or reduced heart-rate variability	12–14,45,47,102, 131–165
Epigenetic changes	Reduced levels of DNA methylation	166,167

CHF = congestive heart failure, CRP = C-reactive protein, ET-1 = endothelin 1, IL = interleukin, MI = myocardial infarction, PAI-1 = plasminogen activator inhibitor 1, PUFA = polyunsaturated fatty acids, sCD40-L = soluble CD40 ligand, TNF-α = tumour necrosis factor α, vWF = von Willebrand factor.

to several days after exposures. Air pollutants have also been associated with increased hospital admissions and deaths due to stroke, predominantly of ischemic etiology.[23,57–59] Studies showing numerous changes in subclinical measures as summarized in Table 2.3 also provide some general insights into the patho-biological mechanisms involved in PM-mediated cardiovascular morbidity and mortality.

2.5 Pathobiological Mechanisms

PM air pollution can induce cardiovascular effects in both an acute (minutes-to-hours) and chronic (months-to-years) fashions. While cumulative epidemiological and clinical evidence has been important to establish the veracity of these associations, experimental work is required to establish causality. Likewise, the detection of subclinical outcomes has helped in elucidating possible pathogenic mechanisms (Figure 2.1), some of which have been validated or are currently being tested in human as well as in cellular and animal studies.

Since the predominant mode of PM exposure is by inhalation, it is intriguing and still unclear how the deposition of particles in the lungs can translate into cardiovascular effects remote from the biological site of exposure. Complete elucidation of the many pathways whereby PM exposure is capable of eliciting cardiovascular diseases is complicated by a large number of variables. Variations in susceptibility (e.g pre-existing diseases, genetic profile), pollution components (*e.g.* specific chemicals with more or less "toxicity"), presence of copollutants (*e.g.* synergistic effects of gases), durations of exposures, pathways and mechanisms evaluated including the specific methodologies employed altogether make this a difficult task. Indeed, it appears likely that there is no single "responsible" mechanism or pathway to which all the effects of air pollution could be attributed. The type of pollutant and its chemistry, as well as the duration of exposure are likely to play important roles in determining the pathway/mechanism activated. The mechanisms are likely to be not mutually exclusive and may overlap and cause complimentary biological effects that could, in theory, synergistically increase the risk for cardiovascular events.

Despite the complexity of this relationship, a review of the literature supports the involvement of three putative "general mediating" pathways as shown in Figure 2.1: (1) autonomic nervous system imbalance, (2) induction of pumonary and thereby systemic inflammation/oxidative stress *via* "spill-over" of mediators (*e.g.* cytokines, activated white cells/platelets) into the systemic circulation, (3) the access of particles or specific chemical constituents to the systemic circulation that thereby cause direct affects upon the heart and vasculature. All three pathways could be potentially involved in the induction of both acute and chronic effects with a degree of overlap that may be important and could influence the timing of effects and dosing required to cause those effects. Several lines of evidence support each of the these pathways. However, even though the feasibility of particle systemic translocation has been shown by employing synthetic particles,[60–62] there is no conclusive evidence (*e.g.* dose or nature of compounds reaching the circulation) about the relevance of the ocurrence of the latter pathway with inhaled ambient particles in humans. It therefore remains unclear whether the lungs are an active player in the induction of cardiovascular effects or whether pollutant particulates or chemical constituents of air pollution can access the systemic circulation in a biologically meaningful manner and thus exert their vascular effects directly, without the need of intermediaries produced by lung-based cells (*e.g.* cytokines). On the other hand, the full extent, nature, cellular sources, and relative importance in mediating the various cardiovascular effects of the numerous blood-borne proinflammatory mediators shown to be associated with PM including cytokines (*e.g.* interleukins 1-beta and 6, tumor necrosis factor alpha), acute phase reactants (*e.g.* C-reactive protein, fibrinogen), molecules with hemodynamic/vascular activity (*e.g.* endothelins), and/or other possible factors (*e.g.* cell-derived microparticles) remain to be described. Finally, the different composition, chemistry, and sizes of inhaled PM or gaseous pollutants may determine the pathways that are subsequently activated, the temporal sequence and degree of overlap among mechanisms induced.

Regardless on the specific intermediary pathway(s), it appears that the ability of PM inhalation to induce cellular oxidative stress and promote inflammatory responses within cardiovascular tissues are central features of the pathogenic mechanisms. Cellular *in vitro*, animal *in vivo* and human experimental evidence supports this notion. For instance, the lung-related proinflammatory effects of diesel-exhaust particles (DEP) and concentrated air particles (CAPs) have been linked to reactive oxygen species (ROS) production[63–65] which includes O_2^- production in macrophages, bronchial epithelial cells and lung microsomes incubated with the particles or their organic extracts.[66,67] This pro-oxidative redox potential is in relation to its content of pro-oxidative and electrophilic chemicals. Therefore, DEP polar and aromatic fractions as well as ultrafine particles, rich in redox active chemicals, concentrate the pro-oxidative effects out of total DEP and ambient particulate, respectively.[9] DEP and ambient PM pro-oxidative effects have also been demonstrated in vascular cells.[68] Indeed, ROS generation can occur as a direct consequence of electrophilic chemicals or by the stimulation of intracellular sources of ROS such as NADPH oxidase,[69,70] mitochondrial perturbation,[71] cytochrome P450 and endothelial nitric oxide synthase. Increased ROS can lead to the activation of pattern-recognition receptors such as toll-like receptors (TLR),[72–76] intracellular pathways such as mitogen-activated protein kinase (MAPK)[71] that may lead to inflammatory transcription factor activation (e.g. NF-kB) and lung infiltration by inflammatory cells such as macrophages, neutrophils, lymphocytes. Systemic proinflammatory mediators can be derived from spillover from lung inflammatory foci or from nonpulmonary sources by unknown mechanisms yet to be elucidated.

At larger sites within cardiovascular tissues, increased oxidative stress and inflammation can lead to several pathological effects that could trigger ischemic events including abnormal vasculomotor function, both endothelial cell dependent and independent, as well as enhanced atherogenesis, as demonstrated in animal models[77–79] and studies with human subjects.[80,81] Similar inflammatory/oxidative responses within the blood, as shown by ambient particle exposure, can also induce prothrombotic effects by increasing platelet aggregation, elevating procoagulant factor production from the liver or vascular cells, decreasing either fibronolitic mechanisms or antithrombotic factors (Table 2.3). Thus, the induction of both proatherosclerotic and prothrombotic effects may lead to the induction of acute ischemic cardiovascular syndromes, largely responsible for most of the PM-induced morbidity and mortality.

Prohypertensive and vasoconstrictive effects appear to occur both in the pulmonary and systemic circulation and seem to include a variety of mechanisms that range from increased ROS generation, endothelial cell dysfunction to neurohormonal activation by increased activity of the sympathetic nervous system as well as enhanced endothelin 1 (ET-1) and angiotensin II receptor pathways, all leading to decreased endothelial-dependent vasodilatation and enhanced vasoconstrictor activity in several studies. While all human studies are not consistent, it does appear that short exposures can result in small elevations of blood pressure, most likely due to autonomic nervous system imbalance, although the other mechanisms cannot be excluded.

The induction of dysrhythmias on the other hand, may appear to be mostly neurally mediated by the activation of receptor-mediated autonomic reflexes in the lung that result in decreased heart rate and heart-rate variability in an acute fashion as suggested by short-term exposure animal studies[82] (Table 2.3). Exposure studies in humans appear to confirm the effects on heart rate and heart-rate variability although the issue of whether pulmonary neural reflexes are activated and responsible for those effects still requires more investigation. It is possible that part of the variability among various studies and lack of reproducibility of some findings may be due to the heterogeneity of the various ambient pollutants under study and large differences in experimental design. Finally, it cannot be excluded that many other biologically active mediators with potentially adverse cardiovascular downstream consequences (*e.g.* microparticles, proinflammatory subtypes of innate immune cells, CD40-ligand, myeloperoxidase, tissue factor) might also be involved in some of the adverse cardiovascular events associated with PM exposures.

2.6 Conclusions and Perspectives

To summarize, accumulating epidemiological and experimental evidence allows us to draw several conclusions about the cardiovascular effects of particulate air pollution. These are: (a) short-term exposure to $PM_{2.5}$ over a period of a few hours-to-weeks triggers both fatal and nonfatal CV events including myocardial ischemia and infarctions, heart failure, arrhythmias and strokes; (b) long-term exposure to $PM_{2.5}$ over a period of months-to-years increases CVD risk and reduces life expectancy within a population by several months to a few years; (c) chronic exposure to $PM_{2.5}$ appears to affect CV mortality to a greater degree than nonfatal events; (d) cardiovascular diseases constitute the largest cause of PM-related deaths; (e) the $PM_{2.5}$ concentration–CV risk relationship appears to be monotonic without a discernible "safe" threshold (even below $15\,\mu g/m^3$); (f) various factors appear to enhance the susceptibility to exposures (elderly age, pre-existent heart/lung diseases and perhaps diabetes and obesity); (g) inflammation and oxidative stress in tissues and at a cellular level are likely to be fundamental mediators of PM toxicity; (h) other pathways, such as the induction of autonomic nervous system imbalance, may also be involved in causing some of the PM-associated events.

It is important to highlight in closing that there are many questions that remain unanswered. Some noteworthy issues include a better identification of individual-level susceptibility factors, gene–environment interactions including the epigenetic mechanisms/factors involved, specific "responsible" PM components/sources and the nature of their chemical toxicity, the potential synergistic or additive roles of copollutants such as gases or volatile organic compounds, insights into the observed differences between geographical locations in health responses, the most important sources of harmful PM pollutants, effects of traffic-related particles and UFP levels in particular, the time frames of exposures required to elicit the majority of the adverse cardiovascular health effects, and many more details pertinent to the molecular mechanisms that mediate CV toxicity and the signal-transduction pathways that initial lung–PM interactions to cardiovascular dysfunction.

Acknowledgment

Writing of this article was supported by the National Institute of Environmental Health Sciences, NIH (RO1 ES016959 to JAA)

References

1. C. A. Pope, M. J. Thun, M. M. Namboodiri, D. W. Dockery, J. S. Evans, F. E. Speizer and C. W. Heath Jr, Particulate air pollution as a predictor of mortality in a prospective study of US adults, *Am. J. Respir. Crit. Care Med.*, 1995, **151**, 669–674.
2. J. M. Samet, F. Dominici, F. C. Curriero, I. Coursac and S. L. Zeger, Fine particulate air pollution and mortality in 20 US cities, 1987–1994, *N. Engl. J. Med.*, 2000, **343**, 1742–1749.
3. D. W. Dockery, C. A. Pope, X. Xu, J. D. Spengler, J. H. Ware, M. E. Fay, B. G. Ferris Jr and F. E. Speizer, An association between air pollution and mortality in six US cities, *N. Engl. J. Med.*, 1993, **329**, 1753–1759.
4. W. H. Organization, World Health Report 2002. Geneva: World Health Organization, 2002., 2002.
5. C. A. Pope 3rd, R. T. Burnett, G. D. Thurston, M. J. Thun, E. E. Calle, D. Krewski and J. J. Godleski, Cardiovascular mortality and long-term

exposure to particulate air pollution: epidemiological evidence of general pathophysiological pathways of disease, *Circulation*, 2004, **109**, 71–77.
6. A. Bhatnagar, Environmental cardiology: studying mechanistic links between pollution and heart disease, *Circ. Res.*, 2006, **99**, 692–705.
7. R. D. Brook, Cardiovascular effects of air pollution, *Clin. Sci. (London)*, 2008, **115**, 175–187.
8. R. D. Brook, B. Franklin, W. Cascio, Y. Hong, G. Howard, M. Lipsett, R. Luepker, M. Mittleman, J. Samet, S. C. Smith Jr and I. Tager, Air pollution and cardiovascular disease: a statement for healthcare professionals from the Expert Panel on Population and Prevention Science of the American Heart Association, *Circulation*, 2004, **109**, 2655–2671.
9. J. A. Araujo and A. E. Nel, Particulate matter and atherosclerosis: role of particle size, composition and oxidative stress, *Part. Fibre Toxicol.*, 2009, **6**, 24.
10. F. Laden, J. Schwartz, F. E. Speizer and D. W. Dockery, Reduction in fine particulate air pollution and mortality: Extended follow-up of the Harvard Six Cities study, *Am. J. Respir. Crit. Care Med.*, 2006, **173**, 667–672.
11. USEPA, Air quality criteria for particulate matter (Final Report, Oct 2004), US Environmental Protection Agency, Washington, DC, EPA 600/P-99/002aF-bF, 2004.
12. C. A. Pope 3rd, R. L. Verrier, E. G. Lovett, A. C. Larson, M. E. Raizenne, R. E. Kanner, J. Schwartz, G. M. Villegas, D. R. Gold and D. W. Dockery, Heart-rate variability associated with particulate air pollution, *Am. Heart J.*, 1999, **138**, 890–899.
13. D. R. Gold, A. Litonjua, J. Schwartz, E. Lovett, A. Larson, B. Nearing, G. Allen, M. Verrier, R. Cherry and R. Verrier, Ambient pollution and heart-rate variability, *Circulation*, 2000, **101**, 1267–1273.
14. A. Peters, S. Perz, A. Doring, J. Stieber, W. Koenig and H. E. Wichmann, Increases in heart rate during an air pollution episode, *Am. J. Epidemiol.*, 1999, **150**, 1094–1098.
15. A. Ibald-Mulli, J. Stieber, H. E. Wichmann, W. Koenig and A. Peters, Effects of air pollution on blood pressure: a population-based approach, *Am. J. Public Health*, 2001, **91**, 571–577.
16. R. D. Brook, J. R. Brook, B. Urch, R. Vincent, S. Rajagopalan and F. Silverman, Inhalation of fine particulate air pollution and ozone causes acute arterial vasoconstriction in healthy adults, *Circulation*, 2002, **105**, 1534–1536.
17. F. Dominici, A. McDermott, M. Daniels, S. L. Zeger and J. M. Samet, Mortality among residents of 90 cities. In *Revised Analyses of Time-Series Studies of Air Pollution and Health*, Health Effects Institute, Boston, MA, 2003, pp. 9–24.
18. A. Analitis, K. Katsouyanni, K. Dimakopoulou, E. Samoli, A. K. Nikoloulopoulos, Y. Petasakis, G. Touloumi, J. Schwartz, H. R. Anderson, K. Cambra, F. Forastiere, D. Zmirou, J. M. Vonk, L. Clancy, B. Kriz, J. Bobvos and J. Pekkanen, Short-term effects of ambient

particles on cardiovascular and respiratory mortality, *Epidemiology*, 2006, **17**, 230–233.
19. C. A. Pope 3rd, J. B. Muhlestein, H. T. May, D. G. Renlund, J. L. Anderson and B. D. Horne, Ischemic heart disease events triggered by short-term exposure to fine particulate air pollution, *Circulation*, 2006, **114**, 2443–2448.
20. A. Zanobetti and J. Schwartz, The effect of fine and coarse particulate air pollution on mortality: a national analysis, *Environ. Health Perspect.*, 2009, **117**, 898–903.
21. F. Dominici, R. D. Peng, M. L. Bell, L. Pham, A. McDermott, S. L. Zeger and J. M. Samet, Fine particulate air pollution and hospital admission for cardiovascular and respiratory diseases, *JAMA*, 2006, **295**, 1127–1134.
22. R. D. Peng, H. H. Chang, M. L. Bell, A. McDermott, S. L. Zeger, J. M. Samet and F. Dominici, Coarse particulate-matter air pollution and hospital admissions for cardiovascular and respiratory diseases among Medicare patients, *JAMA*, 2008, **299**, 2172–2179.
23. K. A. Miller, D. S. Siscovick, L. Sheppard, K. Shepherd, J. H. Sullivan, G. L. Anderson and J. D. Kaufman, Long-term exposure to air pollution and incidence of cardiovascular events in women, *N. Engl. J. Med.*, 2007, **356**, 447–458.
24. Z. Kabir, Decrease in US deaths from coronary disease, *N. Engl. J. Med.*, 2007, **357**, 941; author reply 941.
25. C. A. Pope 3rd, M. Ezzati and D. W. Dockery, Fine-particulate air pollution and life expectancy in the United States, *N. Engl. J. Med.*, 2009, **360**, 376–386.
26. M. Stolzel, S. Breitner, J. Cyrys, M. Pitz, G. Wolke, W. Kreyling, J. Heinrich, H. E. Wichmann and A. Peters, Daily mortality and particulate matter in different size classes in Erfurt, Germany, *J. Expo. Sci. Environ. Epidemiol.*, 2007, **17**, 458–467.
27. H. E. Wichmann, C. Spix, T. Tuch, G. Wolke, A. Peters, J. Heinrich, W. G. Kreyling and J. Heyder, Daily mortality and fine and ultrafine particles in Erfurt, Germany part I: role of particle number and particle mass, *Res. Rep. Health Eff. Inst.*, 2000, 5–86 ; discussion 87–94.
28. A. Peters, S. Von Klot, M. Heier, I. Trentinaglia, J. Cyrys, A. Hormann, M. Hauptmann, H. E. Vichmann and H. Lowel, Air pollution, personal activities, and onset of myocardial infarction in a case-crossover study, Part I, in *In Particulate Air Pollution and Nonfatal Cardiac Events*, Health Effects Institute, Boston, Massachusetts, 2005, p. 124.
29. R. C. Puett, J. Schwartz, J. E. Hart, J. D. Yanosky, F. E. Speizer, H. Suh, C. J. Paciorek, L. M. Neas and F. Laden, Chronic particulate exposure, mortality, and coronary heart disease in the nurses' health study, *Am. J. Epidemiol.*, 2008, **168**, 1161–1168.
30. D. Krewski, Evaluating the effects of ambient air pollution on life expectancy, *N. Engl. J. Med.*, 2009, **360**, 413–415.
31. C. A. Pope 3rd, Particulate pollution and health: a review of the Utah valley experience, *J. Expo. Anal. Environ. Epidemiol.*, 1996, **6**, 23–34.

32. C. Tonne, S. Melly, M. Mittleman, B. Coull, R. Goldberg and J. Schwartz, A case-control analysis of exposure to traffic and acute myocardial infarction, *Environ. Health Perspect.*, 2007, **115**, 53–57.
33. A. Peters, D. W. Dockery, J. E. Muller and M. A. Mittleman, Increased particulate air pollution and the triggering of myocardial infarction, *Circulation*, 2001, **103**, 2810–2815.
34. A. Zanobetti and J. Schwartz, The effect of particulate air pollution on emergency admissions for myocardial infarction: a multicity case-crossover analysis, *Environ. Health Perspect.*, 2005, **113**, 978–982.
35. S. von Klot, A. Peters, P. Aalto, T. Bellander, N. Berglind, D. D'Ippoliti, R. Elosua, A. Hormann, M. Kulmala, T. Lanki, H. Lowel, J. Pekkanen, S. Picciotto, J. Sunyer, F. Forastiere and G. Health, Effects of Particles on Susceptible Subpopulations Study, Ambient air pollution is associated with increased risk of hospital cardiac readmissions of myocardial-infarction survivors in five European cities, *Circulation*, 2005, **112**, 3073–3079.
36. J. Sullivan, L. Sheppard, A. Schreuder, N. Ishikawa, D. Siscovick and J. Kaufman, Relation between short-term fine-particulate matter exposure and onset of myocardial infarction, *Epidemiology*, 2005, **16**, 41–48.
37. D. D'Ippoliti, F. Forastiere, C. Ancona, N. Agabiti, D. Fusco, P. Michelozzi and C. A. Perucci, Air pollution and myocardial infarction in Rome: a case-crossover analysis, *Epidemiology*, 2003, **14**, 528–535.
38. Y. Murakami and M. Ono, Myocardial infarction deaths after high level exposure to particulate matter, *J. Epidemiol. Community Health*, 2006, **60**, 262–266.
39. A. Peters, S. von Klot, M. Heier, I. Trentinaglia, A. Hörmann, H. E. Wichmann and H. Lowel, Cooperative Health Research in the Region of Augsburg Study Group. Exposure to traffic and the onset of myocardial infarction, *N. Engl. J. Med.*, 2004, **351**, 1721–1730.
40. A. Zanobetti, M. J. Canner, P. H. Stone, J. Schwartz, D. Sher, E. Eagan-Bengston, K. A. Gates, L. H. Hartley, H. Suh and D. R. Gold, Ambient pollution and blood pressure in cardiac rehabilitation patients, *Circulation*, 2004, **110**, 2184–2189.
41. A. Le Tertre, S. Medina, E. Samoli, B. Forsberg, P. Michelozzi, A. Boumghar, J. M. Vonk, A. Bellini, R. Atkinson, J. G. Ayres, J. Sunyer, J. Schwartz and K. Katsouyanni, Short-term effects of particulate air pollution on cardiovascular diseases in eight European cities, *J. Epidemiol. Community Health*, 2002, **56**, 773–779.
42. COMEAP, Cardiovascular Disease and Air Pollution. A report by the committee on the medical effects of air pollutant's cardiovascular subgroup., ed. U. K. D. of Health, London, UK, 2006.
43. G. A. Wellenius, J. Schwartz and M. A. Mittleman, Particulate air pollution and hospital admissions for congestive heart failure in seven United States cities, *Am. J. Cardiol.*, 2006, **97**, 404–408.
44. C. A. Pope 3rd, D. G. Renlund, A. G. Kfoury, H. T. May and B. D. Horne, Relation of heart failure hospitalization to exposure to fine particulate air pollution, *Am. J. Cardiol.*, 2008, **102**, 1230–1234.

45. C. A. Pope, M. L. Hansen, R. W. Long, K. R. Nielsen, N. L. Eatough, W. E. Wilson and D. J. Eatough, Ambient particulate air pollution, heart-rate variability, and blood markers of inflammation in a panel of elderly subjects, *Environ. Health Perspect.*, 2004, **112**, 339–345.
46. M. Medina-Ramon, R. Goldberg, S. Melly, M. A. Mittleman and J. Schwartz, Residential exposure to traffic-related air pollution and survival after heart failure, *Environ. Health Perspect.*, 2008, **116**, 481–485.
47. S. K. Park, M. S. O'Neill, P. S. Vokonas, D. Sparrow and J. Schwartz, Effects of air pollution on heart-rate variability: the VA normative aging study, *Environ. Health Perspect.*, 2005, **113**, 304–309.
48. D. W. Dockery, H. Luttmann-Gibson, D. Q. Rich, M. S. Link, M. A. Mittleman, D. R. Gold, P. Koutrakis, J. D. Schwartz and R. L. Verrier, Association of air pollution with increased incidence of ventricular tachyarrhythmias recorded by implanted cardioverter defibrillators, *Environ. Health Perspect.*, 2005, **113**, 670–674.
49. D. Q. Rich, J. Schwartz, M. A. Mittleman, M. Link, H. Luttmann-Gibson, P. J. Catalano, F. E. Speizer and D. W. Dockery, Association of short-term ambient air pollution concentrations and ventricular arrhythmias, *Am. J. Epidemiol.*, 2005, **161**, 1123–1132.
50. A. Berger, W. Zareba, A. Schneider, R. Ruckerl, A. Ibald-Mulli, J. Cyrys, H. E. Wichmann and A. Peters, Runs of ventricular and supraventricular tachycardia triggered by air pollution in patients with coronary heart disease, *J. Occup. Environ. Med.*, 2006, **48**, 1149–1158.
51. D. Q. Rich, M. H. Kim, J. R. Turner, M. A. Mittleman, J. Schwartz, P. J. Catalano and D. W. Dockery, Association of ventricular arrhythmias detected by implantable cardioverter defibrillator and ambient air pollutants in the St Louis, Missouri metropolitan area, *J. Occup. Environ. Med.*, 2006, **63**, 591–596.
52. D. Q. Rich, M. A. Mittleman, M. S. Link, J. Schwartz, H. Luttmann-Gibson, P. J. Catalano, F. E. Speizer, D. R. Gold and D. W. Dockery, Increased risk of paroxysmal atrial fibrillation episodes associated with acute increases in ambient air pollution, *Environ. Health Perspect.*, 2006, **114**, 120–123.
53. S. E. Sarnat, H. H. Suh, B. A. Coull, J. Schwartz, P. H. Stone and D. R. Gold, Ambient particulate air pollution and cardiac arrhythmia in a panel of older adults in Steubenville, Ohio, *J. Occup. Environ. Med.*, 2006, **63**, 700–706.
54. U. P. Santos, M. Terra-Filho, C. A. Lin, L. A. Pereira, T. C. Vieira, P. H. Saldiva and A. L. Braga, Cardiac arrhythmia emergency-room visits and environmental air pollution in Sao Paulo, Brazil, *J. Epidemiol. Community Health*, 2008, **62**, 267–272.
55. F. Forastiere, M. Stafoggia, S. Picciotto, T. Bellander, D. D'Ippoliti, T. Lanki, S. von Klot, F. Nyberg, P. Paatero, A. Peters, J. Pekkanen, J. Sunyer and C. A. Perucci, A case-crossover analysis of out-of-hospital coronary deaths and air pollution in Rome, Italy, *Am. J. Respir. Crit. Care Med.*, 2005, **172**, 1549–1555.

56. F. S. Rosenthal, J. P. Carney and M. L. Olinger, Out-of-hospital cardiac arrest and airborne fine particulate matter: a case-crossover analysis of emergency medical services data in Indianapolis, Indiana, *Environ. Health Perspect.*, 2008, **116**, 631–636.
57. H. Kan, J. Jia and B. Chen, Acute stroke mortality and air pollution: new evidence from Shanghai, China, *J. Occup. Health*, 2003, **45**, 321–323.
58. Y. C. Hong, J. T. Lee, H. Kim and H. J. Kwon, Air pollution: a new risk factor in ischemic stroke mortality, *Stroke*, 2002, **33**, 2165–2169.
59. J. Kettunen, T. Lanki, P. Tiittanen, P. P. Aalto, T. Koskentalo, M. Kulmala, V. Salomaa and J. Pekkanen, Associations of fine and ultrafine particulate air pollution with stroke mortality in an area of low air-pollution levels, *Stroke*, 2007, **38**, 918–922.
60. W. G. Kreyling, M. Semmler, F. Erbe, P. Mayer, S. Takenaka, H. Schulz, G. Oberdorster and A. Ziesenis, Translocation of ultrafine insoluble iridium particles from lung epithelium to extrapulmonary organs is size dependent but very low, *J. Toxicol. Environl. Health-Part A*, 2002, **65**, 1513–1530.
61. A. Nemmar, H. Vanbilloen, M. F. Hoylaerts, P. H. Hoet, A. Verbruggen and B. Nemery, Passage of intratracheally instilled ultrafine particles from the lung into the systemic circulation in hamster, *Am. J. Respir. Crit. Care Med.*, 2001, **164**, 1665–1668.
62. G. Oberdorster, Z. Sharp, V. Atudorei, A. Elder, R. Gelein, A. Lunts, W. Kreyling and C. Cox, Extrapulmonary translocation of ultrafine carbon particles following whole-body inhalation exposure of rats, *J. Toxicol. Environl. Health-Part A*, 2002, **65**, 1531–1543.
63. A. E. Nel, D. Diaz-Sanchez, D. Ng, T. Hiura and A. Saxon, Enhancement of allergic inflammation by the interaction between diesel-exhaust particles and the immune system, *J. Allergy Clin. Immunol.*, 1998, **102**, 539–554.
64. N. Kunzli, S. Medina, R. Kaiser, P. Quenel, F. Horak Jr and M. Studnicka, Assessment of deaths attributable to air pollution: should we use risk estimates based on time series or on cohort studies?, *Am. J. Epidemiol.*, 2001, **153**, 1050–1055.
65. M. Lippmann, M. Frampton, J. Schwartz, D. Dockery, R. Schlesinger, P. Koutrakis, J. Froines, A. Nel, J. Finkelstein, J. Godleski, J. Kaufman, J. Koenig, T. Larson, D. Luchtel, L. J. Liu, G. Oberdorster, A. Peters, J. Sarnat, C. Sioutas, H. Suh, J. Sullivan, M. Utell, E. Wichmann and J. Zelikoff, The US Environmental Protection Agency particulate matter health effects research centers program: A midcourse report of status, progress, and plans, *Environ. Health Persp.*, 2003, **111**, 1074–1092.
66. T. S. Hiura, M. P. Kaszubowski, N. Li and A. E. Nel, Chemicals in diesel-exhaust particles generate reactive oxygen radicals and induce apoptosis in macrophages, *J. Immunol.*, 1999, **163**, 5582–5591.
67. N. Li, M. Wang, T. D. Oberley, J. M. Sempf and A. E. Nel, Comparison of the pro-oxidative and proinflammatory effects of organic diesel-exhaust

particle chemicals in bronchial epithelial cells and macrophages, *J. Immunol.*, 2002, **169**, 4531–4541.
68. K. W. Gong, W. Zhao, N. Li, B. Barajas, M. Kleinman, C. Sioutas, S. Horvath, A. J. Lusis, A. Nel and J. A. Araujo, Air-pollutant chemicals and oxidized lipids exhibit genome-wide synergistic effects on endothelial cells, *Genome Biol.*, 2007, **8**, R149.
69. R. Becher, A. Bucht, J. Ovrevik, J. K. Hongslo, H. J. Dahlman, J. T. Samuelsen and P. E. Schwarze, Involvement of NADPH oxidase and iNOS in rodent pulmonary cytokine responses to urban air and mineral particles, *Inhal. Toxicol.*, 2007, **19**, 645–655.
70. Z. Li, X. Hyseni, J. D. Carter, J. M. Soukup, L. A. Dailey and Y. C. Huang, Pollutant particles enhanced H_2O_2 production from NAD(P)H oxidase and mitochondria in human pulmonary artery endothelial cells, *Am. J. Physiol. Cell Physiol*, 2006, **291**, C357–365.
71. A. Nel, T. Xia, L. Madler and N. Li, Toxic potential of materials at the nanolevel, *Science*, 2006, **311**, 622–627.
72. Z. Li, J. D. Carter, L. A. Dailey and Y. C. Huang, Pollutant particles produce vasoconstriction and enhance MAPK signaling *via* angiotensin type I receptor, *Environ. Health Perspect.*, 2005, **113**, 1009–1014.
73. D. Cao, T. L. Tal, L. M. Graves, I. Gilmour, W. Linak, W. Reed, P. A. Bromberg and J. M. Samet, Diesel-exhaust particulate-induced activation of Stat3 requires activities of EGFR and Src in airway epithelial cells, *Am. J. Physiol. Lung Cell Mol. Physiol.*, 2007, **292**, L422–429.
74. S. Becker, L. Dailey, J. M. Soukup, R. Silbajoris and R. B. Devlin, TLR-2 is involved in airway epithelial cell response to air pollution particles, *Toxicol. Appl. Pharmacol.*, 2005, **203**, 45–52.
75. Y. C. Huang, W. Wu, A. J. Ghio, J. D. Carter, R. Silbajoris, R. B. Devlin and J. M. Samet, Activation of EGF receptors mediates pulmonary vasoconstriction induced by residual oil fly ash, *Exp. Lung Res.*, 2002, **28**, 19–38.
76. J. W. Hollingsworth, D. N. Cook, D. M. Brass, J. K. Walker, D. L. Morgan, W. M. Foster and D. A. Schwartz, The role of Toll-like receptor 4 in environmental airway injury in mice, *Am. J. Respir. Crit. Care Med.*, 2004, **170**, 126–132.
77. Q. Sun, P. Yue, J. A. Deiuliis, C. N. Lumeng, T. Kampfrath, M. B. Mikolaj, Y. Cai, M. C. Ostrowski, B. Lu, S. Parthasarathy, R. D. Brook, S. D. Moffatt-Bruce, L. C. Chen and S. Rajagopalan, Ambient air pollution exaggerates adipose inflammation and insulin resistance in a mouse model of diet induced obesity, *Circulation*, 2009, **119**, 538–546.
78. T. Suwa, J. C. Hogg, K. B. Quinlan, A. Ohgami, R. Vincent and S. F. van Eeden, Particulate air pollution induces progression of atherosclerosis, *J. Am. Coll. Cardiol.*, 2002, **39**, 935–942.
79. J. A. Araujo, B. Barajas, M. Kleinman, X. Wang, B. J. Bennett, K. W. Gong, M. Navab, J. Harkema, C. Sioutas, A. J. Lusis and A. E. Nel, Ambient particulate pollutants in the ultrafine range promote

early atherosclerosis and systemic oxidative stress, *Circ. Res.*, 2008, **102**, 589–596.
80. N. Kunzli, M. Jerrett, W. J. Mack, B. Beckerman, L. LaBree, F. Gilliland, D. Thomas, J. Peters and H. N. Hodis, Ambient air pollution and atherosclerosis in Los Angeles, *Environ. Health Perspect.*, 2005, **113**, 201–206.
81. N. Kunzli, M. Jerrett, R. Garcia-Esteban, X. Basagana, B. Beckermann, F. Gilliland, M. Medina, J. Peters, H. N. Hodis and W. J. Mack, Ambient air pollution and the progression of atherosclerosis in adults, *PLoS One*, **5**, e9096.
82. E. Ghelfi, C. R. Rhoden, G. A. Wellenius, J. Lawrence and B. Gonzalez-Flecha, Cardiac oxidative stress and electrophysiological changes in rats exposed to concentrated ambient particles are mediated by TRP-dependent pulmonary reflexes, *Toxicol. Sci.*, 2008, **102**, 328–336.
83. F. Dominici, S. L. Zeger and J. M. Samet, A measurement error model for time-series studies of air pollution and mortality, *Biostatistics*, 2000, **1**, 157–175.
84. F. Dominici, A. McDermott, S. L. Zeger and J. M. Samet, National maps of the effects of particulate matter on mortality: exploring geographical variation, *Environ. Health Perspect.*, 2003, **111**, 39–44.
85. R. D. Peng, F. Dominici, R. Pastor-Barriuso, S. L. Zeger and J. M. Samet, Seasonal analyses of air pollution and mortality in 100 US cities, *Am. J. Epidemiol.*, 2005, **161**, 585–594.
86. F. Dominici, R. D. Peng, K. Ebisu, S. L. Zeger, J. M. Samet and M. L. Bell, Does the effect of PM_{10} on mortality depend on PM nickel and vanadium content? A reanalysis of the NMMAPS data, *Environ. Health Perspect.*, 2007, **115**, 1701–1703.
87. K. Katsouyanni, Ambient air pollution and health, *Br. Med. Bull.*, 2003, **68**, 143–156.
88. E. Samoli, R. Peng, T. Ramsay, M. Pipikou, G. Touloumi, F. Dominici, R. Burnett, A. Cohen, D. Krewski, J. Samet and K. Katsouyanni, Acute effects of ambient particulate matter on mortality in Europe and North America: results from the APHENA study, *Environ. Health Perspect.*, 2008, **116**, 1480–1486.
89. M. Franklin, A. Zeka and J. Schwartz, Association between $PM_{2.5}$ and all-cause and specific-cause mortality in 27 US communities, *J. Expo. Sci. Environ. Epidemiol.*, 2007, **17**, 279–287.
90. C. C. Chan, K. J. Chuang, L. C. Chien, W. J. Chen and W. T. Chang, Urban air pollution and emergency admissions for cerebrovascular diseases in Taipei, Taiwan, *Eur. Heart J..*, 2006, **27**, 1238–1244.
91. S. S. Tsai, W. B. Goggins, H. F. Chiu and C. Y. Yang, Evidence for an association between air pollution and daily stroke admissions in Kaohsiung, Taiwan, *Stroke*, 2003, **34**, 2612–2616.
92. G. A. Wellenius, J. Schwartz and M. A. Mittleman, Air pollution and hospital admissions for ischemic and hemorrhagic stroke among medicare beneficiaries, *Stroke*, 2005, **36**, 2549–2553.

93. A. Baccarelli, I. Martinelli, A. Zanobetti, P. Grillo, L. F. Hou, P. A. Bertazzi, P. M. Mannucci and J. Schwartz, Exposure to particulate air pollution and risk of deep vein thrombosis, *Arch. Intern. Med.*, 2008, **168**, 920–927.
94. A. Peters, M. Frohlich, A. Doring, T. Immervoll, H. E. Wichmann, W. L. Hutchinson, M. B. Pepys and W. Koenig, Particulate air pollution is associated with an acute phase response in men; results from the MONICA-Augsburg Study, *Eur. Heart J..*, 2001, **22**, 1198–1204.
95. J. Pekkanen, E. J. Brunner, H. R. Anderson, P. Tiittanen and R. W. Atkinson, Daily concentrations of air pollution and plasma fibrinogen in London, *J. Occup. Environ. Med.*, 2000, **57**, 818–822.
96. R. Ruckerl, A. Ibald-Mulli, W. Koenig, A. Schneider, G. Woelke, J. Cyrys, J. Heinrich, V. Marder, M. Frampton, H. E. Wichmann and A. Peters, Air pollution and markers of inflammation and coagulation in patients with coronary heart disease, *Am. J. Respir. Crit. Care Med.*, 2006, **173**, 432–441.
97. K. J. Chuang, C. C. Chan, T. C. Su, C. T. Lee and C. S. Tang, The effect of urban air pollution on inflammation, oxidative stress, coagulation, and autonomic dysfunction in young adults, *Am. J. Respir. Crit. Care Med.*, 2007, **176**, 370–376.
98. A. Zeka, J. R. Sullivan, P. S. Vokonas, D. Sparrow and J. Schwartz, Inflammatory markers and particulate air pollution: characterizing the pathway to disease, *Int. J. Epidemiol.*, 2006, **35**, 1347–1354.
99. L. Calderon-Garciduenas, R. Villarreal-Calderon, G. Valencia-Salazar, C. Henriquez-Roldan, P. Gutierrez-Castrellon, R. Torres-Jardon, N. Osnaya-Brizuela, L. Romero, A. Solt and W. Reed, Systemic inflammation, endothelial dysfunction, and activation in clinically healthy children exposed to air pollutants, *Inhal. Toxicol.*, 2008, **20**, 499–506.
100. M. S. O'Neill, A. Veves, J. A. Sarnat, A. Zanobetti, D. R. Gold, P. A. Economides, E. S. Horton and J. Schwartz, Air pollution and inflammation in type-2 diabetes: a mechanism for susceptibility, *J. Occup. Environ. Med.*, 2007, **64**, 373–379.
101. R. Ruckerl, R. P. Phipps, A. Schneider, M. Frampton, J. Cyrys, G. Oberdorster, H. E. Wichmann and A. Peters, Ultrafine particles and platelet activation in patients with coronary heart disease – results from a prospective panel study, *Part. Fibre Toxicol.*, 2007, **4**, 1.
102. M. Riediker, W. E. Cascio, T. R. Griggs, M. C. Herbst, P. A. Bromberg, L. Neas, R. W. Williams and R. B. Devlin, Particulate matter exposure in cars is associated with cardiovascular effects in healthy young men, *Am. J. Respir. Crit. Care Med.*, 2004, **169**, 934–940.
103. E. V. Brauner, L. Forchhammer, P. Moller, J. Simonsen, M. Glasius, P. Wahlin, O. Raaschou-Nielsen and S. Loft, Exposure to ultrafine particles from ambient air and oxidative stress-induced DNA damage, *Environ. Health Perspect.*, 2007, **115**, 1177–1182.
104. P. S. Vinzents, P. Moller, M. Sorensen, L. E. Knudsen, O. Hertel, F. P. Jensen, B. Schibye and S. Loft, Personal exposure to ultrafine particles

and oxidative DNA damage, *Environ. Health Perspect.*, 2005, **113**, 1485–1490.
105. M. Sorensen, B. Daneshvar, M. Hansen, L. O. Dragsted, O. Hertel, L. Knudsen and S. Loft, Personal $PM_{2.5}$ exposure and markers of oxidative stress in blood, *Environ. Health Perspect.*, 2003, **111**, 161–166.
106. A. Baccarelli, A. Zanobetti, I. Martinelli, P. Grillo, L. Hou, S. Giacomini, M. Bonzini, G. Lanzani, P. M. Mannucci, P. A. Bertazzi and J. Schwartz, Effects of exposure to air pollution on blood coagulation, *J. Thromb. Haemost.*, 2007, **5**, 252–260.
107. S. K. Park, M. S. O'Neill, P. S. Vokonas, D. Sparrow, A. Spiro 3rd, K. L. Tucker, H. Suh, H. Hu and J. Schwartz, Traffic-related particles are associated with elevated homocysteine: the VA normative aging study, *Am. J. Respir. Crit. Care Med.*, 2008, **178**, 283–289.
108. I. Romieu, R. Garcia-Esteban, J. Sunyer, C. Rios, M. Alcaraz-Zubeldia, S. R. Velasco and F. Holguin, The effect of supplementation with omega-3 polyunsaturated fatty acids on markers of oxidative stress in elderly exposed to PM(2.5), *Environ. Health Perspect.*, 2008, **116**, 1237–1242.
109. A. V. Diez Roux, A. H. Auchincloss, T. G. Franklin, T. Raghunathan, R. G. Barr, J. Kaufman, B. Astor and J. Keeler, Long-term exposure to ambient particulate matter and prevalence of subclinical atherosclerosis in the Multi-Ethnic Study of Atherosclerosis, *Am. J. Epidemiol.*, 2008, **167**, 667–675.
110. B. Hoffmann, S. Moebus, S. Mohlenkamp, A. Stang, N. Lehmann, N. Dragano, A. Schmermund, M. Memmesheimer, K. Mann, R. Erbel, K. H. Jockel and G. Heinz, Nixdorf Recall Study Investigative, Residential exposure to traffic is associated with coronary atherosclerosis, *Circulation*, 2007, **116**, 489–496.
111. R. W. Allen, M. H. Criqui, A. V. Diez Roux, M. Allison, S. Shea, R. Detrano, L. Sheppard, N. D. Wong, K. H. Stukovsky and J. D. Kaufman, Fine particulate matter air pollution, proximity to traffic, and aortic atherosclerosis, *Epidemiology*, 2009, 254–264.
112. M. S. O'Neill, A. Veves, A. Zanobetti, J. A. Sarnat, D. R. Gold, P. A. Economides, E. S. Horton and J. Schwartz, Diabetes enhances vulnerability to particulate air pollution-associated impairment in vascular reactivity and endothelial function, *Circulation*, 2005, **111**, 2913–2920.
113. K. W. Rundell, J. R. Hoffman, R. Caviston, R. Bulbulian and A. M. Hollenbach, Inhalation of ultrafine and fine particulate matter disrupts systemic vascular function, *Inhal. Toxicol.*, 2007, **19**, 133–140.
114. R. Dales, L. Liu, M. Szyszkowicz, M. Dalipaj, J. Willey, R. Kulka and T. D. Ruddy, Particulate air pollution and vascular reactivity: the bus stop study, *Int. Arch. Occup. Environ. Health*, 2007, **81**, 159–164.
115. M. Briet, C. Collin, S. Laurent, A. Tan, M. Azizi, M. Agharazii, X. Jeunemaitre, F. Alhenc-Gelas and P. Boutouyrie, Endothelial function and chronic exposure to air pollution in normal male subjects, *Hypertension*, 2007, **50**, 970–976.

116. A. Schneider, L. Neas, M. C. Herbst, M. Case, R. W. Williams, W. Cascio, A. Hinderliter, F. Holguin, J. B. Buse, K. Dungan, M. Styner, A. Peters and R. B. Devlin, Endothelial dysfunction: associations with exposure to ambient fine particles in diabetic individuals, *Environ. Health Perspect.*, 2008, **116**, 1666–1674.
117. K. J. Chuang, C. C. Chan, G. M. Shiao and T. C. Su, Associations between submicrometer particles exposures and blood pressure and heart rate in patients with lung function impairments, *J. Occup. Environ. Med.*, 2005, **47**, 1093–1098.
118. J. H. Choi, Q. S. Xu, S. Y. Park, J. H. Kim, S. S. Hwang, K. H. Lee, H. J. Lee and Y. C. Hong, Seasonal variation of effect of air pollution on blood pressure, *J. Epidemiol. Community Health*, 2007, **61**, 314–318.
119. A. H. Auchincloss, A. V. Roux, J. T. Dvonch, P. L. Brown, R. G. Barr, M. L. Daviglus, D. C. Goff, J. D. Kaufman and M. S. O'Neill, Associations between recent exposure to ambient fine particulate matter and blood pressure in the Multi-ethnic Study of Atherosclerosis (MESA), *Environ. Health Perspect.*, 2008, **116**, 486–491.
120. L. Liu, T. D. Ruddy, M. Dalipaj, M. Szyszkowicz, H. You, R. Poon, A. Wheeler and R. Dales, Influence of personal exposure to particulate air pollution on cardiovascular physiology and biomarkers of inflammation and oxidative stress in subjects with diabetes, *J. Occup. Environ. Med.*, 2007, **49**, 258–265.
121. J. T. Dvonch, S. Kannan, A. J. Schulz, G. J. Keeler, G. Mentz, J. House, A. Benjamin, P. Max, R. L. Bard and R. D. Brook, Acute effects of ambient particulate matter on blood pressure. differential effects across urban communities, *Hypertension*, 2009, **53**, 853–859.
122. D. Q. Rich, R. S. Freudenberger, P. Ohman-Strickland, Y. Cho and H. M. Kipen, Right heart pressure increases after acute increases in ambient particulate concentration, *Environ. Health Perspect.*, 2008, **116**, 1167–1171.
123. L. Calderon-Garciduenas, R. Vincent, A. Mora-Tiscareno, M. Franco-Lira, C. Henriquez-Roldan, G. Barragan-Mejia, L. Garrido-Garcia, L. Camacho-Reyes, G. Valencia-Salazar, R. Paredes, L. Romero, H. Osnaya, R. Villarreal-Calderon, R. Torres-Jardon, M. J. Hazucha and W. Reed, Elevated plasma endothelin-1 and pulmonary arterial pressure in children exposed to air pollution, *Environ. Health Perspect.*, 2007, **115**, 1248–1253.
124. D. Liao, G. Heiss, V. M. Chinchilli, Y. Duan, A. R. Folsom, H. M. Lin and V. Salomaa, Association of criteria pollutants with plasma hemostatic/inflammatory markers: a population-based study, *J. Expo. Anal. Environ. Epidemiol.*, 2005, **15**, 319–328.
125. T. C. Su, C. C. Chan, C. S. Liau, L. Y. Lin, H. L. Kao and K. J. Chuang, Urban air pollution increases plasma fibrinogen and plasminogen activator inhibitor-1 levels in susceptible patients, *Eur. J. Cardiovasc. Prev. Rehabil.*, 2006, **13**, 849–852.

126. M. Riediker, Cardiovascular effects of fine particulate matter components in highway patrol officers, *Inhal. Toxicol.*, 2007, **19**(Suppl 1), 99–105.
127. J. Pekkanen, A. Peters, G. Hoek, P. Tiittanen, B. Brunekreef, J. de Hartog, J. Heinrich, A. Ibald-Mulli, W. G. Kreyling, T. Lanki, K. L. Timonen and E. Vanninen, Particulate air pollution and risk of ST-segment depression during repeated submaximal exercise tests among subjects with coronary heart disease: the Exposure and Risk Assessment for Fine and Ultrafine Particles in Ambient Air (ULTRA) study, *Circulation*, 2002, **106**, 933–938.
128. T. Lanki, J. J. de Hartog, J. Heinrich, G. Hoek, N. A. Janssen, A. Peters, M. Stolzel, K. L. Timonen, M. Vallius, E. Vanninen and J. Pekkanen, Can we identify sources of fine particles responsible for exercise-induced ischemia on days with elevated air pollution? The ULTRA study, *Environ. Health Perspect.*, 2006, **114**, 655–660.
129. T. Lanki, G. Hoek, K. L. Timonen, A. Peters, P. Tiittanen, E. Vanninen and J. Pekkanen, Hourly variation in fine particle exposure is associated with transiently increased risk of ST segment depression, *J. Occup. Environ. Med.*, 2008, **65**, 782–786.
130. D. R. Gold, A. A. Litonjua, A. Zanobetti, B. A. Coull, J. Schwartz, G. MacCallum, R. L. Verrier, B. D. Nearing, M. J. Canner, H. Suh and P. H. Stone, Air pollution and ST-segment depression in elderly subjects, *Environ. Health Perspect.*, 2005, **113**, 883–887.
131. J. M. Cavallari, S. C. Fang, E. A. Eisen, J. Schwartz, R. Hauser, R. F. Herrick and D. C. Christiani, Time course of heart-rate variability decline following particulate matter exposures in an occupational cohort, *Inhal. Toxicol.*, 2008, **20**, 415–422.
132. T. Chahine, A. Baccarelli, A. Litonjua, R. O. Wright, H. Suh, D. R. Gold, D. Sparrow, P. Vokonas and J. Schwartz, Particulate air pollution, oxidative stress genes, and heart-rate variability in an elderly cohort, *Environ. Health Perspect.*, 2007, **115**, 1617–1622.
133. S. K. Park, M. S. O'Neill, P. S. Vokonas, D. Sparrow, R. O. Wright, B. Coull, H. Nie, H. Hu and J. Schwartz, Air pollution and heart-rate variability: effect modification by chronic lead exposure, *Epidemiology*, 2008, **19**, 111–120.
134. A. Peretz, J. D. Kaufman, C. A. Trenga, J. Allen, C. Carlsten, M. R. Aulet, S. D. Adar and J. H. Sullivan, Effects of diesel exhaust inhalation on heart-rate variability in human volunteers, *Environ. Res.*, 2008, **107**, 178–184.
135. K. B. Min, J. Y. Min, S. I. Cho and D. Paek, The relationship between air pollutants and heart-rate variability among community residents in Korea, *Inhal. Toxicol.*, 2008, **20**, 435–444.
136. J. M. Cavallari, E. A. Eisen, S. C. Fang, J. Schwartz, R. Hauser, R. F. Herrick and D. C. Christiani, $PM_{2.5}$ metal exposures and nocturnal heart-rate variability: a panel study of boilermaker construction workers, *Environ. Health*, 2008, **7**, 36.

137. M. Cardenas, M. Vallejo, P. Romano-Riquer, S. Ruiz-Velasco, A. D. Ferreira-Vidal and A. G. Hermosillo, Personal exposure to $PM_{2.5}$ air pollution and heart-rate variability in subjects with positive or negative head-up tilt test, *Environ. Res.*, 2008, **108**, 1–6.
138. J. C. Chen, J. M. Cavallari, P. H. Stone and D. C. Christiani, Obesity is a modifier of autonomic cardiac responses to fine metal particulates, *Environ. Health Perspect.*, 2007, **115**, 1002–1006.
139. K. J. Chuang, C. C. Chan, T. C. Su, L. Y. Lin and C. T. Lee, Associations between particulate sulfate and organic carbon exposures and heart-rate variability in patients with or at risk for cardiovascular diseases, *J. Occup. Environ. Med.*, 2007, **49**, 610–617.
140. J. M. Cavallari, E. A. Eisen, J. C. Chen, S. C. Fang, C. B. Dobson, J. Schwartz and D. C. Christiani, Night heart-rate variability and particulate exposures among boilermaker construction workers, *Environ. Health Perspect.*, 2007, **115**, 1046–1051.
141. S. D. Adar, D. R. Gold, B. A. Coull, J. Schwartz, P. H. Stone and H. Suh, Focused exposures to airborne traffic particles and heart-rate variability in the elderly, *Epidemiology*, 2007, **18**, 95–103.
142. A. Wheeler, A. Zanobetti, D. R. Gold, J. Schwartz, P. Stone and H. H. Suh, The relationship between ambient air pollution and heart-rate variability differs for individuals with heart and pulmonary disease, *Environ. Health Perspect.*, 2006, **114**, 560–566.
143. M. Vallejo, S. Ruiz, A. G. Hermosillo, V. H. Borja-Aburto and M. Cardenas, Ambient fine particles modify heart-rate variability in young healthy adults, *J. Expo. Sci. Environ. Epidemiol.*, 2006, **16**, 125–130.
144. K. L. Timonen, E. Vanninen, J. de Hartog, A. Ibald-Mulli, B. Brunekreef, D. R. Gold, J. Heinrich, G. Hoek, T. Lanki, A. Peters, T. Tarkiainen, P. Tiittanen, W. Kreyling and J. Pekkanen, Effects of ultrafine and fine particulate and gaseous air pollution on cardiac autonomic control in subjects with coronary artery disease: the ULTRA study, *J. Expo. Sci. Environ. Epidemiol.*, 2006, **16**, 332–341.
145. H. Riojas-Rodriguez, J. A. Escamilla-Cejudo, J. A. Gonzalez-Hermosillo, M. M. Tellez-Rojo, M. Vallejo, C. Santos-Burgoa and L. Rojas-Bracho, Personal $PM_{2.5}$ and CO exposures and heart-rate variability in subjects with known ischemic heart disease in Mexico City, *J. Expo. Sci. Environ. Epidemiol.*, 2006, **16**, 131–137.
146. H. Luttmann-Gibson, H. H. Suh, B. A. Coull, D. W. Dockery, S. E. Sarnat, J. Schwartz, P. H. Stone and D. R. Gold, Short-term effects of air pollution on heart-rate variability in senior adults in Steubenville, *Ohio, J. Occup. Environ. Med.*, 2006, **48**, 780–788.
147. M. J. Lipsett, F. C. Tsai, L. Roger, M. Woo and B. D. Ostro, Coarse particles and heart-rate variability among older adults with coronary artery disease in the Coachella Valley, California, *Environ. Health Perspect.*, 2006, **114**, 1215–1220.

148. J. C. Chen, P. H. Stone, R. L. Verrier, B. D. Nearing, G. MacCallum, J. Y. Kim, R. F. Herrick, J. You, H. Zhou and D. C. Christiani, Personal coronary risk profiles modify autonomic nervous system responses to air pollution, *J. Occup. Environ. Med.*, 2006, **48**, 1133–1142.
149. A. Henneberger, W. Zareba, A. Ibald-Mulli, R. Ruckerl, J. Cyrys, J. P. Couderc, B. Mykins, G. Woelke, H. E. Wichmann and A. Peters, Repolarization changes induced by air pollution in ischemic heart disease patients, *Environ. Health Perspect.*, 2005, **113**, 440–446.
150. J. H. Sullivan, A. B. Schreuder, C. A. Trenga, S. L. Liu, T. V. Larson, J. Q. Koenig and J. D. Kaufman, Association between short term exposure to fine particulate matter and heart-rate variability in older subjects with and without heart disease, *Thorax*, 2005, **60**, 462–466.
151. J. Schwartz, A. Litonjua, H. Suh, M. Verrier, A. Zanobetti, M. Syring, B. Nearing, R. Verrier, P. Stone, G. MacCallum, F. E. Speizer and D. R. Gold, Traffic related pollution and heart-rate variability in a panel of elderly subjects, *Thorax*, 2005, **60**, 455–461.
152. I. Romieu, M. M. Tellez-Rojo, M. Lazo, A. Manzano-Patino, M. Cortez-Lugo, P. Julien, M. C. Belanger, M. Hernandez-Avila and F. Holguin, Omega-3 fatty acid prevents heart-rate variability reductions associated with particulate matter, *Am. J. Respir. Crit. Care Med.*, 2005, **172**, 1534–1540.
153. K. J. Chuang, C. C. Chan, N. T. Chen, T. C. Su and L. Y. Lin, Effects of particle size fractions on reducing heart-rate variability in cardiac and hypertensive patients, *Environ. Health Perspect.*, 2005, **113**, 1693–1697.
154. C. C. Chan, K. J. Chuang, G. M. Shiao and L. Y. Lin, Personal exposure to submicrometer particles and heart-rate variability in human subjects, *Environ. Health Perspect.*, 2004, **112**, 1063–1067.
155. C. C. Chan, K. J. Chuang, T. C. Su and L. Y. Lin, Association between nitrogen dioxide and heart-rate variability in a susceptible population, *Eur. J. Cardiovasc. Prev. Rehabil.*, 2005, **12**, 580–586.
156. D. Liao, Y. Duan, E. A. Whitsel, Z. J. Zheng, G. Heiss, V. M. Chinchilli and H. M. Lin, Association of higher levels of ambient criteria pollutants with impaired cardiac autonomic control: a population-based study, *Am. J. Epidemiol.*, 2004, **159**, 768–777.
157. F. Holguin, M. M. Tellez-Rojo, M. Hernandez, M. Cortez, J. C. Chow, J. G. Watson, D. Mannino and I. Romieu, Air pollution and heart-rate variability among the elderly in Mexico City, *Epidemiology*, 2003, **14**, 521–527.
158. R. B. Devlin, A. J. Ghio, H. Kehrl, G. Sanders and W. Cascio, Elderly humans exposed to concentrated air pollution particles have decreased heart-rate variability, *Eur. Respir. J. Suppl.*, 2003, **40**, 76s–80s.
159. S. R. Magari, R. Hauser, J. Schwartz, P. L. Williams, T. J. Smith and D. C. Christiani, Association of heart-rate variability with occupational and environmental exposure to particulate air pollution, *Circulation*, 2001, **104**, 986–991.

160. S. R. Magari, J. Schwartz, P. L. Williams, R. Hauser, T. J. Smith and D. C. Christiani, The association between personal measurements of environmental exposure to particulates and heart-rate variability, *Epidemiology*, 2002, **13**, 305–310.
161. C. A. Pope, D. J. Eatough, D. R. Gold, Y. Pang, K. R. Nielsen, P. Nath, R. L. Verrier and R. E. Kanner, Acute exposure to environmental tobacco smoke and heart-rate variability, *Environ. Health Perspect.*, 2001, **109**, 711–716.
162. J. Creason, L. Neas, D. Walsh, R. Williams, L. Sheldon, D. Liao and C. Shy, Particulate matter and heart-rate variability among elderly retirees: the Baltimore 1998 PM study, *J. Expo. Anal. Environ. Epidemiol.*, 2001, **11**, 116–122.
163. A. Peters, E. Liu, R. L. Verrier, J. Schwartz, D. R. Gold, M. Mittleman, J. Baliff, J. A. Oh, G. Allen, K. Monahan and D. W. Dockery, Air pollution and incidence of cardiac arrhythmia, *Epidemiology*, 2000, **11**, 11–17.
164. D. Liao, J. Creason, C. Shy, R. Williams, R. Watts and R. Zweidinger, Daily variation of particulate air pollution and poor cardiac autonomic control in the elderly, *Environ. Health Perspect.*, 1999, **107**, 521–525.
165. D. W. Dockery, C. A. Pope 3rd, R. E. Kanner, G. Martin Villegas and J. Schwartz, Daily changes in oxygen saturation and pulse rate associated with particulate air pollution and barometric pressure, *Res. Rep. Health Eff. Inst.*, 1999, 1–19;discussion 21–18..
166. A. Baccarelli, R. O. Wright, V. Bollati, L. Tarantini, A. A. Litonjua, H. H. Suh, A. Zanobetti, D. Sparrow, P. S. Vokonas and J. Schwartz, Rapid DNA methylation changes after exposure to traffic particles, *Am. J. Respir. Crit. Care Med.*, 2009.
167. L. Tarantini, M. Bonzini, P. Apostoli, V. Pegoraro, V. Bollati, B. Marinelli, L. Cantone, G. Rizzo, L. Hou, J. Schwartz, P. A. Bertazzi and A. Baccarelli, Effects of particulate matter on genomic DNA methylation content and iNOS promoter methylation, *Environ. Health Perspect.*, 2009, **117**, 217–222.

CHAPTER 3
Air Pollution and Atherosclerosis: Epidemiologic Studies

V. C. VAN HEE[1] AND J. D. KAUFMAN[2]

[1] University of Washington, Department of Internal Medicine, Harborview Medical Center, 325 Ninth Avenue, Seattle, WA 98104, USA; [2] University of Washington, Department of Environmental and Occupational Health Sciences, 4225 Roosevelt Way NE, Seattle, WA, 98105, USA

3.1 Introduction

A large and growing body of epidemiologic studies has demonstrated increased cardiovascular morbidity and mortality associated with increased air pollution levels, but the pathogenesis of air pollution-related deaths remains uncertain. Although both acute elevations[1,2] in particulate-matter air pollution (PM) and increased long-term average[3–6] PM or traffic-related pollutants more generally have been associated with overall cardiovascular mortality and events such as myocardial infarction and sudden cardiac death, whether pollutants cause only acute triggering of cardiovascular events or also induce slow progression of atherosclerotic disease burden is not known. Toxicologic studies have revealed that inhalation of fine particles accelerates atherogenesis in animals,[7,8] but the human epidemiologic studies characterizing the role of PM in atherosclerosis have only begun to examine this hypothesis in detail. Most human epidemiologic studies have focused on outcomes (such as mortality and acute myocardial infarction) that do not distinguish the extent of atherosclerotic burden in exposed individuals, and therefore cannot assess whether long-term air-pollution exposure is associated with an accumulation of atherosclerotic disease

---

Issues in Toxicology No. 8
Environmental Cardiology: Pollution and Heart Disease
Edited by Aruni Bhatnagar
© The Royal Society of Chemistry 2011
Published by the Royal Society of Chemistry, www.rsc.org

burden or simply triggering of acute cardiovascular events. To date, few epidemiologic studies have directly addressed the relationship between air pollution (whether PM specifically or traffic-related pollutants in general) and atherosclerotic disease.

3.2 Atherosclerosis: A Chronic, Inflammatory Disease Leading to Acute Cardiac Events

Atherosclerosis results from long-term systemic inflammatory changes that impact the vasculature, causing gradual alterations in the vessel wall that lead to increased susceptibility to myocardial infarction.[9,10] Atherosclerosis begins as early as childhood, with the formation of "fatty streaks" consisting of accumulations of lipid-laden smooth muscle cells within the vessel intima. Over time, fatty streaks progress to fibrous plaques, protrusions within the intima consisting of connective tissue and smooth muscle cells surrounding an inner lipid-rich core that contains inflammatory cells including macrophages and T cells. Although atherosclerosis itself can lead to progressive arterial obstruction and ischemia, its association with plaque rupture followed by thrombosis produces the acute mortality that accounts for the greatest proportion of cardiac deaths. Plaque rupture leads to an inflammatory and thrombotic cascade that produces acute arterial obstruction, causing ischemia and myocardial cell death. Prior to the development of symptomatic disease, however, atherogenesis has a long, preclinical phase during which time it can only be detected through the use of specific, targeted methods.

3.3 Subclinical Atherosclerosis: Measurement Methods

To determine whether PM exposures lead to long-term alterations in atherosclerosis, several methods that quantify subclinical disease burden have been employed in epidemiologic studies (and several show great promise for future studies). Although coronary atherosclerosis itself is impractical to measure in living human subjects, several useful, noninvasive measurement methods that are highly predictive of atherosclerosis and incident cardiovascular events have been developed. Some of these methods are used routinely in the clinical setting to stratify individuals by cardiac risk, while others have been used thus far almost exclusively in the research setting.

In epidemiological studies it is generally desirable to use methods with minimal risk to the participants. Noninvasive measures that can be used to detect subclinical disease include ultrasound carotid artery intima-media thickness (CIMT) determination and plaque detection, Doppler readings of ankle-brachial index (ABI), pressure recordings of aortic pulse-wave velocity (PWV), computed tomography (CT) detection of aortic and coronary artery calcium, and magnetic resonance imaging (MRI) characterization of aortic and coronary artery plaque. Although all methods have demonstrated varying degrees of association with cardiovascular events, they each have advantages and disadvantages and appear to measure different aspects of cardiovascular disease risk.[11] With the exception of methods that directly image the coronary arteries, these metrics generally depend upon the diffuseness of atherosclerotic disease for their ability to predict coronary artery disease (CAD). Atherosclerosis in one vascular bed must determine atherosclerosis in the coronary arteries for these methods to be useful in predicting risk. When evaluating epidemiologic studies that use these measures to understand the relationship between air pollution and atherosclerosis, it is important to understand their individual characteristics, strengths, and weaknesses.

3.3.1 Carotid Intima-Media Thickness (CIMT)

CIMT, as evaluated by B-mode ultrasound, is a direct measure of carotid atherosclerosis that corresponds to disease of the carotid artery, as demonstrated in autopsy studies[12] and on angiography.[13] Additionally, multiple epidemiologic studies have demonstrated its association with cardiovascular events.[14,15] Reliability and reproducibility of this method depend to a large extent on sonographer technique, however, necessitating rigorous procedures to standardize results if participants are to be evaluated by multiple operators.[16] Because atherosclerotic disease is not completely homogenous throughout vascular beds, measurements of the thickness of the carotid depend upon both the location of measurement (common carotid, bifurcation, or internal carotid) and whether local plaque is present at the measured location. Procedures for IMT measurement that take into account either averages or maximal thickness of multiple cross sections of carotid artery are associated with widely varying levels of risk, such that focal intrusive plaque or calcification is linked to substantially higher 10-year MI risk than increased average overall IMT or maximum common carotid IMT.[17] Measuring CIMT at a single location risks missing potentially important data on plaque burden or size. Overall, CIMT has the benefit of providing a continuous outcome measure, free of radiation exposure, which correlates well with clinical disease.

3.3.2 Coronary Artery Calcium (CAC)

CAC, measured by electron beam or multidetector computed tomography, has been demonstrated in numerous studies to be associated with CAD risk and to add significantly to models that predict clinical coronary artery disease (such as the clinically popular Framingham Risk Score).[18] Atherosclerosis is frequently associated with arterial calcium deposition, which increases as disease progresses. Measurement of that calcium in the particular vascular bed of interest (the coronary arteries) represents presumably the most relevant location for the cardiac disease of primary interest. CAC is frequently quantified through the use of the Agatston scoring system, which estimates the amount of calcium in the coronary arteries by examining the intensity, number and area of bright regions on CT imaging within the arteries.[19] The Agatston score is designed to maximize sensitivity and specificity in identifying those with more advanced atherosclerosis. For epidemiological purposes, one disadvantage of the Agatston scoring system is that it uses a relatively high threshold for providing a score, so that in many individuals (most younger people and nearly all premenopausal women), an Agatston score of zero is found. This necessitates statistical transformation of the outcome (often logarithmic transformation of CAC + 1) in studies with many young or female individuals and does not capture the important extent of early atherosclerosis in many individuals. In the case of individuals with zero CAC, measurement of CIMT may be advantageous, detecting gradations of subclinical disease in the absence of detectable calcium.[20] In contrast to CIMT, CAC is generally operator independent, although differences between individual machines may affect comparability, and radiation exposure is a concern.

3.3.3 Aortic Calcium

Measurement of aortic calcium uses identical methods to those used for CAC, albeit in a different vascular bed, again relying on the known pathophysiology of atherosclerosis as a systemic disease that affects vessels throughout the body. There appears to be good correlation between aortic and coronary artery calcium, suggesting that aortic calcium may be a good proxy for coronary artery disease.[21] Calcium in the aorta can also be quantified through the use of the Agatston scoring system. Similar to CAC, although to a lesser extent, aortic calcium has been associated with risk of cardiovascular disease events.[22]

3.3.4 Ankle-Arm Index (AAI)

AAI (also known as ankle-brachial index, or ABI) is a measure of restrictive atherosclerotic disease in the peripheral arteries. This measure is the simple ratio between the systolic blood pressure (SBP) obtained at the ankle (posterior tibial or dorsalis pedis artery) and that obtained at the arm (brachial artery). Under normal conditions, SBP in the legs is slightly higher than systolic pressure in the arms. With progressive narrowing of the peripheral arteries of legs due to atherosclerosis, the ratio of ankle to arm SBP is reduced. Decreased AAI has been associated with cardiovascular mortality and appears to correlate with overall atherosclerotic burden including CIMT particularly of the carotid bifurcation and femoral IMT.[23] Some individuals with extensive arterial calcification may have normal or elevated ABI despite significant atherosclerotic disease, although this scenario appears to be relatively rare. For this reason, a very high ABI (> 1.4) also confers increased risk of cardiovascular mortality. In general, although ABI is probably a less precise measure of the extent of atherosclerosis than imaging methods that can take into account non-obstructive disease, AAI is relatively inexpensive to measure and has low operator dependence.

3.3.5 Other Methods

Additional methods to detect atherosclerosis that have not yet been used in epidemiologic studies examining the relation between air pollution and atherosclerosis may provide further insight into the effects of air pollution on CVD. Doppler assessment of aortic PWV and MRI assessment of aortic distensibility[24,25] are methods that may provide further information specifically about the aortic vascular bed. Ultrasound and MRI assessment of plaque burden in the carotids or the coronary arteries may also be useful to achieve a measure of atherosclerosis that does not rely on generalized wall thickening or the presence of calcium.[26,27] Because overall effect sizes examining the relationship between air pollutants and atherosclerosis are likely to be small compared with traditional risk factors such as smoking, and to develop slowly over years, very precise methods to assess subclinical disease are needed.

3.4 Epidemiologic Studies Addressing the Relationship Between Air Pollutants and Atherosclerosis

To date, few epidemiologic studies in three populations have examined the link between air-pollution exposures and atherosclerosis using a subset of the subclinical measures described above (Table 3.1). Although results have not been entirely consistent from study to study, they have begun to confirm toxicologic evidence, demonstrating a link between air-pollutant exposures (whether general traffic-related pollutants or PM specifically) and the extent of atherosclerosis.

3.4.1 Particulate-Matter Air Pollution and CIMT in Los Angeles

In the first study to examine a link between air pollution and atherosclerosis, Künzli et al.[28] studied cross-sectional data from 798 healthy, nondiabetic adult participants collected in 2 randomized controlled trials[29,30] in Los Angeles County of individuals with elevated LDL or homocysteine. Using data from 23 state and local monitoring stations during the year of the examination (2000), mean particulate matter less than 2.5 μm in diameter ($PM_{2.5}$) concentrations were interpolated to each participant's home address using geostatistical methods (kriging). Carotid IMT was measured using B-mode ultrasound imaging of the right common carotid artery (average of 70–100 measurements along a 1-cm length distal to the carotid artery bulb).

For each 10 μg/m³ increase in annual average $PM_{2.5}$ exposure, a 4.2% higher CIMT [95% CI: –0.2,8.9] was observed. Although the overall effect size in this study did not reach statistical significance, the effect was robust to adjustment for multiple well-measured potential confounders. In addition, participants presumed to have better measurement of exposure, such as older individuals who tend to stay at home more frequently and women, showed larger effect

Table 3.1 Epidemiologic studies assessing the relation between air pollution exposure and atherosclerotic disease.

Study	Population	Design	Exposure location/geocoding method	Exposure	Outcome	Adjusted main effect	Susceptible groups	Confounders/Covariates
Künzli, et al.[28]	798 healthy participants from 2 RCTs in LA	Cross-sectional	Single home address	Year of exam (2000) average interpolated $PM_{2.5}$	CIMT	4.2% [95% CI: −0.2,8.9] per 10 μg/m^3	Older women, participants on lipid-lowering therapy	Age, sex, education, income, ETS, smoking, MVI, EtOH use
Hoffmann, et al. (2007)[32]	4494 participants from a population-based study in 3 German cities	Cross-sectional	Single home address	$PM_{2.5}$ based on EURAD dispersion model	log(CAC +1)	17.2% [95% CI: 5.6,45.5] per 3.91 μg/m^3 (interdecile range)	Women	Distance to roadway or $PM_{2.5}$, city, area of residence, age, sex, education, ETS, smoking, physical activity, WHR, DM, BP, LDL, HDL, TG
				Proximity to major roadways	log(CAC +1)	7.0% [95% CI: 0.1,14.4] per reduction of distance by half	Younger men	
Diez-Roux, et al.[34]	5172 participants without CVD in 6 US regions	Cross-sectional	Home residential history	PM_{10} (20 year imputed, modeled mean)	log(CIMT)	2% [95% CI: 0.3] per 21 μg/m^3 (interdecile range)	None	Age, sex, race / ethnicity, SES, BMI, HTN, LDL, HDL, smoking, DM, diet, physical activity
					Presence of CAC (>0)	Relative prevalence 1.02 [95% CI: 0.96,1.08] per 21 μg/m^3		
					ABI	0.001 [95% CI: −0.006,0.009] per 21 μg/m^3		
				$PM_{2.5}$ (20 year imputed, modeled mean)	log(CIMT)	1% [95% CI: 0.2] per 12.5 μg/m^3 (interdecile range)		
					Presence of CAC (>0)	Relative prevalence 1.01 [95% CI: 0.96,1.06] per 12.5 μg/m^3		
					ABI	−0.001 [95% CI: −0.006,0.006] per 12.5 μg/m^3		

Study	Participants	Design	Exposure assessment	Exposure measure	Outcome	Result	Effect modification	Covariates
Allen, et al.[35]	1147 participants without CVD in 6 US regions	Cross-sectional	Single home address	2000–2002 average interpolated $PM_{2.5}$	Presence of aortic calcium (>0)	RR 1.06 [95% CI: 0.96,1.16] per 10 $\mu g/m^3$	Hispanics, women, on lipid-lowering therapy, older individuals	CT scanner, age, gender, race, BMI, smoking, DM, education, income, LDL, HDL, LLT, BP, antihypertensive medication
				Proximity to major roadways	Presence of aortic calcium (>0)	Not reported; no significant effect noted	Higher income	
Hoffmann, et al. (2009)[36]	4348 participants from a population-based study in 3 German cities	Cross-sectional	Single home address	$PM_{2.5}$ based on EURAD dispersion model	ABI	0.0 [95% CI: −0.014,0.014] per 3.91 $\mu g/m^3$ (interdecile range)	Little evidence of effect modification	Distance to roadway or $PM_{2.5}$, city, area of residence, age, sex, SES, ETS, smoking, physical inactivity, DM, BP, BMI
					ABI<0.9 or PAD	OR 0.87 [95% CI: 0.57,1.34] per 3.91 $\mu g/m^3$		
				Proximity to major roadway (≤50 m compared to >200 m)	ABI	−0.024 [95% CI: −0.047, −0.001]	Women, smokers, elderly, nondiabetics, on LLT, obese	
					ABI<0.9 or PAD	OR 1.77 [95% CI: 1.01, 2.1]		

Definitions: CAC, coronary artery calcium; CIMT, carotid intima-media thickness; CVD, cardiovascular disease; EtOH, alcohol; ETS, environmental tobacco smoke; LA, Los Angeles; MVI, multivitamin; RCT, randomized-controlled trial; WHR, waist-to-hip ratio; BP, blood pressure; DM, diabetes mellitus; LDL, low-density lipoprotein cholesterol; HDL, high-density lipoprotein cholesterol; TG, triglycerides; BMI, body mass index; SES, socioeconomic status; HTN, hypertension; LLT, lipid-lowering therapy; PAD, peripheral artery disease estimates. Based upon prior studies of the relationship between CIMT and risk of myocardial infarction, the magnitude of this effect was thought to be comparable to a 3–6% increase in risk of MI per 10 μg/m^3 of PM$_{2.5}$ exposure,[31] or a portion of the PM-related CAD risk observed in multiple cohort studies described above.[3–6] In addition to older individuals and women, participants on lipid-lowering therapy also demonstrated greater effects.

3.4.2 PM, Traffic-Related Air Pollution, and CAC in Three German Cities

In a cross-sectional study, Hoffman et al.[32] examined 4494 adults (45–74 years old) from three large urban areas in Germany in the Heinz Nixdorf Recall study. Background mean PM$_{2.5}$ air-pollution levels in 2002 were assessed at participant home addresses using the EURAD dispersion model,[33] which estimates levels based upon emissions inventories, topographical data, and meteorology. Residential distance to major roadways (defined as a mean daily vehicle count 10 000 to 130 000) was calculated using geographic information systems (GIS). CAC was measured at the baseline examination (2000–2003) using EBCT, and the Agatston calcium score was computed. Detailed information on potential confounders was utilized.

Urban background PM$_{2.5}$ exposures were not significantly associated with CAC, although the effect estimates were large and positive, suggesting increases in CAC with increasing background PM$_{2.5}$ levels. The study was limited by a relatively small variation in pollution levels. Looking at a subset of participants with more time spent at home (nonworking individuals and women), effect estimates were noted to be larger. Additionally, examining the impact of near-roadway exposures beyond PM$_{2.5}$, a reduction in distance to major roadways by half was associated with a 7.0% [95% CI: 0.1,14.4] increase in CAC. The relationship between roadway proximity and CAC was not modified overall by various risk factors, although younger participants, men, and individuals with lower socioeconomic status (SES) had slightly larger effect sizes.

This study suggested that beyond particulate-matter air pollution, which had been the primary pollutant of interest, traffic-related air pollution in particular might be associated with larger impacts on atherosclerotic disease. It also was the first to use CAC, a highly sensitive marker of subclinical atherosclerotic disease in the vascular bed of primary interest, as an outcome.

3.4.3 PM$_{2.5}$, PM$_{10}$, and Multiple Subclinical Measures in Six US Cities

Diez-Rouz et al.[34] studied 5172 adults (age 44–84) without clinical cardiovascular disease in six US cities in the Multi-Ethnic Study of Atherosclerosis (MESA). Exposures to both PM$_{2.5}$ and PM$_{10}$ were calculated at participant residential addresses using two primary methods—imputation (modeling methods incorporating geographic and temporal covariates) and using the average value of the nearest monitor to participant residence. Unlike the other three studies, this study took advantage of a complete 20-year residential address history rather than a single address, and thus was able to assess average exposures over time. This study also evaluated multiple measures of subclinical disease, including CIMT (as in the Künzli study), CAC (as in the Hoffmann study), and ABI. CIMT was measured by B-mode ultrasound, calculating the mean of all maximum wall thicknesses (near and far walls) of both the right and left common carotid. CAC was measured by EBCT (3 centers) or MDCT (3 centers). The mean Agatston score of 2 scans per participant was calculated. As in all five of the studies described here, good control of potential confounders was possible.

Multiple regression analysis demonstrated a significant relationship between PM_{10}, $PM_{2.5}$, and CIMT, but not CAC or ABI. Per $12.5\,\mu g/m^3$ increase in $PM_{2.5}$ exposure (the interdecile range), there was a 1% [95% CI: 0,2] increase in median CIMT. Although somewhat smaller in magnitude than the effect previously observed in the study by Künzli *et al.*, this estimate of the effect of PM on CIMT lay within the confidence intervals of the prior single-city study. The relationship between CAC and traffic exposures was not investigated, as in the Hoffmann study. No evidence of effect modification by important demographic variables or risk factors was noted. Results were not sensitive to the utilization of alternate exposure estimation methods or participant-selection criteria.

3.4.4 PM, Traffic-Related Pollution, and Aortic Calcium in Six US Cities

Allen *et al.*[35] investigated a subset of 1147 of the MESA participants described above. Using 2-year average exposures to $PM_{2.5}$ (2000–2002) interpolated to participant baseline home addresses, proximity to major roadways as a binary indicator of living within 100 meters of a highway or 50 meters of a major arterial, and aortic calcium Agatston score measured by EBCT or MDCT, cross-sectional relationships were examined. After adjustment for multiple potential confounders (Table 3.1), neither $PM_{2.5}$ nor proximity to major roadways was associated with aortic calcium, although older individuals, Hispanics, women, and those on lipid-lowering therapy had larger effect estimates for the $PM_{2.5}$–aortic calcium relationship. Additionally, those with presumably better exposure estimates (individuals living >20 years at home address, those living nearer to monitoring stations) showed larger effect estimates.

3.4.5 PM, Traffic-Related Air Pollution, and ABI in Three German Cities

In a second cross-sectional study in the German cohort (Heinz Nixdorf Recall Study) described above, Hoffmann *et al.*[36] examined 4348 participants with

ABI determination. Exposure measurements at baseline residential address were performed as above (using the EURAD dispersion model for $PM_{2.5}$ and GIS-coded distance to major roadway for assessment of traffic exposure). ABI was calculated as the ratio of the highest SBP in the posterior tibial and dorsalis pedis artery of both legs to the highest SBP obtained in the right and left arm. Participants with an ABI > 1.3 were excluded from the study. Cases considered to have peripheral arterial disease were defined as individuals with either an ABI of less than 0.9 or a history of surgical treatment for peripheral arterial disease. Control for potential confounders was similar to the prior Heinz Nixdorf study (Table 3.1).

Compared with living more than 200 meters away from a major roadway, living within 50 meters of a major roadway in this study was associated with an OR of 1.77 (95% CI: 1.01,2.1) for peripheral arterial disease. The strongest associations between ABI (modeled as a linear term) and roadway proximity were also seen nearest to major roadways, with only small effects seen at distances greater than 50 m. Women were noted to have larger proximity-PAD associations than men, as were obese individuals, smokers, those on lipid-lowering therapy, nondiabetics, and the elderly. No significant associations between $PM_{2.5}$ and peripheral arterial disease or ABI were noted.

This study again showed more consistent associations between measures of subclinical atherosclerosis and traffic exposure than $PM_{2.5}$.

All of the above studies had good control of potential confounders, using detailed characterization of participants' socioeconomic status and other important risk factors that may also be associated with pollution exposures.

3.5 Consistency between Relationships Observed in Current Studies

Based upon the limited epidemiologic evidence described above, there appears to be a relationship between air-pollution exposures and atherosclerosis in human populations, with the most consistent associations observed between (1) PM and CIMT, and (2) traffic exposures and several measures of subclinical atherosclerosis. Although the majority of the studies suggest that women may be a particularly sensitive subgroup, the existence of susceptible subgroups varies somewhat from study to study. There are several potential reasons for significant differences in effect sizes, significance, and subgroup analysis observed in the above studies.

First, the differences in effect sizes and effect modification results between the studies may relate to differing populations with differing degrees of measurement error. As noted above, four of the five studies showed stronger effects in either women,[28,32,36] nonworking individuals,[32,35,36] or individuals living closer to $PM_{2.5}$ monitors.[35] All of these particular characteristics would be associated with reduced measurement error, given that all exposures were characterized at participant homes and that these groups either stay at home more frequently or have more accurately measured pollution concentrations due to their proximity to monitoring stations.

Differences may also be related to the very different exposure and outcome measures used from study to study. As discussed above, although all of the subclinical measures assessed do predict CVD, they relate to different vascular beds and different stages of disease, and they are sometimes only minimally correlated.[11] The only exposures and outcomes with relatively direct overlap in these studies are $PM_{2.5}$/carotid IMT (Künzli and Diez-Roux), $PM_{2.5}$/CAC (Hoffmann 2007 and Diez-Roux), and $PM_{2.5}$/ABI (Diez-Roux, Hoffmann 2009). Looking at these overlapping studies specifically, both Künzli et al. and Diez Roux et al. showed a cross-sectional relationship between $PM_{2.5}$ exposures and log (CIMT), with effect estimates of similar orders of magnitude (although precision was significantly greater in the latter study due to larger exposure contrasts across the six cities and the increased study size). Both Diez-Roux and Hoffmann saw no significant evidence of a relationship between $PM_{2.5}$ exposures and CAC, although effect estimates were positive and large (though imprecise) for both studies. Additionally, both Diez-Roux and Hoffmann (2009) saw no relationship between $PM_{2.5}$ exposures and ABI. Hoffmann et al.'s assessment of a significant relationship between traffic-exposure surrogates and CAC/ABI could not be replicated in Diez-Roux et al.'s study due to a lack of traffic-exposure estimates.

Although all of the five studies either look broadly at $PM_{2.5}/PM_{10}$ alone or both $PM_{2.5}$ and roadway proximity for exposure estimates, the methods for estimating exposures and precision of estimates vary widely. For example, Allen et al. used a simple binary measure of roadway proximity to estimate traffic exposure, whereas Hoffmann et al. (2007 and 2009) used finer categories of distance to major roadway to estimate exposures. Varied methods of PM exposure assessment included dispersion modeling on a very large scale (25 km grid),[32,36] geostatistical methods of interpolation,[28,35] and modeling using both spatial and temporal covariates.[34] All PM exposure estimates in the studies described could generally be said to ignore small area spatial variability in pollutant concentrations, operating on a larger scale, and none of these estimates took into account regional variations in particle composition (which could also account for differences in health-effect estimates).

Finally, interstudy differences may relate to the cross-sectional nature of these studies, which cannot capture the time course of the associations. If each study captured participants at a different stage in the disease process, this could affect cross-sectional estimates of effect, particularly if the relationship between PM and atherosclerosis is modified by time.

3.6 Air-Pollution Exposure and Atherosclerosis: Ancillary Epidemiologic Evidence

In addition to the above five studies that report subclinical atherosclerosis, there is suggestive evidence implicating longer-term acceleration of atherosclerotic disease burden rather than acute "triggering" of events. Time-series studies in particular have attempted to capture the public-health impact of air-pollution exposures by examining the extent to which air pollution causes a "mortality displacement" or "harvesting" effect. This effect refers to the potential of exposures to advance the date of death by only a few days or weeks. If true, this effect would imply that the cardiovascular impacts of air pollution have little overall public health significance, because they would only reduce the number of years of potential life lost by a small fraction. Additionally, if air pollution only advanced the onset of cardiovascular morbidity and mortality by a short time, it would be less plausible that air pollution could be associated with significant atherosclerotic disease.

Based upon several studies that demonstrated no decrease in mortality following episodes of air-pollution-related mortality, it appears that PM is not associated with a mortality displacement effect.[37] This observation lends weight to the argument that PM might induce longer-term changes in cardiovascular disease risk (such as those associated with atherosclerosis) that impact mortality over longer timescales. Although it is possible that these mortality effects could stem from very early "triggering" of acute MI, it seems more plausible to attribute this significant advancement of the date of death to acceleration of atherosclerosis, a phenomenon that advances gradually over time and is strongly associated with cardiovascular morbidity and mortality. Several studies have supported this hypothesis by showing increased cardiovascular morbidity and mortality effects with increasing duration of air-pollution exposure.[5,37–39]

Further epidemiologic evidence to support a role of PM in the development of atherosclerosis is the large disparity between the magnitude of risk of cardiovascular death seen in cohort studies compared to that which is seen in time series. Time-series analyses, which examine impacts of only short-term deviations from temporally smoothed mean pollutant levels, may underestimate mortality effects through an inability to account for chronic atherosclerotic changes over many years. A recent study performed in the WHI cohort showed a cardiovascular mortality HR of 1.76 [95% CI: 1.25,2.47] associated with an increase in long-term average $PM_{2.5}$ of $10\,\mu g/m^3$.[6] This magnitude of effect contrasts rather dramatically with large time-series studies like the National Morbidity, Mortality, and Air Pollution Study (NMMAPS), which showed an increase in cardiovascular mortality risk of 0.4% or a RR of 1.004 associated with $20\,\mu g/m^3$ short-term elevations in PM_{10}.[2] Although these somewhat dramatic differences in effect magnitude could relate partly to differences in the type and quantity of measurement error or confounding, the longer timescales and larger effect sizes seen in cohort studies likely relate to the improved ability of longer-term estimates of effect to pick up chronic phenomenon—such as acceleration of atherosclerotic disease over time—beyond a simple "triggering" effect.

To summarize, there is growing epidemiologic evidence that, in addition to causing cardiac morbidity and mortality through either long- or short-term effects or both, chronic air-pollution exposures lead to acceleration or initiation of atherosclerotic processes. Because the epidemiologic studies described above have all been cross-sectional in design, causality is difficult to assess at present, and significant questions remain about the time course of the observed associations. Exposure-assessment methods to date have relied primarily on widely dispersed monitoring stations that fail to account for local variability in pollutant levels, choosing instead to use coarse measures of traffic proximity as surrogates for better specific pollutant measurement. Although several studies implicate PM and traffic-related air pollution, the specific components of these heterogeneous mixtures responsible for the cardiovascular effects of air pollution have yet to be delineated. As is the case for the understanding of the relationship between PM and cardiac morbidity and mortality more generally, knowledge about the causal pathways involved in air-pollution-related atherosclerosis and its pathophysiologic mechanisms is limited. In addition to ongoing toxicologic work, further exploration of the genetic factors that modify these associations in epidemiologic studies may lead to a better understanding of causal mechanisms that mediate the cardiovascular effects of air pollution. Although the above studies show a possible atherosclerotic effect on the carotid arteries (for $PM_{2.5}$) and the coronaries and peripheral vasculature (for traffic-related pollutants), the relative susceptibility of different vascular beds has not yet been determined. Similarly, although the studies suggest that certain populations (such as women) may be more at risk of pollution-related atherosclerosis, the population subgroups at greatest risk remain poorly defined. In addition to replicating the preliminary findings observed in these few epidemiologic studies, future studies that incorporate longitudinal assessments of multiple vascular beds and conduct richer exposure assessment with particular emphasis on local scale variability in pollutant concentrations, traffic-related pollutants, and specific components of air pollution, will begin to address these questions in greater detail.

References

1. A. Peters, S. von Klot, M. Heier, I. Trentinaglia, A. Hörmann, H. E. Wichmann and H. Löwel, Cooperative Health Research in the Region of Augsburg Study Group, Exposure to traffic the onset of myocardial infarction, *N. Engl. J. Med.*, 2004, **351**, 1721–1730.
2. F. Dominici, A. McDermott, M. Daniels, S. L. Zeger and J. M. Samet, Revised analyses of the National Morbidity, Mortality, and Air Pollution Study: mortality among residents of 90 cities, *J. Toxicol. Environ. Health. A.*, 2005, **68**, 1071–1092.
3. F. Laden, L. M. Neas, D. W. Dockery and J. Schwartz, Association of fine particulate matter from different sources with daily mortality in six US cities, *Environ. Health. Perspect.*, 2000, **108**, 941–947.
4. C. A. Pope, R. T. Burnett, G. D. Thurston, M. J. Thun, E. E. Calle, D. Krewski and J. J. Godleski, Cardiovascular mortality and long-term exposure to particulate air pollution: epidemiological evidence of general pathophysiological pathways of disease, *Circulation*, 2004, **109**, 71–77.

5. F. Laden, J. Schwartz, F. E. Speizer and D. W. Dockery, Reduction in fine particulate air pollution and mortality: Extended follow-up of the Harvard Six Cities study, *Am. J. Respir. Crit. Care. Med.*, 2006, **173**, 667–672.
6. K. A. Miller, D. S. Siscovick, L. Sheppard, K. Shepherd, J. H. Sullivan, G. L. Anderson and J. D. Kaufman, Long-term exposure to air pollution and incidence of cardiovascular events in women, *N. Engl. J. Med.*, 2007, **356**, 447–458.
7. N. L. Mills, H. Törnqvist, S. D. Robinson, M. C. Gonzalez, S. Söderberg, T. Sandström, A. Blomberg, D. E. Newby and K. Donaldson, Air pollution and atherothrombosis, *Inhal. Toxicol.*, 2007, **19**(Suppl 1), 81–89.
8. J. A. Araujo, B. Barajas, M. Kleinman, X. Wang, B. J. Bennett, K. W. Gong, M. Navab, J. Harkema, C. Sioutas, A. J. Lusis and A. E. Nel, Ambient particulate pollutants in the ultrafine range promote early atherosclerosis and systemic oxidative stress, *Circ. Res.*, 2008, **102**, 589–596.
9. R. Ross, Atherosclerosis – an inflammatory disease, *N. Engl. J. Med.*, 1999, **340**, 115–126.
10. P. Libby, Inflammation in atherosclerosis, *Nature*, 2002, **420**, 868–874.
11. S. Kathiresan, M. G. Larson, M. J. Keyes, J. F. Polak, P. A. Wolf, R. B. D'Agostino, F. A. Jaffer, M. E. Clouse, D. Levy, W. J. Manning and C. J. O'Donnell, Assessment by cardiovascular magnetic resonance, electron beam computed tomography, and carotid ultrasonography of the distribution of subclinical atherosclerosis across Framingham risk strata, *Am. J. Cardiol.*, 2007, **99**, 310–314.
12. P. Pignoli, E. Tremoli, A. Poli, P. Oreste and R. Paoletti, Intimal plus medial thickness of the arterial wall: a direct measurement with ultrasound imaging, *Circulation*, 1986, **74**, 1399–1406.
13. W. J. Mack, L. LaBree, C. Liu, R. H. Selzer and H. N. Hodis, Correlations between measures of atherosclerosis change using carotid ultrasonography and coronary angiography, *Atherosclerosis*, 2000, **150**, 371–379.
14. M. L. Bots and D. E. Grobbee, Intima media thickness as a surrogate marker for generalised atherosclerosis, *Cardiovasc. Drugs. Ther.*, 2002, **16**, 341–351.
15. M. L. Bots, D. Baldassarre, A. Simon, E. de Groot, D. H. O'Leary, W. Riley, J. J. Kastelein and D. E. Grobbee, Carotid intima-media thickness and coronary atherosclerosis: weak or strong relations?, *Eur. Heart. J.*, 2007, **28**, 398–406.
16. R. Tang, M. Hennig, M. G. Bond, R. Hollweck, G. Mancia and A. Zanchetti, ELSA Investigators, Quality control of B-mode ultrasonic measurement of carotid artery intima-media thickness: the European Lacidipine Study on Atherosclerosis, *J. Hypertens.*, 2005, **23**, 1047–1054.
17. A. Simon, G. Chironi and J. Levenson, Comparative performance of subclinical atherosclerosis tests in predicting coronary heart disease in asymptomatic individuals, *Eur. Heart. J.*, 2007, **28**, 2967–2971.
18. M. J. Budoff and K. M. Gul, Expert review on coronary calcium, *Vasc. Health. Risk. Manag.*, 2008, **4**, 315–324.

19. A. S. Agatston, W. R. Janowitz, F. J. Hildner, N. R. Zusmer, M. Viamonte and R. Detrano, Quantification of coronary artery calcium using ultrafast computed tomography, *J. Am. Coll. Cardiol.*, 1990, **15**, 827–832.
20. S. J. Lester, M. F. Eleid, B. K. Khandheria and R. T. Hurst, Carotid intima-media thickness and coronary artery calcium score as indications of subclinical atherosclerosis, *Mayo Clin. Proc.*, 2009, **84**, 229–233.
21. J. Takasu, M. J. Budoff, K. D. O'Brien, D. M. Shavelle, J. L. Probstfield, J. J. Carr and R. Katz, Relationship between coronary artery and descending thoracic aortic calcification as detected by computed tomography: the Multi-Ethnic Study of Atherosclerosis, *Atherosclerosis*, 2009, **204**, 440–446.
22. A. Eisen, A. Tenenbaum, N. Koren-Morag, D. Tanne, J. Shemesh, M. Imazio, E. Z. Fisman, M. Motro, E. Schwammenthal and Y. Adler, Calcification of the thoracic aorta as detected by spiral computed tomography among stable angina pectoris patients: association with cardiovascular events and death, *Circulation*, 2008, **118**, 1328–1334.
23. C. M. Papamichael, J. P. Lekakis, K. S. Stamatelopoulos, T. G. Papaioannou, M. K. Alevizaki, A. T. Cimponeriu, J. E. Kanakakis, A. Papapanagiotou, A. T. Kalofoutis and S. F. Stamatelopoulos, Ankle-brachial index as a predictor of the extent of coronary atherosclerosis and cardiovascular events in patients with coronary artery disease, *Am. J. Cardiol.*, 2000, **86**, 615–618.
24. G. M. London and J. N. Cohn, Prognostic application of arterial stiffness: task forces, *Am. J. Hypertens.*, 2002, **15**, 754–758.
25. A. A. Malayeri, S. Natori, H. Bahrami, A. G. Bertoni, R. Kronmal, J. A. Lima and D. A. Bluemke, Relation of aortic wall thickness and distensibility to cardiovascular risk factors (from the Multi-Ethnic Study of Atherosclerosis [MESA]), *Am. J. Cardiol.*, 2008, **102**, 491–496.
26. I. M. van der Meer, M. L. Bots, A. Hofman, A. I. del Sol, D. A. van der Kuip and J. C. Witteman, Predictive value of noninvasive measures of atherosclerosis for incident myocardial infarction: the Rotterdam Study, *Circulation*, 2004, **109**, 1089–1094.
27. B. Kantor, E. Nagel, P. Schoenhagen, J. Barkhausen and T. C. Gerber, Coronary computed tomography and magnetic resonance imaging, *Curr. Probl. Cardiol.*, 2009, **34**, 145–217.
28. N. Künzli, M. Jerrett, W. J. Mack, B. Beckerman, L. LaBree, F. Gilliland, D. Thomas, J. Peters and H. N. Hodis, Ambient air pollution and atherosclerosis in Los Angeles, *Environ. Health. Perspect.*, 2005, **113**, 201–206.
29. H. N. Hodis, W. J. Mack, L. LaBree, P. R. Mahrer, A. Sevanian, C. R. Liu, C. H. Liu, J. Hwang, R. H. Selzer and S. P. Azen, VEAPS Research Group, Alpha-tocopherol supplementation in healthy individuals reduces low-density lipoprotein oxidation but not atherosclerosis: the Vitamin E Atherosclerosis Prevention Study (VEAPS), *Circulation*, 2002, **106**, 1453–1459.
30. H. N. Hodis, W. J. Mack, L. Dustin, P. R. Mahrer, S. P. Azen, R. Detrano, J. Selhub, P. Alaupovic, C. R. Liu, C. H. Liu, J. Hwang, A. G. Wilcox and

R. H. Selzer, BVAIT Research Group, High-dose B vitamin supplementation, progression of subclinical atherosclerosis: a randomized controlled trial, *Stroke*, 2009, **40**, 730–736.
31. D. H. O'Leary, J. F. Polak, R. A. Kronmal, T. A. Manolio, G. L. Burke and S. K. Wolfson, Carotid-artery intima and media thickness as a risk factor for myocardial infarction and stroke in older adults. Cardiovascular Health Study Collaborative Research Group, *N. Engl. J. Med.*, 1999, **340**, 14–22.
32. B. Hoffmann, S. Moebus, S. Möhlenkamp, A. Stang, N. Lehmann, N. Dragano, A. Schmermund, M. Memmesheimer, K. Mann, R. Erbel and K. H. Jöckel, Heinz Nixdorf Recall Study Investigative Group, Residential exposure to traffic is associated with coronary atherosclerosis, *Circulation*, 2007, **116**, 489–496.
33. M. Memmesheimer, E. Friese, A. Ebel, H. J. Jakobs, H. Feldmann, C. Kessler and G. Piekorz, Long-term simulations of particulate matter in Europe on different scales using sequential nesting of a regional model, *Int. J. Environ. Pollut.*, 2004, **22**, 108–132.
34. A. V. Diez Roux, A. H. Auchincloss, T. G. Franklin, T. Raghunathan, R. G. Barr, J. Kaufman, B. Astor and J. Keeler, Long-term exposure to ambient particulate matter and prevalence of subclinical atherosclerosis in the Multi-Ethnic Study of Atherosclerosis, *Am. J. Epidemiol.*, 2008, **167**, 667–675.
35. R. W. Allen, M. H. Criqui, A. V. Diez Roux, M. Allison, S. Shea, R. Detrano, L. Sheppard, N. D. Wong, K. H. Stukovsky and J. D. Kaufman, Fine particulate-matter air pollution, proximity to traffic, and aortic atherosclerosis, *Epidemiology*, 2009, **20**, 254–264.
36. B. Hoffmann, S. Moebus, K. Kröger, A. Stang, S. Möhlenkamp, N. Dragano, A. Schmermund, M. Memmesheimer, R. Erbel and K. H. Jöckel, Residential exposure to urban air pollution, ankle-brachial index, and peripheral arterial disease, *Epidemiology*, 2009, **20**, 280–288.
37. C. A. Pope, Mortality effects of longer term exposures to fine particulate air pollution: review of recent epidemiological evidence, *Inhal. Toxicol.*, 2007, **19**(Suppl 1), 33–38.
38. C. A. Pope, J. Schwartz and M. R. Ransom, Daily mortality and PM10 pollution in Utah Valley, *Arch. Environ. Health.*, 1992, **47**, 211–217.
39. J. Schwartz, D. W. Dockery and L. M. Neas, Is daily mortality associated specifically with fine particles?, *J. Air. Waste. Manag. Assoc.*, 1996, **46**, 927–939.

CHAPTER 4
Hypertension and Vascular Toxicity of PM

Z. YING[1] AND S. RAJAGOPALAN[2]

[1] Department of Physiology, Medical College of Georgia, Augusta, Georgia, USA; [2] Davis Heart and Lung Research Institute, Ohio State University, 473 W 12th Avenue, Rm 110, Columbus, OH 43210-1252, USA

4.1 Introduction

A growing body of data implicates particulate-matter air pollution (PM) as an important factor in the pathogenesis of cardiovascular disease.[1] PM influences susceptibility to adverse cardiovascular events and may be particularly harmful in individuals with pre-existing cardiovascular disease (CVD) risk factors such as diabetics, hypertensives and smokers.[2] The synergistic interaction of PM with other conventional risk factors is internally consistent with our current understanding of how risk factors mediate complex diseases such as hypertension, whereby the pathways involved in mediating the adverse effects often converge on final common mechanisms. A number of studies have now demonstrated rapid effects of inhaled particulates on cardiovascular variables, such as vascular tone and function, arguing for mechanisms transducing PM signals within minutes to hours. There is also good evidence to suggest subacute and chronic effects of PM with persuasive lines of evidence to support a role for reactive oxygen species (ROS) dependent mechanisms. The extent of these effects depends on the source, composition and duration of exposure to PM and the underlying susceptibility of the individual or the animal. Our understanding of the locus of generation of these mediators (lung *versus* extra-pulmonary), the source(s), time course of release and the relative contribution to various vascular effects continues to evolve. In this section we will discuss recent studies that have enhanced our understanding of PM-mediated vascular effects with a focus on hypertension, and we will discuss the relative biological significance of these findings.

4.2 Current Evidence from Animal and Toxicological Studies

Oxidative stress and inflammation are common etiopathogenic mechanisms that underlie risk factors such as hyperlipidemia, diabetes and hypertension. PM is a prototypical stimulus for ROS generation. The constituents of PM include transition metals, various reactive hydrocarbons and other intermediates that are facile generators of oxidant stress and also play a role in the propagation of free-radical chain reactions. Indeed, a number of studies have demonstrated the prowess of PM to induce ROS in a variety of cell types in the lung and the extrapulmonary vasculature including inflammatory cells such as monocytes and macrophages.

4.2.1 Systemic Oxidative Stress and Endothelial Function

Although a number of *in vitro* studies have demonstrated clear activation of a variety of ROS pathways in cultured pulmonary and vascular cells such as NADPH oxidases, mitochondrial, cytochrome P450 enzymes including eNOS,[3–15] it is only recently that ambient exposure experiments have extended these findings to an *in vivo* context. The ROS-generating pathway depends on a variety of factors specific to the type, duration and host susceptibility and in reality may involve more than one system. Both short-term and long-term studies in animal models have demonstrated an important effect of PM in increasing systemic and cardiovascular oxidant stress. Nurkiewicz *et al.* have shown a relationship between the enhanced constriction in spinotrapezius microvessels and the release of myeloperoxidase (MPO) from leukocytes at this site.[16] Using a model of instilled ROFA (partially soluble), the authors demonstrated dose-dependent impairment in calcium ionophore-induced dilation and increased adherence and "rolling" of leukocytes in the systemic vasculature, indicative of an activated endothelium. These findings were associated with increased vascular deposition of MPO. Interestingly in their study, an insoluble particle (TiO_2) induced very similar vascular effects. The effects of PM on ROS generation may also be synergistic to other risk factors such as hypertension. Sun *et al.*, have demonstrated that 10 weeks of exposure to $PM_{2.5}$ increased superoxide production in response to the short-term administration of angiotensin II (7 days) and resulted in upregulation of NAD(P)H oxidase subunits and depletion of tetrahydrobiopterin (cofactor for eNOS) in the vasculature.[17] These effects had functional consequences in terms of endothelial dysfunction and increase in blood pressure. In another investigation involving apoE$^{-/-}$ mice fed a high-fat diet, concomitant chronic exposure to $PM_{2.5}$ (particulate matter <2.5 µm) exacerbated vascular oxidant stress and vascular endothelial dysfunction.[18] Araujo et al., compared the proatherogenic effects of ambient ultrafine with $PM_{2.5}$ in apoE$^{-/-}$ mice in a mobile facility close to a Los Angeles freeway. Exposure to ultrafine particles resulted in an inhibition of the anti-inflammatory capacity of plasma high-density lipoprotein and greater systemic oxidative stress as evidenced by increased hepatic malondialdehyde and upregulation of Nrf2-regulated antioxidant genes.[19] Rhoden et al., tested the role of the autonomic nervous system in driving PM-induced cardiac oxidative stress and found that hexamethonium, a pharmacological inhibitor of the autonomic nervous system, could significantly reduce the chemiluminescence in the heart following exposure.[20] More recently, an upstream modulator, the TRPV1 receptor, was identified as central to the inhaled CAP (concentrated airborne particulates)-mediated induction of cardiac chemiluminescence.[21] In these studies, the TRPV1 inhibitor, capsazapine, was able to abrogate electrocardiographic alterations in rats during the 5-h exposure. Controlled exposure studies with diesel-exhaust particles in rodent studies of pulmonary exposure to diesel-exhaust particles (DEP) indicate increased levels of oxidative DNA damage in lung tissue.[22] A study by Ito et al., has shown that Wistar Kyoto male rats exposed to CAP (0.6–1.5 mg/m^3) in Yokohama for 4 days upregulated endothelin A receptor expression in the heart and that this finding weakly correlated with increase in blood pressure.[23] Ultrafine particles have been shown to activate downstream pathways such as phosphorylation of ERK1/2 and p38 MAPK in endothelial cells, an effect antagonized by losartan suggesting a role for the angiotensin II type-1 receptor. The soluble and insoluble fractions of the ultrafine PM were capable of inducing the phosphorylation of ERK1/2 and p38 MAPK in cultured cells, and Cu and Zn, both significant components of the PM sample, could replicate these effects.[24] Thus, there is evidence of downstream activation of signaling pathways relevant for the maintenance of vascular smooth muscle constriction and consequent increase in vascular tone.

4.2.2 Autonomic Tone and Function

Studies on PM-mediated effects on autonomic tone in the experimental and human context have been limited to analysis of heart-rate variability (HRV). The earliest indication that there may be systemic effects of PM came from ECG studies in rats[25] and have been confirmed by other groups that have also extended these findings to other susceptible models.[26–30] Decrease in heart rate and changes in HRV indices have been reported to be pronounced in senescent mice, indicating that aging may be a susceptibility factor.[31] Change in rMSSD (influenced by vagal tone) was greater while the LF/HF ratio was depressed in senescent mice. Using an anesthetized model of postinfarction myocardium sensitivity, Wellenius and colleagues tested the effects of acute exposure to concentrated PM (CAP for one hour) on heart rate and the induction of spontaneous ventricular arrhythmias such as ventricular premature beats. CAP exposure did not increase VPB frequency or change the heart rate in this study.[32]

In a post-MI heart failure model (3 months post-MI) in Sprague–Dawley rats when exposed to diesel-exhaust emissions demonstrated a reduction in rMSSD in both healthy and heart-failure groups, along with prominent increases in the incidence of premature ventricular contractions. Recent studies of ECG alterations in mice have indicated a role for PM-associated transition metals in driving the cardiovascular effects.[30] It has been recently reported that PM-induced EKG changes in rats could be prevented by inhibiting the Transient Receptor Potential Vanilloid Receptor 1 (TRPV1), indicating that changes in cardiac rhythm may be secondary to the activation of receptor-mediated autonomic reflexes in the lung.[21] Whether similar mechanisms can account for autonomic changes in humans remains to be established.

4.2.3 Pulmonary and Systemic Inflammation

Elevations in systemic and pulmonary levels of IL-6 and TNF-α have been observed following PM exposure, typically coincident with pulmonary inflammation.[33–40] Moreover, there is some evidence showing that the degree of pulmonary inflammation correlates with the elevation of systemic cytokines and systemic vascular dysfunction.[40] In a 4-week inhalation exposure to freshly generated diesel exhaust, IL-6 null mice did not demonstrate increased cellular inflammation in BAL fluid, implicating a role for IL-6.[41] Consistent with these finding Mutlu *et al.*, have demonstrated that acute intratracheal exposure to PM_{10} resulted in an increase in IL-6, TNF-α and IFN-γ in the bronchio-alveolar lavage fluid.[42] In this acute single-dose study with coarse particles, the cytokines appeared to originate from macrophages, since depletion of these cells in the lung abolished increases in some of the cytokine(s). The source of IL-6 and TNF-α and their involvement in the systemic inflammatory response following PM exposure have not been fully elucidated although these and other experiments appear to suggest that at least with PM_{10} particles, alveolar macrophages play a dominant role.[36,40,42] IL-6 null mice had control levels of TNF-α following diesel-exhaust exposure, unlike the wild-type animals, indicating a linkage between IL-6 and TNF-α.[41] However, another study employing IL-6 null mice exposed intratracheally to ambient PM_{10} showed roughly the same levels of TNF-α in BAL fluid compared with the wild-type mice.[42] Interestingly, IFN-γ was decreased to control values in the IL-6 null mice following PM exposure.[42] The upstream signaling pathways responsible for the recognition of PM components resulting in inflammation remain to be determined, although there is some evidence with other particulates and experimental models of lung injury that ROS generated by NADPH oxidase and pattern-recognition receptors may modulate some of these responses.[38,43,44] The NADPH-oxidase-null mice demonstrated a significantly lower IL-6 and MIP-2 responses to collected particulate matter (SRM-1648) compared to wild-type mice.[38] In a recent study by Sun *et al.*, $PM_{2.5}$ exposure in a model of diet-induced obesity in C57BL/6 mice for a duration of 24 weeks resulted in elevations in TNF-α and IL-6. In addition, there were increases noted in adipokines such as resistin and PAI-1.[45] While the source of these cytokines was not investigated, the elevation in cytokines thought to be derived from adipose sources in addition to findings of adipose inflammation in the study raises the possibility of systemic (nonpulmonary sources) sources of such cytokines.

4.2.4 Integrated Animal Studies Supporting a Role in Hypertension

Earlier studies have suggested small and inconsistent effects of PM on blood pressure.[46–48] A potential explanation for these inconsistent results may lie in the variations in the experimental protocol used to address this question, including differences in delivery, duration and composition of exposure. Table 4.1 summarizes studies that have demonstrated a positive relationship between blood pressure and PM exposure. Chang *et al.*, noted increases in rat BP (5–10 mm Hg

Table 4.1 Blood pressure and exposure to PM in controlled animal studies.

Author	Animals	PM exposures	BP method	Findings
Chang[27]	Spontaneously hypertensive rats	Several days intermittent exposures to fine PM	Implanted Radiotelemetry	CAP significantly elevated by 8.7 mm Hg in the spring months. Summer month exposures yielded smaller effects.
Sun[17]	Sprague–Dawley rats	Fine PM exposures for 9 weeks followed by angiotensin II	Implanted Radiotelemetry	No increase in basal BP. CAP increased BP response to angiotensin II infusion compared to filtered-air exposed rats
Ying[83]	C57Bl/6 mice	Fine PM exposures for 12 weeks followed by angiotensin II infusion	Noninvasive cuff. Technique	CAP increased BP response to angiotensin II, prevented by fasudil. CAP increased LVH and LV fibrosis.
Bartoli[52]	Dogs	Intermittent fine PM exposures for 5 h	Implanted Radiotelemetry	CAP increased BP by 2.6/4.7 mm Hg compared to filtered air during 5-h exposure. Alpha blockade mitigated the BP increase

during exposures) when exposed to $\sim 200\,\mu g/m^3$ CAP during spring months. However, during summer months, when the PM level decreased to $140\,\mu g/m^3$, this effect was not observed.[49] It was unclear, therefore, whether the effects were seasonal or dose-related. Sun *et al.*, investigated the BP effect of more prolonged fine CAP exposures over 9 weeks in healthy Sprague–Dawley rats.[50] Basal mean arterial pressure was not affected by $PM_{2.5}$ exposure. Nevertheless, starting at 24 h the increase in aortic mean BP during angiotensin II infusion was amplified in the CAP exposed rats compared with a filtered-air control group. During a 7-day long infusion, average BP was higher in rats exposed to CAP. Their thoracic aortic segments exhibited enhanced constriction to an α-agonist (phenylephrine), blunted vasodilatation to an endothelial-dependent agent (acetylcholine), and exaggerated relaxation in response to a Rho-kinase inhibitor (Y-27632). Aortic vessel superoxide production was greatly increased in the $PM_{2.5}$ exposed rats. This response was prevented by an inhibitor of nicotinamide adenine dinucleotide phosphate-oxidase (NAD(P)H oxidase) (apocynin) and of NO synthase. $PM_{2.5}$ also reduced tetrahydobiopterin (BH_4) levels in the heart, mesenteric vasculature, and liver, consistent with systemic effects of $PM_{2.5}$ exposure on depletion of an essential cofactor for endothelial NO synthase.

In aortic tissues, mRNA levels of the subunits of NAD(P)H oxidase and Rho-kinase 1 were increased. *In vitro* exposure of aortic smooth muscle cells to ultrafine particles and $PM_{2.5}$ caused a reactive oxygen-dependent activation of rho-kinase and phosphorylation of myosin light chain. Together, these findings demonstrate a mechanism whereby the inhalation of $PM_{2.5}$ for several weeks can sensitize the vasculature to the provasoconstrictive effects of endogenous pressor hormones. Through upregulation of ROS-producing enzymes such as NAD(P)H oxidase, oxidative stress lowers the bioavailability of NO. Excess ROS was likely responsible for the lowering of BH_4 levels that lead to the uncoupling of NO synthase, thus transforming it into a ROS-producing enzyme. The marked increase in oxidative stress from multiple sources activates the Rho/Rho-kinase pathway. The ensuing increase in calcium sensitization of the smooth muscle contractile apparatus could further potentiate vasoconstriction. Collectively, these responses lead to a significant imbalance in vasomotor tone, favoring the development of vasoconstriction and hypertension.

In a subsequent experiment, the relevance of the previously demonstrated activation of Rho-kinase by $PM_{2.5}$ was tested in mice. Several studies have demonstrated a critical role for the RhoA/Rho-kinase pathway in regulating systemic vascular responses and in the pathogenesis of hypertension. In these experiments, treatment with Fasudil blunted the increase in blood pressure by $PM_{2.5}$ in the presence of angiotensin II (Ang II).[51] Ang II was infused over a 14-day period in these experiments with treatment with Fasudil being initiated 1 day following Ang II infusion. Both cardiac and vascular RhoA activation was enhanced by CAP exposure along with increased expression of the guanine exchange factors, PDZ-RhoGEF and p115 RhoGEF. Parallel with increased RhoA activation, CAP exposure increased Ang-II-induced cardiac hypertrophy and collagen deposition, with these increases being normalized by fasudil treatment. These results directly implicate the Rho-ROCK pathway as a pathophysiologically relevant mechanism in PM-mediated effects, at least in rodent models.

In experiments performed in a conscious canine model with implanted blood-pressure catheters repeated inhalation exposure to $PM_{2.5}$ ($358 \pm 307\,\mu g/m^3$) in a crossover protocol, increased systolic, diastolic and mean arterial blood pressure, when compared with filtered air.[52] Interestingly, the changes between $PM_{2.5}$ and the filtered-air group were abolished in a subset of animals by Prazosin, an alpha adrenergic antagonist consistent with the heightened responsiveness to alpha-adrenergic responses to phenlyephrine noted by Sun and Ying et al.

4.3 Evidence to Support Vascular Effects of Inhaled Particles in Humans

Controlled human studies typically involve exposure to particles in an exposure chamber connected to an ambient source (concentrated or nonconcentrated) or exposure to collected particles such as those collected on filters (given by nebulization). The protocols used in different studies vary considerably in terms of duration of exposure, concentration of particles and differences in gaseous components including ozone, a factor that must be taken into consideration when interpreting results.

4.3.1 Systemic Oxidative Stress and Endothelial Dysfunction

Barregard et al., found an increase in urinary excretion of free 8-iso-prostaglandin2α among healthy adults following a 4-h exposure to concentrated wood smoke exposed in a chamber.[53] The particles were comprised of both $PM_{2.5}$ and ultrafines. This finding provides evidence of a PM-induced increase in free-radical-mediated lipid peroxidation. Potential evidence for systemic oxidative stress has also been observed following controlled human exposures to diesel exhaust. Tornqvist et al., reported an increase in plasma antioxidant capacity [Trolox equivalent antioxidant capacity] 24 h after exposure to diesel exhaust [300 µg/m3] in a group of healthy volunteers after a 1-h exposure.[39] The investigators suggested that systemic oxidative stress occurring following exposure may have caused this upregulation in antioxidant defense.[39] Peretz et al., observed significant differences in expression of genes involved in oxidative stress pathways between exposure to diesel exhaust ($200\,\mu g/m^3$ $PM_{2.5}$) and filtered air.[54] However, the conclusions of this investigation were limited by a small number of subjects and require validation in future studies. Based on the results of these studies, it appears that exposure to ambient levels of PM is capable of inducing acute systemic oxidative stress in human subjects.

Acute effects of concentrated inhaled particulates include steady-state vasoconstriction in peripheral conduit vessels, as previously reported by Brook et al.[55] In this study, no effect on endothelium-dependent vasodilatation using reactive hyperemia as a stimulus was noted. Mills et al., using exposure to dilute diesel-exhaust particles demonstrated reduction in peripheral resistance vessel responses to acetylcholine, bradykinin and nitroprusside 6 h following exposure.[56] Tornquist et al., employing an identical protocol in healthy adults, showed that responses to acetylcholine were impaired 24 h following exposure.[39] In contrast, bradykinin- and sodium-nitroprusside-mediated vasodilatation or bradykinin-induced acute plasma tissue plasminogen activator release, a key regulator of endogenous fibrinolysis, were no different 24 h following exposure. In an extension of these findings to patients with stable coronary artery disease, subjects exposed to dilute diesel exhaust (300 μg/m^3, particulate concentration) for one hour during intermittent exercise demonstrated reduced release tissue plasminogen activator, compared to filtered-air exposed patients.[57] This effect predominated 6 h following exposure. Interestingly in this study, microvascular endothelial function assessed by agonist responsiveness, was no different between filtered air and diesel exhaust, perhaps related to pre-existing endothelial dysfunction in patients with coronary artery disease. During the exposure, myocardial ischemia was quantified by ST-segment analysis using continuous 12-lead electrocardiography. Exercise-induced ST-segment depression was present in all patients, but there was a 3-fold greater increase in ST segment depression and ischemic burden during exposure to diesel exhaust. The implications of these results are that reductions in coronary flow reserve owing to rapid alterations in coronary microvascular function may have contributed to myocardial ischemia in this at-risk population.

In contrast to studies demonstrating impaired endothelial function in healthy adults with diesel-exhaust exposure, studies with acute exposure to concentrated ambient $PM_{2.5}$ have not demonstrated an acute impairment in endothelial function. In a study in patients with stable coronary artery disease, 12 patients and 12 healthy volunteers were exposed to concentrated ambient fine particles or filtered air for 2 h using a randomized, double-blind crossover study design. Peripheral microvascular function was assessed by responses to endothelium-dependent and -independent agonists in conjunction with tPA release to bradykinin, and inflammatory variables – including circulating leukocytes, serum C-reactive protein, and exhaled breath 8-isoprostane and nitrotyrosine 6–8 h following exposure. No changes in microvascular function were noted, although exposure to $PM_{2.5}$ resulted in an increase in exhaled 8-isoprostane concentrations after exposure.[58] These results are consistent with other studies demonstrating no effects on conduit endothelial function in response to reactive hyperemic stimulus, following acute exposure to $PM_{2.5}$.[55,58] Unique aspects of this protocol included substantially lower concentrations of elemental carbon, indicating that combustion sources were not a major source of ambient particulates and pretreatment of the diseased population with drugs such as statins, ASA and ACE inhibitors all of which may preclude appreciation of differences between cases and controls. The additional implication of this study is that differences in particle size, numbers and concentration between the exposures may also impact the results. Diesel-exhaust particles are dominated by ultrafine particles rich in combustion-derived carbon nanoparticulate with substantially higher surface area and a greater effect per unit mass compared to larger particles.[59,60]

In the study by Peretz et al.,[27] adult volunteers (10 healthy and 17 with metabolic syndrome) were exposed in randomized order to filtered air (FA) and each of two levels of diluted DE (100 or 200 µg/m^3 of fine particulate matter) in 2-h sessions.[61] Before and after each exposure, endothelial function by ultrasound assessment of flow-mediated dilation was evaluated. Diesel exhaust had no effects on flow-mediated dilation, but did dose dependently constrict the brachial artery along with elevations in plasma levels of ET-1 when compared with filtered air. In this study the hypothesis that metabolic syndrome patients, an "at risk" population being more susceptible to the effects of air-pollution-mediated vascular dysfunction was not borne out by the data that on the contrary seemed to suggest greater vasoconstriction after exposure to DE in the healthy participants. The findings were limited by the small sample size and the baseline variability of the measurements.

4.3.2 Autonomic Tone and Function

The results of several new controlled human-exposure studies provide limited evidence to suggest that acute exposure to near-ambient levels of particulate matter may be associated with small changes in HRV. There are at least 4 studies to support this. In the first study, healthy elderly individuals experienced significant decreases in HRV immediately following exposure.[62] Some of these changes persisted for at least 24 h. Gong et al., studied healthy and asthmatic adults exposed to coarse CAP with intermittent exercise.[63] HRV was not affected immediately following the exposure, but decreased in both groups at 4 and 22 h after the end of the exposure, greater responses were observed in nonasthmatics.[63] Another study involved healthy elderly subjects and patients with COPD exposed to approximately 200 µg/m^3 CAP or filtered air for 2 h with intermittent mild exercise. Heart-rate variability over multihour intervals was lower after CAP than after FA in healthy elderly subjects but not in COPD subjects. A significant negative effect of CAP on ectopic heartbeats during/after CAP exposure relative to FA was noted in the healthy subjects, while the COPD group experiencing an improvement during/after CAP relative to FA.[64]

Samet et al., recently compared the effects of a 2-h exposure with intermittent exercise to ultrafine (average concentration 47 µg/m^3), fine (average concentration 120 µg/m^3), and coarse (average concentration 89 µg/m^3) CAP among healthy subjects.[65] In both the ultrafine and thoracic coarse studies, a crossover design was used in which each subject was exposed to both PM and filtered air. In the case of the fine PM study, subjects did not serve as their own control, but were exposed to either PM or filtered air. Thoracic coarse fraction CAP produced a statistically significant decrease in SDNN, 20 h after exposure compared with filtered air. No statistically significant effects on HRV were observed following exposure to ultrafine PM as measured during controlled 5-min intervals. However, the authors did observe a significant decrease in

SDNN following exposure to ultrafine PM based on an analysis of the 24-h measurements. No differences were reported in HRV with fine PM exposures. Definitive studies examining other susceptible populations with ambient exposures have not yet been reported, although PM exposure appears to result in a decrease in HRV more consistently in healthy older adults.[62,64]

4.3.3 Systemic Inflammation

Several controlled human-exposure studies have measured CRP, IL-6 and TNF-α to evaluate systemic inflammation following exposure to PM and diesel exhaust. In some of these studies, no statistically significant changes in levels of these cytokines have been observed following controlled CAP exposures,[58,65,66] or diesel exhaust.[46,56,57,67] In contrast, Tornqvist et al., observed increases in IL-6 and TNF-α 24 h following exposure to diesel exhaust in healthy adults.[39] Several controlled human-exposure studies have shown no increases in CRP following controlled exposures to ultrafine, fine, or thoracic coarse CAP or diesel exhaust.[39,56,57,67] The lack of an association between the vast majority of acute exposure studies and inflammatory markers does not preclude an effect with chronic exposure nor does it preclude an effect on other cytokine pathways.

The exposure of humans to high levels of ambient particles stimulates the bone marrow and the release of neutrophils, band cells, and monocytes into the circulation.[68,69] Two studies with controlled exposures have shown increases in white cell indices following exposure. In one study increased peripheral basophils in healthy older adults, 4 h following a 2-h exposure to $PM_{2.5}$ CAP was noted.[64] In another, increased white blood cell counts in healthy young adults 12 h following a 2-h exposure to $PM_{2.5}$ CAP was noted.[66] Frampton et al., reported decreases in blood monocytes, basophils, and eosinophils following exposure to ultrafine carbon (10–50 µg/m^3) among exercising asthmatics and healthy adults.[70] Particle exposure also reduced the expression of adhesion molecules CD54 and CD18 on monocytes. These results may be interpreted as increased sequestration of these cells in tissue compartments such as the lung or vasculature where there may be selective expression of the corresponding receptors for these ligands.[71] However, other recent human clinical studies have found no association between peripheral blood-cell counts and exposure to fine or ultrafine particles such as zinc oxide,[72] ultrafine carbon,[73] or diesel exhaust.[39,56,57]

In the only study to date to examine the effects on specific subsets of peripheral cells, Peretz et al., evaluated changes in gene expression using an expression array in monocytes following 4 h of exposure to diesel exhaust.[54] Ten genes involved in the inflammatory response were modulated in response to $PM_{2.5}$ exposure (8 upregulated, 2 downregulated). These findings will need to be reproduced in larger studies and raise the possibility that functional changes in inflammatory cells may occur without discernible changes in counts of these cells in the peripheral circulation.[54] Moreover, with the expanding complexity of

4.3.4 Integrated Hemodynamic Studies in Humans

In general, the studies that have attempted to discern an association between PM exposure and changes in systemic arterial pressure have been diverse with significant differences in exposure protocols. In almost all studies blood pressure was one of many secondary endpoints collected and was not the primary outcome variable of interest. In one of the earliest studies $PM_{2.5}$ (174 µg/m^3) increased systolic blood pressure in healthy subjects but decreased it in asthmatics.[76] Three other controlled studies among healthy adults where blood pressure was not a primary endpoint, 2-h exposures to resuspended diesel-exhaust particles,[46] CAP + ozone[55] or ultrafine exposure.[77] In a subsequent reanalysis in the study by Brook et al., that involved examination of blood pressure immediately following the exposure rather than 10 min following the exposure, the authors reported a significant increase in diastolic blood pressure following 2-h exposure.[78] This increase in blood pressure was associated with the organic carbon content.[79]

The effect of inhaled particulates on blood pressure has been further investigated in several recent controlled human-exposure studies. Two new studies have assessed blood-pressure changes following a 1-h exposure to diesel exhaust with a particle concentration of 300 µg/m.^{3} Mills et al., evaluated changes in blood pressure 2 h following exposure to diesel exhaust and found a 6-mm Hg increase in diastolic blood pressure of marginal statistical significance ($p = 0.08$) compared with filtered air control.[56] Using a similar study design and duration of exposure, Tornqvist et al., reported no changes in blood pressure 24 h following exposure to diesel exhaust compared with filtered air control.[56] Thus, the data to date appears to suggest a rapid effect of $PM_{2.5}$ or diesel exhaust on blood pressure.

To elucidate the underlying biological mechanisms, Brook et al., designed a subsequent study with the objective of investigating the effect of 4 different 2-h long controlled air-pollution exposures on BP.[80] During randomized, blinded exposures conducted in downtown Toronto, Ontario, diastolic BP linearly increased during CAP-containing exposures only. CAP (~150 µg/m^3) and CAP + ozone (120 ppb) exposures caused a significant 2.9- and 3.6-mm Hg elevation after 2 h, respectively. BP levels were not altered by filtered air or ozone. The diastolic BP slope for the 2-h period during all exposures pooled was associated with reductions in time-domain measures of heart-rate variability indicated that altered ANS balance favoring sympathetic over parasympathetic CV tone may be causally involved. Moreover, the 2-h integrated $PM_{2.5}$ gravimetric $PM_{2.5}$ mass concentration predicted the magnitude of diastolic BP change ($\mu = 1.6$ mm Hg per 100 µg/m^3 of $PM_{2.5}$, $p = 0.01$). A concomitant experiment in Ann Arbor, Michigan tested potential operative mechanisms. Subjects received randomized blinded pretreatments of vitamin C 2 g, Bosentan 250 mg, and placebo prior to undergoing 2-h long CAP + ozone exposures. Neither high-dose antioxidant treatment nor endothelin antagonism obviated the prohypertensive response during $PM_{2.5}$ inhalation. These results imply that the underlying mechanisms responsible for the acute effect are unlikely to involve these biological pathways. Therefore, the totality of findings supports that short-term exposure to high levels of $PM_{2.5}$, at least derived from relatively urban ambient air, is capable of rapidly raising diastolic BP over a 2-h period. The most likely underlying mechanism is related to a rapid alteration in ANS balance, favoring relative sympathetic activation.

4.4 Summary of Biological Mechanisms

Based on the collective body of evidence supporting systemic effects of inhaled particulates, four putative pathways can be proposed: (1) autonomic mechanisms (*e.g.* parasympathetic nervous system withdrawal and/or sympathetic nervous system activation; (2) release of proinflammatory mediators (*e.g.* cytokines and activated immune cells such as monocytes) into the systemic circulation; (3) nanoscale particles and/or soluble PM constituents translocating into the systemic circulation (Figure 4.1). The 3 general pathways are not

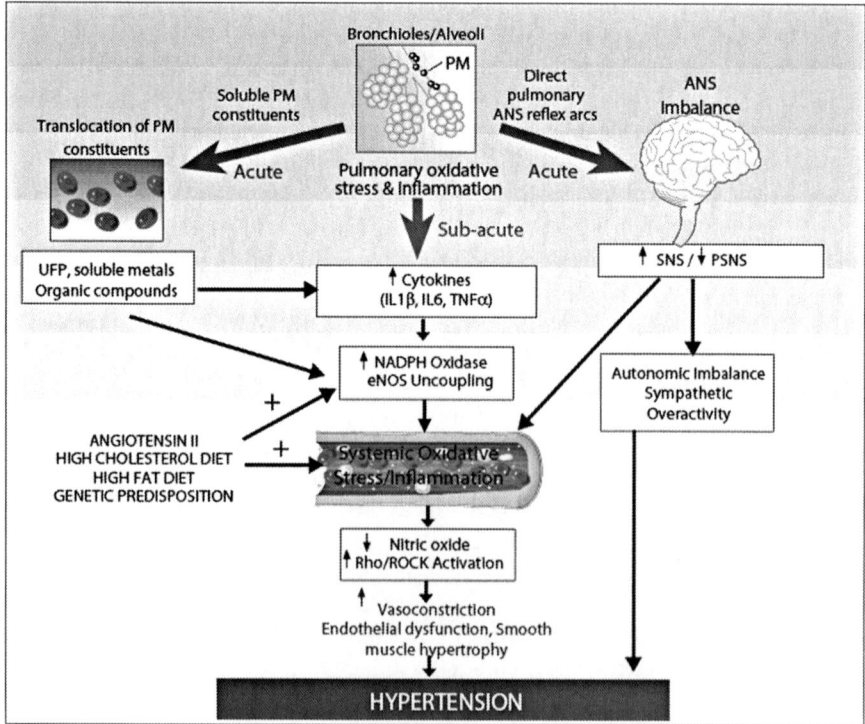

Figure 4.1 Mechanisms of PM-mediated hypertension. SNS = Sympathetic nervous system; PSNS = parasympathetic nervous system.

Table 4.2 Controlled exposure studies in humans that reported increased blood pressure.

Author	Human subjects	PM exposures	BP method	Findings
Urch[78]	23 healthy nonsmoking adults. Age 32 ± 10 years	2 h fine CAP ($147 \pm 27\,\mu g/m^3$) + ozone (121 ± 3 ppb) and filtered air exposures (crossover study) with intermittent exercise	Automated left arm seated BP taken once every 30 min during exposures	Median diastolic BP ↑ 6 (0–11) mm Hg during the CAP exposure period versus a ↑ 1 (–2 to 4) mm Hg response during the filtered air period ($p = 0.017$ for difference). No SBP change
Mills[56]	15 healthy nonsmoking men. Age 20–38 years	1-h diesel exhaust (PM levels $300\,\mu g/m^3$) and filtered air exposures (crossover study) with intermittent exercise.	Supine automated BP measured 2 h postexposure	Borderline SBP ↑ of 8 mm Hg ($p = 0.13$) and DBP ↑ of 6 mm Hg ($p = 0.08$) 2 h after diesel exhaust compared to after filtered air exposures.
Gong[63]	12 healthy and asthmatic subjects. Age 18–45 years	2-h fine CAP ($174\,\mu g/m^3$) and filtered air exposures (crossover study) with intermittent exercise	BP immediately after exposures	Systolic BP ↑ in healthy subjects immediately, 4 h, and 2 days post-CAP. Asthmatic patients ↓ SBP post and 2 days after CAP.
Fang[84]	26 male welders. Age 41 ± 12 years	Per $460\,\mu g/m^3$ ↑ of personal $PM_{2.5}$ during 6-h welding and non-welding sessions	Radial pulse wave analysis before, after and 24 h postsession.	Augmentation index (% of aortic systolic pressure derived from arterial wave reflection) ↑ of 3.5 (1.2–5.6)% after welding
Brook[80]	50 healthy subjects exposed 3 times for 2 h each time with 3 differing pre-treatments (Ann Arbor)	Ann Arbor: CAP + ozone; pretreated with placebo, vitamin C, or bosentan (crossover study)	Automated BP monitoring during and after exposures	Ann Arbor: DBP significantly increased during exposures between 3–4 mm Hg regardless of pre-treatment.

Table 4.2 (*Continued*)

Author	Human subjects	PM exposures	BP method	Findings
	31 healthy subjects exposed 4 times for 2 h to 4 differing conditions (Toronto)	Toronto: 4 exposures of CAP + ozone, CAP, ozone, and filtered air (crossover study)	Additional 24-h BP monitoring (Ann Arbor)	Toronto: DBP significantly increased during all exposures containing CAP between 3–4 mm Hg. Filtered air and ozone did not increase BP.

mutually exclusive. They may overlap temporally and/or be principally activated at differing time points. For instance, based on the studies examining blood pressure in humans (Table 4.2) it is apparent that the temporal nature of blood-pressure increase (rapid) would implicate rapid mechanisms that are likely autonomic or dependent on translocation of particles. The biological mechanisms responsible for more delayed responses such as sustained hypertension or progression of atherosclerosis, may dependent on slower-onset changes.

The types and sizes of pollutants inhaled may also determine their toxicity and the relative importance of the pathways. Larger fine or coarse PM cannot be transported into the circulation and will require secondary neural or proinflammatory responses to mediate extrapulmonary actions, while ultrafine PM (or soluble constituents of larger particles) might directly enter the blood stream and thus may exert their effects in the absence of a substantive pulmonary inflammatory component. The idea that oxidative stress is both a cause and consequence of PM-mediated effects has a sound experimental basis.[61–68] The generation of ROS likely plays a critical role in altering vasomotor balance and promoting arterial vasoconstriction and a propensity for an elevation in BP through a variety of mechanisms including activation of calcium-sensitization pathways such as Rho-kinase. The activation of ROS-sensitive signal-transduction pathways (*e.g.* NF-kB and mitogen-activated protein kinase pathways) could play a pivotal role in aggravation of the local vascular inflammatory responses and oxidative stress that could facilitate longer-term mechanisms that may play a role in sustenance of increases in blood pressure. The sensitivity to the effect of PM could also vary depending on the physiological and genetic mileu.[18,81] This is supported by studies demonstrating a synergistic effect of PM with hyperlipidemia, high-fat diet and Ang II in animals and in population studies demonstrating a differential effect of subjects with polymorphisms to key antioxidant enzyme systems.[82] There is considerably more work that needs to be done to characterize the initial source of the systemic proinflammatory reaction and the detailed mechanisms by which it is conveyed throughout the body following PM exposure.

References

1. R. D. Brook, B. Franklin, W. Cascio, Y. Hong, G. Howard, M. Lipsett, R. Luepker, M. Mittleman, J. Samet, S. C. Smith Jr and I. Tager, Air pollution and cardiovascular disease: a statement for healthcare professionals from the Expert Panel on Population and Prevention Science of the American Heart Association, *Circulation*, 2004, **109**, 2655–2671.
2. C. A. Pope 3rd, R. T. Burnett, G. D. Thurston, M. J. Thun, E. E. Calle, D. Krewski and J. J. Godleski, Cardiovascular mortality and long-term exposure to particulate air pollution: epidemiological evidence of general pathophysiological pathways of disease, *Circulation*, 2004, **109**, 71–77.
3. A. Shukla, C. Timblin, K. BeruBe, T. Gordon, W. McKinney, K. Driscoll, P. Vacek and B. T. Mossman, Inhaled particulate matter causes expression of nuclear factor (NF)-kappaB-related genes and oxidant-dependent NF-kappaB activation *in vitro*, *Am. J. Respir. Cell Molec. Biol.*, 2000, **23**, 182–187.
4. J. Y. Han, K. Takeshita and H. Utsumi, Noninvasive detection of hydroxyl radical generation in lung by diesel-exhaust particles, *Free Rad. Biol. Med.*, 2001, **30**, 516–525.
5. G. L. Squadrito, R. Cueto, B. Dellinger and W. A. Pryor, Quinoid redox cycling as a mechanism for sustained free radical generation by inhaled airborne particulate matter, *Free Rad. Biol. Med.*, 2001, **31**, 1132–1138.
6. A. Baulig, M. Garlatti, V. Bonvallot, A. Marchand, R. Barouki, F. Marano and A. Baeza-Squiban, Involvement of reactive oxygen species in the metabolic pathways triggered by diesel-exhaust particles in human airway epithelial cells, *Am. J. Physiol. Lung Cell Mol. Physiol.*, 2003, **285**, L671–679.
7. K. Donaldson, V. Stone, P. J. Borm, L. A. Jimenez, P. S. Gilmour, R. P. Schins, A. M. Knaapen, I. Rahman, S. P. Faux, D. M. Brown and W. MacNee, Oxidative stress and calcium signaling in the adverse effects of environmental particles (PM10), *Free Rad. Biol. Med.*, 2003, **34**, 1369–1382.
8. I. Beck-Speier, N. Dayal, E. Karg, K. L. Maier, G. Schumann, H. Schulz, M. Semmler, S. Takenaka, K. Stettmaier, W. Bors, A. Ghio, J. M. Samet and J. Heyder, Oxidative stress and lipid mediators induced in alveolar macrophages by ultrafine particles, *Free Rad. Biol. Med.*, 2005, **38**, 1080–1092.
9. Z. Li, X. Hyseni, J. D. Carter, J. M. Soukup, L. A. Dailey and Y. C. Huang, Pollutant particles enhanced H2O2 production from NAD(P)H oxidase and mitochondria in human pulmonary artery endothelial cells, *Am. J. Physiol. Cell Physiol.*, 2006, **291**, C357–365.
10. A. K. Prahalad, J. M. Soukup, J. Inmon, R. Willis, A. J. Ghio, S. Becker and J. E. Gallagher, Ambient air particles: effects on cellular oxidant radical generation in relation to particulate elemental chemistry, *Toxicol. Appl. Pharmacol.*, 1999, **158**, 81–91.
11. B. Dellinger, W. A. Pryor, R. Cueto, G. L. Squadrito, V. Hegde and W. A. Deutsch, Role of free radicals in the toxicity of airborne fine particulate matter, *Chem. Res. Toxicol.*, 2001, **14**, 1371–1377.

12. B. Gonzalez-Flecha, Oxidant mechanisms in response to ambient air particles, *Mol. Aspects Med.*, 2004, **25**, 169–182.
13. M. Ikeda, K. Watarai, M. Suzuki, T. Ito, H. Yamasaki, M. Sagai and T. Tomita, Mechanism of pathophysiological effects of diesel-exhaust particles on endothelial cells, *Environ. Tox. Pharm.*, 1998, **6**, 117–123.
14. Y. Bai, A. K. Suzuki and M. Sagai, The cytotoxic effects of diesel-exhaust particles on human pulmonary artery endothelial cells *in vitro*: role of active oxygen species, *Free Rad. Biol. Med.*, 2001, **30**, 555–562.
15. J. Y. Han, K. Takeshita and H. Utsumi, Noninvasive detection of hydroxyl radical generation in lung by diesel-exhaust particles, *Free Rad. Biol. Med.*, 2001, **30**, 516–525.
16. T. R. Nurkiewicz, D. W. Porter, M. Barger, L. Millecchia, K. M. Rao, P. J. Marvar, A. F. Hubbs, V. Castranova and M. A. Boegehold, Systemic microvascular dysfunction and inflammation after pulmonary particulate matter exposure, *Environ. Health Pers.*, 2006, **114**, 412–419.
17. Q. Sun, P. Yue, Z. Ying, A. J. Cardounel, R. D. Brook, R. Devlin, J. S. Hwang, J. L. Zweier, L. C. Chen and S. Rajagopalan, Air-pollution exposure potentiates hypertension through reactive oxygen species-mediated activation of Rho/ROCK, *Arterioscler. Thromb. Vasc. Biol.*, 2008, **28**, 1760–1766.
18. Q. Sun, A. Wang, X. Jin, A. Natanzon, D. Duquaine, R. D. Brook, J. G. Aguinaldo, Z. A. Fayad, V. Fuster, M. Lippmann, L. C. Chen and S. Rajagopalan, Long-term air-pollution exposure and acceleration of atherosclerosis and vascular inflammation in an animal model, *JAMA*, 2005, **294**, 3003–3010.
19. J. A. Araujo, B. Barajas, M. Kleinman, X. Wang, B. J. Bennett, K. W. Gong, M. Navab, J. Harkema, C. Sioutas, A. J. Lusis and A. E. Nel, Ambient particulate pollutants in the ultrafine range promote early atherosclerosis and systemic oxidative stress, *Circ. Res.*, 2008, **102**, 589–596.
20. C. R. Rhoden, E. Ghelfi and B. Gonzalez-Flecha, Pulmonary inflammation by ambient air particles is mediated by superoxide anion, *Inhal. Toxicol.*, 2008, **20**, 11–15.
21. E. Ghelfi, C. R. Rhoden, G. A. Wellenius, J. Lawrence and B. Gonzalez-Flecha, Cardiac oxidative stress and electrophysiological changes in rats exposed to concentrated ambient particles are mediated by TRP-dependent pulmonary reflexes, *Toxicol. Sci.*, 2008, **102**, 328–336.
22. L. Risom, P. Møller and S. Loft, Oxidative stress-induced DNA damage by particulate air pollution, *Mutat. Res./Fundam. Molec. Mechan. Mutagen.*, 2005, **592**, 119–137.
23. T. Ito, T. Suzuki, K. Tamura, T. Nezu, K. Honda and T. Kobayashi, Examination of mRNA expression in rat hearts and lungs for analysis of effects of exposure to concentrated ambient particles on cardiovascular function, *Toxicology*, 2008, **243**, 271–283.

24. Z. Li, J. D. Carter, L. A. Dailey and Y. C. Huang, Pollutant particles produce vasoconstriction and enhance MAPK signaling *via* angiotensin type I receptor, *Environ. Health Pers.*, 2005, **113**, 1009–1014.
25. W. P. Watkinson, M. J. Campen and D. L. Costa, Cardiac arrhythmia induction after exposure to residual oil fly ash particles in a rodent model of pulmonary hypertension, *Toxicol. Sci.*, 1998, **41**, 209–216.
26. G. A. Wellenius, P. H. Saldiva, J. R. Batalha, G. G. Krishna Murthy, B. A. Coull, R. L. Verrier and J. J. Godleski, Electrocardiographic changes during exposure to residual oil fly ash (ROFA) particles in a rat model of myocardial infarction, *Toxicol. Sci.*, 2002, **66**, 327–335.
27. C. C. Chang, J. S. Hwang, C. C. Chan, P. Y. Wang, T. H. Hu and T. J. Cheng, Effects of concentrated ambient particles on heart-rate variability in spontaneously hypertensive rats, *J. Occupat. Health*, 2005, **47**, 471–480.
28. C. C. Chang, J. S. Hwang, C. C. Chan and T. J. Cheng, Interaction effects of ultrafine carbon black with iron and nickel on heart-rate variability in spontaneously hypertensive rats, *Environ. Health Pers.*, 2007, **115**, 1012–1017.
29. L. C. Chen and J. S. Hwang, Effects of subchronic exposures to concentrated ambient particles (CAPs) in mice. IV. Characterization of acute and chronic effects of ambient air fine particulate matter exposures on heart-rate variability, *Inhal. Toxicol.*, 2005, **17**, 209–216.
30. F. Anselme, S. Loriot, J. P. Henry, F. Dionnet, J. G. Napoleoni, C. Thuillez and J. P. Morin, Inhalation of diluted diesel-engine emission impacts heart-rate variability and arrhythmia occurrence in a rat model of chronic ischemic heart failure, *Arch. Toxicol.*, 2007, **81**, 299–307.
31. C. G. Tankersley, H. C. Champion, E. Takimoto, K. Gabrielson, D. Bedja, V. Misra, H. El-Haddad, R. Rabold and W. Mitzner, Exposure to inhaled particulate matter impairs cardiac function in senescent mice, *Am. J. Physiol. Regul. Integr. Comp. Physiol.*, 2008, **295**, R252–263.
32. G. A. Wellenius, B. A. Coull, J. R. Batalha, E. A. Diaz, J. Lawrence and J. J. Godleski, Effects of ambient particles and carbon monoxide on supraventricular arrhythmias in a rat model of myocardial infarction, *Inhal. Toxicol.*, 2006, **18**, 1077–1082.
33. J. L. Quay, W. Reed, J. Samet and R. B. Devlin, Air Pollution Particles Induce IL-6 Gene Expression in Human Airway Epithelial Cells *via* NF-kappa B Activation, *Am. J. Respir. Cell Mol. Biol.*, 1998, **19**, 98–106.
34. S. Boland, A. Baeza-Squiban, T. Fournier, O. Houcine, M. C. Gendron, M. Chevrier, G. Jouvenot, A. Coste, M. Aubier and F. Marano, Diesel-exhaust particles are taken up by human airway epithelial cells *in vitro* and alter cytokine production, *Am. J. Physiol.*, 1999, **276**, L604–613.
35. A. Shukla, C. Timblin, K. BeruBe, T. Gordon, W. McKinney, K. Driscoll, P. Vacek and B. T. Mossman, Inhaled particulate matter causes expression of nuclear factor (NF)-kappaB-related genes and oxidant-dependent NF-kappaB activation *in vitro*, *Am. J. Respir. Cell Molec. Biol.*, 2000, **23**, 182–187.

36. S. F. van Eeden, W. C. Tan, T. Suwa, H. Mukae, T. Terashima, T. Fujii, D. Qui, R. Vincent and J. C. Hogg, Cytokines involved in the systemic inflammatory response induced by exposure to particulate-matter air pollutants (PM(10)), *Am. J. Respir. Crit. Care Med.*, 2001, **164**, 826–830.
37. T. Fujii, S. Hayashi, J. C. Hogg, H. Mukae, T. Suwa, Y. Goto, R. Vincent and S. F. van Eeden, Interaction of alveolar macrophages and airway epithelial cells following exposure to particulate matter produces mediators that stimulate the bone marrow, *Am. J. Respir. Cell Molec. Biol.*, 2002, **27**, 34–41.
38. R. Becher, A. Bucht, J. Ovrevik, J. K. Hongslo, H. J. Dahlman, J. T. Samuelsen and P. E. Schwarze, Involvement of NADPH oxidase and iNOS in rodent pulmonary cytokine responses to urban air and mineral particles, *Inhal. Toxicol.*, 2007, **19**, 645–655.
39. H. Tornqvist, N. L. Mills, M. Gonzalez, M. R. Miller, S. D. Robinson, I. L. Megson, W. Macnee, K. Donaldson, S. Soderberg, D. E. Newby, T. Sandstrom and A. Blomberg, Persistent endothelial dysfunction in humans after diesel exhaust inhalation, *Am. J. Respir. Crit. Care Med.*, 2007, **176**, 395–400.
40. E. Tamagawa, N. Bai, K. Morimoto, C. Gray, T. Mui, K. Yatera, X. Zhang, L. Xing, Y. Li, I. Laher, D. D. Sin, S. F. Man and S. F. van Eeden, Particulate matter exposure induces persistent lung inflammation and endothelial dysfunction, *Am. J. Physiol. Lung Cell Mol. Physiol.*, 2008, **295**, L79–85.
41. H. Fujimaki, Y. Kurokawa, S. Yamamoto and M. Satoh, Distinct requirements for interleukin-6 in airway inflammation induced by diesel exhaust in mice, *Immunopharm. Immunotoxicol.*, 2006, **28**, 703–714.
42. G. M. Mutlu, D. Green, A. Bellmeyer, C. M. Baker, Z. Burgess, N. Rajamannan, J. W. Christman, N. Foiles, D. W. Kamp, A. J. Ghio, N. S. Chandel, D. A. Dean, J. I. Sznajder and G. R. Budinger, Ambient particulate matter accelerates coagulation *via* an IL-6-dependent pathway, *J. Clin. Investig.*, 2007, **117**, 2952–2961.
43. C. Dostert, V. Petrilli, R. Van Bruggen, C. Steele, B. T. Mossman and J. Tschopp, Innate immune activation through Nalp3 inflammasome sensing of asbestos and silica, *Science (New York, N.Y.)*, 2008, **320**, 674–677.
44. J. W. Hollingsworth 2nd, D. N. Cook, D. M. Brass, J. K. Walker, D. L. Morgan, W. M. Foster and D. A. Schwartz, The role of Toll-like receptor 4 in environmental airway injury in mice, *Am. J. Respir. Crit. Care Med.*, 2004, **170**, 126–132.
45. Q. Sun, P. Yue, J. A. Deiuliis, C. N. Lumeng, T. Kampfrath, M. B. Mikolaj, Y. Cai, M. C. Ostrowski, B. Lu, S. Parthasarathy, R. D. Brook, S. D. Moffatt-Bruce, L. C. Chen and S. Rajagopalan, Ambient air pollution exaggerates adipose inflammation and insulin resistance in a mouse model of diet induced obesity, *Circulation*, 2009, **118**.
46. J. A. Nightingale, R. Maggs, P. Cullinan, L. E. Donnelly, D. F. Rogers, R. Kinnersley, K. F. Chung, P. J. Barnes, M. Ashmore and A. Newman-Taylor, Airway inflammation after controlled exposure to diesel-exhaust particulates, *Am. J. Respir. Crit. Care Med.*, 2000, **162**, 161–166.

47. L. B. Wichers, J. P. Nolan, D. W. Winsett, A. D. Ledbetter, U. P. Kodavanti, M. C. Schladweiler, D. L. Costa and W. P. Watkinson, Effects of instilled combustion-derived particles in spontaneously hypertensive rats. Part I: Cardiovascular responses, *Inhal. Toxicol.*, 2004, **16**, 391–405.
48. T. J. Cheng, J. S. Hwang, P. Y. Wang, C. F. Tsai, C. Y. Chen, S. H. Lin and C. C. Chan, Effects of concentrated ambient particles on heart rate and blood pressure in pulmonary hypertensive rats, *Environ. Health Pers.*, 2003, **111**, 147–150.
49. C. C. Chang, J. S. Hwang, C. C. Chan, P. Y. Wang, T. H. Hu and T. J. Cheng, Effects of concentrated ambient particles on heart rate, blood pressure, and cardiac contractility in spontaneously hypertensive rats, *Inhal. Toxicol.*, 2004, **16**, 421–429.
50. Q. Sun, P. Yue, Z. Ying, A. J. Cardounal, R. D. Brook, R. Devlin, J. S. Hwang, J. L. Zweier, L. C. Chen and S. Rajagopalan, Air-pollution exposure potentiates hypertension through reactive oxygen species mediated activation of Rho/ROCK. *Arterioscler. Thromb. Vasc. Biol.*, 2008, **28**, 1760–1766.
51. Z. Ying, T. Kampfrath, G. Thurston, B. Farrar, M. Lippmann, A. Wang, Q. Sun, L. C. Chen and S. Rajagopalan, Ambient Particulates Alter Vascular Function through Induction of Reactive Oxygen and Nitrogen Species, *Toxicol. Sci.*, 2009, **111**, 80–88.
52. C. R. Bartoli, G. A. Wellenius, E. A. Diaz, J. Lawrence, B. A. Coull, I. Akiyama, L. M. Lee, K. Okabe, R. L. Verrier and J. J. Godleski, Mechanisms of inhaled fine particulate air pollution-induced arterial blood pressure changes, *Environ. Health Pers.*, 2008, **117**, 361–366.
53. L. Barregard, G. Sallsten, P. Gustafson, L. Andersson, L. Johansson, S. Basu and L. Stigendal, Experimental exposure to wood-smoke particles in healthy humans: effects on markers of inflammation, coagulation, and lipid peroxidation, *Inhal. Toxicol.*, 2006, **18**, 845–853.
54. A. Peretz, E. C. Peck, T. K. Bammler, R. P. Beyer, J. H. Sullivan, C. A. Trenga, S. Srinouanprachnah, F. M. Farin and J. D. Kaufman, Diesel exhaust inhalation and assessment of peripheral blood mononuclear cell gene transcription effects: an exploratory study of healthy human volunteers, *Inhal. Toxicol.*, 2007, **19**, 1107–1119.
55. R. D. Brook, J. R. Brook, B. Urch, R. Vincent, S. Rajagopalan and F. Silverman, Inhalation of fine particulate air pollution and ozone causes acute arterial vasoconstriction in healthy adults, *Circulation*, 2002, **105**, 1534–1536.
56. N. L. Mills, H. Tornqvist, S. D. Robinson, M. Gonzalez, K. Darnley, W. MacNee, N. A. Boon, K. Donaldson, A. Blomberg, T. Sandstrom and D. E. Newby, Diesel exhaust inhalation causes vascular dysfunction and impaired endogenous fibrinolysis, *Circulation*, 2005, **112**, 3930–3936.
57. N. L. Mills, H. Tornqvist, M. C. Gonzalez, E. Vink, S. D. Robinson, S. Soderberg, N. A. Boon, K. Donaldson, T. Sandstrom, A. Blomberg and D. E. Newby, Ischemic and thrombotic effects of dilute diesel-exhaust inhalation in men with coronary heart disease, *New Engl. J. Med.*, 2007, **357**, 1075–1082.

58. N. L. Mills, S. D. Robinson, P. H. Fokkens, D. L. Leseman, M. R. Miller, D. Anderson, E. J. Freney, M. R. Heal, R. J. Donovan, A. Blomberg, T. Sandstrom, W. MacNee, N. A. Boon, K. Donaldson, D. E. Newby and F. R. Cassee, Exposure to concentrated ambient particles does not affect vascular function in patients with coronary heart disease, *Environ. Health Pers.*, 2008, **116**, 709–715.
59. R. Duffin, L. Tran, D. Brown, V. Stone and K. Donaldson, Proinflammogenic effects of low-toxicity and metal nanoparticles *in vivo* and *in vitro*: highlighting the role of particle surface area and surface reactivity, *Inhal. Toxicol.*, 2007, **19**, 849–856.
60. G. Oberdorster, E. Oberdorster and J. Oberdorster, Nanotoxicology: an emerging discipline evolving from studies of ultrafine particles, *Environ. Health Pers.*, 2005, **113**, 823–839.
61. A. Peretz, J. H. Sullivan, D. F. Leotta, C. A. Trenga, F. N. Sands, J. Allen, C. Carlsten, C. W. Wilkinson, E. A. Gill and J. D. Kaufman, Diesel exhaust inhalation elicits acute vasoconstriction *in vivo*, *Environ. Health Pers.*, 2008, **116**, 937–942.
62. R. B. Devlin, A. J. Ghio, H. Kehrl, G. Sanders and W. Cascio, Elderly humans exposed to concentrated air pollution particles have decreased heart-rate variability, *Eur. Respir. J.*, 2003, **40**, 76s–80s.
63. H. Gong Jr, W. S. Linn, S. L. Terrell, K. W. Clark, M. D. Geller, K. R. Anderson, W. E. Cascio and C. Sioutas, Altered heart-rate variability in asthmatic and healthy volunteers exposed to concentrated ambient coarse particles, *Inhal. Toxicol.*, 2004, **16**, 335–343.
64. H. Gong, W. S. Linn, S. L. Terrell, K. R. Anderson, K. W. Clark, C. Sioutas, W. E. Cascio, N. Alexis and R. B. Devlin, Exposures of elderly volunteers with and without chronic obstructive pulmonary disease (COPD) to concentrated ambient fine particulate pollution, *Inhal. Toxicol.*, 2004, **16**, 731–744.
65. J. M. Samet, D. Graff, J. Berntsen, A. J. Ghio, Y. C. Huang and R. B. Devlin, A comparison of studies on the effects of controlled exposure to fine, coarse and ultrafine ambient particulate matter from a single location, *Inhal. Toxicol.*, 2007, **19**(Suppl 1), 29–32.
66. A. J. Ghio, A. Hall, M. A. Bassett, W. E. Cascio and R. B. Devlin, Exposure to concentrated ambient air particles alters hematologic indices in humans, *Inhal. Toxicol.*, 2003, **15**, 1465–1478.
67. C. Carlsten, J. D. Kaufman, A. Peretz, C. A. Trenga, L. Sheppard and J. H. Sullivan, Coagulation markers in healthy human subjects exposed to diesel exhaust, *Thromb. Res.*, 2007, **120**, 849–855.
68. W. C. Tan, D. Qiu, B. L. Liam, T. P. Ng, S. H. Lee, S. F. van Eeden, Y. D'Yachkova and J. C. Hogg, The human bone marrow response to acute air pollution caused by forest fires, *Am. J. Respir. Crit. Care Med.*, 2000, **161**, 1213–1217.
69. S. F. van Eeden, A. Yeung, K. Quinlam and J. C. Hogg, Systemic response to ambient particulate matter: relevance to chronic obstructive pulmonary disease, *Proc. Am. Thoracic Soc.*, 2005, **2**, 61–67.

70. M. W. Frampton, J. C. Stewart, G. Oberdorster, P. E. Morrow, D. Chalupa, A. P. Pietropaoli, L. M. Frasier, D. M. Speers, C. Cox, L. S. Huang and M. J. Utell, Inhalation of ultrafine particles alters blood leukocyte expression of adhesion molecules in humans, *Environ. Health Pers.*, 2006, **114**, 51–58.
71. K. Yatera, J. Hsieh, J. C. Hogg, E. Tranfield, H. Suzuki, C. H. Shih, A. R. Behzad, R. Vincent and S. F. van Eeden, Particulate matter air-pollution exposure promotes recruitment of monocytes into atherosclerotic plaques, *Am. J. Physiol.*, 2008, **294**, H944–953.
72. W. S. Beckett, D. F. Chalupa, A. Pauly-Brown, D. M. Speers, J. C. Stewart, M. W. Frampton, M. J. Utell, L. S. Huang, C. Cox, W. Zareba and G. Oberdorster, Comparing inhaled ultrafine versus fine zinc oxide particles in healthy adults: a human inhalation study, *Am. J. Respir. Crit. Care Med.*, 2005, **171**, 1129–1135.
73. H. C. Routledge, S. Manney, R. M. Harrison, J. G. Ayres and J. N. Townend, Effect of inhaled sulphur dioxide and carbon particles on heart-rate variability and markers of inflammation and coagulation in human subjects, *Heart (British Cardiac Society)*, 2006, **92**, 220–227.
74. I. F. Charo, Macrophage polarization and insulin resistance: PPARgamma in control, *Cell Metab.*, 2007, **6**, 96–98.
75. F. Geissmann, C. Auffray, R. Palframan, C. Wirrig, A. Ciocca, L. Campisi, E. Narni-Mancinelli and G. Lauvau, Blood monocytes: distinct subsets, how they relate to dendritic cells, and their possible roles in the regulation of T-cell responses, *Immunol. Cell Biol.*, 2008, **86**, 398–408.
76. H. Gong Jr, W. S. Linn, C. Sioutas, S. L. Terrell, K. W. Clark, K. R. Anderson and L. L. Terrell, Controlled exposures of healthy and asthmatic volunteers to concentrated ambient fine particles in Los Angeles, *Inhal. Toxicol.*, 2003, **15**, 305–325.
77. M. W. Frampton, Systemic and cardiovascular effects of airway injury and inflammation: ultrafine particle exposure in humans, *Environ. Health Pers.*, 2001, **109**(Suppl 4), 529–532.
78. B. Urch, F. Silverman, P. Corey, J. R. Brook, K. Z. Lukic, S. Rajagopalan and R. D. Brook, Acute blood pressure responses in healthy adults during controlled air-pollution exposures, *Environ. Health Pers.*, 2005, **113**, 1052–1055.
79. B. Urch, J. R. Brook, D. Wasserstein, R. D. Brook, S. Rajagopalan, P. Corey and F. Silverman, Relative contributions of $PM_{2.5}$ chemical constituents to acute arterial vasoconstriction in humans, *Inhal. Toxicol.*, 2004, **16**, 345–352.
80. R. D. Brook, B. Urch, J. T. Dvonch, R. L. Bard, M. Speck, G. Keeler, M. Morishita, F. J. Marsik, A. S. Kamal, N. Kaciroti, J. Harkema, P. Corey, F. Silverman, D. R. Gold, G. Wellenius, M. A. Mittleman, S. Rajagopalan and J. R. Brook, Insights into the mechanisms and mediators of the effects of air-pollution exposure on blood pressure and vascular function in healthy humans, *Hypertension*, 2009, **54**, 659–667.
81. Q. Sun, P. Yue, J. A. Deiuliis, C. N. Lumeng, T. Kampfrath, M. B. Mikolaj, Y. Cai, M. C. Ostrowski, B. Lu, S. Parthasarathy, R. D. Brook,

S. D. Moffatt-Bruce, L. C. Chen and S. Rajagopalan, Ambient air pollution exaggerates adipose inflammation and insulin resistance in a mouse model of diet-induced obesity, *Circulation*, 2009, **119**, 538–546.
82. J. Schwartz, S. K. Park, M. S. O'Neill, P. S. Vokonas, D. Sparrow, S. Weiss and K. Kelsey, Glutathione-S-transferase M1, obesity, statins, and autonomic effects of particles: gene-by-drug-by-environment interaction, *Am. J. Respir. Crit. Care Med.*, 2005, **172**, 1529–1533.

CHAPTER 5
Air Pollution and Diabetes

E. H. WILKER[1,2] AND J. D. SCHWARTZ[2,3]

[1] Cardiovascular Epidemiology Research Unit, Beth Israel Deaconess Medical Center, 375 Longwood Ave, Boston MA, 02215, USA; [2] Harvard School of Public Health, Department of Environmental Health, 665 Huntington Ave Boston MA, 02215, USA; [3] Channing Laboratory, Department of Medicine, Brigham and Women's Hospital, 181 Longwood Avenue, Boston, MA, 02115, USA

5.1 Introduction

Type-2 diabetes is a complex medical condition defined by insulin resistance and increased hepatic glucose production, but also closely associated with adiposity, hypertension, and hypercholesterolemia.[1] Together, these factors have contributed to substantial health and economical consequences. A recent study estimates that the world-wide prevalence of diabetes among adults (aged 20–79 years) in 2010 will be 6.4%, affecting 285 million individuals, and is expected to increase to 7.7%, and 439 million adults by 2030.[2] Particulate air pollution has been associated with a number of cardiovascular health effects[3] including mortality[4] and hospital admissions[5] as well as autonomic dysfunction[6] and inflammation.[7] Although differences in effects may vary across populations, a number of studies have consistently demonstrated that the cardiotoxic effects of air pollution are particularly strong among persons diagnosed with type-2 diabetes. Diabetes is known to be a predictor of cardiovascular disease[8,9] and atherosclerosis is one of the primary causes of diabetes morbidity and mortality.[10] It has been hypothesized that diabetics and other persons with similar characteristics of higher baseline inflammation may be particularly susceptible to the effects of ambient air-pollution exposures. Nevertheless, the underlying mechanisms of the relationship remain unclear.[11]

Evaluating how air pollution may influence susceptibility to diabetes is also complicated by the heterogeneous composition of ambient particulates, which may include secondary aerosols derived from gases (such as sulfates and nitrates), silicates, carbon compounds, metal oxides as well as other biological materials. Variation observed across studies may be in part due to heterogeneity in the composition of ambient exposures resulting from different mixtures of pollutants. Recent evidence suggests that certain components in air pollution are more toxic than others[12-14] and that differences across studies may arise because of variation in sources, composition, mixtures as well as the time windows examined.

The purpose of this chapter is to provide a summary of major findings from key areas of research related to the association between air pollution and diabetes. We focus primarily on the role in which diabetes has been hypothesized to contribute to effect measure modification of the relationship between air pollution and cardiovascular outcomes, examining evidence from administrative data, autonomic and endothelial responses, and biomarkers of systemic inflammation and haemostatic function. Given the extent to which diabetes affects the health and well-being of individuals globally as well as the growing understanding of how air pollutants contribute to influencing both acute and chronic disease, improving understanding of this association is critical to prevention and treatment strategies for public health.

5.2 Evidence from Administrative Data Sources

5.2.1 Mortality

Numerous health studies have shown that short-[15,16] and long-term[17] ambient particle exposures are associated with early death, particularly from cardiovascular disease. In an extension to these observed all-cause mortality relationships, some studies have attempted to identify the subpopulations that may be more susceptible to these effects. One of the first studies to suggest that diabetic individuals may be particularly vulnerable to air pollution utilized data from Montreal, Quebec, where the mean percentage change in daily mortality from diabetes evaluated for an interquartile range change in pollutant levels averaged over the day of death and the preceding 2 days was 7.59% (95% confidence interval (CI): 2.36–13.09%) for predicted $PM_{2.5}$ and 4.48% (95% CI: 1.08–7.99%) for sulfates, which represent regional pollution resulting from combustion sources.[18] These results were followed by an additional investigation within the same population that expanded upon the previous study to include all ambient air pollutants monitored routinely in Montreal as well as Medicare data to identify subjects over the age of 65 years as having diabetes and other comorbid conditions before death. The second study found that persons with diabetes are at an elevated risk of death when levels of air pollution increase and that these effects are not restricted to pollutants generated from combustion sources, suggesting that mixture and composition may be important in determining the effects of pollution on mortality. These data also suggested that diabetics with comorbidities (*i.e.* cardiovascular disease, airways disease, cancer) had higher rates of death when air-pollution levels were higher than those with diabetes alone.[19]

A case-crossover study in Cook County, Illinois showed a 2.0-fold higher risk (95% CI = –1.5 to 5.5) of mortality for a 10 µg/m^3 increase in PM_{10} among diabetics when compared to participants with no diabetes, myocardial infarction congestive heart failure, chronic obstructive pulmonary disease or conduction disorders.[20] Similarly, stronger associations of daily fine particulate matter ($PM_{2.5}$) concentrations with mortality for diabetics were observed in a study of 9 California counties, where 2.4% (95% CI 0.6 to 4.2) changes in mortality per 10 µg/m^3 increment in $PM_{2.5}$[21] were observed, compared with 0.6% (95% CI: 0.2 to 1.0%) for mortality for all persons. A recent study conducted within the Boston area also examined effects of black carbon, a marker of local traffic, and sulfate (SO_4) and found a much larger association with diabetes deaths (5.7%, 95% CI: –1.7, 13.7) compared with all-cause mortality (2.3% increase 95% CI: 1.2, 3.4%), but noted these deaths accounted for only 2.5% of all deaths within the Boston metropolitan area from 1995–2002.[22] While most studies are from North America, studies from other parts of the world report similar findings. For example, a recent study of 9 Italian cities found that the effects of PM_{10} on mortality were stronger for diabetics than nondiabetics,[23] and in Shanghai 10 µg/m^3 changes in PM_{10}, SO_2, or NO_2 were found to correspond to small but consistently higher relative risk of diabetes mortality.[24]

5.2.2 Hospital Admissions and Acute Events

Large studies utilizing administrative data sources have also investigated whether diabetics are not just more susceptible to dying during periods of higher air pollution, but whether they undergo hospitalization in response to higher air pollution at greater rates. Some of the earliest evidence comes from the Los Angeles area, where cardiovascular disease admissions were analyzed separately for diabetics in a study examining pollutants including ozone (O_3), nitric oxide (NO_2), carbon monoxide (CO), and PM_{10}. In both year-round analyses, as well as season-specific estimates for several pollutants, steeper slopes were observed for diabetics, although the associations did not reach statistical significance.[25] Subsequently, in a study in Cook County, IL, greater increases in admission were observed for heart disease among diabetics, and diabetics were identified as an important risk group.[26] In this study, a 10-µg/m^3 increase in PM_{10} was associated with a 2.01% (95% CI: 1.40–2.62%) higher rate in admissions for heart disease with diabetes, but only a 0.94% (95% CI: 0.61–1.28%) higher rate in persons without diabetes. Further investigations from a follow-up study included data from Chicago, IL, Detroit, MI, Pittsburgh, PA, and Seattle, WA, for the years 1988–1994. These studies showed a doubling of the risk of a PM_{10}-associated cardiovascular admission compared with nondiabetics.[27]

More recent studies have provided additional evidence to support these associations for increases in hospital admissions among diabetics. A study from the KORA Registry in Augsburg Germany found that the odds of myocardial infarction (MI) within 1 h after exposure to traffic was 4.63 (2.57–8.33) in sensitivity models restricted to diabetic participants, but in primary models odds were only 2.92 (2.22–3.83).[28] In a study examining emergency department visits in Atlanta, evidence of effect modification by comorbid hypertension anddiabetes was observed in relation to PM_{10}, NO_2, and CO exposures. In particular, there was evidence that the association of dysrhythmia visits in relation to NO_2 among patients with diabetes was stronger than those without (per 20 parts per billion (ppb): odds ratio (OR) = 1.158, 95% CI: 1.046, 1.282 vs. OR = 1.014, 95% CI: 0.988, 1.040).[29] When Chiu et al. examined effect modification for potentially sensitive subgroups of Taipei area residents presenting at the emergency department for arrhythmias, statistically significant effect modification for diabetics was not observed, however, the authors acknowledge that diabetics comprised only 2% of their visits between 2000 and 2006.[30] Lack of statistical power may be one reason that some other studies have failed to observe differences in associations among subpopulations of diabetics.[31]

5.3 Evidence from Measurements of Physiologic Outcomes

5.3.1 Heart-Rate Variability

Given the associations observed between cardiovascular and diabetic conditions, investigators have examined whether preclinical markers of cardiovascular disease are influenced by air pollution and also whether diabetics are more susceptible to these effects. A number of studies have examined associations between particulate air pollution and heart-rate variability (HRV).[32–35] HRV is used widely as a quantitative marker of autonomic function reflecting rhythmic activity of the sinus node.[36] For example, in the Atherosclerosis Risk in Communities Study (ARIC), diabetes was found to be associated with more rapid secular decreases in HRV, suggesting the presence of cardiac autonomic impairment at early stages of diabetes. The study also indicates that in diabetics, autonomic cardiac function worsens over the long-term (9 years of follow-up).[37]

In participants of the Normative Aging Study, comprised of older males residing in the Boston area, Park and colleagues observed that exposures to $PM_{2.5}$ and O_3 were associated with decreased HRV. In this study, associations between standard deviation of normal-to-normal beats (SDNN) and low frequency (LF) with $PM_{2.5}$ were stronger in people with diabetes[38] and associated with almost 4-fold higher percentage changes than those observed among persons without diabetes. However, diabetes did not modify the effect of O_3 on HRV. A more recent study investigated associations between particulate exposures and HRV in the Womens' Health Initiative (WHI) observed strong inverse associations between particulate matter (PM_{10}) and root mean square of successive differences in normal-to-normal RR intervals (rMMSD). The study found that the effects were not limited to individuals with diabetes diagnosis and that impaired fasting glucose was also associated with changes in HRV among otherwise healthy female participants.[39] Gender differences may be due to the fact that women's coronary arteries are smaller in size and women's microvessels appear to be more frequently dysfunctional than those of men;[40] however, this is an area that requires further investigation. Electrocardiogram measures from Holter monitors have also been used to measure air-pollution responses in more controlled settings.[41] In a study of ST segment depression (an indicator of myocardial ischemia) patients with coronary artery disease and diabetes showed higher response to increased levels of $PM_{2.5}$ (P for interaction < 0.001).[42]

5.3.2 Brachial-Artery Diameter and Flow-Mediated Dilation

Other physiologic metrics provide information not just about autonomic function, but about endothelial activity as well. For example, brachial flow-mediated dilation (FMD) is a noninvasive physiological parameter that can measure endothelial function and has been validated as a predictor of cardiovascular events in older adults.[43] Literature on the effects of ambient pollutants and FMD is limited, but a few small controlled experiments have examined FMD association with air pollution, with mixed results. Diabetes enhanced vulnerability to impairment of vascular reactivity and endothelial function in a cross-sectional study using baseline data from 270 patients with or at risk of diabetes who participated in clinical trials.[44] Interquartile range (IQR) changes in black carbon were associated with decreased flow-mediated vascular reactivity (−12.6%; 95% CI: −21.7 to −2.4), and $PM_{2.5}$ was associated with nitroglycerin-mediated reactivity (−7.6%; 95% CI: −12.8 to −2.1), which may indicate nonendothelial and endothelial pollution responses. This study also found that IQR changes in SO_4 were associated with lower flow-mediated (−10.7%; 95% CI: −17.3 to −3.5) as well as nitroglycerin-mediated (−5.4%; 95% CI: −10.5 to −0.1) vascular reactivity among those with diabetes. In a recent study, personal exposure to PM_{10} was positively and consistently associated with brachial artery FMD but showed an inverse association with basal arterial diameter and flow in diabetics.[45] In this study, PM_{10} was also significantly positively associated with blood pressure, a classic measure of cardiovascular function, but inversely associated with artery flow. Two other studies used controlled exposures of volunteers. One, using diesel exhaust, found reduced brachial artery diameter, but no significant effect on FMD.[46] The other, using concentrated $PM_{2.5}$ and ozone, again found effects on brachial diameter, but not FMD.[47] However, another study, in which volunteers were stationed at bus stops for 2 h, reported associations of $PM_{2.5}$ with FMD.[48] None of these studies examined effect modification by diabetes.

5.3.3 Evidence from Biomarkers

Endothelial dysfunction is a hallmark of the early stages of atherosclerosis and has also been assessed by directly using information measured directly from serum biomarkers. Cell-adhesion molecules play important roles levels of endothelial biomarkers and both intercellular adhesion molecule-1 (ICAM-1) and (vascular cellular adhesion molecule (VCAM-1) have been shown to be higher among diabetics.[49] O'Neill et al. measured plasma levels of soluble ICAM-1, VCAM-1, and Von Willibrand Factor (vWF) as well as particle exposure in 92 Boston patients with type-2 diabetes.[50] Air-pollutant exposure measures showed consistently positive point estimates of association with these inflammatory markers, with particularly strong associations observed between $PM_{2.5}$ and VCAM-1. Madrigano et al. examined repeated measurements of ICAM-1 and VCAM-1 in 809 participants of the Normative aging study, and reported that black carbon was associated with VCAM, with a significant interaction with obesity.[51] Increased ICAM-1 has been reported in response to air pollution among asthmatic children,[52] and Delfino and coworkers reported positive, but not significant associations of multiple particle measures with both ICAM and VCAM.[53] These findings are supported by toxicological studies, showing that instillation of PM in Watanabe rabbits results in increased expression of ICAM and VCAM in atherosclerotic plaques.[54]

Particles have been associated with thrombotic and inflammatory factors in a number of studies.[55,56] Within the ARIC Study, which measured inflammatory and hemostatic factors, a standard deviation (SD) increment of PM_{10} (12.8 µg/m^3) was significantly associated with 3.93% higher levels of Willebrand Factor (vWF) among diabetics, suggesting that the strongest associations in the upper range of the pollutant distributions, and in persons with a positive history of diabetes and CHD.[57] C-reactive protein (CRP) has also been shown to be a predictor of new-onset diabetes among[58] and in a small study of 44 senior citizens that examined repeated measures of inflammatory markers and particulate matter, CRP, interleukin 6 (IL-6) and white blood cells, associations between $PM_{2.5}$ and CRP were consistently and often significantly elevated among the 8 individuals with diabetes. The strongest associations were between CRP and $PM_{2.5}$ at 4-day moving averages.[59]

5.3.4 Toxicology Studies

Much of the recent work that has informed our understanding of the potential mechanisms underlying associations between air pollution and diabetes comes from animal models. The positive associations that have been observed, as well as the consistency across studies regarding the enhanced susceptibility among subjects with diabetes and its risk factors are supported by the growing body of evidence from toxicologic studies that have began to elucidate the mechanisms of chronic PM effects on CVD. In Watanabe heritable hyperlipidemic rabbits,[60] instillation with PM_{10} (collected from outdoor air in Ottawa, Ontario, Canada) for 4 weeks was found to accelerate the progression of coronary atherosclerosis.

PM$_{10}$ also increased plaque-cell turnover and extracellular-lipid pools in both coronary and aortic atherosclerotic lesions. In one study, exposure of apolipoprotein E-null mice to PM$_{2.5}$ for 6 months (with equivalent concentration of 15.2 µg/m^3, collected in Tuxedo, NY), was found to result in a 1.58-fold increase in percentage plaque area in the abdominal aorta in mice maintained on high-fat, but not regular, chow.[61] Further analyses indicated that vascular inflammation was elevated in the atherosclerotic plaques of PM-exposed mice. A recent study has also demonstrated that PM$_{2.5}$ exposure exaggerates insulin resistance and visceral inflammation and adiposity.[62] Collectively, these studies further support the associations observed in cohort studies and underscore the need for further work in this area.

5.3.5 Potential Mechanisms

The etiology of the diabetes is complex and involves both genetic and environmental factors,[63] but recent evidence has demonstrated that some lifestyle factors can reduce the risk of developing the disease. It is well established that diabetes is associated with inflammatory processes[64] and oxidative stress,[65] as well as endothelial and vascular dysfunction.[66] Inhalation and immediate contact with pollutants *via* respiration is thought to be associated with downstream cardiovascular effects.[67] Oxidative stress is also a key mechanism by which air pollution is thought to induce its harmful effects.[68,69] Exposure of animals to concentrated air particles has been demonstrated to produce systemic and cardiac oxidative stress,[68,70–72] and studies of genetic variations in oxidative defenses in humans have confirmed the role of oxidative stress in mediating the cardiovascular effects of PM.[51,73–75] The combined effects of inflammation and oxidative stress may also lead to disturbances in nitric-oxide levels.[76]

Current evidence also suggests that obesity, a risk factor for type-2 diabetes development, modifies the association between air pollution and cardiovascular disease. It is clear that adipose tissue secretes leptin, hormones, adipokines and cytokines and can act as an endocrine organ.[77,78] Body mass index (BMI) is an important risk factor for type-2 diabetes. Obesity can also lead to changes in the adipokine profiles. Several studies suggest that obesity affects susceptibility to air pollution. Results from the Normative Aging Study suggest that obesity is a significant modifier of the relationship between PM exposure and cardiovascular disease. For example, a study that examined effect modification by glutathione-s transferase mu-1 (*GSTM1*) found an association between PM$_{2.5}$ and reduced high frequency (HF) is only evident in persons missing the allele for *GSTM1*, or in persons likely to have greater than average baseline systemic inflammation and oxidative stress, such as obese individuals.[75] In the same population, obesity was found to be a significant susceptibility factor for the acute effects of O$_3$ on lung function; the estimated decrease in forced expiratory volume (FEV1) due to O$_3$ was twice as much for obese subjects than non-obese.[79] NAS investigators have also observed a greater effect of traffic-related particles on inflammatory markers in the obese.[80] Another recent study found that larger effects of black carbon on VCAM-1 were seen in the obese ($p=0.007$) and in subjects missing the allele for *GSTM1* ($p=0.02$).[51]

It has been hypothesized that exposure to ultrafine particles may invoke alveolar inflammation, release inflammatory mediators, exacerbate lung conditions, and increase the coagulability of blood, thereby leading to acute episodes of cardiovascular disease.[81] Additionally, it has been shown that inhalation of urban particles in animals without a structural lung injury increases the circulating levels of endothelins, which are potent vasoconstrictors.[82] There is some evidence that diabetics have abnormal plasma levels of endothelins.[83] Additionally, certain factors may be initiated or activated as a result of hyperglycemia and may lead to endothelin dysfunction.[84]

5.4 Does Air Pollution Cause Diabetes?

While thus far we have highlighted evidence that suggests that diabetics are more susceptible to the effects of air pollution, a more controversial question is whether the opposite is true: does exposure to PM predispose people to develop insulin resistance and diabetes? A growing body of literature is beginning to address this question. Exposure to NO_2, a marker of traffic-related air pollutants, was associated with type-2 diabetes prevalence among women.[85] However, no positive association with men was observed, suggesting that there may be some gender differences in susceptibility as well, or that there are gender differences in exposure misclassification. A case-control study performed in California suggests that cumulative exposure to O_3 and SO_4 in ambient air may predispose children to the development of type-1 diabetes.[86] The authors hypothesize that oxidative stress induced by exposure to O_3 and SO_4 may accelerate B-cell destruction, known to be a hallmark of type-1 diabetes. They also cite evidence that air pollution may alter cellular and humoral immunity. Although type-I and type-II diabetes differ in their etiology, they share commonalities in their clinical presentation. Thus, increased understanding of the influence of ambient pollutants on type-1 diabetes diagnosed in childhood may help to develop more targeted prevention strategies.

While particulate matter has not specifically been shown to be associated with type-2 diabetes, some metals, as well as organic compounds, have been associated with diabetes prevalence. Particularly strong effects have been observed with arsenic. For instance, a recent study found that high levels of total urine arsenic were positively associated with the prevalence of type-2 diabetes and with levels of glycated hemoglobin.[87] This association persisted even after adjusting for diabetes risk factors and markers of seafood intake (which could be a major source of arsenic). This work also supports that the high prevalence of diabetes in some parts of the world, *e.g.*, Taiwan, Bangladesh, and Mexico may be related to the high levels of arsenic in these regions.[88] Studies from the National Health and Nutrition Examination Survey have also shown associations between dioxins and diabetes.[89,90] Organic compounds like bisphenol

A have been associated with insulin resistance and profiles of metabolic syndrome.[91] Currently, this is a growing area of research that is likely to generate a better understanding of the source-specific relationships between pollutant exposure and diabetes prevalence.

5.5 Conclusions

Interpreting the associations between diabetes and air pollution still remains a challenge. A growing body of evidence suggests that diabetes influences vulnerability to air pollution, but it remains unclear which components of the air pollution are driving this interaction and whether other seasonal, temporal or, biological factors may further alter these associations. Both air pollution and diabetes influence cardiovascular disease through the common pathways of inflammation, oxidative stress, and endothelial dysfunction, and this may be the physiologic basis for the interaction between the two risk factors. However, further work is clearly needed to identify the nature of this synergy. Despite the decline in ambient air-pollution levels in recent years, studies continue to show that some populations are particularly susceptible to air-pollution-related health effects. Additionally, an important challenge for future research will be to disentangle the effects of air pollution from lifestyle issues such as diet and exercise. Undiagnosed diabetes and abnormal glucose tolerance are also of substantial clinical importance. As obesity continues to rise, and comorbid conditions become increasingly common, it has become even more important to understand with unprecedented clarity the role of environment risk factors in increasing CVD risk in vulnerable populations. Because environmental exposures are readily modifiable risk factors, minimizing their impact could provide significant public-health benefit. Work in this direction has already begun. Several pollution-controlling devices have been developed to prevent air pollution from several sources. Increasing use of these devices and other similar technology could substantially and expeditiously decrease the impact of air pollution on cardiovascular health and disease.

References

1. S. M. Haffner and H. Miettinen, Insulin resistance implications for type II diabetes mellitus and coronary heart disease, *Am. J. Med.*, 1997, **103**, 152–162.
2. J. E. Shaw, R. A. Sicree and P. Z. Zimmet, Global estimates of the prevalence of diabetes for 2010 and 2030, *Diabetes Res. Clin. Pract.*, 2009, **87**, 4–14.
3. C. A. Pope 3rd and D. W. Dockery, Health effects of fine particulate air pollution: lines that connect, *J. Air Waste Manag. Assoc.*, 2006, **56**, 709–742.

4. D. W. Dockery, C. A. Pope, X. Xu, J. D. Spengler, J. H. Ware, M. E. Fay, B. G. Ferris and F. E. Speizer, An association between air pollution and mortality in six US cities, *N. Engl. J. Med.*, 1993, **329**, 1753–1759.
5. J. Schwartz, Air pollution and hospital admissions for heart disease in eight US counties, *Epidemiology*, 1999, **10**, 17–22.
6. J. Schwartz, A. Litonjua, H. Suh, M. Verrier, A. Zanobetti, M. Syring, B. Nearing, R. Verrier, P. Stone, G. MacCallum, F. E. Speizer and D. R. Gold, Traffic related pollution and heart-rate variability in a panel of elderly subjects, *Thorax*, 2005, **60**, 455–461.
7. C. A. Pope 3rd, R. T. Burnett, G. D. Thurston, M. J. Thun, E. E. Calle, D. Krewski and J. J. Godleski, Cardiovascular mortality and long-term exposure to particulate air pollution: epidemiological evidence of general pathophysiological pathways of disease, *Circulation*, 2004, **109**, 71–77.
8. S. M. Grundy, I. J. Benjamin, G. L. Burke, A. Chait, R. H. Eckel, B. V. Howard, W. Mitch, S. C. Smith Jr and J. R. Sowers, Diabetes and cardiovascular disease: a statement for healthcare professionals from the American Heart Association, *Circulation*, 1999, **100**, 1134–1146.
9. W. B. Kannel and D. L. McGee, Diabetes and cardiovascular disease. The Framingham study, *JAMA*, 1979, **241**, 2035–2038.
10. E. Cersosimo and R. A. DeFronzo, Insulin resistance and endothelial dysfunction: the road map to cardiovascular diseases, *Diabetes Metab. Res. Rev.*, 2006, **22**, 423–436.
11. J. E. Kanter, F. Johansson, R. C. LeBoeuf and K. E. Bornfeldt, Do glucose and lipids exert independent effects on atherosclerotic lesion initiation or progression to advanced plaques?, *Circ. Res.*, 2007, **100**, 769–781.
12. F. Laden, Association of fine particulate matter from different sources with daily mortality in six US cities, *Environ. Health Perspect.*, 2000, **108**, 941–941.
13. M. S. Goldberg, On the interpretation of epidemiological studies of ambient air pollution, *J. Expo. Sci. Environ. Epidemiol.*, 2007, **17**(Suppl 2), S66–70.
14. M. Franklin, P. Koutrakis and P. Schwartz, The role of particle composition on the association between $PM_{2.5}$ and mortality, *Epidemiology*, 2008, **19**, 680–689.
15. P. L. Kinney and H. Ozkaynak, Associations of daily mortality and air pollution in Los Angeles County, *Environ. Res.*, 1991, **54**, 99–120.
16. J. Schwartz and D. W. Dockery, Increased mortality in Philadelphia associated with daily air pollution concentrations, *Am. Rev. Respir. Dis.*, 1992, **145**, 600–604.
17. C. A. Pope 3rd, J. Schwartz and M. R. Ransom, Daily mortality and PM10 pollution in Utah Valley, *Arch. Environ. Health*, 1992, **47**, 211–217.
18. M. S. Goldberg, R. T. Burnett, J. C. Bailar 3rd, R. Tamblyn, P. Ernst, K. Flegel, J. Brook, Y. Bonvalot, R. Singh, M. F. Valois and R. Vincent, Identification of persons with cardiorespiratory conditions who are at risk

of dying from the acute effects of ambient air particles, *Environ. Health Perspect.*, 2001, **109**(Suppl 4), 487–494.
19. M. S. Goldberg, R. T. Burnett, J. F. Yale, M. F. Valois and J. R. Brook, Associations between ambient air pollution and daily mortality among persons with diabetes and cardiovascular disease, *Environ. Res.*, 2006, **100**, 255–267.
20. T. F. Bateson and J. Schwartz, Who is sensitive to the effects of particulate air pollution on mortality? A case-crossover analysis of effect modifiers, *Epidemiology*, 2004, **15**, 143–149.
21. B. Ostro, R. Broadwin, S. Green, W. Y. Feng and M. Lipsett, Fine particulate air pollution and mortality in nine California counties: results from CALFINE, *Environ. Health Perspect.*, 2006, **114**, 29–33.
22. D. Maynard, B. A. Coull, A. Gryparis and J. Schwartz, Mortality risk associated with short-term exposure to traffic particles and sulfates, *Environ. Health Perspect.*, 2007, **115**, 751–755.
23. F. Forastiere, M. Stafoggia, G. Berti, L. Bisanti, A. Cernigliaro, M. Chiusolo, S. Mallone, R. Miglio, P. Pandolfi, M. Rognoni, M. Serinelli, R. Tessari, M. Vigotti and C. A. Perucci, SISTI Group, Particulate matter, daily mortality: a case-crossover analysis of individual effect modifiers, *Epidemiology*, 2008, **19**, 571–580.
24. H. Kan, J. Jia and B. Chen, The association of daily diabetes mortality and outdoor air pollution in Shanghai, *China, J. Environ. Health*, 2004, **67**, 21–26.
25. W. S. Linn, Y. Szlachcic, H. Gong Jr, P. L. Kinney and K. T. Berhane, Air pollution and daily hospital admissions in metropolitan Los Angeles, *Environ. Health Perspect.*, 2000, **108**, 427–434.
26. A. Zanobetti and J. Schwartz, Are diabetics more susceptible to the health effects of airborne particles?, *Am. J. Respir. Crit. Care Med.*, 2001, **164**, 831–833.
27. A. Zanobetti and J. Schwartz, Cardiovascular damage by airborne particles: are diabetics more susceptible?, *Epidemiology*, 2002, **13**, 588–592.
28. A. Peters, S. von Klot, M. Heier, I. Trentinaglia, A. Hormann, H. E. Wichmann and H. Lowel, Cooperative Health Research in the Region of Augsburg Study Group, Exposure to traffic, the onset of myocardial infarction, *N. Engl. J. Med.*, 2004, **351**, 1721–1730.
29. J. L. Peel, K. B. Metzger, M. Klein, W. D. Flanders, J. A. Mulholland and P. E. Tolbert, Ambient air pollution and cardiovascular emergency department visits in potentially sensitive groups, *Am. J. Epidemiol.*, 2007, **165**, 625–633.
30. H. F. Chiu and C. Y. Yang, Air pollution and emergency-room visits for arrhythmias: are there potentially sensitive groups?, *J. Toxicol. Environ. Health A*, 2009, **72**, 817–823.
31. J. Sullivan, L. Sheppard, A. Schreuder, N. Ishikawa, D. Siscovick and J. Kaufman, Relation between short-term fine-particulate matter exposure and onset of myocardial infarction, *Epidemiology*, 2005, **16**, 41–48.

32. C. A. Pope 3rd, R. L. Verrier, E. G. Lovett, A. C. Larson, M. E. Raizenne, R. E. Kanner, J. Schwartz, G. M. Villegas, D. R. Gold and D. W. Dockery, Heart-rate variability associated with particulate air pollution, *Am. Heart J.*, 1999, **138**, 890–899.
33. R. B. Devlin, A. J. Ghio, H. Kehrl, G. Sanders and W. Cascio, Elderly humans exposed to concentrated air pollution particles have decreased heart-rate variability, *Eur. Respir. J. Suppl.*, 2003, **40**, 76s–80s.
34. F. Holguin, M. M. Tellez-Rojo, M. Hernandez, M. Cortez, J. C. Chow, J. G. Watson, D. Mannino and I. Romieu, Air pollution and heart-rate variability among the elderly in Mexico City, *Epidemiology*, 2003, **14**, 521–527.
35. D. R. Gold, A. Litonjua, J. Schwartz, E. Lovett, A. Larson, B. Nearing, G. Allen, M. Verrier, R. Cherry and R. Verrier, Ambient pollution and heart rate variability, *Circulation*, 2000, **101**, 1267–1273.
36. C. M. van Ravenswaaij-Arts, L. A. Kollee, J. C. Hopman, G. B. Stoelinga and H. P. van Geijn, Heart-rate variability, *Ann. Intern. Med.*, 1993, **118**, 436–447.
37. E. B. Schroeder, L. E. Chambless, D. Liao, R. J. Prineas, G. W. Evans, W. D. Rosamond and G. Heiss, Atherosclerosis Risk in Communities (ARIC) study, Diabetes, glucose, insulin,, heart-rate variability: the Atherosclerosis Risk in Communities (ARIC) study, *Diabetes Care*, 2005, **28**, 668–674.
38. S. K. Park, M. S. O'Neill, P. S. Vokonas, D. Sparrow and J. Schwartz, Effects of air pollution on heart-rate variability: the VA normative aging study, *Environ. Health Perspect.*, 2005, **113**, 304–309.
39. E. A. Whitsel, P. M. Quibrera, S. L. Christ, D. Liao, R. J. Prineas, G. L. Anderson and G. Heiss, Heart-rate variability, ambient particulate-matter air pollution, and glucose homeostasis: the environmental epidemiology of arrhythmogenesis in the women's health initiative, *Am. J. Epidemiol.*, 2009, **169**, 693–703.
40. C. J. Pepine, R. A. Kerensky, C. R. Lambert, K. M. Smith, G. O. von Mering, G. Sopko and C. N. Bairey Merz, Some thoughts on the vasculopathy of women with ischemic heart disease, *J. Am. Coll. Cardiol.*, 2006, **47**, S30–5.
41. D. R. Gold, A. A. Litonjua, A. Zanobetti, B. A. Coull, J. Schwartz, G. MacCallum, R. L. Verrier, B. D. Nearing, M. J. Canner, H. Suh and P. H. Stone, Air pollution and ST-segment depression in elderly subjects, *Environ. Health Perspect.*, 2005, **113**, 883–887.
42. K. J. Chuang, B. A. Coull, A. Zanobetti, H. Suh, J. Schwartz, P. H. Stone, A. Litonjua, F. E. Speizer and D. R. Gold, Particulate air pollution as a risk factor for ST-segment depression in patients with coronary artery disease, *Circulation*, 2008, **118**, 1314–1320.
43. J. Yeboah, J. R. Crouse, F. C. Hsu, G. L. Burke and D. M. Herrington, Brachial flow-mediated dilation predicts incident cardiovascular events in older adults: the Cardiovascular Health Study, *Circulation*, 2007, **115**, 2390–2397.

44. M. S. O'Neill, A. Veves, A. Zanobetti, J. A. Sarnat, D. R. Gold, P. A. Economides, E. S. Horton and J. Schwartz, Diabetes enhances vulnerability to particulate air pollution-associated impairment in vascular reactivity and endothelial function, *Circulation*, 2005, **111**, 2913–2920.
45. L. Liu, T. D. Ruddy, M. Dalipaj, M. Szyszkowicz, H. You, R. Poon, A. Wheeler and R. Dales, Influence of personal exposure to particulate air pollution on cardiovascular physiology and biomarkers of inflammation and oxidative stress in subjects with diabetes, *J. Occup. Environ. Med.*, 2007, **49**, 258–265.
46. A. Peretz, J. H. Sullivan, D. F. Leotta, C. A. Trenga, F. N. Sands, J. Allen, C. Carlsten, C. W. Wilkinson, E. A. Gill and J. D. Kaufman, Diesel exhaust inhalation elicits acute vasoconstriction *in vivo*, *Environ. Health Perspect.*, 2008, **116**, 937–942.
47. R. D. Brook, J. R. Brook, B. Urch, R. Vincent, S. Rajagopalan and F. Silverman, Inhalation of fine particulate air pollution and ozone causes acute arterial vasoconstriction in healthy adults, *Circulation*, 2002, **105**, 1534–1536.
48. R. Dales, L. Liu, M. Szyszkowicz, M. Dalipaj, J. Willey, R. Kulka and T. D. Ruddy, Particulate air pollution and vascular reactivity: the bus stop study, *Int. Arch. Occup. Environ. Health*, 2007, **81**, 159–164.
49. J. B. Meigs, F. B. Hu, N. Rifai and J. E. Manson, Biomarkers of endothelial dysfunction and risk of type-2 diabetes mellitus, *JAMA*, 2004, **291**, 1978–1986.
50. M. S. O'Neill, A. Veves, J. A. Sarnat, A. Zanobetti, D. R. Gold, P. A. Economides, E. S. Horton and J. Schwartz, Air pollution and inflammation in type-2 diabetes: a mechanism for susceptibility, *Occup. Environ. Med.*, 2007, **64**, 373–379.
51. J. Madrigano, A. Baccarelli, R. Wright, H. Suh, D. Sparrow, P. Vokonas and J. Schwartz, Air pollution, obesity, genes, and cellular adhesion molecules, *Occup. Environ. Med.*, 2009.
52. M. Ando, M. Shima, M. Adachi and Y. Tsunetoshi, The role of intercellular adhesion molecule-1 (ICAM-1), vascular cell-adhesion molecule-1 (VCAM-1), and regulated on activation, normal T-cell expressed and secreted (RANTES) in the relationship between air pollution and asthma among children, *Arch. Environ. Health*, 2001, **56**, 227–233.
53. R. J. Delfino, N. Staimer, T. Tjoa, A. Polidori, M. Arhami, D. L. Gillen, M. T. Kleinman, N. D. Vaziri, J. Longhurst, F. Zaldivar and C. Sioutas, Circulating biomarkers of inflammation, antioxidant activity, and platelet activation are associated with primary combustion aerosols in subjects with coronary artery disease, *Environ. Health Perspect.*, 2008, **116**, 898–906.
54. K. Yatera, J. Hsieh, J. C. Hogg, E. Tranfield, H. Suzuki, C. H. Shih, A. R. Behzad, R. Vincent and S. F. van Eeden, Particulate matter air-pollution exposure promotes recruitment of monocytes into atherosclerotic plaques, *Am. J. Physiol. Heart Circ. Physiol.*, 2008, **294**, H944–53.

55. A. Peters, D. W. Dockery, J. E. Muller and M. A. Mittleman, Increased particulate air pollution and the triggering of myocardial infarction, *Circulation*, 2001, **103**, 2810–2815.
56. K. Donaldson, V. Stone, A. Seaton and W. MacNee, Ambient particle inhalation and the cardiovascular system: potential mechanisms, *Environ. Health Perspect.*, 2001, **109**, 523–527.
57. D. Liao, G. Heiss, V. M. Chinchilli, Y. Duan, A. R. Folsom, H. M. Lin and V. Salomaa, Association of criteria pollutants with plasma hemostatic/inflammatory markers: a population-based study, *J. Expo. Anal. Environ. Epidemiol.*, 2005, **15**, 319–328.
58. T. H. Chen, P. Gona, P. A. Sutherland, E. J. Benjamin, P. W. Wilson, M. G. Larson, R. S. Vasan and S. J. Robins, Long-term C-reactive protein variability and prediction of metabolic risk, *Am. J. Med.*, 2009, **122**, 53–61.
59. S. D. Dubowsky, H. Suh, J. Schwartz, B. A. Coull and D. R. Gold, Diabetes, obesity, and hypertension may enhance associations between air pollution and markers of systemic inflammation, *Environ. Health Perspect.*, 2006, **114**, 992–998.
60. T. Suwa, J. C. Hogg, K. B. Quinlan, A. Ohgami, R. Vincent and S. F. van Eeden, Particulate air pollution induces progression of atherosclerosis, *J. Am. Coll. Cardiol.*, 2002, **39**, 935–942.
61. Q. Sun, A. Wang, X. Jin, A. Natanzon, D. Duquaine, R. D. Brook, J. G. Aguinaldo, Z. A. Fayad, V. Fuster, M. Lippmann, L. C. Chen and S. Rajagopalan, Long-term air-pollution exposure and acceleration of atherosclerosis and vascular inflammation in an animal model, *JAMA*, 2005, **294**, 3003–3010.
62. Q. Sun, P. Yue, J. A. Deiuliis, C. N. Lumeng, T. Kampfrath, M. B. Mikolaj, Y. Cai, M. C. Ostrowski, B. Lu, S. Parthasarathy, R. D. Brook, S. D. Moffatt-Bruce, L. C. Chen and S. Rajagopalan, Ambient air pollution exaggerates adipose inflammation and insulin resistance in a mouse model of diet-induced obesity, *Circulation*, 2009, **119**, 538–546.
63. R. Sladek, G. Rocheleau, J. Rung, C. Dina, L. Shen, D. Serre, P. Boutin, D. Vincent, A. Belisle, S. Hadjadj, B. Balkau, B. Heude, G. Charpentier, T. J. Hudson, A. Montpetit, A. V. Pshezhetsky, M. Prentki, B. I. Posner, D. J. Balding, D. Meyre, C. Polychronakos and P. Froguel, A genome-wide association study identifies novel risk loci for type-2 diabetes, *Nature*, 2007, **445**, 881–885.
64. A. Sjoholm and T. Nystrom, Inflammation and the etiology of type-2 diabetes, *Diabetes Metab. Res. Rev.*, 2006, **22**, 4–10.
65. J. L. Evans, I. D. Goldfine, B. A. Maddux and G. M. Grodsky, Oxidative stress and stress-activated signaling pathways: a unifying hypothesis of type-2 diabetes, *Endocr. Rev.*, 2002, **23**, 599–622.
66. M. I. Schmidt, B. B. Duncan, A. R. Sharrett, G. Lindberg, P. J. Savage, S. Offenbacher, M. I. Azambuja, R. P. Tracy and G. Heiss, Markers of inflammation and prediction of diabetes mellitus in adults (Atherosclerosis Risk in Communities study): a cohort study, *Lancet*, 1999, **353**, 1649–1652.

67. R. D. Brook, J. R. Brook and S. Rajagopalan, Air pollution: the "Heart" of the problem, *Curr. Hypertens. Rep.*, 2003, **5**, 32–39.
68. J. A. Araujo, B. Barajas, M. Kleinman, X. Wang, B. J. Bennett, K. W. Gong, M. Navab, J. Harkema, C. Sioutas, A. J. Lusis and A. E. Nel, Ambient particulate pollutants in the ultrafine range promote early atherosclerosis and systemic oxidative stress, *Circ. Res.*, 2008, **102**, 589–596.
69. F. J. Kelly, Oxidative stress: its role in air pollution and adverse health effects, *Occup. Environ. Med.*, 2003, **60**, 612–616.
70. C. R. Rhoden, G. A. Wellenius, E. Ghelfi, J. Lawrence and B. González-Flecha, PM-induced cardiac oxidative stress and dysfunction are mediated by autonomic stimulation, *BBA-General Subjects*, 2005, **1725**, 305–313.
71. E. Ghelfi, C. R. Rhoden, G. A. Wellenius, J. Lawrence and B. Gonzalez-Flecha, Cardiac oxidative stress and electrophysiological changes in rats exposed to concentrated ambient particles are mediated by TRP-dependent pulmonary reflexes, *Toxicol. Sci.*, 2008, **102**, 328–336.
72. S. A. Gurgueira, J. Lawrence, B. Coull, G. G. K. Murthy and B. Gonzalez-Flecha, Rapid increases in the steady-state concentration of reactive oxygen species in the lungs and heart after particulate air pollution inhalation, *Environ. Health Perspect.*, 2002, **110**, 749–755.
73. T. Chahine, A. Baccarelli, A. Litonjua, R. O. Wright, H. Suh, D. R. Gold, D. Sparrow, P. Vokonas and J. Schwartz, Particulate air pollution, oxidative stress genes, and heart-rate variability in an elderly cohort, *Environ. Health Perspect.*, 2007, **115**, 1617–1622.
74. I. Romieu, M. Ramirez-Aguilar, J. J. Sienra-Monge, H. Moreno-Macias, B. E. del Rio-Navarro, G. David, J. Marzec, M. Hernandez-Avila and S. London, GSTM1 and GSTP1 and respiratory health in asthmatic children exposed to ozone, *Eur. Respir. J.*, 2006, **28**, 953–959.
75. J. Schwartz, S. K. Park, M. S. O'Neill, P. S. Vokonas, D. Sparrow, S. Weiss and K. Kelsey, Glutathione-S-Transferase M1, Obesity, statins, and autonomic effects of particles: gene-by-drug-by-environment interaction, *Am. J. Respir. Crit. Care Med.*, 2005, **172**, 1529–1529.
76. M. L. Honing, P. J. Morrison, J. D. Banga, E. S. Stroes and T. J. Rabelink, Nitric oxide availability in diabetes mellitus, *Diabetes Metab. Rev.*, 1998, **14**, 241–249.
77. R. S. Ahima, Y. Qi and N. S. Singhal, Adipokines that link obesity and diabetes to the hypothalamus, *Prog. Brain Res.*, 2006, **153**, 155–174.
78. M. A. Lazar, How obesity causes diabetes: not a tall tale, *Science*, 2005, **307**, 373–375.
79. S. E. Alexeeff, A. A. Litonjua, H. Suh, D. Sparrow, P. S. Vokonas and J. Schwartz, Ozone exposure and lung function: effect modified by obesity and airways hyperresponsiveness in the VA normative aging study, *Chest*, 2007, **132**, 1890–1897.
80. A. Zeka, J. R. Sullivan, P. S. Vokonas, D. Sparrow and J. Schwartz, Inflammatory markers and particulate air pollution: characterizing the pathway to disease, *Int. J. Epidemiol.*, 2006, **35**, 1347–1347.

81. A. Seaton, A. Soutar, V. Crawford, R. Elton, S. McNerlan, J. Cherrie, M. Watt, R. Agius and R. Stout, Particulate air pollution and the blood, *Thorax*, 1999, **54**, 1027–1032.
82. L. Bouthillier, R. Vincent, P. Goegan, I. Y. Adamson, S. Bjarnason, M. Stewart, J. Guenette, M. Potvin and P. Kumarathasan, Acute effects of inhaled urban particles and ozone: lung morphology, macrophage activity, and plasma endothelin-1, *Am. J. Pathol.*, 1998, **153**, 1873–1884.
83. A. S. De Vriese, T. J. Verbeuren, J. Van de Voorde, N. H. Lameire and P. M. Vanhoutte, Endothelial dysfunction in diabetes, *Br. J. Pharmacol.*, 2000, **130**, 963–974.
84. Z. A. Khan and S. Chakrabarti, Endothelins in chronic diabetic complications, *Can. J. Physiol. Pharmacol.*, 2003, **81**, 622–634.
85. R. D. Brook, M. Jerrett, J. R. Brook, R. L. Bard and M. M. Finkelstein, The relationship between diabetes mellitus and traffic-related air pollution, *J. Occup. Environ. Med.*, 2008, **50**, 32–38.
86. E. H. Hathout, W. L. Beeson, M. Ischander, R. Rao and J. W. Mace, Air pollution and type-1 diabetes in children, *Pediatr. Diabetes*, 2006, **7**, 81–87.
87. A. Navas-Acien, E. K. Silbergeld, R. Pastor-Barriuso and E. Guallar, Arsenic exposure and prevalence of type-2 diabetes in US adults, *JAMA*, 2008, **300**, 814–822.
88. A. Navas-Acien, E. Guallar, E. K. Silbergeld and S. J. Rothenberg, Lead exposure and cardiovascular disease – a systematic review, *Environ. Health Perspect.*, 2007, **115**, 472–482.
89. D. H. Lee, I. K. Lee, K. Song, M. Steffes, W. Toscano, B. A. Baker and D. R. Jacobs Jr, A strong dose–response relation between serum concentrations of persistent organic pollutants and diabetes: results from the National Health and Examination Survey 1999–2002, *Diabetes Care*, 2006, **29**, 1638–1644.
90. D. H. Lee, I. K. Lee, M. Porta, M. Steffes and D. R. Jacobs Jr, Relationship between serum concentrations of persistent organic pollutants and the prevalence of metabolic syndrome among non-diabetic adults: results from the National Health and Nutrition Examination Survey 1999–2002, *Diabetologia*, 2007, **50**, 1841–1851.
91. N. Ben-Jonathan, E. R. Hugo and T. D. Brandebourg, Effects of bisphenol A on adipokine release from human adipose tissue: Implications for the metabolic syndrome, *Mol. Cell. Endocrinol.*, 2009, **304**, 49–54.

CHAPTER 6
Ambient Particulate Matter and the Risk of Stroke

G. A. WELLENIUS,[1] D. R. GOLD[2] AND M. A. MITTLEMAN[3]

[1] Brown University, Department of Community Health, Center for Environmental Health and Technology, 121 S. Main Street, Providence, RI, 02912, USA; [2] Harvard Medical School, Brigham and Women's Hospital, Channing Laboratory, 181 Longwood Avenue, Boston, MA, 02115, USA; [3] Harvard Medical School, Cardiovascular Epidemiology Research Unit, Beth Israel Deaconess Medical Center, 375 Longwood Ave, Boston, MA, 02215, USA

6.1 Stroke is a Public-Health Problem

Stroke is a leading cause of long-term disability and the third leading cause of death after heart disease and cancer both in the United States[1] and globally.[2] Each year, almost 800 000 people in the US experience a new or recurrent stroke, resulting in more than 140 000 deaths and almost 900 000 hospital admissions per year.[1] Health-care expenditures for stroke have been estimated at over $32 billion annually in direct costs, and an additional $21 billion in lost productivity. Because of the potential for long-term disability, the true cost due to diminished productivity and reduced quality of life among disabled survivors of stroke and their caregivers is likely even greater. Hence, identifying strategies for stroke prevention is an important public-health goal.

The World Health Organization (WHO) defines stroke as "rapidly developed clinical signs of focal (or global) disturbance of cerebral function, lasting more than 24 h or leading to death, with no apparent cause other than of vascular origin".[3] The most common type of stroke is ischemic stroke (80%), followed by primary intracerebral hemorrhage (15%) and subarachnoid hemorrhage

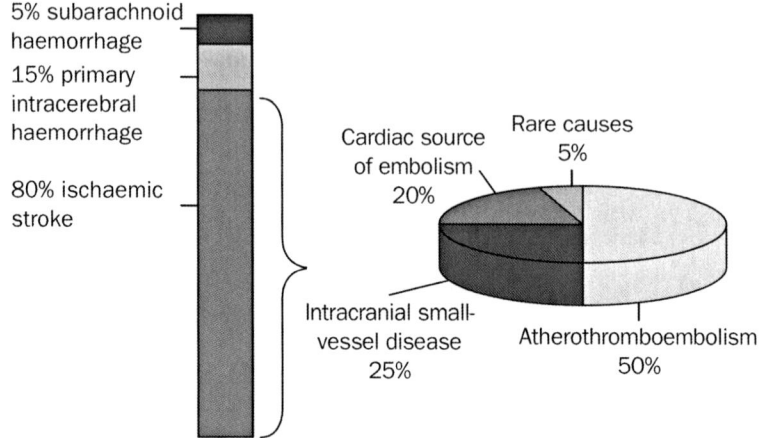

Figure 6.1 Approximate frequency of three main pathological types of stroke (in white populations) and of main subtypes of ischemic stroke as shown by population-based studies (Reproduced from: Warlow C, Sudlow C, Dennis M, Wardlaw J, Sandercock P, *Lancet*, 2003; **362**(9391): 1211–1224.)

(5%) (Figure 6.1).[2] Ischemic stroke can be further classified according to presumed mechanistic subtypes, including cardioembolism, large-vessel atherothromboembolism, and small-vessel (lacunar) causes of stroke. Traditionally, patients with clinical signs lasting less than 24 h are considered to have suffered a transient ischemic attack (TIA) with no permanent brain injury. However, more recent studies using computed tomography (CT) or magnetic resonance imaging (MRI) techniques have shown that a substantial percentage of patients who were classified as having TIAs according to symptom duration in fact had permanent ischemic brain injuries.[4] Increasingly, patients with clinical evidence of cerebral infarction are considered to have suffered a stroke, regardless of the duration of symptoms.[4]

6.2 Cardiovascular Health Effects of Ambient Particulate Matter

A large number of epidemiologic studies have found an association between short-term increases in particulate air pollution and cardiovascular morbidity and mortality.[5–11] Respirable particulate matter (PM_{10}, particulate matter with aerodynamic diameter < 10 μm) or fine particulate matter ($PM_{2.5}$, aerodynamic diameter < 2.5 μm) has been specifically implicated in the triggering of myocardial infarction (MI),[12–13] cardiac arrhythmias,[14–16] heart-failure exacerbations,[17–19] and the exacerbation of myocardial ischemia.[20–22]

Relatively fewer studies have evaluated the effects of long-term exposure to ambient air pollution. The Harvard Six Cities study was the first study to show an association between long-term exposure to PM and increased mortality.[23] The findings of this study have been confirmed and extended in analyses of data from nearly 300 000 participants in the American Cancer Society (ACS) Cancer Prevention Study II,[24,25] a cohort of 6338 nonsmoking California Seventh-day Adventists,[26] and among 58 000 participants from the Women's Health Initiative (WHI) Observational Study.[27] The findings of the Harvard Six Cities and ACS studies were verified by an independent reanalysis team sponsored by the Health Effects Institute.[28] Cohort studies in Europe provide additional evidence.[29]

While the original Six Cities study examined only the effects on total mortality, subsequent analyses found the strongest associations for cardiovascular mortality.[30] Similarly, extended analyses of the ACS study found the strongest associations for cardiovascular mortality in general and deaths due to coronary heart disease in particular.[31] The WHI study focused on cardiovascular outcomes and found the strongest associations with death from coronary heart disease.[27] When considering both fatal and nonfatal cardiovascular events together, the strongest associations were observed for cerebrovascular disease and stroke.

The distinction made above between studies of the effects of short-term exposure (over hours to days) *versus* long-term exposure (over months to years) to ambient air pollution is somewhat artificial as the true effects of PM are likely to occur over a continuous timescale. Nevertheless, this distinction is useful because the effects and/or mechanisms of effects likely differ according to the timescale of interest. Moreover, exposure assessment strategies, study designs, potential confounders, and statistical methods differ depending on the timescale of interest. In the remainder of this chapter we review the evidence related to the effects on cerebrovascular morbidity and mortality of short-term and long-term exposure to ambient PM.

6.3 Effects of Short-Term PM Exposure on Cerebrovascular Hospitalizations

Time-series studies evaluating the link between short-term increases in ambient PM levels and risk of hospitalization for cerebrovascular disease have been inconsistent, with a minority of studies reporting statistically significant positive associations,[32–34] and several studies reporting null or inverse associations.[9,35–41]

The largest of these is the US Medicare Air Pollution Study (MCAPS) study that evaluated this hypothesis using data on hospital admissions among Medicare beneficiaries aged 65 years and older in 204 US counties.[33] The MCAPS investigators found a 0.8% (95% CI: 0.3, 1.4) excess relative risk of hospitalization for cerebrovascular disease per $10\,\mu g/m^3$ increase in same-day $PM_{2.5}$.[33] Investigators from the French National Program on Air Pollution Health Effects (PSAS) reported a 0.8% (–0.9, 2.5%) excess relative risk per $10\,\mu g/m^3$ increase in PM_{10} among patients aged ≥ 65 years and a 0.2%

(−1.6, 1.9%) excess relative risk among all patients hospitalized for cerebrovascular disease in 8 French cities.[9] Although neither estimate was statistically significant, the estimated excess relative risk among the elderly is similar to that observed in the US MCAPS study. In contrast, the large APHEA study, reported a 0.0% (−0.3, 0.3%) excess relative risk of hospitalization for cerebrovascular disease per $10\,\mu g/m^3$ increase in the 2-day moving average of PM_{10} in 8 European cities.[36] Barnett et al.[37] examined this hypothesis in 7 cities in New Zealand and Australia and reported finding no association, but did not report point estimates or confidence intervals.

All of the above studies have identified cases of cerebrovascular disease based on primary discharge diagnoses as specified by the International Classification of Disease, 9th Revision (ICD-9) or 10th Revision (ICD-10). An approach common to these studies is to define cerebrovascular hospitalizations using ICD-9 codes 430–438 or ICD-10 codes I60–I69. However, these ranges of ICD codes include diagnoses of ischemic stroke, hemorrhagic stroke, transient ischemic attack (TIA), several poorly defined forms of acute neurological events and the late sequelae of acute events (Table 6.1). It is plausible that ambient PM levels may be unrelated to some of these endpoints or that ambient PM has different effects on different outcomes.

Table 6.1 Summary of ICD-9 and ICD-10 codes for cerebrovascular disease.

ICD-9		ICD-10	
Code	Description	Code	Description
430	Subarachnoid hemorrhage	I60	Subarachnoid hemorrhage
431	Intracerebral hemorrhage	I61	Intracerebral hemorrhage
432	Other and unspecified intracranial hemorrhage	I62	Other nontraumatic intracranial hemorrhage
433	Occlusion and stenosis of pre-cerebral arteries	I63	Cerebral infarction
434	Occlusion of cerebral arteries	I64	Stroke, not specified as hemorrhage or infarction
435	Transient cerebral ischemia	I65	Occlusion and stenosis of pre-cerebral arteries, not resulting in cerebral infarction
436	Acute but ill-defined cerebrovascular disease	I66	Occlusion and stenosis of cerebral arteries, not resulting in cerebral infarction
437	Other and ill-defined cerebrovascular disease	I67	Other cerebrovascular diseases
438	Late effects of cerebrovascular disease	I68	Cerebrovascular disorders in diseases classified elsewhere
		I69	Sequelae of cerebrovascular disease
		G45	Transient cerebral ischemic attacks and related syndromes

6.4 Ischemic Stroke and Transient Ischemic Attack (TIA)

An increasing number of studies have specifically evaluated the association between PM and the risk of hospitalization for ischemic stroke.[34,40,42–46] Linn et al.[45] found a 0.1% (95% CI: –0.4, 0.6%) excess relative risk of hospitalization for ischemic stroke in the Los Angeles metropolitan area. Wellenius et al.[42] reported a statistically significant 0.4% (0.0, 0.9%) excess relative risk per 10 µg/m^3 increase in same-day PM$_{10}$ among elderly Medicare beneficiaries in 9 US cities. In Kaohsiung, Taiwan, Tsai et al.[44] found a 5.9% (4.3, 7.4%) excess relative risk of hospitalization for ischemic stroke per 10 µg/m^3 increase in PM$_{10}$ after excluding days with mean daily temperature < 20 °C. Meanwhile, in Taipei, Taiwan, Chan et al.[34] found a 1.6% (–0.8, 3.9%) and 3.0% (–0.8, 6.6%) excess relative risk per 10 µg/m^3 increase in PM$_{10}$ and PM$_{2.5}$, respectively. Villeneuve et al.[40] found no association between either PM$_{2.5}$ or PM$_{10}$ and emergency department visits for acute ischemic stroke in Edmonton, Canada.

Two recent studies are particularly noteworthy given the high specificity of the outcome definition. Henrotin et al.[43] used data on 1432 confirmed cases of ischemic stroke from the French Dijon Stroke Register and found 0.9% (–7.0, 9.4%) excess relative risk of ischemic stroke per 10 µg/m^3 increase in PM$_{10}$ on the same day and a 1.1% (–6.6, 9.4%) excess relative risk on the previous day (lag 1 day). Lisabeth et al.[46] used data on 2350 confirmed cases of ischemic stroke and 1158 cases of TIA from the Brain Attack Surveillance in Corpus Christi Project (BASIC), a population-based stroke-surveillance project designed to capture all strokes in Nueces County, Texas. The authors found a 3% (95% CI: –1%, 7%) and 3% (–0%, 7%) excess relative risk of ischemic stroke/TIA per 5.1 µg/m^3 increase in PM$_{2.5}$ on the same day and previous day, respectively.

As noted earlier, ischemic stroke can be categorized according to pathophysiologic mechanism as small-vessel (lacunar), large-vessel atherothromboembolic, or cardioembolic stroke. If short-term exposure to elevated levels of PM truly increases the risk of ischemic stroke, it is likely that the effect varies in magnitude according to ischemic stroke type. For example, if we hypothesize that short-term exposure to PM increases the risk of ischemic stroke primarily through a mechanism that includes destabilizing vulnerable atherosclerotic plaques, we would not expect to find an association between long-term PM exposure and cardioembolic stroke. Only one published study to date has evaluated the effects of ambient PM by ischemic stroke subtype. In the context of the French Dijon Stroke Register, Henrotin et al.[43] found no association between PM$_{10}$ and any ischemic stroke subtype, but the study was underpowered to detect effects within subtype. Additional studies with information of stroke etiology are clearly needed.

Most studies of PM and cerebrovascular disease include TIA in the definition of the outcome, while studies specifically looking at ischemic stroke usually exclude TIA. Few studies have specifically examined the association between short-term changes in PM and TIA. Villeneuve et al.[40] specifically evaluated the association between ambient PM and the risk of TIAs, but failed to find any evidence of an association with either $PM_{2.5}$ or PM_{10}. Henrotin et al.[43] similarly found no association between PM_{10} and TIAs, but this study was underpowered.

TIAs share many of the same pathophysiologic and clinical features as ischemic strokes and some investigators have been inclined to evaluate the two together. However, TIAs are more difficult to diagnose than ischemic strokes, as they depend almost entirely on the patient history rather than clinical signs or neuroimaging. Thus, diagnoses of TIA may be more likely to be misclassified as compared to ischemic stroke. Additionally, patients classified as having a TIA often undergo a less thorough clinical workup to determine etiology. The absence of a documented etiology precludes the possibility of gaining insights into the mechanisms of PM-related strokes.

6.5 Hemorrhagic Stroke

Many of the studies in the preceding section also evaluated the association between ambient PM and the risk of hemorrhagic stroke.[34,40,42–44] In Kaohsiung, Taiwan, Tsai et al.[44] noted a 6.7% (4.2, 9.4%) excess relative risk of hospitalization for intracerebral hemorrhages per $10\,\mu g/m^3$ increase in PM_{10}, after excluding days where the mean temperature was $<20\,°C$. However, in the US, Wellenius et al.[42] failed to find any association between ambient PM levels and risk of hospitalization with intracerebral hemorrhages among elderly Medicare beneficiaries in 9 US cities. Similarly, others have found no evidence of an association between emergency department visits for hemorrhagic stroke (intracerebral and subarachnoid hemorrhages combined) and either PM_{10} or $PM_{2.5}$ levels in Edmonton, Canada[40] or Taipei, Taiwan,[34] or PM_{10} levels in Dijon, France.[43] However, because hemorrhagic strokes are far less common than ischemic strokes, most of these studies might have had sufficient statistical power to detect only very large effects.

6.6 Effects of Short-Term PM Exposure on Cerebrovascular Mortality

Most,[47–53] but not all studies,[54–56] suggest that short-term increases in levels of PM_{10} and/or $PM_{2.5}$ may increase the risk of death from cerebrovascular disease. The largest of these studies examined the association between cerebrovascular death and daily levels of ambient fine and course particles in 112 US cities.[53] The authors found a 1.78% (95% CI: 0.96, 2.62%) increased risk of cerebrovascular death per $10\,\mu g/m^3$ increase in $PM_{2.5}$. In a subset of 47 cities where data on course particulate matter were available, the investigators found a 0.84% (0.07, 1.62%) increased risk of cerebrovascular death per $10\,\mu g/m^3$ increase in $PM_{10-2.5}$.

Few studies have specifically evaluated the association between PM and ischemic or hemorrhagic stroke mortality. Hong et al.[57] carried out a time-series study in Seoul, Korea, and found a 3% (95% CI: 0, 6%) increased risk of ischemic stroke death and a 4% (2, 7%) increased risk of hemorrhagic stroke death per interquartile range increase in total suspended particulates (TSP). A later analysis from Seoul reported a positive association between ischemic stroke death and PM_{10}, but only when considering a 6-day distributed lag model.[49] A study covering 13 metropolitan regions in Japan found an association between hemorrhagic stroke death and particulate matter with aerodynamic diameter $<7\,\mu m$ (PM_7) 2-h prior to death, but only during the warm months.[58] No association was observed for hemorrhagic stroke for 24-h averages of PM_7 and no significant associations were seen for ischemic stroke deaths.

6.7 Effects of Long-Term PM Exposure on Cerebrovascular Morbidity and Mortality

The earliest evidence supporting an association between long-term exposure to ambient pollutants and cererbrovascular disease came from small-area ecological studies. Maheswaran and Elliott[59] related cerebrovascular mortality rates to distance to major roadway using data from 113 465 census enumeration districts in England and Wales. In this analysis, the authors used the distance from the enumeration district centroid to the nearest major roadway as a surrogate for long-term exposure to elevated levels of pollution from traffic. Adjusting for potential confounders, the authors found that cerebrovascular mortality was 7% (95% CI: 4, 9%) higher in men and 4% (2, 6%) higher in women living within 200 m of a main road compared with those living ≥ 1000 m away. Subsequently, Maheswaran et al.[60] related cerebrovascular mortality and hospitalization rates to modeled air-pollution levels in 1030 census enumeration districts in Sheffield, England. The authors found that adjusting for potential confounders, cerebrovascular mortality was 37% (95% CI: 19, 57%) and 33% (14, 56%) higher when comparing the highest versus the lowest quintiles of modeled 5-year mean NO_x and PM_{10}. Associations for cerebrovascular hospitalizations were more modest. Using a similar ecological approach, Hu et al.[61] reported an association between annual average daily traffic counts and cerebrovascular mortality in Northwest Florida.

While these studies suggest that long-term exposure to ambient PM and/or pollution from traffic sources increases the risk of cerebrovascular events, the ecological design used is susceptible to potentially important biases. Therefore, this hypothesis is better assessed in the setting of a cohort study. Of the large cohort studies evaluating the health effects of long-term exposure to ambient PM, only the WHI study specifically evaluated the association with cerebrovascular endpoints. The authors of the WHI study found a 35% (95% CI: 8, 68%) increase in risk of cerebrovascular disease (including stroke and death from cerebrovascular disease) per $10\,\mu g/m^3$ increase in $PM_{2.5}$.[27] The excess relative risk was similar for stroke and more than twice as large for cerebrovascular death. The Netherlands Cohort Study on Diet and Cancer followed for 11 years 117 000 men and women aged 55–69 at baseline and found a 41% (1, 97%) increase in risk of cerebrovascular death associated with a 10-µg/m^3 increase in modeled average residential levels of black smoke after controlling for potential confounders including indicators of noise pollution.[27] Taken together, the findings from both the ecological and cohort studies suggest that long-term exposure to ambient PM is associated with increased risk of cerebrovascular morbidity and mortality.

6.8 Potential Mechanisms

Short-term increases in ambient PM are thought to increase the risk of cardiovascular events by a combination of at least four separate but related mechanisms. First, PM-related changes in autonomic nervous system activity, as assessed by heart-rate variability, have been observed in experimental animals[62] and in human panel studies.[63–66] Results from these studies are consistent with sympathetic activation or reduction of parasympathetic (vagal) tone following exposure to PM. Secondly, PM-related changes in hematological parameters have been reported, including decreased red blood cell indices,[67] increased blood viscosity,[68] and enhanced peripheral arterial thrombosis.[69] Thirdly, there is evidence that short-term exposure to PM can induce an acute systemic inflammatory response with increased number of circulating neutrophils[70] and increased levels of plasma C-reactive protein.[67,71] Fourthly, short-term exposure to ambient PM can promote endothelial cell injury and impair endothelial cell function, as evidenced by PM-related increases in plasma markers of endothelial injury,[72–78] changes in systemic hemodynamics,[79–85] impaired endothelium-dependent vasodilation,[86–88] and reduced brachial artery diameter.[89] These findings suggest that autonomic, hemostatic, inflammatory, and endothelial disturbances with consequent changes in cardiac and vascular function may underlie the particulate-related increased risk of cardiovascular events.

Given these putative mechanisms, it is plausible that short-term increases in PM levels could similarly increase the risk of other circulatory diseases including stroke. Specifically, acute PM-related changes in blood pressure, atherosclerotic plaque stability, vascular sheer stress, and hemostatic balance could increase the risk of atherothrombotic strokes; acute elevations in blood pressure or vascular endothelial cell activation could trigger small-vessel (lacunar) strokes; and episodic atrial fibrillation could lead to increased risk of cardioembolic stroke. Transient elevations in arterial blood pressure could similarly increase the risk of both primary intracerebral hemorrhage and subarachnoid hemorrhage. It seems likely that depending on the pathophysiological actions of PM, short-term exposure could have different effects on ischemic stroke versus hemorrhagic stroke, and even on different subtypes of ischemic stroke. Moreover, it is plausible that the effects on each stroke type or subtype differ by PM constituents or hypothesized delay time between exposure and stroke onset. These are questions that need to be addressed in future studies with high-quality clinical data and sufficient statistical power.

Similarly, clinical characteristics that potentially place some individuals at greater risk of pollution-related strokes have yet to be evaluated in detail. For example, the pathophysiologic effects of PM may be more pronounced in the setting of the microvascular and macrovascular injury associated with diabetes mellitus and insulin resistance. Studies of the short-term effects of PM on cardiovascular morbidity and mortality have identified several subgroups of the population that may be at increased risk of PM-related effects, including the elderly[36–37] and those with diabetes mellitus, heart failure, a history of myocardial infarction, pulmonary disease, or concurrent respiratory infections.[50,90,91] It is important to note that the identification of susceptibility factors is still an active area of research and that few of these proposed factors have been studied in detail. The strongest evidence for susceptibility exists for diabetes; people with diabetes appear to be at greater risk of PM-related changes in vascular function and acute cardiac events.[50,87,90,92,93] People with diabetes may similarly be at increased risk of PM-related cerebrovascular events,[50] but this hypothesis has not been evaluated in detail. Whether the presence of other established risk factors for stroke (*e.g.*: atrial fibrillation, hypercholesterolemia, hypertension, smoking) confer additional susceptibility also warrants further evaluation. Similarly, the potentially protective effects of treatments for these conditions has not been examined in the context of PM and stroke.

The pathophysiologic mechanisms by which long-term exposure to ambient PM increases cardiovascular morbidity and mortality have not been fully elucidated. However, several studies in humans have found associations between long-term exposure to air pollution and noninvasive measures of subclinical atherosclerosis.[94–96] This is supported by animal studies that have found that long-term exposure to ambient aerosols promotes atherosclerosis in susceptible animals.[97–99] Such changes would be expected to increase the risk of stroke. Long-term exposure to ambient PM may also increase stroke risk through changes in inflammation, oxidative stress, hemostatic factors or hemodynamic parameters, although evidence for such effects is still largely lacking.

6.9 Summary

In summary, large studies from the US, Europe, and Australia/New Zealand provide inconsistent support for an association between short-term increases in ambient levels of PM_{10} and $PM_{2.5}$ and risk of cerebrovascular disease morbidity or mortality. The evidence in support of an association is more consistent among studies that have specifically evaluated hospitalization for ischemic stroke, although again, considerable heterogeneity across studies is observed. Importantly, the sample sizes of several of the studies reviewed were too small to detect statistically significant associations of the expected magnitude.

Some of the heterogeneity observed across studies of short-term changes in ambient PM is likely attributable to differences in the sensitivity and specificity of the outcome definitions used, in addition to differences in population characteristics and PM sources and composition. The putative mechanisms of PM health effects raise the possibility that there may be important differences in the effects of PM according to stroke type or even between ischemic stroke subtypes. Additional studies with large enough sample sizes and detailed data on stroke etiology will help answer these questions.

In contrast to the heterogeneity seen in studies of short-term effects, findings from both ecological and cohort studies consistently show that long-term exposure to ambient PM is associated with increased risk of cerebrovascular morbidity and mortality. Whether important differences exist between ischemic and hemorrhagic stroke remains unknown. Similarly, which sources or components of PM may be most deleterious has not been explored in detail, although the ecological studies suggest an important role for traffic-related pollution.

Given the well-documented associations between ambient PM exposure and cardiac disease as well as the putative mechanisms of these effects, it is plausible that ambient particles increase the risk of ischemic and/or hemorrhagic stroke. Studies to date broadly support this hypothesis, but additional studies are clearly needed. Even though the magnitude of any potential health effect is expected to be small, given the large number of people simultaneously at risk for stroke and exposed to urban air pollution, the attributable risk may be considerable.

References

1. D. Lloyd-Jones, R. Adams, M. Carnethon, G. De Simone, T. B. Ferguson, K. Flegal, E. Ford, K. Furie, A. Go, K. Greenlund, N. Haase, S. Hailpern, M. Ho, V. Howard, B. Kissela, S. Kittner, D. Lackland, L. Lisabeth, A. Marelli, M. McDermott, J. Meigs, D. Mozaffarian, G. Nichol, C. O'Donnell, V. Roger, W. Rosamond, R. Sacco, P. Sorlie, R. Stafford, J. Steinberger, T. Thom, S. Wasserthiel-Smoller, N. Wong, J. Wylie-Rosett and Y. Hong, Heart disease and stroke statistics – 2009 update: a report from the American Heart Association Statistics Committee and Stroke Statistics Subcommittee, *Circulation*, 2009, **119**, e21–181.
2. C. Warlow, C. Sudlow, M. Dennis, J. Wardlaw and P. Sandercock, Stroke, *Lancet*, 2003, **362**, 1211–1224.
3. K. Aho, P. Harmsen, S. Hatano, J. Marquardsen, V. E. Smirnov and T. Strasser, Cerebrovascular disease in the community: results of a WHO collaborative study, *Bull. World Health Organ.*, 1980, **58**, 113–130.
4. G. W. Albers, L. R. Caplan, J. D. Easton, P. B. Fayad, J. P. Mohr, J. L. Saver and D. G. Sherman, Transient ischemic attack – proposal for a new definition, *New Engl. J. Med.*, 2002, **347**, 1713–1716.
5. J. M. Samet, F. Dominici, F. C. Curriero, I. Coursac and S. L. Zeger, Fine particulate air pollution and mortality in 20 US cities, 1987–1994, *N. Engl. J. Med.*, 2000, **343**, 1742–1749.

6. J. M. Samet, S. L. Zeger, F. Dominici, F. Curriero, I. Coursac, D. W. Dockery, J. Schwartz and A. Zanobetti, The National Morbidity, Mortality, and Air Pollution Study. Part II: Morbidity and mortality from air pollution in the United States, *Res. Rep. Health Eff. Inst.*, 2000, **94**, 5–70; discussion 71–79.
7. J. Schwartz, Air pollution and hospital admissions for heart disease in eight US counties, *Epidemiology*, 1999, **10**, 17–22.
8. C. A. Pope 3rd, J. B. Muhlestein, H. T. May, D. G. Renlund, J. L. Anderson and B. D. Horne, Ischemic heart disease events triggered by short-term exposure to fine particulate air pollution, *Circulation*, 2006, **114**, 2443–2448.
9. S. Larrieu, J. F. Jusot, M. Blanchard, H. Prouvost, C. Declercq, P. Fabre, L. Pascal, A. L. Tertre, V. Wagner, S. Riviere, B. Chardon, D. Borrelli, S. Cassadou, D. Eilstein and A. Lefranc, Short term effects of air pollution on hospitalizations for cardiovascular diseases in eight French cities: the PSAS program, *Sci Total Environ*, 2007, **387**, 105–112.
10. J. L. Peel, K. B. Metzger, M. Klein, W. D. Flanders, J. A. Mulholland and P. E. Tolbert, Ambient air pollution and cardiovascular emergency department visits in potentially sensitive groups, *Am. J. Epidemiol.*, 2007, **165**, 625–633.
11. A. Zanobetti and J. Schwartz, Air pollution and emergency admissions in Boston, MA, *J. Epidemiol. Community Health*, 2006, **60**, 890–895.
12. A. Peters, D. W. Dockery, J. E. Muller and M. A. Mittleman, Increased particulate air pollution and the triggering of myocardial infarction, *Circulation*, 2001, **103**, 2810–2815.
13. D. D'Ippoliti, F. Forastiere, C. Ancona, N. Agabiti, D. Fusco, P. Michelozzi and C. A. Perucci, Air pollution and myocardial infarction in Rome: A case-crossover analysis., *Epidemiology*, 2003, **14**, 528–535.
14. A. Peters, E. Liu, R. L. Verrier, J. Schwartz, D. R. Gold, M. Mittleman, J. Baliff, J. A. Oh, G. Allen, K. Monahan and D. W. Dockery, Air pollution and incidence of cardiac arrhythmia, *Epidemiology*, 2000, **11**, 11–17.
15. H. Luttmann-Gibson, D. R. Gold, M. Link, M. A. Mittleman, D. Q. Rich, J. Schwartz, R. L. Verrier and D. W. Dockery, Daily air pollution effects on cardiac arrhythmias in patients with implanted cardioverter defibrillators., *Epidemiology*, 2002, **13**, S170.
16. D. W. Dockery, H. Luttmann-Gibson, D. Q. Rich, M. S. Link, M. A. Mittleman, D. R. Gold, P. Koutrakis, J. D. Schwartz and R. L. Verrier, Association of air pollution with increased incidence of ventricular tachyarrhythmias recorded by implanted cardioverter defibrillators, *Environ. Health Perspect.*, 2005, **113**, 670–674.
17. J. Schwartz and R. Morris, Air pollution and hospital admissions for cardiovascular disease in Detroit, *Michigan, Am. J. Epidemiol.*, 1995, **142**, 23–35.
18. R. D. Morris and E. N. Naumova, Carbon monoxide and hospital admissions for congestive heart failure: evidence of an increased effect at low temperatures, *Environ. Health Perspect.*, 1998, **106**, 649–653.

19. G. A. Wellenius, T. F. Bateson, M. A. Mittleman and J. Schwartz, Particulate Air Pollution is Associated with Hospital Admissions for Congestive Heart Failure among the Elderly in Pittsburgh, PA, *Epidemiology*, 2003, **14**, S83.
20. G. A. Wellenius, B. A. Coull, J. J. Godleski, P. Koutrakis, K. Okabe, S. T. Savage, J. E. Lawrence, G. G. Murthy and R. L. Verrier, Inhalation of concentrated ambient air particles exacerbates myocardial ischemia in conscious dogs, *Environ. Health Perspect.*, 2003, **111**, 402–408.
21. J. Pekkanen, A. Peters, G. Hoek, P. Tiittanen, B. Brunekreef, J. de Hartog, J. Heinrich, A. Ibald-Mulli, W. G. Kreyling, T. Lanki, K. L. Timonen and E. Vanninen, Particulate air pollution and risk of ST-segment depression during repeated submaximal exercise tests among subjects with coronary heart disease: the Exposure and Risk Assessment for Fine and Ultrafine Particles in Ambient Air (ULTRA) study, *Circulation*, 2002, **106**, 933–938.
22. N. L. Mills, H. Tornqvist, M. C. Gonzalez, E. Vink, S. D. Robinson, S. Soderberg, N. A. Boon, K. Donaldson, T. Sandstrom, A. Blomberg and D. E. Newby, Ischemic and thrombotic effects of dilute diesel-exhaust inhalation in men with coronary heart disease, *N. Engl. J. Med.*, 2007, **357**, 1075–1082.
23. D. W. Dockery, C. A. Pope 3rd, X. Xu, J. D. Spengler, J. H. Ware, M. E. Fay, B. G. Ferris Jr and F. E. Speizer, An association between air pollution and mortality in six US cities, *N. Engl. J. Med.*, 1993, **329**, 1753–1759.
24. C. A. Pope 3rd, M. J. Thun, M. M. Namboodiri, D. W. Dockery, J. S. Evans, F. E. Speizer and C. W. Heath Jr, Particulate air pollution as a predictor of mortality in a prospective study of US adults, *Am. J. Respir. Crit. Care Med.*, 1995, **151**, 669–674.
25. C. A. Pope, R. T. Burnett, M. J. Thun, E. E. Calle, D. Krewski, K. Ito and G. D. Thurston, Lung cancer, cardiopulmonary mortality, and long-term exposure to fine particulate air pollution, *JAMA*, 2002, **287**, 1132–1141.
26. D. E. Abbey, N. Nishino, W. F. McDonnell, R. J. Burchette, S. F. Knutsen, W. Lawrence Beeson and J. X. Yang, Long-term inhalable particles and other air pollutants related to mortality in nonsmokers, *Am. J. Respir. Crit. Care Med.*, 1999, **159**, 373–382.
27. K. A. Miller, D. S. Siscovick, L. Sheppard, K. Shepherd, J. H. Sullivan, G. L. Anderson and J. D. Kaufman, Long-term exposure to air pollution and incidence of cardiovascular events in women, *N. Engl. J. Med.*, 2007, **356**, 447–458.
28. D. Krewski, R. T. Burnett, M. S. Goldberg, B. K. Hoover, J. Siemiatycki, M. Jerrett, M. Abrahamowicz and W. H. White, Overview of the reanalysis of the Harvard Six Cities Study and American Cancer Society Study of Particulate Air Pollution and Mortality, *J. Toxicol. Environ. Health A*, 2003, **66**, 1507–1551.
29. R. Beelen, G. Hoek, P. A. van den Brandt, R. A. Goldbohm, P. Fischer, L. J. Schouten, M. Jerrett, E. Hughes, B. Armstrong and B. Brunekreef, Long-term effects of traffic-related air pollution on mortality in a Dutch cohort (NLCS-AIR study), *Environ. Health Perspect.*, 2008, **116**, 196–202.

30. F. Laden, J. Schwartz, F. E. Speizer and D. W. Dockery, Reduction in fine particulate air pollution and mortality: Extended follow-up of the Harvard Six Cities study, *Am. J. Respir. Crit. Care Med.*, 2006, **173**, 667–672.
31. C. A. Pope 3rd, R. T. Burnett, G. D. Thurston, M. J. Thun, E. E. Calle, D. Krewski and J. J. Godleski, Cardiovascular mortality and long-term exposure to particulate air pollution: epidemiological evidence of general pathophysiological pathways of disease, *Circulation*, 2004, **109**, 71–77.
32. J. Wordley, S. Walters and J. G. Ayres, Short term variations in hospital admissions and mortality and particulate air pollution, *J. Occup. Environ. Med.*, 1997, **54**, 108–116.
33. F. Dominici, R. D. Peng, M. L. Bell, L. Pham, A. McDermott, S. L. Zeger and J. M. Samet, Fine particulate air pollution and hospital admission for cardiovascular and respiratory diseases, *JAMA*, 2006, **295**, 1127–1134.
34. C. C. Chan, K. J. Chuang, L. C. Chien, W. J. Chen and W. T. Chang, Urban air pollution and emergency admissions for cerebrovascular diseases in Taipei, *Taiwan, Eur. Heart J.*., 2006, **27**, 1238–1244.
35. T. W. Wong, T. S. Lau, T. S. Yu, A. Neller, S. L. Wong, W. Tam and S. W. Pang, Air pollution and hospital admissions for respiratory and cardiovascular diseases in Hong Kong, *J. Occup. Environ. Med.*, 1999, **56**, 679–683.
36. A. Le Tertre, S. Medina, E. Samoli, B. Forsberg, P. Michelozzi, A. Boumghar, J. M. Vonk, A. Bellini, R. Atkinson, J. G. Ayres, J. Sunyer, J. Schwartz and K. Katsouyanni, Short-term effects of particulate air pollution on cardiovascular diseases in eight European cities, *J. Epidemiol. Community Health*, 2002, **56**, 773–779.
37. A. G. Barnett, G. M. Williams, J. Schwartz, T. L. Best, A. H. Neller, A. L. Petroeschevsky and R. W. Simpson, The effects of air pollution on hospitalizations for cardiovascular disease in elderly people in Australian and New Zealand cities, *Environ. Health Pers.*, 2006, **114**, 1018–1023.
38. H. R. Anderson, S. A. Bremner, R. W. Atkinson, R. M. Harrison and S. Walters, Particulate matter and daily mortality and hospital admissions in the west midlands conurbation of the United Kingdom: associations with fine and coarse particles, black smoke and sulphate, *J. Occup. Environ. Med.*, 2001, **58**, 504–510.
39. B. Jalaludin, G. Morgan, D. Lincoln, V. Sheppeard, R. Simpson and S. Corbett, Associations between ambient air pollution and daily emergency department attendances for cardiovascular disease in the elderly (65+ years), Sydney, Australia, *J. Exp. Sci. Environ. Epidemiol.*, 2006, **16**, 225–237.
40. P. J. Villeneuve, L. Chen, D. Stieb and B. H. Rowe, Associations between outdoor air pollution and emergency department visits for stroke in Edmonton, *Canada, Eur. J. Epidemiol.*, 2006, **21**, 689–700.
41. R. T. Burnett, M. Smith-Doiron, D. Stieb, S. Cakmak and J. R. Brook, Effects of particulate and gaseous air pollution on cardiorespiratory hospitalizations, *Arch. Environ. Health*, 1999, **54**, 130–139.

42. G. A. Wellenius, J. Schwartz and M. A. Mittleman, Air pollution and hospital admissions for ischemic and hemorrhagic stroke among medicare beneficiaries, *Stroke; J. Cereb. Circ.*, 2005, **36**, 2549–2553.
43. J. B. Henrotin, J. P. Besancenot, Y. Bejot and M. Giroud, Short-term effects of ozone air pollution on ischaemic stroke occurrence: a case-crossover analysis from a 10-year population-based study in Dijon, France, *J. Occup. Environ. Med.*, 2007, **64**, 439–445.
44. S. S. Tsai, W. B. Goggins, H. F. Chiu and C. Y. Yang, Evidence for an association between air pollution and daily stroke admissions in Kaohsiung, Taiwan, *Stroke; J. Cereb. Circ.*, 2003, **34**, 2612–2616.
45. W. S. Linn, Y. Szlachcic, H. Gong Jr, P. L. Kinney and K. T. Berhane, Air pollution and daily hospital admissions in metropolitan Los Angeles, *Environ. Health Pers.*, 2000, **108**, 427–434.
46. L. D. Lisabeth, J. D. Escobar, J. T. Dvonch, B. N. Sanchez, J. J. Majersik, D. L. Brown, M. A. Smith and L. B. Morgenstern, Ambient air pollution and risk for ischemic stroke and transient ischemic attack, *Ann. Neurol.*, 2008, **64**, 53–59.
47. Y. C. Hong, J. T. Lee, H. Kim, E. H. Ha, J. Schwartz and D. C. Christiani, Effects of air pollutants on acute stroke mortality, *Environ. Health Perspect.*, 2002, **110**, 187–191.
48. H. Kan, J. Jia and B. Chen, Acute stroke mortality and air pollution: new evidence from Shanghai, *China, J. Occup. Health*, 2003, **45**, 321–323.
49. H. Kim, Y. Kim and Y. C. Hong, The lag-effect pattern in the relationship of particulate air pollution to daily mortality in Seoul, Korea, *Int. J. Biometeorol.*, 2003, **48**, 25–30.
50. A. Zeka, A. Zanobetti and J. Schwartz, Individual-level modifiers of the effects of particulate matter on daily mortality, *Am. J. Epidemiol.*, 2006, **163**, 849–859.
51. Z. Qian, Q. He, H. M. Lin, L. Kong, D. Liao, J. Dan, C. M. Bentley and B. Wang, Association of daily cause-specific mortality with ambient particle air pollution in Wuhan, China, *Environ. Res.*, 2007, **105**, 380–389.
52. L. Perez, M. Medina-Ramon, N. Kunzli, A. Alastuey, J. Pey, N. Perez, R. Garcia, A. Tobias, X. Querol and J. Sunyer, Size fractionate particulate matter, vehicle traffic, and case-specific daily mortality in Barcelona, Spain, *Environ. Sci. Technol.*, 2009, **43**, 4707–4714.
53. A. Zanobetti and J. Schwartz, The effect of fine and coarse particulate air pollution on mortality: a national analysis, *Environ. Health Perspect.*, 2009, **117**, 898–903.
54. S. H. Moolgavkar, Air pollution and daily mortality in three US counties, *Environ. Health Perspect.*, 2000, **108**, 777–784.
55. G. Hoek, B. Brunekreef, P. Fischer and J. van Wijnen, The association between air pollution and heart failure, arrhythmia, embolism, thrombosis, and other cardiovascular causes of death in a time series study, *Epidemiology*, 2001, **12**, 355–357.
56. T. W. Wong, W. S. Tam, T. S. Yu and A. H. Wong, Associations between daily mortalities from respiratory and cardiovascular diseases

and air pollution in Hong Kong, China, *J. Occup. Environ. Med.*, 2002, **59**, 30–35.
57. Y. C. Hong, J. T. Lee, H. Kim and H. J. Kwon, Air pollution: a new risk factor in ischemic stroke mortality, *Stroke; J. Cereb. Circ.*, 2002, **33**, 2165–2169.
58. S. Yamazaki, H. Nitta, M. Ono, J. Green and S. Fukuhara, Intracerebral haemorrhage associated with hourly concentration of ambient particulate matter: case-crossover analysis, *J. Occup. Environ. Med.*, 2007, **64**, 17–24.
59. R. Maheswaran and P. Elliott, Stroke mortality associated with living near main roads in England and wales: a geographical study, *Stroke; J. Cereb. Circ.*, 2003, **34**, 2776–2780.
60. R. Maheswaran, R. P. Haining, P. Brindley, J. Law, T. Pearson, P. R. Fryers, S. Wise and M. J. Campbell, Outdoor air pollution and stroke in Sheffield, United Kingdom: a small-area level geographical study, *Stroke; J. Cereb. Circ.*, 2005, **36**, 239–243.
61. Z. Hu, J. Liebens and K. R. Rao, Linking stroke mortality with air pollution, income, and greenness in northwest Florida: an ecological geographical study, *Int. J. Health Geogr.*, 2008, **7**, 20.
62. J. J. Godleski, R. L. Verrier, P. Koutrakis, P. Catalano, B. Coull, U. Reinisch, E. G. Lovett, J. Lawrence, G. G. Murthy, J. M. Wolfson, R. W. Clarke, B. D. Nearing and C. Killingsworth, Mechanisms of morbidity and mortality from exposure to ambient air particles, *Res. Rep. Health Eff. Inst.*, 2000, 5–88; discussion 89–103.
63. D. Liao, J. Creason, C. Shy, R. Williams, R. Watts and R. Zweidinger, Daily variation of particulate air pollution and poor cardiac autonomic control in the elderly, *Environ. Health Perspect.*, 1999, **107**, 521–525.
64. C. A. Pope 3rd, R. L. Verrier, E. G. Lovett, A. C. Larson, M. E. Raizenne, R. E. Kanner, J. Schwartz, G. M. Villegas, D. R. Gold and D. W. Dockery, Heart-rate variability associated with particulate air pollution, *Am. Heart J.*, 1999, **138**, 890–899.
65. D. R. Gold, A. Litonjua, J. Schwartz, E. Lovett, A. Larson, B. Nearing, G. Allen, M. Verrier, R. Cherry and R. Verrier, Ambient pollution and heart-rate variability, *Circulation*, 2000, **101**, 1267–1273.
66. J. Creason, L. Neas, D. Walsh, R. Williams, L. Sheldon, D. Liao and C. Shy, Particulate matter and heart-rate variability among elderly retirees: the Baltimore 1998 PM study, *J. Expo. Anal. Environ. Epidemiol.*, 2001, **11**, 116–122.
67. A. Seaton, A. Soutar, V. Crawford, R. Elton, S. McNerlan, J. Cherrie, M. Watt, R. Agius and R. Stout, Particulate air pollution and the blood, *Thorax*, 1999, **54**, 1027–1032.
68. A. Peters, A. Doring, H. E. Wichmann and W. Koenig, Increased plasma viscosity during an air pollution episode: a link to mortality?, *Lancet*, 1997, **349**, 1582–1587.
69. A. Nemmar, P. H. Hoet, D. Dinsdale, J. Vermylen, M. F. Hoylaerts and B. Nemery, Diesel-exhaust particles in lung acutely enhance experimental peripheral thrombosis, *Circulation*, 2003, **107**, 1202–1208.

70. S. F. van Eeden and J. C. Hogg, Systemic inflammatory response induced by particulate-matter air pollution: the importance of bone-marrow stimulation, *J. Toxicol. Environ. Health A*, 2002, **65**, 1597–1613.
71. A. Peters, M. Frohlich, A. Doring, T. Immervoll, H. E. Wichmann, W. L. Hutchinson, M. B. Pepys and W. Koenig, Particulate air pollution is associated with an acute phase response in men; results from the MONICA-Augsburg Study, *Eur. Heart J.*, 2001, **22**, 1198–1204.
72. L. Barregard, G. Sallsten, P. Gustafson, L. Andersson, L. Johansson, S. Basu and L. Stigendal, Experimental exposure to wood-smoke particles in healthy humans: effects on markers of inflammation, coagulation, and lipid peroxidation, *Inhal. Toxicol.*, 2006, **18**, 845–853.
73. D. Liao, G. Heiss, V. M. Chinchilli, Y. Duan, A. R. Folsom, H. M. Lin and V. Salomaa, Association of criteria pollutants with plasma hemostatic/inflammatory markers: a population-based study, *J. Expo. Anal. Environ. Epidemiol.*, 2005, **15**, 319–328.
74. L. Bouthillier, R. Vincent, P. Goegan, I. Y. Adamson, S. Bjarnason, M. Stewart, J. Guenette, M. Potvin and P. Kumarathasan, Acute effects of inhaled urban particles and ozone: lung morphology, macrophage activity, and plasma endothelin-1, *Am J. Pathol*, 1998, **153**, 1873–1884.
75. R. Vincent, P. Kumarathasan, B. Mukherjee, C. Gravel, S. Bjarnason, B. Urch, M. Speck, J. Brook, S. Tarlo, B. Zimmerman and F. Silverman, Exposure to urban particles (PM2.5) causes elevations of the plasma vasopeptides endothelin ET-1 and ET-3 in humans [Abstract], *Am. J. Respir. Crit. Care Med.*, 2001, **163**, A313.
76. R. Vincent, P. Kumarathasan, P. Goegan, S. G. Bjarnason, J. Guenette, D. Berube, I. Y. Adamson, S. Desjardins, R. T. Burnett, F. J. Miller and B. Battistini, Inhalation toxicology of urban ambient particulate matter: acute cardiovascular effects in rats, *Res. Rep. Health Eff. Inst.*, 2001, 5–54; discussion 55–62.
77. M. Ando, M. Shima, M. Adachi and Y. Tsunetoshi, The role of intercellular adhesion molecule-1 (ICAM-1), vascular cell-adhesion molecule-1 (VCAM-1), and regulated on activation, normal T-cell expressed and secreted (RANTES) in the relationship between air pollution and asthma among children, *Arch. Environ. Health*, 2001, **56**, 227–233.
78. S. Salvi, A. Blomberg, B. Rudell, F. Kelly, T. Sandstrom, S. T. Holgate and A. Frew, Acute inflammatory responses in the airways and peripheral blood after short-term exposure to diesel exhaust in healthy human volunteers, *Am. J. Respir. Crit. Care Med.*, 1999, **159**, 702–709.
79. B. Urch, F. Silverman, P. Corey, J. R. Brook, K. Z. Lukic, S. Rajagopalan and R. D. Brook, Acute blood pressure responses in healthy adults during controlled air-pollution exposures, *Environ. Health Perspect.*, 2005, **113**, 1052–1055.
80. A. Zanobetti, M. J. Canner, P. H. Stone, J. Schwartz, D. Sher, E. Eagan-Bengston, K. A. Gates, L. H. Hartley, H. Suh and D. R. Gold, Ambient pollution and blood pressure in cardiac rehabilitation patients, *Circulation*, 2004, **110**, 2184–2189.

81. A. Ibald-Mulli, J. Stieber, H. E. Wichmann, W. Koenig and A. Peters, Effects of air pollution on blood pressure: a population-based approach, *Am. J. Public Health*, 2001, **91**, 571–577.
82. W. S. Linn, H. Gong Jr, K. W. Clark and K. R. Anderson, Day-to-day particulate exposures and health changes in Los Angeles area residents with severe lung disease, *J. Air Waste Manag Assoc*, 1999, **49**, 108–115.
83. V. C. Van Hee, S. D. Adar, A. A. Szpiro, R. G. Barr, D. A. Bluemke, A. V. Diez Roux, E. A. Gill, L. Sheppard and J. D. Kaufman, Exposure to traffic and left ventricular mass and function: the multi-ethnic study of atherosclerosis, *Am. J. Respir. Crit. Care Med.*, 2009.
84. A. H. Auchincloss, A. V. Roux, J. T. Dvonch, P. L. Brown, R. G. Barr, M. L. Daviglus, D. C. Goff, J. D. Kaufman and M. S. O'Neill, Associations between recent exposure to ambient fine particulate matter and blood pressure in the Multi-Ethnic Study of Atherosclerosis (MESA), *Environ. Health Perspect.*, 2008, **116**, 486–491.
85. C. R. Bartoli, G. A. Wellenius, E. A. Diaz, J. Lawrence, B. A. Coull, I. Akiyama, L. M. Lee, K. Okabe, R. L. Verrier and J. J. Godleski, Mechanisms of inhaled fine particulate air pollution-induced arterial blood pressure changes, *Environ. Health Perspect.*, in press.
86. H. Tornqvist, N. L. Mills, M. Gonzalez, M. R. Miller, S. D. Robinson, I. L. Megson, W. Macnee, K. Donaldson, S. Soderberg, D. E. Newby, T. Sandstrom and A. Blomberg, Persistent endothelial dysfunction in humans after diesel exhaust inhalation, *Am. J. Respir. Crit. Care Med.*, 2007, **176**, 395–400.
87. M. S. O'Neill, A. Veves, A. Zanobetti, J. A. Sarnat, D. R. Gold, P. A. Economides, E. S. Horton and J. Schwartz, Diabetes enhances vulnerability to particulate air pollution-associated impairment in vascular reactivity and endothelial function, *Circulation*, 2005, **111**, 2913–2920.
88. R. Dales, L. Liu, M. Szyszkowicz, M. Dalipaj, J. Willey, R. Kulka and T. D. Ruddy, Particulate air pollution and vascular reactivity: the bus stop study, *Int. Arch. Occup. Environ. Health*, 2007, **81**, 159–164.
89. R. D. Brook, J. R. Brook, B. Urch, R. Vincent, S. Rajagopalan and F. Silverman, Inhalation of fine particulate air pollution and ozone causes acute arterial vasoconstriction in healthy adults, *Circulation*, 2002, **105**, 1534–1536.
90. T. F. Bateson and J. Schwartz, Who is sensitive to the effects of particulate air pollution on mortality? A case-crossover analysis of effect modifiers, *Epidemiology*, 2004, **15**, 143–149.
91. A. Zanobetti and J. Schwartz, The effect of particulate air pollution on emergency admissions for myocardial infarction: a multicity case-crossover analysis, *Environ. Health Perspect.*, 2005, **113**, 978–982.
92. A. Zanobetti and J. Schwartz, Are diabetics more susceptible to the health effects of airborne particles?, *Am. J. Respir. Crit. Care Med.*, 2001, **164**, 831–833.
93. A. Zanobetti and J. Schwartz, Cardiovascular damage by airborne particles: are diabetics more susceptible?, *Epidemiology*, 2002, **13**, 588–592.

94. N. Kunzli, M. Jerrett, W. J. Mack, B. Beckerman, L. LaBree, F. Gilliland, D. Thomas, J. Peters and H. N. Hodis, Ambient air pollution and atherosclerosis in Los Angeles, *Environ. Health Perspect.*, 2005, **113**, 201–206.
95. B. Hoffmann, S. Moebus, S. Mohlenkamp, A. Stang, N. Lehmann, N. Dragano, A. Schmermund, M. Memmesheimer, K. Mann, R. Erbel and K. H. Jockel, Residential exposure to traffic is associated with coronary atherosclerosis, *Circulation*, 2007, **116**, 489–496.
96. R. W. Allen, M. H. Criqui, A. V. Diez Roux, M. Allison, S. Shea, R. Detrano, L. Sheppard, N. D. Wong, K. H. Stukovsky and J. D. Kaufman, Fine particulate-matter air pollution, proximity to traffic, and aortic atherosclerosis, *Epidemiology*, 2009, **20**, 254–264.
97. T. Suwa, J. C. Hogg, K. B. Quinlan, A. Ohgami, R. Vincent and S. F. van Eeden, Particulate air pollution induces progression of atherosclerosis, *J. Am. Coll. Cardiol.*, 2002, **39**, 935–942.
98. Q. Sun, A. Wang, X. Jin, A. Natanzon, D. Duquaine, R. D. Brook, J. G. Aguinaldo, Z. A. Fayad, V. Fuster, M. Lippmann, L. C. Chen and S. Rajagopalan, Long-term air-pollution exposure and acceleration of atherosclerosis and vascular inflammation in an animal model, *JAMA*, 2005, **294**, 3003–3010.
99. J. A. Araujo, B. Barajas, M. Kleinman, X. Wang, B. J. Bennett, K. W. Gong, M. Navab, J. Harkema, C. Sioutas, A. J. Lusis and A. E. Nel, Ambient particulate pollutants in the ultrafine range promote early atherosclerosis and systemic oxidative stress, *Circ. Res.*, 2008, **102**, 589–596.

CHAPTER 7
Environmental Pollutants and Heart Failure

S. D. PRABHU[1,2]

[1] University of Louisville, Institute of Molecular Cardiology, 580 S. Preston Street, Louisville, KY, 40202, USA; [2] Louisville VA Medical Center, 800 Zorn Avenue, Louisville, KY, 40206, USA

7.1 Introduction

Urban air pollution is primarily derived from the combustion of fossil fuels and is comprised of a complex mixture of particulate matter (PM) and gases such as sulfur dioxide (SO_2), carbon monoxide (CO), nitrogen dioxide (NO_2), and ozone.[1–5] While both PM and gaseous pollutants have been implicated in adverse effects on human health, compelling evidence suggests that PM is the most relevant fraction.[1,2] PM is a heterogeneous mixture of solid and liquid particles varying in size and chemical composition to include elemental carbon, organic carbon and hydrocarbon compounds (*e.g.*, aldehydes), transition metals and metal oxides, biological material (*e.g.*, endotoxin), and sulfates and nitrates.[2–5] PM is categorized in terms of aerodynamic diameter: thoracic particles < 10 µm diameter (PM_{10}), fine particles < 2.5 µm diameter ($PM_{2.5}$), and ultrafine particles < 0.1 µm diameter ($PM_{0.1}$). $PM_{2.5}$ has received the greatest scientific attention as they can reach the small airways and alveoli in the lung.[1] The short-lived ultrafine particles are also likely to be of considerable biological importance as these particles account for the largest number of particles per volume air and present the largest surface area for transporting particulate matter into the pulmonary tree and circulation.[2,5]

Substantial epidemiological data over the last two decades indicate that exposure to air pollution is associated with adverse effects on cardiovascular

---

Issues in Toxicology No. 8
Environmental Cardiology: Pollution and Heart Disease
Edited by Aruni Bhatnagar
© The Royal Society of Chemistry 2011
Published by the Royal Society of Chemistry, www.rsc.org

health and increased death and hospitalizations from cardiovascular disease.[1–5] Pope[6] has estimated that for each 50 μg/m^3 increase in $PM_{2.5}$ exposure there is a 7% and 11% increase in all-cause and cardiovascular daily mortality, respectively; the percentage of excess deaths due to PM exposure is attributable primarily to cardiovascular disease. Moreover, long-term exposure to PM has been suggested to reduce life expectancy by 1.8–3.1 years, again primarily due to excess cardiovascular mortality.[1,2,6,7] Due to statistical considerations (*e.g.*, limited number of daily events or sample size) or methodological constraints (*e.g.*, accuracy of death certification for mortality data), most human studies examining the health effects of pollution have reported effect sizes relevant for broad or grouped diagnostic categories of cardiovascular disease, and studies examining pathology-specific causes have been fewer in number.[4]

However, such distinctions are of particular importance for the identification of high-risk patient populations, so that therapeutic, educational, and regulatory interventions can be more effectively targeted. Accordingly, this chapter will examine the available studies linking pollution to morbidity and mortality for the disease subset of heart failure (HF).

7.2 Clinical and Pathological Characteristics of HF

HF is of high significance from a public health standpoint.[8] HF is common; more than 5 million Americans are afflicted, and the lifetime risk of developing HF is 1 in 5. HF is increasing in prevalence, especially in the elderly (>65 years), in whom the disease incidence approaches 1 in 100. HF is expensive with an estimated health care cost of over $35 billion annually. Moreover, HF also carries a high mortality; the 5-year survival for patients with systolic HF is only 40%, despite advances in pharmacotherapy.[9] Given these characteristics, the recognition and identification of environmental factors that precipitate and/or exacerbate HF is of paramount importance at both an individual and societal level.

HF occurs when the heart cannot maintain adequate output to the peripheral tissues or can do so only at elevated filling pressure.[10] This results in clinical manifestations of low cardiac output (fatigue, effort intolerance, reduced tissue perfusion) and pulmonary and/or systemic congestion (cough, exertional dyspnea, orthopnea, paroxysmal nocturnal dyspnea, nocturia, and peripheral edema). The underlying basis for HF is pathological structural remodeling and hypertrophy of the left ventricle (LV), resulting in systolic and/or diastolic dysfunction.[10–12] The process of LV remodeling (Figure 7.1) is triggered by an index insult such as myocardial infarction, chronic hypertension, and valvular lesions and is perpetuated over the long-term by factors that include augmented mechanical load and the concomitant activation of neurohormonal and inflammatory systems (*e.g.*, adrenergic nervous system, renin–angiotensin system, arginine vasopressin, proinflammatory cytokines) that have deleterious biological effects in the heart such as hypertrophy, fibrosis, fetal gene expression, and apoptotic cell death. This complex interplay ultimately results in progressive

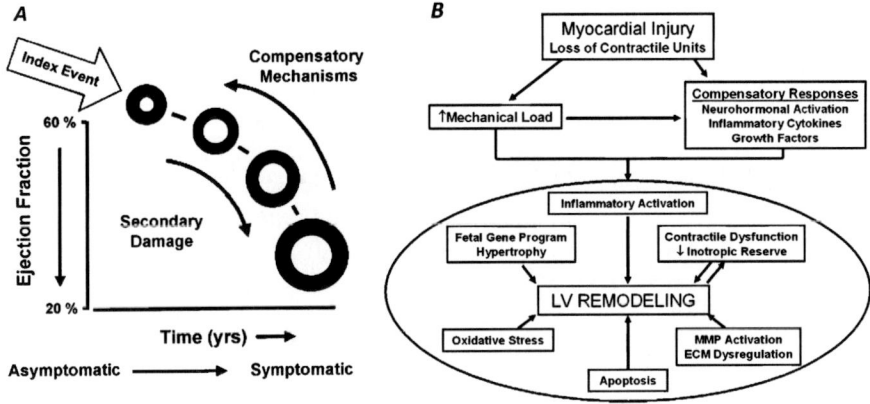

Figure 7.1 (A) Time course of left ventricular (LV) remodeling. After an index injury such as myocardial infarction, compensatory mechanisms (*e.g.*, adrenergic and renin–angiotensin–aldosterone systems) are activated to maintain cardiac performance. Sustained activation of these systems induces toxicity and secondary damage that leads to progressive LV enlargement and dysfunction (reduced ejection fraction) over time and, ultimately, symptomatic heart failure (reproduced from reference 12, Mann DL, Circulation 1999, with permission from the American Heart Association). (B) Pathophysiology of LV remodeling. Following myocardial injury, augmented cardiac mechanical load together with the aforementioned activation of compensatory neurohormonal systems induce a range of cellular and tissue responses in the heart (*e.g.*, apoptosis, inflammation, hypertrophy, contractile dysfunction, oxidative stress) that cause the progression of LV remodeling and the heart failure phenotype (reproduced from reference 11, Prabhu SD, Journal of Molecular and Cellular Cardiology 2005, with permission from Elsevier Inc.). MMP, matrix metalloproteinase; ECM, extracellular matrix.

LV dilatation and dysfunction over time (months to years) and the signs and symptoms of HF.[12] Hence, when considering pollution-induced responses, those effects that adversely impact either structural cardiac remodeling (and its triggers) or relevant extracardiac factors (*e.g.*, load, neurohormonal systems, and inflammation) can potentially influence clinical outcomes in persons with HF.

7.3 Pollution and Heart Failure: Short-Term Effects

Epidemiological studies have demonstrated an association between short-term exposures to air pollution (time-series or case-crossover analyses over a few days of exposure) and increases in: (1) the incidence of select cardiorespiratory symptoms and signs,[13–16] (2) mortality related to HF,[17–23] and (3) HF hospitalizations and emergency-room visits.[24–32] Taken together, these studies indicate a clinically discernible and important role for acute pollutant exposure on disease exacerbations and deaths from HF. Moreover, these effects are uniformly reported to be much more pronounced in the elderly population.

7.3.1 HF Symptoms and Signs

Small group studies have attempted to dissect the relationship between short-term exposure and worsening of cardiorespiratory symptoms. The ULTRA study[13] evaluated symptomatic response with changes in daily mass concentrations of PM_{10}, $PM_{2.5}$, and $PM_{0.01-0.1}$ in three panels of elderly subjects with stable coronary heart disease from three different European countries. Pollutant concentrations were measured at fixed monitoring sites representing urban background levels. The authors found that a 10-μg/m^3 short-term increase in $PM_{2.5}$ was associated with a ~10% increase in the incidence of dyspnea and effort intolerance. A 10 000-particles/cm^3 increase in $PM_{0.01-0.1}$ was also associated with effort intolerance. Interestingly, there was no association between air particulates and chest pain. Goldberg and colleagues[14] reported that in 31 subjects with HF, oxygen saturation by pulse oximetry was inversely associated with concentrations of $PM_{2.5}$, ozone, and SO_2, whereas heart rate was increased in relation to $PM_{2.5}$, NO_2, and SO_2. In this same group of subjects, increased concurrent day ozone levels predicted poorer self-perceived general health.[15] Importantly, in a separate study, Wellenius *et al.* reported that blood levels of B-type natriuretic peptide (BNP, a HF biomarker that correlates with filling pressure) in subjects with chronic HF were substantially variable and not associated with any number of pollutants, including $PM_{2.5}$, CO, ozone, SO_2, NO_2, and carbon black.[16] Overall, these studies suggested that symptoms and functional status in subjects with HF can be influenced by short-term variations in ambient air pollutants. However, this effect is not reliably detected by changes in circulating BNP.

7.3.2 HF Mortality

The demonstration of air pollution-related HF mortality from large-scale time-series studies has provided more definitive evidence of an important association between short-term exposure and disease.[17–23] A 10-year study of the residents of Milan, Italy (~1.5 million people) revealed a 7% increase in concurrent-day HF mortality and a 3.3% increase in all-cause mortality for each 100-μg/m^3 increase in total particulates.[18] An 8-year study evaluated the associations between daily variations in air pollution and mortality for the entire population of the Netherlands (~15 million people).[19,20] Daily mortality was significantly associated with the concentration of all air pollutants studied (PM_{10}, black smoke, ozone, CO, SO_2, and NO_2).[19] Further analysis indicated that the effect estimates were most pronounced for deaths due to HF than for total cardiovascular mortality and that the excess relative risks for HF were ~3 times higher for all pollutants except ozone.[20] The authors estimated that HF deaths comprised 10% of all cardiovascular deaths but accounted for 30% of the cardiovascular deaths related to PM. Moreover, smaller studies with more limited statistical power have nonetheless demonstrated that daily mortality in patient cohorts with known HF increases linearly with the concentration of particulates, SO_2, and NO_2 in the air and that these associations are more pronounced in the elderly (≥ 65 years).[21,22] The estimated effects on mortality were 2.5–4.1 times higher in patients with HF than in the general population.[23] Taken together, these studies establish that persons with pre-existing HF are particularly vulnerable to the harmful health effects of pollution and are at greater risk of dying.

7.3.3 HF Hospital Visits and Admissions

Although mortality is a hard endpoint, hospital admissions and emergency-room visits for HF exacerbation may be more sensitive indicators of the impact of pollutants on disease. Indeed, such studies have revealed statistically robust associations between a wide array of pollutants and HF. Several multicity and single-region studies have demonstrated significant positive associations between both ambient PM and gaseous pollutants and HF hospitalizations and emergency-room visits.[24–32] A 7-city study of Medicare recipients (≥ 65 years of age) examining more than 290 000 HF hospital admissions found that a 10-μg/m^3 increase in PM$_{10}$ was associated with a 0.72% increase in HF hospitalizations on the same day.[24] In a very large study of 11.5 million Medicare enrollees (>65 years) from 204 urban US counties, Dominici et al.[25] found a 1.28% increase in HF hospitalization rate for every 10-μg/m^3 increase in same-day PM$_{2.5}$. This was the largest association for any cardiovascular diagnosis that was studied (Figure 7.2). Interestingly, the degree of risk exhibited marked geographical variability and tended to be higher in Eastern US counties. The authors suggested that this spatial heterogeneity may be secondary to known differences in the source mix for PM$_{2.5}$ and differences in particle composition.

Importantly, a more recent study by Pope et al.,[26] though markedly smaller in size (examining 2628 HF hospitalizations in Utah), found that the strongest PM$_{2.5}$ association with HF hospitalization occurred 2–3 weeks after exposure (Figure 7.3). A 14-day lagged average increase of 10-μg/m^3 PM$_{2.5}$ was linked to a 13.1% increase in HF admissions, representing a 16-fold greater effect than for concurrent-day PM$_{2.5}$ concentrations. This delayed effect for the effects of PM on HF exacerbations is in distinct contrast to the pattern of PM-related risk for ischemic heart disease events such as myocardial infarction, which typically occur during a much shorter time window (generally 0–1 d).[33–35] This pathology-specific difference suggests that for HF disease exacerbations, cumulative PM exposure may be of greater importance than proximity in time of exposure. Alternatively, it may suggest underlying biological mechanisms that are distinct for HF as opposed to ischemic and atherosclerotic disease.

Along with PM, short-term exposure to gaseous pollutants has also been demonstrated to have a detrimental impact on HF exacerbations and hospitalizations.[27–32] In general, the most consistent and robust associations with HF have been for CO and NO$_2$,[27–30] with somewhat weaker associations for SO$_2$ and ozone.[29,31,32] A large case-crossover study of seven cities in Australia and New Zealand (total population >12 million) examined the relationship between daily levels of PM$_{2.5}$, PM$_{10}$, NO$_2$, CO, and ozone and five categories of

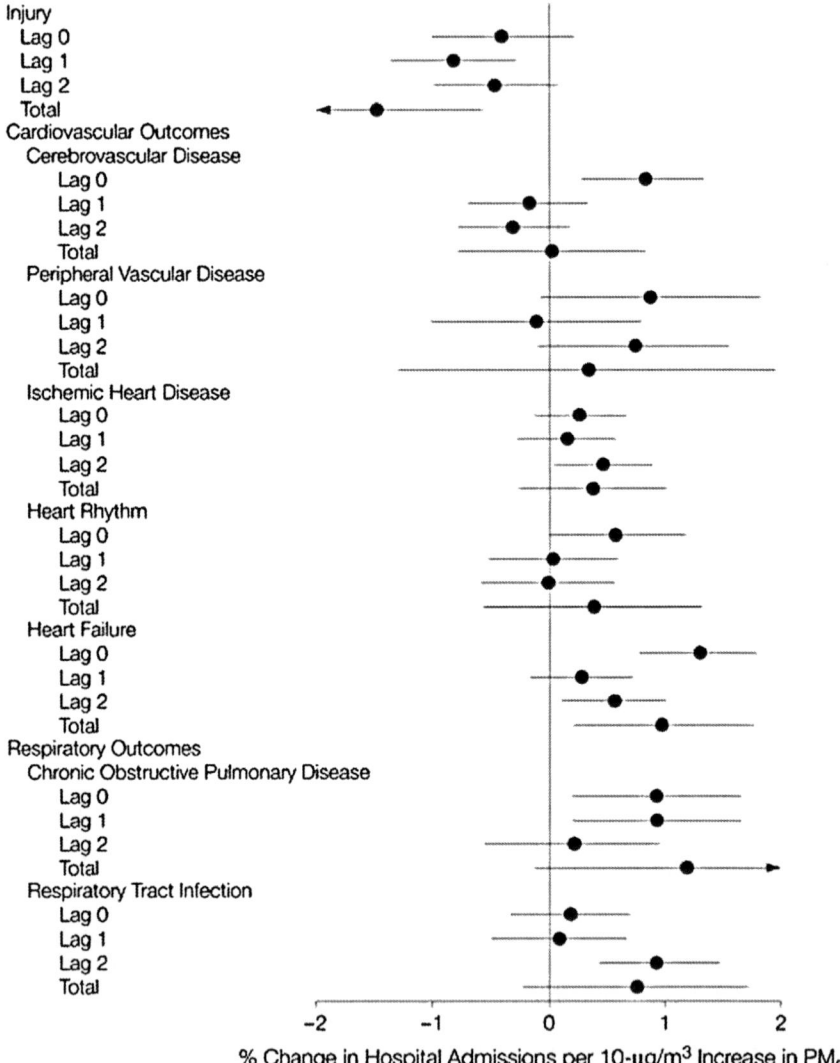

Figure 7.2 Percentage change in rate of hospitalization for cardiovascular and respiratory causes per 10 μg/m³ increase in PM$_{2.5}$ for 204 US counties from the study by Dominici et al., JAMA 2006 (reference 25). Point estimates and 95% posterior intervals are shown. The HF hospitalization rate increased by 1.28%, the largest association for cardiovascular causes. Reproduced with permission from the American Medical Association.

cardiovascular disease admissions, including cardiac failure.[27] PM, CO, and NO$_2$ (but not ozone) exhibited significant associations with HF admissions, an effect that was much greater in the elderly (>65 years). Moreover, when considering all cardiovascular disease categories, the effects in the elderly were

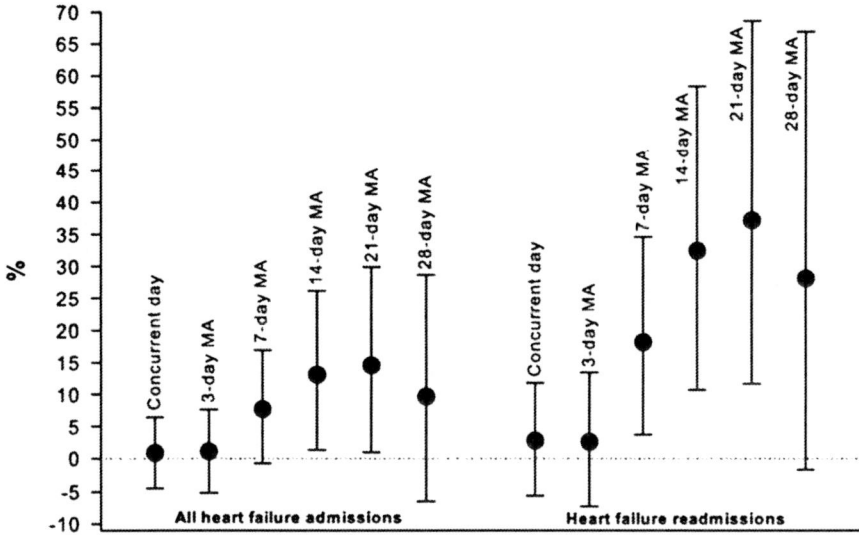

Figure 7.3 Data from reference 26 (Pope et al., American Journal of Cardiology, 2008) illustrating the percentage increase in HF hospitalization risk (and 95% confidence intervals) associated with a 10-μg/m^3 increase in PM$_{2.5}$. The largest increase in risk was seen with a 2–3 week lagged moving average exposure. Reproduced with permission from Elsevier Inc.

most pronounced for HF, with a 6.0% increase for a 0.9-ppm increase in CO and a 6.9% increase for a 5.1-ppb increase in NO$_2$. Importantly, these effects were found at concentrations below air-quality health guidelines and suggested susceptibility to air pollution from common emission sources for PM, CO, and NO$_2$, such as motor-vehicle exhaust. Other studies have similarly shown consistent associations between CO and NO$_2$ and HF exacerbations.[28–32]

Short-term exposure studies also indicate the potential impact of disease comorbidities on the link between pollutants and HF. Specifically, patients with recent myocardial infarction are at greater risk of PM-related HF admission.[36,37] Conversely, patients with an underlying diagnosis of HF are more vulnerable to ischemic heart disease hospital admissions related to ambient CO and NO$_2$ exposure.[38] There is also apparent increased susceptibility to pollution-related HF emergency-room visits in subjects with comorbid hypertension,[39] although this has not been uniformly reported.[40]

7.4 Pollution and Heart Failure: Long-Term Effects

In contrast to short-term studies, the long-term effects of pollutants on HF have been studied less frequently.[7,41–44] Nonetheless, the available data point to a substantial health risk of chronic PM exposure on HF disease progression and mortality. Whether this health risk represents a summation of acute exposure risks over time or a true independent effect of longer-term exposures is not entirely clear, although generally the disease risk delineated in chronic studies tends to be higher than that reported in acute studies.[2]

7.4.1 Particulate Exposure

Pope *et al.* analyzed $PM_{2.5}$-related cardiac and pulmonary mortality from a large cohort of subjects from all 50 states in the US participating in the American Cancer Society Cancer Prevention Study (analytical cohort ranging from 319 000 to 500 000 subjects).[7] During the overall follow-up period of 16 years, nearly half of the deaths were secondary to cardiovascular disease, with statistically robust associations between $PM_{2.5}$ exposure and cardiovascular mortality. The strongest associations were for mortality related to ischemic heart disease, dysrhythmias, heart failure, and cardiac arrest. For the category of dysrhythmias, heart failure, and cardiac arrest, a 10-μg/m^3 elevation in $PM_{2.5}$ was associated with a 13% increased mortality risk. The authors felt that these data were most consistent with putative mechanisms of PM-mediated systemic inflammation and endothelial dysfunction and altered cardiac autonomic tone (see below). Extending these observations, a more recent study evaluated nearly 200 000 elderly (>65 years) survivors of acute myocardial infarction from a broad cross section of the US (21 cities).[41] The authors found that over an average follow-up of 3.7–5.1 years, a 10-μg/m^3 elevation in PM_{10} conferred a 30% increased risk of mortality and a 40% increased risk for new-onset HF. As the development of new-onset HF following myocardial infarction is thought to be related to the progression of underlying left ventricular remodeling over time (Figure 7.1), these intriguing data suggest that chronic particulate exposure can exacerbate underlying structural myocardial disease (remodeling) and thereby contribute to the progression of heart failure.

7.4.2 Motor-Vehicle Traffic Exposure

Motor traffic is an important source of pollutants, including PM and gaseous pollutants. Traffic exposure may be of particular importance with regard to health effects, as fresh emissions contain PM enriched with ultrafine particles, which can penetrate deep into the pulmonary tree[1,5] together with aldehydes, CO, NO_2, and elemental carbon.[42,43] Moreover, in contrast to city-level measurement of particulates and other pollutants that are global indexes, traffic proximity can provide an index of effects closer to an individual level. Traffic exposure in individuals, for example, increases the risk of myocardial infarction by nearly threefold in the hour following exposure.[44] In view of these considerations, several studies have explored the association between residential proximity to traffic, as an index of chronic pollution exposure, and HF. Over a five-year follow-up of 1389 patients with established HF, residential proximity within 100 m of a major roadway or 50 m of a bus route increased mortality risk by 30% (Figure 7.4).[45] Importantly, the HF mortality associated with traffic intensity was not explained by traffic noise,[46] suggesting that residential

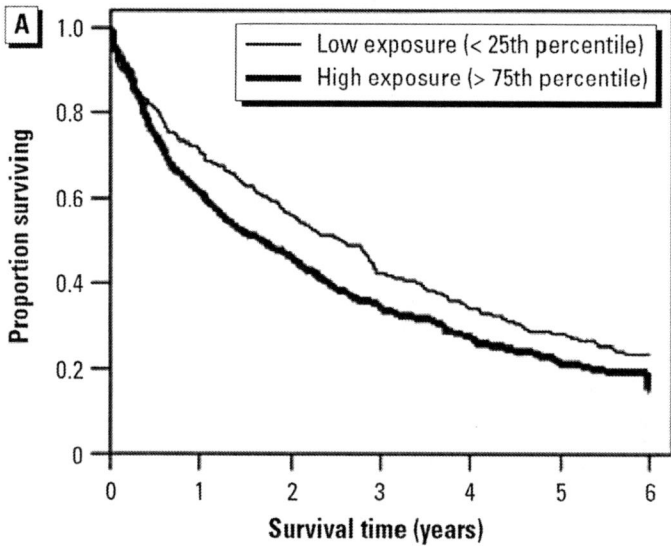

Figure 7.4 Long-term survival rates in patients with HF for low and high traffic-related pollution exposure groups as indexed by proximity to daily traffic within 100 m. There was a significant ($p = 0.017$) increase in mortality in the high-exposure group over a six-year period. Reproduced from reference 45, Medina-Ramón et al., Environmental Health Perspective 2008, with permission from the National Institute of Environmental Health Sciences.

exposure to traffic-mediated pollution was the responsible factor. Residential proximity to traffic has also been shown to correspond to greater left ventricular mass, as determined by magnetic resonance imaging, an effect that appeared to be independent of differences in systolic or diastolic blood pressure.[47] These data also support a potential pathophysiological link between pollutant exposure and LV remodeling.

Taken together, the available human epidemiological studies suggest both short-term and long-term deleterious effects of air pollution on patients with HF, primarily in the elderly, at both a population level and at an individual level. These are comprised of worsening symptoms and signs, significant disease exacerbation requiring emergency-room visits and hospital admissions, increased disease mortality, and potentially detrimental remodeling responses that influence long-term disease progression. These associations have been demonstrated for both PM and gaseous pollutants, but especially CO and NO_2, suggesting that motor vehicle emissions may be an important source of HF risk. The role of motor-vehicle exhaust in triggering detrimental cardiovascular responses in individuals is also inferred from studies examining traffic-related effects. This compelling epidemiological evidence raises the important (but as of yet not definitively answered) question as to what are the relevant biological mechanisms underlying these observations.

7.5 Pathophysiological Mechanisms of Pollution-Related HF Risk

Based on the results of a large number of human and animal studies, three general biological mechanisms have been proposed to explain how inhaled pollutants can adversely affect the cardiovascular system.[1-5] These include: (1) alterations in autonomic tone, triggered by pollutant-induced activation of pulmonary neural afferent reflexes, which result in sympathetic activation and parasympathetic withdrawal; (2) the elaboration of proinflammatory, pro-oxidant, and vasoactive substances from the lungs that induce systemic inflammation and oxidative stress, thereby promoting endothelial and vascular dysfunction, hypertension, and atherothrombosis; and (3) physical translocation of soluble constituents of inhaled PM into the circulation that have direct effects on the heart and vasculature.

In principle, all these putative mechanisms can produce adverse effects in subjects with HF. Reduced heart-rate variability (HRV), reflecting disturbances in autonomic tone resulting from parasympathetic withdrawal and/or excessive sympathetic (adrenergic) activation, is a hallmark of HF and is an independent predictor of mortality in such patients.[48-50] Moreover, sustained adrenergic activation is an important mediator of myocardial dysfunction and remodeling in HF, whereas beta-adrenergic blockade improves remodeling and clinical outcomes and is considered standard therapy for HF.[10-12,51,52]

Human studies have demonstrated that exposure to particulates and gaseous pollutants reduces HRV and that in the elderly, such effects are more pronounced for traffic particles.[53,54] Therefore, pathological alterations in autonomic tone may fundamentally contribute to the adverse effects induced by pollutants on remodeling and ventricular function in patients with HF. Patients with HF are also exquisitely susceptible to changes in myocardial load, such as those induced by elevated blood pressure. In this regard, PM has been demonstrated to induce arterial vasoconstriction[55] and increase mean arterial blood pressure at rest and during exercise.[56] Alterations in autonomic tone may significantly impact cardiac load. Such pathophysiological responses can contribute to symptomatic decompensation and effort intolerance and, ultimately, acute clinical exacerbations of HF.

HF is also characterized by chronic inflammation that is thought to contribute to LV remodeling and disease progression.[57] Several studies have demonstrated that particulate exposure increases the systemic elaboration of inflammatory mediators such as proinflammatory cytokines and C-reactive protein.[1,58,59] This would suggest that promotion of chronic inflammation can also contribute to pollution-mediated exacerbation and/or progression of HF. However, a recent human study[60] indicated that PM-related augmentation of proinflammatory mediators in survivors of myocardial infarction was quite modest (*i.e.* only a 2.7% increase in IL-6, 0.6% increase in fibrinogen, and no consistent change in CRP) raising questions about the physiological significance of these observations. Furthermore, this hypothesis needs to be tested directly in animal models.

The aforementioned pathophysiological mechanisms are indicative of a generalized stress response to pollutant-induced injury involving, neural, humoral, and inflammatory systems. In contrast, much less is known about specific responses to individual particulate and gaseous constituents. Moreover, it is unclear as to whether, and by which cellular mechanisms, specific soluble particulate constituents (*e.g.*, volatile organic compounds such as aldehydes, transition metals, elemental carbon, nitrates, sulfates) induce direct effects on the heart and vasculature. Epidemiological clues, however, suggest that determining the role of individual particle constituents is of pressing importance. In their study of 11.5 million Medicare enrollees, Dominici *et al.*[25] found substantial geographic variation in $PM_{2.5}$-related cardiovascular disease hospitalization, suggesting that known differences in the source mix for $PM_{2.5}$ (and consequently particle composition) in different regions of the US could be responsible for these observations. Moreover, a spatial analysis of air pollution in a single city revealed important intraurban gradients in $PM_{2.5}$ exposure, suggesting variations in source and associated differences in particle characteristics.[3,61]

7.6 Aldehydes Impart Significant Cardiotoxic Effects

In view of the above mechanistic considerations, it is evident that attention should be focused on determining the biological effects uniquely referable to specific PM components and defining those particulate characteristics that are ultimately responsible for toxicity in cardiovascular disorders. Accordingly, our laboratory has been interested in delineating the cardiac effects engendered by exposure to reactive aldehydes,[62–64] which are important copollutants with particulate matter and the pollutant mix derived from other environmental sources, particularly automobile emission and smoke.[65,66] High levels of aldehydes are also present in food and water. Indeed, more than 300 different aldehydes have been identified in various foods and at least 36 different aldehydes are present in water,[63,65,66] and with the exception of metals, aldehydes are considered to be the major pollutants in drinking water.[66] However, toxicological profiles of the complex aldehyde mixtures present in environmental sources (*e.g.* food, water, smoke, ambient air) are difficult to establish. Hence, to assess prototypic aldehyde toxicity, we studied the cardiac effects of acrolein, a highly reactive C3 α,β-unsaturated aldehyde ($CH_2=CH-CHO$) classified by the Environmental Protection Agency as a high-priority air and water toxic,[67] at concentrations documented in human disease or at estimated human oral exposure levels.[62–64] In the investigations summarized below, we sought to derive proof-of-concept and potential mechanisms for cardiotoxicity induced by acrolein in the mouse model in order to gain insights into the potential cardiac consequences of environmental aldehyde exposure in humans.

Acrolein, at concentrations observed clinically in human disease, induced profound, rapid, and to a large extent, reversible cardiac contractile dysfunction, and studies in isolated mouse heart revealed that the contractile dysfunction was due to impaired myofibrillar Ca^{2+} responsiveness, rather than changes in Ca^{2+} cycling (Figure 7.5), a response similar to that seen during myocardial stunning (postischemic reversible myocardial dysfunction).[62] These effects were in part related to acrolein–sulfhydryl interactions and the formation of acrolein–protein adducts (Figure 7.6), as both the contractile dysfunction and adduct formation were markedly attenuated by thiol donors. Importantly, the acrolein–protein modifications were selective and predominantly affected contractile/cytoskeletal proteins (cardiac α-actin, desmin, myosin light polypeptide 3) and mitochondrial energy metabolism proteins

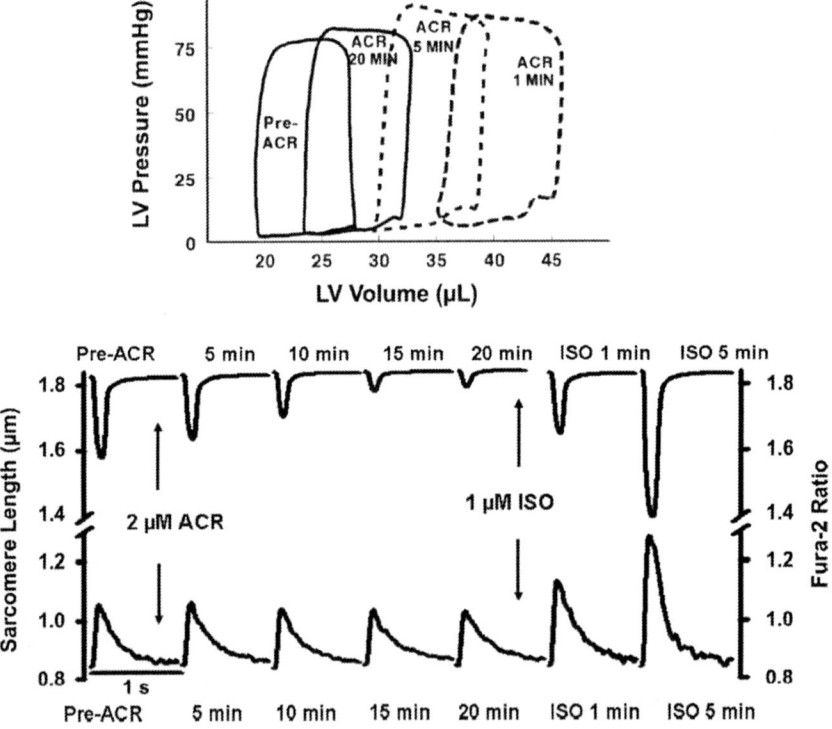

Figure 7.5 Acrolein depresses cardiac performance. Top, LV pressure-volume loops were measured in anesthetized mice at baseline and after 0.5 mg/kg intravenous acrolein. The pressure–volume loop demonstrated a rapid rightward shift and elevation of LV end-diastolic pressure, indicating significant chamber dilatation and mechanical dysfunction. The effect was reversible as there was a return toward baseline by 20 min. Bottom, Sarcomere shortening (top trace) and Ca^{2+} transients (bottom trace, fura-2 ratio) were measured in adult mouse cardiomyocytes upon myocyte exposure to 2 μM acrolein (ACR). Acrolein induced progressive depression of shortening without significant change in the Ca^{2+} transients, suggesting underlying changes in myofibrillar Ca^{2+} responsiveness. The contractile and Ca^{2+} responses to 1 μM isoproterenol (ISO) remained robust.

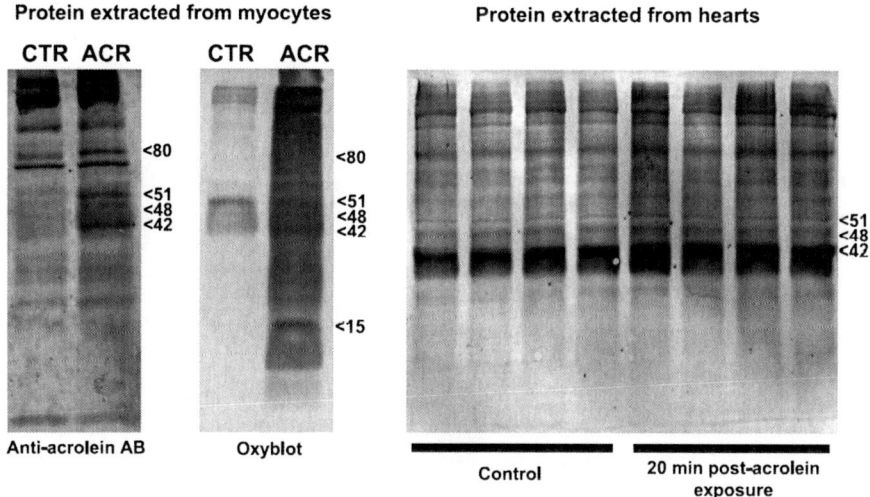

Figure 7.6 Acrolein exposure increases the formation of acrolein–protein adducts. Left, abundance of acrolein–protein adducts by Western blot and protein-carbonyls by Oxyblot in control (CTRL) and acrolein-treated (ACR 10 µmol/L, 20 min) mouse cardiomyocytes. ACR-exposed myocytes exhibited increased immunoreactivity with both techniques over a broad molecular weight range. Right, protein–carbonyl abundance (Oxyblot) in hearts 20 min after 0.5 µg/kg intravenous acrolein. There was a 1.8-fold increased protein–carbonyl abundance over control consistent with acrolein–protein adduct formation *in vivo*. Reproduced from reference 62, Luo *et al.*, American Journal of Physiology: Heart and Circulatory Physiology 2007, with permission from the American Physiological Society.

(mitochondrial creatine kinase 2, ATP synthase). These findings suggested that acrolein may impart analogous, detrimental effects on the heart after environmental acrolein exposure.

To extend these studies, we further evaluated the cardiac effects of acute and chronic oral acrolein exposure. We estimated the approximate daily consumption of unsaturated aldehydes (such as acrolein) in humans to be 5 mg/kg, whereas that of saturated aldehydes (such as formaldehyde and acetaldehyde) to be 2 mg/kg.[63] Based on these estimates a 5 mg/kg dose of acrolein, representing the expected unsaturated aldehyde intake, was chosen for acute exposure studies.[63] Interestingly, we found that that oral acrolein consumption increased myocardial ischemia-reperfusion injury and abolished the cardioprotective effects of nitric oxide during ischemia-reperfusion (Figure 7.7), suggesting that exposure to acrolein and related aldehydes could enhance the susceptibility of the heart to ischemia. The signaling mechanisms by which these changes occur appear to involve attenuated translocation of the cardioprotective enzyme protein kinase Cε (PKCε), decreased mitochondrial PKCε expression, and the formation of mitochondrial acrolein–PKCε adducts, suggesting that the mitochondria are highly susceptible to aldehyde toxicity.

190 Chapter 7

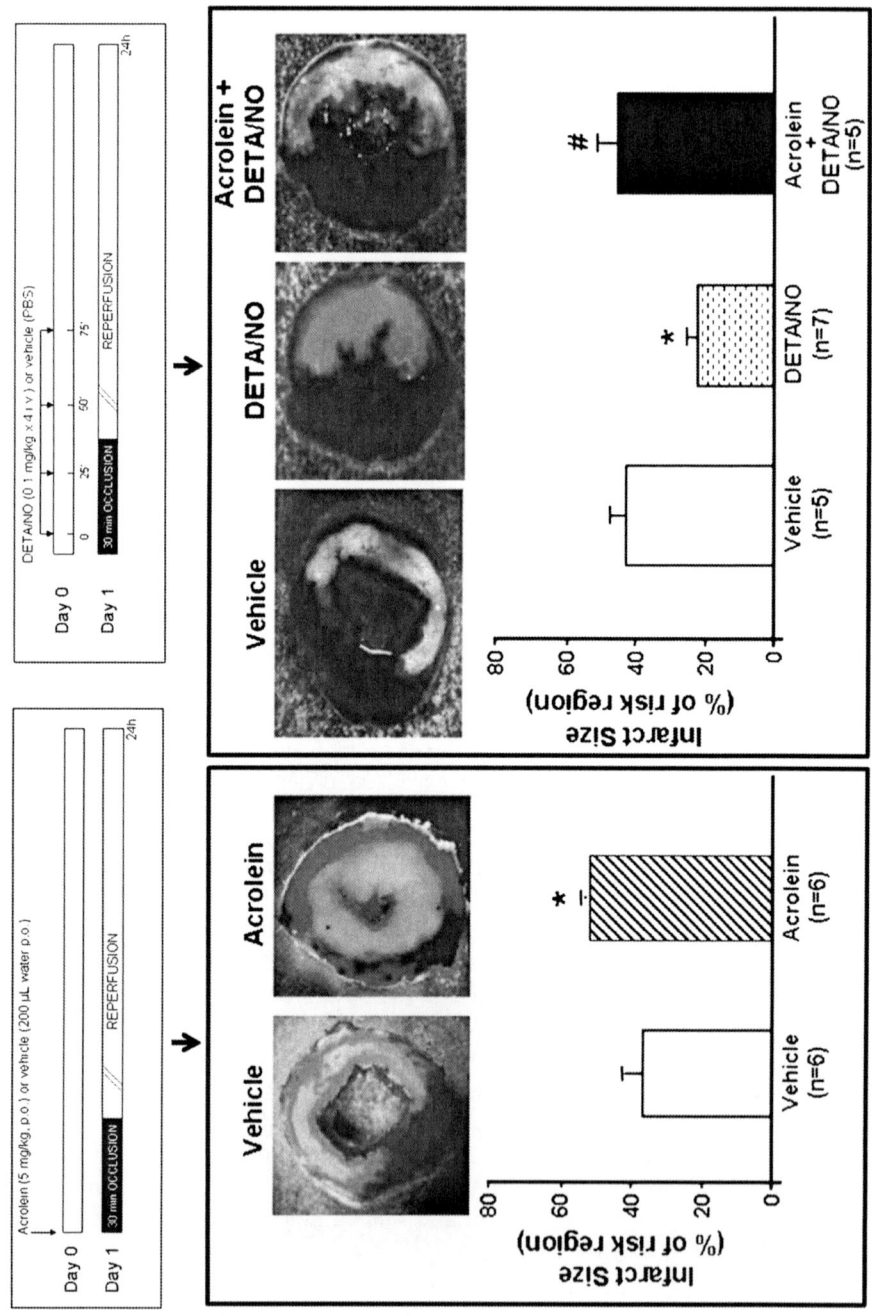

Overall, our studies suggest a direct link between exposure to a specific aldehyde pollutant and the sensitivity of the heart to ischemic injury, further underscoring the unique vulnerability of the heart to pollutant exposure. Importantly, chronic (7 weeks) oral exposure to acrolein at a five-fold lower daily dose (1 mg/kg/day) induced myocardial inflammation and dilated cardiomyopathy.[64]

Taken together, these studies indicated that exposure to acrolein, a prototypical unsaturated aldehyde, is sufficient to induce myocardial dysfunction, augment the sensitivity of the heart to ischemic injury, and trigger pathological ventricular remodeling. Therefore, it is possible that analogous exposure to acrolein and related aldehydes *via* inhalation may significantly contribute to pollutant-associated cardiovascular dysfunction and HF risk. This hypothesis requires direct testing *via* studies of inhalation exposure to acrolein- and aldehyde-containing pollutants in animal models. Nonetheless, these investigations highlight the potential importance of specific chemical pollutants in the exposure mix in the development of untoward cardiac effects in general and the progression of cardiac dysfunction and heart failure in particular. Moreover, our findings provide a strong rationale for future human investigations into the biological mechanisms of pollution-mediated cardiovascular risk to focus on the health impact of specific chemical constituents, including volatile organic compounds such as aldehydes.

Figure 7.7 Acrolein exposure increases myocardial vulnerability to ischemia-reperfusion injury. Left, on day 0, male ICR mice were fed acrolein (5 mg/kg, p.o.) or vehicle by gavage. Twenty four hours later (day 1), the mice underwent 30 min of coronary occlusion followed by 24 of reperfusion, after which time the heart was excised and perfused postmortem and infarct size was determined. The infarcted region was delineated by perfusion with 2,3,5-triphenyltetrazolium chloride (TTC). Noninfarcted tissue takes up TTC (red), whereas infarcted regions are TTC-free (white). Also, to delineate the occluded-reperfused coronary vascular bed, the coronary artery was tied at the site of the previous occlusion and the aortic root perfused with phthalo blue dye. The nonischemic region (region not at risk) is stained blue, whereas the risk region is not stained blue. Infarct size (white area) was measured by videoplanimetry with NIH Image software and expressed as a percentage of the region at risk (red plus white area). Acrolein exposure increased infarct size in naïve ($*p<0.05$ *vs.* vehicle). Right, The same coronary occlusion and reperfusion protocol was performed except that on day 0, male ICR mice were given 4 intravenous boluses of the nitric oxide (NO) donor diethylenetriamine/NO (DETA/NO, 0.1 mg/kg) every 25 min (total dose, 0.4 mg/kg) to induce pharmacological preconditioning. Control mice were administered PBS (vehicle). As expected, preconditioning with DETA/NO significantly reduced infarct size ($*p<0.05$ *vs.* vehicle). However, this protection was blocked by administration of acrolein 2 h prior to DETA/NO treatment ($\#p<0.05$ *vs.* DETA/NO), indicating that NO-mediated preconditioning and subsequent cardioprotection was blocked by acrolein. Reproduced from reference 63, Wang *et al.*, Journal of Molecular and Cellular Cardiology, 2008, with permission from Elsevier Inc.

Acknowledgement

This work was supported by NIH grant ES-11860.

References

1. R. D. Brook, B. Franklin, W. Cascio, Y. Hong, G. Howard, M. Lipsett, R. Luepker, M. Mittleman, J. Samet, S. C. Smith Jr and I. Tager, Air pollution and cardiovascular disease: a statement for healthcare professionals from the Expert Panel on Population and Prevention Science of the American Heart Association, *Circulation*, 2004, **109**, 2655–71.
2. R. D. Brook, Cardiovascular effects of air pollution, *Clin. Sci. (London)*, 2008, **115**, 175–87.
3. R. D. Brook, Is air pollution a cause of cardiovascular disease? Updated review and controversies, *Rev. Environ. Health*, 2007, **22**, 115–37.
4. H. C. Routledge and J. G. Ayres, Air pollution and the heart, *Occup. Med. (London)*, 2005, **55**, 439–47.
5. A. Bhatnagar, Environmental cardiology: studying mechanistic links between pollution and heart disease, *Circ. Res.*, 2006, **99**, 692–705.
6. C. A. Pope 3rd, Epidemiology of fine particulate air pollution and human health: biologic mechanisms and who's at risk?, *Environ. Health Perspect.*, 2000, **108**(Suppl 4), 713–23.
7. C. A. Pope 3rd, R. T. Burnett, G. D. Thurston, M. J. Thun, E. E. Calle, D. Krewski and J. J. Godleski, Cardiovascular mortality and long-term exposure to particulate air pollution: epidemiological evidence of general pathophysiological pathways of disease, *Circulation*, 2004, **109**, 71–7.
8. W. Rosamond, K. Flegal, K. Furie, A. Go, K. Greenlund, N. Haase, S. M. Hailpern, M. Ho, V. Howard, B. Kissela, S. Kittner, D. Lloyd-Jones, M. McDermott, J. Meigs, C. Moy, G. Nichol, C. O'Donnell, V. Roger, P. Sorlie, J. Steinberger, T. Thom, M. Wilson and Y. Hong, American Heart Association Statistics Committee and Stroke Statistics Subcommittee. Heart disease and stroke statistics – 2008 update: a report from the American Heart Association Statistics Committee and Stroke Statistics Subcommittee, *Circulation*, 2008, **117**, e25–146.
9. T. E. Owan, D. O. Hodge, R. M. Herges, S. J. Jacobsen, V. L. Roger and M. M. Redfield, Trends in prevalence and outcome of heart failure with preserved ejection fraction, *N. Engl. J. Med.*, 2006, **355**, 251–9.
10. M. Jessup and S. Brozena, Heart failure, *N. Engl. J. Med.*, 2003, **348**, 2007–18.
11. S. D. Prabhu, Post-infarction ventricular remodeling: an array of molecular events, *J. Mol. Cell Cardiol.*, 2005, **38**, 547–50.
12. D. L. Mann, Mechanisms and models in heart failure: a combinatorial approach, *Circulation*, 1999, **100**, 999–1008.
13. J. J. de Hartog, G. Hoek, A. Peters, K. L. Timonen, A. Ibald-Mulli, B. Brunekreef, J. Heinrich, P. Tiittanen, J. H. van Wijnen, W. Kreyling, M. Kulmala and J. Pekkanen, Effects of fine and ultrafine particles on

cardiorespiratory symptoms in elderly subjects with coronary heart disease: the ULTRA study, *Am. J. Epidemiol.*, 2003, **157**, 613–23.
14. M. S. Goldberg, N. Giannetti, R. T. Burnett, N. E. Mayo, M. F. Valois and J. M. Brophy, A panel study in congestive heart failure to estimate the short-term effects from personal factors and environmental conditions on oxygen saturation and pulse rate, *Occup. Environ. Med*, 2008, **65**, 659–66.
15. M. S. Goldberg, N. Giannetti, R. T. Burnett, N. E. Mayo, M. F. Valois and J. M. Brophy, Shortness of breath at night and health status in congestive heart failure: effects of environmental conditions and health-related and dietary factors, *Environ. Res.*, 2009, **109**, 166–74.
16. G. A. Wellenius, G. Y. Yeh, B. A. Coull, H. H. Suh, R. S. Phillips and M. A. Mittleman, Effects of ambient air pollution on functional status in patients with chronic congestive heart failure: a repeated-measures study, *Environ. Health*, 2007, **6**, 26.
17. H. E. Wichmann, W. Mueller, P. Allhoff, M. Beckmann, N. Bocter, M. J. Csicsaky, M. Jung, B. Molik and G. Schoeneberg, Health effects during a smog episode in West Germany in 1985, *Environ. Health Perspect.*, 1989, **79**, 89–99.
18. G. Rossi, M. A. Vigotti, A. Zanobetti, F. Repetto, V. Gianelle and J. Schwartz, Air pollution and cause-specific mortality in Milan, Italy, 1980–1989, *Arch. Environ. Health*, 1999, **54**, 158–64.
19. G. Hoek, B. Brunekreef, A. Verhoeff, J. van Wijnen and P. Fischer, Daily mortality and air pollution in The Netherlands, *J. Air Waste Manag. Assoc.*, 2000, **50**, 1380–9.
20. G. Hoek, B. Brunekreef, P. Fischer and J. van Wijnen, The association between air pollution and heart failure, arrhythmia, embolism, thrombosis, and other cardiovascular causes of death in a time series study, *Epidemiology*, 2001, **12**, 355–7.
21. M. S. Goldberg, R. T. Burnett, J. C. Bailar 3rd, R. Tamblyn, P. Ernst, K. Flegel, J. Brook, Y. Bonvalot, R. Singh, M. F. Valois and R. Vincent, Identification of persons with cardiorespiratory conditions who are at risk of dying from the acute effects of ambient air particles, *Environ. Health Perspect.*, 2001, **109**(Suppl 4), 487–94.
22. M. S. Goldberg, R. T. Burnett, M. F. Valois, K. Flegel, J. C. Bailar 3rd, J. Brook, R. Vincent and K. Radon, Associations between ambient air pollution and daily mortality among persons with congestive heart failure, *Environ. Res.*, 2003, **91**, 8–20.
23. H. J. Kwon, S. H. Cho, F. Nyberg and G. Pershagen, Effects of ambient air pollution on daily mortality in a cohort of patients with congestive heart failure, *Epidemiology*, 2001, **12**, 413–9.
24. G. A. Wellenius, J. Schwartz and M. A. Mittleman, Particulate air pollution and hospital admissions for congestive heart failure in seven United States cities, *Am. J. Cardiol.*, 2006, **97**, 404–8.
25. F. Dominici, R. D. Peng, M. L. Bell, L. Pham, A. McDermott, S. L. Zeger and J. M. Samet, Fine particulate air pollution and hospital admission for cardiovascular and respiratory diseases, *JAMA*, 2006, **295**, 1127–34.

26. C. A. Pope 3rd, D. G. Renlund, A. G. Kfoury, H. T. May and B. D. Horne, Relation of heart failure hospitalization to exposure to fine particulate air pollution, *Am. J. Cardiol.*, 2008, **102**, 1230–4.
27. A. G. Barnett, G. M. Williams, J. Schwartz, T. L. Best, A. H. Neller, A. L. Petroeschevsky and R. W. Simpson, The effects of air pollution on hospitalizations for cardiovascular disease in elderly people in Australian and New Zealand cities, *Environ. Health Perspect.*, 2006, **114**, 1018–23.
28. D. M. Stieb, M. Szyszkowicz, B. H. Rowe and J. A. Leech, Air pollution and emergency department visits for cardiac and respiratory conditions: a multi-city time-series analysis, *Environ. Health*, 2009, **8**, 25.
29. R. T. Burnett, R. E. Dales, J. R. Brook, M. E. Raizenne and D. Krewski, Association between ambient carbon monoxide levels and hospitalizations for congestive heart failure in the elderly in 10 Canadian cities, *Epidemiology*, 1997, **8**, 162–7.
30. R. D. Morris, E. N. Naumova and R. L. Munasinghe, Ambient air pollution and hospitalization for congestive heart failure among elderly people in seven large US cities, *Am. J. Public Health*, 1995, **85**, 1361–5.
31. C. Y. Yang, Air pollution and hospital admissions for congestive heart failure in a subtropical city: Taipei, Taiwan, *J. Toxicol. Environ. Health A*, 2008, **71**, 1085–90.
32. I. M. Lee, S. S. Tsai, C. K. Ho, H. F. Chiu and C. Y. Yang, Air pollution and hospital admissions for congestive heart failure in a tropical city: Kaohsiung, Taiwan, *Inhal. Toxicol.*, 2007, **19**, 899–904.
33. C. A. Pope 3rd, J. B. Muhlestein, H. T. May, D. G. Renlund, J. L. Anderson and B. D. Horne, Ischemic heart disease events triggered by short-term exposure to fine particulate air pollution, *Circulation*, 2006, **114**, 2443–8.
34. A. Peters, D. W. Dockery, J. E. Muller and M. A. Mittleman, Increased particulate air pollution and the triggering of myocardial infarction, *Circulation*, 2001, **103**, 2810–5.
35. J. B. Ruidavets, M. Cournot, S. Cassadou, M. Giroux, M. Meybeck and J. Ferrières, Ozone air pollution is associated with acute myocardial infarction, *Circulation*, 2005, **111**, 563–9.
36. G. A. Wellenius, T. F. Bateson, M. A. Mittleman and J. Schwartz, Particulate air pollution and the rate of hospitalization for congestive heart failure among medicare beneficiaries in Pittsburgh, Pennsylvania, *Am. J. Epidemiol.*, 2005, **161**, 1030–6.
37. S. von Klot, A. Peters, P. Aalto, T. Bellander, N. Berglind, D. D'Ippoliti, R. Elosua, A. Hörmann, M. Kulmala, T. Lanki, H. Löwel, J. Pekkanen, S. Picciotto, J. Sunyer and F. Forastiere, Health Effects of Particles on Susceptible Subpopulations (HEAPSS) Study Group. Ambient air pollution is associated with increased risk of hospital cardiac readmissions of myocardial-infarction survivors in five European cities, *Circulation*, 2005, **112**, 3073–9.
38. J. K. Mann, I. B. Tager, F. Lurmann, M. Segal, C. P. Quesenberry Jr, M. M. Lugg, J. Shan and S. K. Van Den Eeden, Air pollution and hospital

admissions for ischemic heart disease in persons with congestive heart failure or arrhythmia, *Environ. Health Perspect.*, 2002, **110**, 1247–52.
39. J. L. Peel, K. B. Metzger, M. Klein, W. D. Flanders, J. A. Mulholland and P. E. Tolbert, Ambient air pollution and cardiovascular emergency department visits in potentially sensitive groups, *Am. J. Epidemiol.*, 2007, **165**, 625–33.
40. I. M. Lee, S. S. Tsai, C. K. Ho, H. F. Chiu, T. N. Wu and C. Y. Yang, Air pollution and hospital admissions for congestive heart failure: are there potentially sensitive groups?, *Environ. Res.*, 2008, **108**, 348–53.
41. A. Zanobetti and J. Schwartz, Particulate air pollution, progression, and survival after myocardial infarction, *Environ. Health Perspect.*, 2007, **115**, 769–75.
42. A. Nel, Atmosphere. Air pollution-related illness: effects of particles, *Science*, 2005, **308**, 804–6.
43. A. Sydbom, A. Blomberg, S. Parnia, N. Stenfors, T. Sandström and S. E. Dahlén, Health effects of diesel-exhaust emissions, *Eur. Respir. J.*, 2001, **17**, 733–46.
44. A. Peters, S. von Klot, M. Heier, I. TrentinagliaI, A. Hörmann, H. E. Wichmann and H. Löwel, Cooperative Health Research in the Region of Augsburg Study Group. Exposure to traffic and the onset of myocardial infarction, *N. Engl. J. Med.*, 2004, **351**, 1721–30.
45. M. Medina-Ramón, R. Goldberg, S. Melly, M. A. Mittleman and J. Schwartz, Residential exposure to traffic-related air pollution and survival after heart failure, *Environ. Health Perspect.*, 2008, **116**, 481–5.
46. R. Beelen, G. Hoek, D. Houthuijs, P. A. van den Brandt, R. A. Goldbohm, P. Fischer, L. J. Schouten, B. Armstrong and B. Brunekreef, The joint association of air pollution and noise from road traffic with cardiovascular mortality in a cohort study, *Occup. Environ. Med.*, 2009, **66**, 243–50.
47. V. C. Van Hee, S. D. Adar, A. A. Szpiro, R. G. Barr, D. A. Bluemke, A. V. Diez Roux, E. A. Gill, L. Sheppard and J. D. Kaufman, Exposure to traffic and left ventricular mass and function: the Multi-Ethnic Study of Atherosclerosis, *Am. J. Respir. Crit. Care Med.*, 2009, **179**, 827–34.
48. P. Ponikowski, S. D. Anker, T. P. Chua, R. Szelemej, M. Piepoli, S. Adamopoulos, K. Webb-Peploe, D. Harrington, W. Banasiak, K. Wrabec and A. J. Coats, Depressed heart-rate variability as an independent predictor of death in chronic congestive heart failure secondary to ischemic or idiopathic dilated cardiomyopathy, *Am. J. Cardiol.*, 1997, **79**, 1645–50.
49. J. Nolan, P. D. Batin, R. Andrews, S. J. Lindsay, P. Brooksby, M. Mullen, W. Baig, A. D. Flapan, A. Cowley, R. J. Prescott, J. M. Neilson and K. A. Fox, Prospective study of heart-rate variability and mortality in chronic heart failure: results of the United Kingdom heart failure evaluation and assessment of risk trial (UK-heart), *Circulation*, 1998, **98**, 1510–6.
50. G. R. Sandercock and D. A. Brodie, The role of heart-rate variability in prognosis for different modes of death in chronic heart failure, *Pacing Clin. Electrophysiol.*, 2006, **29**, 892–904.

51. S. D. Prabhu, B. Chandrasekar, D. R. Murray and G. L. Freeman, Beta-adrenergic blockade in developing heart failure: effects on myocardial inflammatory cytokines, nitric oxide, and remodeling, *Circulation*, 2000, **101**, 2103–9.
52. M. R. Bristow, Beta-adrenergic receptor blockade in chronic heart failure, *Circulation*, 2000, **101**, 558–69.
53. D. R. Gold, A. Litonjua, J. Schwartz, E. Lovett, A. Larson, B. Nearing, G. Allen, M. Verrier, R. Cherry and R. Verrier, Ambient pollution and heart-rate variability, *Circulation*, 2000, **101**, 1267–73.
54. J. Schwartz, A. Litonjua, H. Suh, M. Verrier, A. Zanobetti, M. Syring, B. Nearing, R. Verrier, P. Stone, G. MacCallum, F. E. Speizer and D. R. Gold, Traffic related pollution and heart-rate variability in a panel of elderly subjects, *Thorax*, 2005, **60**, 455–61.
55. R. D. Brook, J. R. Brook, B. Urch, R. Vincent, S. Rajagopalan and F. Silverman, Inhalation of fine particulate air pollution and ozone causes acute arterial vasoconstriction in healthy adults, *Circulation*, 2002, **105**, 1534–6.
56. Zanobetti, M. J. Canner, P. H. Stone, J. Schwartz, D. Sher, E. Eagan-Bengston, K. A. Gates, L. H. Hartley, H. Suh, D. R. Gold, Ambient pollution and blood pressure in cardiac rehabilitation patients, *Circulation*, 2004, **110**, 2184–9.
57. D. L. Mann, Inflammatory mediators and the failing heart: past, present, and the foreseeable future, *Circ. Res.*, 2002, **91**, 988–98.
58. S. F. van Eeden, W. C. Tan, T. Suwa, H. Mukae, T. Terashima, T. Fujii, D. Qui, R. Vincent and J. C. Hogg, Cytokines involved in the systemic inflammatory response induced by exposure to particulate-matter air pollutants (PM(10)), *Am. J. Respir. Crit. Care Med.*, 2001, **164**, 826–30.
59. A. Peters, M. Fröhlich, A. Döring, T. Immervoll, H. E. Wichmann, W. L. Hutchinson, M. B. Pepys and W. Koenig, Particulate air pollution is associated with an acute phase response in men; results from the MONICA-Augsburg Study, *Eur. Heart J.*., 2001, **22**, 1198–204.
60. R. Rückerl, S. Greven, P. Ljungman, P. Aalto, C. Antoniades, T. Bellander, N. Berglind, C. Chrysohoou, F. Forastiere, B. Jacquemin, S. von Klot, W. Koenig, H. Küchenhoff, T. Lanki, J. Pekkanen, C. A. Perucci, A. Schneider, J. Sunyer and A. Peters, AIRGENE Study Group. Air pollution and inflammation (interleukin-6, C-reactive protein, fibrinogen) in myocardial-infarction survivors, *Environ. Health Perspect.*, 2007, **115**, 1072–80.
61. M. Jerrett, R. T. Burnett, R. Ma, C. A. Pope 3rd, D. Krewski, K. B. Newbold, G. Thurston, Y. Shi, N. Finkelstein, E. E. Calle and M. J. Thun, Spatial analysis of air pollution and mortality in Los Angeles, *Epidemiology*, 2005, **16**, 727–36.
62. J. Luo, B. G. Hill, Y. Gu, J. Cai, S. Srivastava, A. Bhatnagar and S. D. Prabhu, Mechanisms of acrolein-induced myocardial dysfunction: implications for environmental and endogenous aldehyde exposure, *Am. J. Physiol. Heart Circ. Physiol.*, 2007, **293**, H3673–84.

63. G. W. Wang, Y. Guo, T. M. Vondriska, J. Zhang, S. Zhang, L. L. Tsai, N. C. Zong, R. Bolli, A. Bhatnagar and S. D. Prabhu, Acrolein consumption exacerbates myocardial ischemic injury and blocks nitric oxide-induced PKCepsilon signaling and cardioprotection, *J. Mol. Cell Cardiol.*, 2008, **44**, 1016–22.
64. S. D. Prabhu, S. Srivastava, B. Chandrasekar, G. Yan, D. Bolanowski, R. Ortines, P. P. Ping and A. Bhatnagar, Chronic exposure to acrolein, an environmental Aldehydic pollutant, causes myocardial oxidative stress, inflammation and dilated cardiomyopathy (abstract), *Circulation*, 2003, **108**(17 Suppl), 58.
65. V. J. Feron, H. P. Til, F. de Vrijer, R. A. Woutersen, F. R. Cassee and P. J. van Bladeren, Aldehydes: occurrence, carcinogenic potential, mechanism of action and risk assessment, *Mutat. Res.*, 1991, **259**, 363–385.
66. Committee on Aldehydes. Formaldehyde and other aldehydes. Board on toxicology and environmental health hazards, Assembly of Life Sciences. Washington, D.C.: National Academy Press, 1981, p. 234–41.
67. R. S. G. M. DeWoskin, W. Pepelko and J. Strickland, *Toxicological Review of Acrolein*, US Environmental Protection Agency, Washington, DC, 2003, p. 1–99, (CAS No. 107-02-08).

CHAPTER 8
Ultrafine Particles and Atherosclerosis

J. A. ARAUJO

Division of Cardiology, Department of Medicine, David Geffen School of Medicine, University of California, Los Angeles, CA 90095, USA

8.1 Introduction

Extensive epidemiological evidence supports the association of air pollution with adverse health effects.[1–3] It is increasingly being recognized that such effects lead to enhanced cardiovascular morbidity and mortality,[4] largely due to the exacerbation of acute ischemic events.[5] Indeed, in addition to classical cardiovascular risk factors such as high LDL cholesterol, low HDL cholesterol, hypertension, smoking, diabetes and obesity, among others, recent data support the association of exposure to air particulate matter with atherosclerosis to the extent that PM exposure can be regarded as an additional cardiovascular risk factor. This has been the subject of extensive reviews,[6–8] and a consensus statement from the American Heart Association.[9]

While air pollution is a complex mixture of gaseous (ozone, CO and nitrogen oxides) and particulate components,[9] the strongest evidence coming from a large body of epidemiological literature links the adverse health affects of ambient air pollution with the particulates.[9–15] Particulate matter (PM) is comprised of heterogenous compounds varying in size, number, chemical composition, surface area, concentration and source.[9,10] Among these physicochemical variations, a growing body of evidence suggests a correlation of a smaller particle size with larger adverse cardiopulmonary effects. Particulates < 100 nm, so called ultrafine particles (UFP), are capable of penetrating deeper into the lungs where they can be retained with a high degree of efficiency

---

Issues in Toxicology No. 8
Environmental Cardiology: Pollution and Heart Disease
Edited by Aruni Bhatnagar
© The Royal Society of Chemistry 2011
Published by the Royal Society of Chemistry, www.rsc.org

and be capable of making their high content of combustion-related organic chemicals bioavailable to catalyze the production of reactive oxygen species (ROS), leading to pro-oxidative and proinflammatory effects in the lungs and the cardiovascular system. However, while UFP exhibit the largest proinflammatory and proatherogenic potentials in experimental cellular and animal studies, there is an urgent need for clinical and epidemiological studies to demonstrate whether the same relationship is true for humans. This chapter reviews the biological evidence and discusses the possible factors and potential mechanisms that may explain the enhanced toxicity of ultrafine particles.

8.2 Particulate Matter of the Smallest Size Has the Greatest Pro-Oxidative Potential

PM from different size fractions exhibit significant differences that go beyond their size and number properties, which include the chemical composition, the source that they derive from, the surface area and the bioavailability of their components as important variables that could determine their biological impact. Ultrafine particles are typically derived from fossil-fuel combustion and result from the condensation of vapors. They are typically present in the highest number close to the source of emission (*e.g.* freeways or other driving roads). The newly formed particles in the so-called nucleation mode can undergo coagulation (Aitken mode) and may ultimately become associated with larger particle types. UFP differ from the bigger particles in the fine-particle fraction ($PM_{0.1-2.5}$), where particles grow in size and "accumulate" by coagulation (two particles combining to form one) or by condensation (gas molecules condensing on a particle), and from the coarse particles ($PM_{2.5-10}$), typically derived from soil, road dust, construction debris and mainly formed by the mechanical breakdown (crushing, grinding, abrasion of surfaces) of crustal material, minerals and organic debris.[10]

It appears that a key element that mediates the cardiovascular effects of ambient PM is the ability of PM to trigger and/or enhance free-radical reactions. Various potential pathogenic mechanisms involve free-radical production (by transition metals and organic compounds), oxidative stress, cytokine release, inflammation and covalent modification of key cellular molecules, among others.[6,7,16,17] Indeed, the use of diesel-exhaust particles (DEP) has shown that PM can promote the generation of reactive oxygen species (ROS) in macrophages, bronquial epithelial cells and lung microsomes incubated with the particles or their organic extracts.[18,19] In addition, DEP induce oxidative stress and cellular effects in target vascular cells such as pulmonary artery endothelial cells,[20] rat heart microvessel endothelial cells,[21] human microvascular endothelial cells[22] and human aortic endothelial cells.[23] DEP in the ultrafine size range stimulates superoxide production in human aortic endothelial cells, partly mediated by JNK activation.[23] Interestingly, ultrafine DEP generated by the same engine under different driving cycles can result in particles with different chemical compositions and different abilities to promote oxidative stress and inflammatory events in vascular endothelial cells.[24] Moreover, DEP pro-oxidant effects on endothelial cells can lead to the activation of proinflammatory pathways and unfolded protein response signals in a synergistic fashion with oxidized phospholipids that are generated with the oxidation of low-density lipoproteins.[22]

DEP pro-oxidant effects have been ascribed to their aromatic and polar constituents as these fractions have been shown to be the most active in the induction of antioxidant genes such as heme oxygenase 1 (HO-1), glutathione S-transferase (GST) and other phase-II enzymes that offer protection against oxidative stress in macrophages and epithelial cells.[25] This line of antioxidant defense is regulated by the transcription factor p45-NFE2 related transcription factor 2 (Nrf2) *via* modulation of its proteasomal degradation in the cytosol and translocation into the nucleus.[25] Likewise, the redox potential of ambient PM and its ability to trigger Nrf2-regulated genes vary significantly among particles of different sizes, partly due to its different content of pro-oxidative and electrophilic chemicals. UFP have been shown to exhibit greater redox potential than larger particles.[25,26] Indeed, UFP samples from the Los Angeles basin are more potent than $PM_{2.5}$ or coarse PM at inducing oxidative stress as measured by the dithiothreitol (DTT) assay, leading to a higher degree of induction of HO-1, larger reduction in glutathione to oxidized glutathione (GSH/GSSG) ratios and greater ability to cause mitochondrial damage.[26] UFP larger pro-oxidative effects are associated with their greater content of PM organic carbon and polycyclic aromatic hydrocarbons (PAH), suggesting a role for these organic substances in generating redox activity.[26–28] In addition, there is an excellent correlation between the PAH content of UFP and their ability to engage in redox cycling reactions *in vitro*. UFP collected from various locations and seasons differ significantly in their ability to trigger Nrf2-regulated responses, also in strong relation to their chemical composition and pro-oxidant potential.[26] This could be explained by the variance in PAH and organic carbon content; several of these chemical species are semivolatile and undergo atmospheric changes that could influence the pro-oxidant effects of the smaller particles. Thus, it is possible that various PM sizes and composition resulting in different pro-oxidative potentials could determine their differential abilities to induce cardiovascular proatherogenic effects *in vivo*.

8.3 UFP Activate Proinflammatory Pathways in Vascular Cells

The ability of PM to induce ROS generation and promote pro-oxidant and proinflammatory effects could be key in the promotion of atherogenesis, given the importance of both oxidation and inflammation in the development of atherosclerotic plaques. Indeed, atherosclerosis is a vascular inflammatory process where lipid deposition and oxidation in the artery wall constitute a hallmark of the disease[29–34] (Figure 8.1). Infiltrating lipids come from low-density lipoprotein (LDL) particles that travel into the arterial wall and get

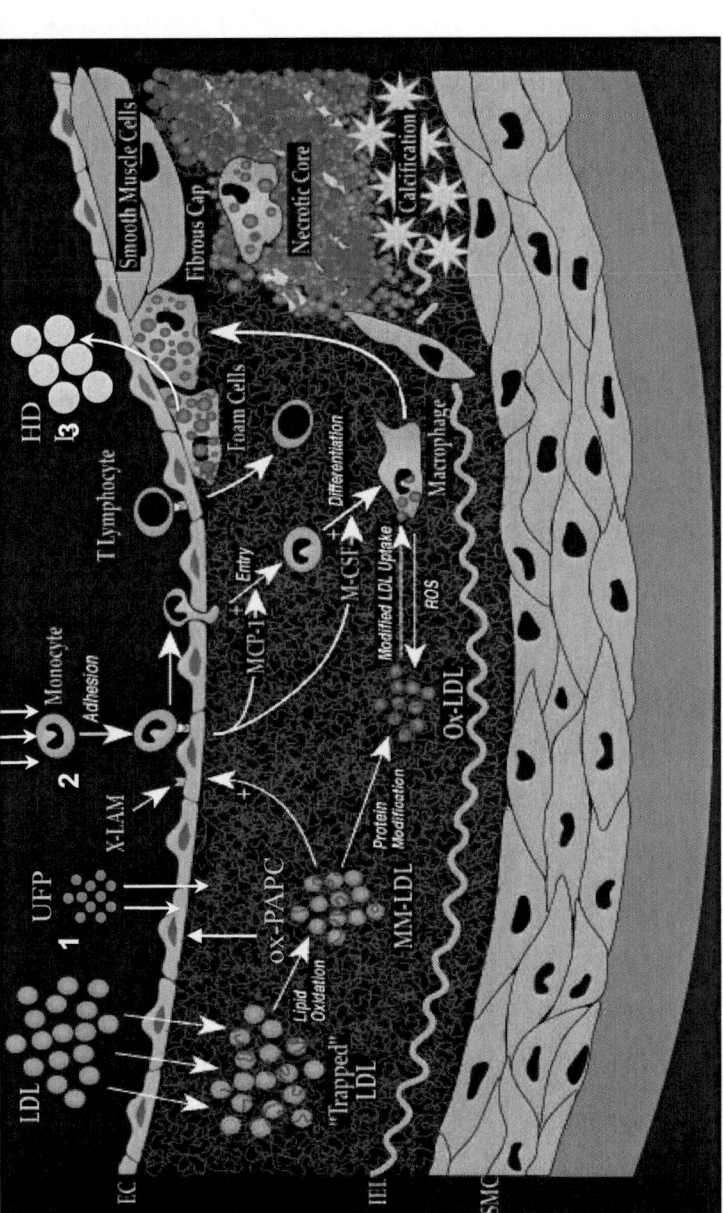

Figure 8.1 UFP and the pathogenesis of atherosclerosis. Lipid infiltration of the artery wall originating from circulating LDL followed by oxidative modification in the subendothelial space, monocyte chemotaxis and foam-cell formation are among the earliest events in atherogenesis. Monocytes differentiate into macrophages, followed by release of inflammatory mediators and a vicious cycle of inflammation. More advanced stages of the disease include smooth muscle cell proliferation, formation of fibrous caps, necrotic cores, calcification, rupture, hemorrhage and thrombosis. Possible mechanisms of how UFP enhances atherosclerosis include: (1) Systemically translocated UFP or their chemical constituents may synergize with ox-PAPC generated within ox-LDL in the activation of proatherogenic molecular pathways in endothelial cells, (2) inflammatory mediators released from the lungs may promote monocyte chemotaxis into the vessels, (3) UFP may induce HDL dysfunction with loss of its anti-inflammatory properties. Taken from Araujo and Nel.[8]

trapped in a three-dimensional cage-work of extracellular fibers and fibrils in the subendothelial space,[35,36] where they are subject to oxidative modifications[37–39] leading to the generation of "minimally modified" LDL (mm-LDL). Such oxidized LDL is capable of activating the overlying endothelial cells to produce proinflammatory molecules such as cell-adhesion molecules, macrophage colony-stimulating factor (M-CSF) and monocyte chemotactic protein-1 (MCP-1)[40–42] that contribute to atherogenesis by recruiting additional monocytes and inducing macrophage differentiation.[30,34,43] Later stages of the disease involve the proliferation of smooth muscle cells, formation of fibrous caps, induction of apoptosis, hemorrhage and calcification,[34] all steps that are the result of a vicious cycle of inflammation where ROS are continuously generated and play a role in disease progression.

The interplay between pro-oxidant and antioxidant elements in the vascular wall may constitute a determining factor in disease pathogenesis. We have shown that factors that increase a vascular pro-oxidative burden such as an iron overload[44] or decreased HO-1 antioxidant protection[45,46] lead to increased proinflammatory effects in vascular cells and enhanced atherosclerosis. There is growing evidence that ambient PM can exert pro-oxidative effects in the vascular wall as well. PM has been shown to promote ROS generation in key cell types involved in atherogenesis, such as endothelial cells,[20–23] macrophages[18,19] and possibly smooth muscle cells.[47] Pro-oxidative effects in vascular endothelial cells can lead to cytotoxic injury[48] and inflammatory events like increased expression of cell-adhesion molecules such as E-and P-selectins, intracellular adhesion molecule-1 (ICAM-1), vascular cell-adhesion molecule-1 (VCAM-1)[49] and increased secretion of proinflammatory cytokines such as MCP-1.[48]

Although the actual contribution of air pollutants to atherogenesis might be small, these effects could be exacerbated by synergizing with other known proatherogenic mediators. Such interactions could be based on synergistic effects with molecular pathways known to mediate disease pathogenesis. Thus, PM could exacerbate the biological activity of well-known pro-oxidative and proatherogenic factors such as ox-LDL. We tested this hypothesis by evaluating the antioxidant response of microvascular endothelial cells to DEP in the absence and presence of oxidized PAPC (1-palmitoyl-2-arachidonyl-sn-glycero-3-phosphorylcholine), one of the key pro-oxidative components generated in LDL particles. While both DEP and ox-PAPC have been shown to exert pro-oxidative effects in vascular cells,[20,21] the combination of both stimuli synergized in increasing antioxidant gene expression, including HO-1.[22] In addition, analysis of the genomic profiles of cells subjected to the various treatments unveiled a large number of genes that reflected the synergy between DEP and ox-PAPC. These genes were grouped into clusters that were enriched in proinflammatory, apoptotic and unfolded protein response (UPR) pathways, all important players in atherogenesis. Some representative examples included proinflammatory genes such as interleukin 8 (IL-8) and chemokine (C-X-C motif) ligand 1 (CXCL1), immune response genes such as Interleukin 11 (IL-11) and UPR genes such as activating transcription factor 4 (ATF 4), heat-shock 70 kDa protein 8 (HSPA8) and X-box binding protein 1 (XBP1).

Importantly, some of these same genes were found to be preferentially upregulated in the livers of apoE$^{-/-}$ mice exposed to UFP, supporting the validity of *in vitro* generated data to predict relevant *in vivo* outcomes. In addition, parallel PM$_{2.5}$ exposures were incapable of inducing gene effects in the livers of the exposed mice,[22] suggesting that the smaller particle size could portray the greatest toxicity *in vivo*. Subsequent sections will discuss the possible explanations for this ultrafine effect.

8.4 UFP Exert Largest PM Proatherogenic Effects

Work with experimental animal models has tested the hypothesis that exposure to PM could lead to enhanced atherosclerosis, either using intratraqueal instillations of PM, inhalation of polluted air, concentrated ambient particles (CAPs) or diesel-exhaust emissions, summarized in Table 8.1. The first evidence was from work at the University of British Columbia in Ontario, Canada.[50] Suwa *et al.* exposed female Watanabe heritable hyperlipidemic rabbits to biweekly intrapharingeal instillations of urban air PM$_{10}$. The degree of aortic and coronary atherosclerosis was evaluated by estimating the atherosclerotic lesion volumes in sections from the various vessels. PM$_{10}$ administration resulted in a 71% increase in the relative volume of atherosclerotic lesions in the coronary arteries and a 62% increase in the relative volume of extracellular lipids in the aorta. A more recent study from the same group showed that the PM$_{10}$-mediated promotion of atherosclerotic plaques in the aorta (Table 8.1) is likely the result of enhanced recruitment of circulating monocytes into atherosclerotic plaques.[51]

Long-term exposures to inhaled concentrated PM$_{2.5}$ also result in increased aortic atherosclerosis as determined by five studies where apoE null mice have been exposed to concentrated ambient particles (CAPs) at Sterling Forest, New York[47,52–54] or in Manhattan, New York.[55] Chen and Nadziejko reported in the first study in 2005 that 39–41-week-old apoE$^{-/-}$ mice fed a chow diet and exposed to 10×ambient concentrations of PM$_{2.5}$ for 6 h per day, 5 days per week for 5 months at an exposure facility in Sterling Forest, exhibited a 57% increase in the percentage of atherosclerotic plaque area in the aortic root[52] (Table 8.1). Sun *et al.* reported in another study published the same year that younger 6-week-old apoE$^{-/-}$ mice fed a chow diet and exposed to similar conditions for 6 months displayed a 45% increase in the percentage of aortic atherosclerotic plaque area, that was not statistically significant, however.[54] The feeding of a high-fat diet resulted in a greater promotion of atherosclerosis that was statistically significant, as PM$_{2.5}$-exposed mice displayed a 58% increase in aortic root plaque area,[54] together with impaired vasomotor response. In another study, Sun *et al.* reported that apoE$^{-/-}$ mice fed a high-fat diet and exposed to concentrated PM$_{2.5}$ also resulted in enhancement of aortic atherosclerosis as assessed by the percentage of the plaque area in the aortic arch by ultrasound biomicroscopy.[47] PM$_{2.5}$ exposures led to atherosclerotic plaques that were also richer in tissue factor,[47] which could play a causative role

Table 8.1 Animal studies evaluating the effect of air pollution on atherosclerosis.

Study	PM fraction (mode of administration)	Animal model	Diet	Assessment of atherosclerosis (method)	Effect of PM on atherosclerosis
Suwa et al. 2002[50]	PM_{10} (I.T.)	Watanabe rabbits	Chow	% lesional volume in coronary arteries and aorta (histology)	Increase
Chen and Nadziejko 2005[52]	$PM_{2.5}$ (Inhaled CAPs)	apoE$^{-/-}$, LDL$^{-/-}$ mice & apoE$^{-/-}$ mice	Chow	% lesional area in whole aorta (histology)	No change (apoE$^{-/-}$, LDL$^{-/-}$ mice) Increase (apoE$^{-/-}$ mice)
Sun et al. 2005[54]	$PM_{2.5}$ (Inhaled CAPs)	apoE$^{-/-}$ mice	Chow or CED	% lesional area in cross sections of aorta (histology)	N.S. increase (chow-fed mice) Increase (CED-fed mice)
Niwa et al. 2007[88]	Carbon black (I.T.)	LDLR$^{-/-}$ mice	CED	% lesional area in whole aorta (histology)	Increase
Sun et al. 2008[47]	$PM_{2.5}$ (Inhaled CAPs)	apoE$^{-/-}$ mice	Chow CED	% lesional area in aorta (ultrasound)	N.S. increase (chow-fed mice) Increase (CED-fed mice)
Yatera et al. 2008[51]	PM_{10} (I.T.)	Watanabe rabbits	Chow	% lesional volume and % lesional area in the aorta (histology)	Increase

Study	Exposure	Model	Diet	Endpoint	Result
Araujo et al. 2008[56]	PM$_{2.5}$ & UFP (Inhaled CAPs)	apoE$^{-/-}$ mice	Chow	Mean lesional area in aortic root (histology)	N.S. increase (PM$_{2.5}$-exposed mice) Increase (UFP-exposed mice)
Soares et al. 2009[89]	PM (inhaled polluted air)	LDLR$^{-/-}$ mice	Chow	% lesional area in aortic root	No change
			CED	Aortic wall thickness (histology)	Increase
Yin et al. 2009[55]	PM$_{2.5}$ (Inhaled CAPs)	apoE$^{-/-}$ mice	CED	% lesional area in cross sections of aorta (histology)	Increase
Campen et al. 2010[57]	PM + gases (Inhaled DE)	apoE$^{-/-}$ mice	CED	% lesional area in the aortic root	N.S. increase.
				Plaque composition	Increase in macrophage content
Chen et al. 2010[53]	PM$_{2.5}$ (Inhaled CAPs)	apoE$^{-/-}$ mice	Chow	% lesional area in brachio-cephalic and left common arteries (ultrasound)	Increase

Studies are shown in chronological order based on the year of publication. I.T. = Intratracheal, CAPs = Concentrated ambient particles, DE = Diesel emissions, CED = Cholesterol enriched diet, N.S. = Not significant.

or simply be an indicator of greater atherosclerotic plaque burden. When compared to inhaled sidestream tobacco smoke, concentrated $PM_{2.5}$ appears to exert greater promotion of plaque formation as assessed by ultrasound biomicroscopy.[53] Importantly, concentrated $PM_{2.5}$ from a different location, such as an exposure laboratory in Manhattan, also resulted in similar proatherogenic effects.[55] Despite the use of a relatively small number of sex-matched animals ($n<10$/group) in the assessment of vascular pathology that has a large phenotypic variance such as aortic atherosclerosis, the consistent reproducibility of $PM_{2.5}$ proatherogenic effects in all these studies decrease the potential for type-I error.

As the PM_{10} and $PM_{2.5}$ fractions contain a large number of ultrafine particles and based on the greater redox potential of UFP as well as their greater ability to induce synergistic effects with cholesterol-enriched diets *in vivo* as discussed in the previous section, we hypothesized that the UFP fraction could concentrate some of the PM proatherogenic effects and could be more proatherogenic than $PM_{2.5}$. We tested this notion in an animal study at the University of California in Los Angeles and Southern California Particle Center (SCPC). Araujo *et al.* exposed apoE$^{-/-}$ mice fed a chow diet to concentrated $PM_{2.5}$, UFP or filtered air for 5 h/day, 3 days per week for 5 weeks. UFP-exposed mice developed 25% and 55% greater aortic atherosclerotic plaques, assessed by the mean atherosclerotic lesional area in the aortic root, as compared to $PM_{2.5}$ or FA-exposed mice, respectively.[56] Although $PM_{2.5}$ exposures tended to enhance lesion formation over FA controls, resulting in a 23% increment in lesional area, these effects were relatively smaller than in the previous studies, which could be due to the shorter duration of exposures. The magnitude of proatherogenic effects was enhanced by UFP exposures, which led to a comparable degree of atherogenesis enhancement as 5–6-month-long $PM_{2.5}$ exposures.[47,52,54] One limitation of our study is that the concentrator technology employed (VACES: Versatile Aerosol Concentration Enrichment System) generates overlapping CAPs aerosols, which makes it difficult to estimate the true relative proatherogenic strength of UFP *vs.* $PM_{2.5}$. It is certainly possible that the UFP fraction could concentrate the PM proatherogenic effects, but a clear demonstration of this notion will require the straight comparison of UFP to the accumulation mode particles in the 0.1–2.5 µm range, which would warrant a different concentration technology.[8] Additional studies are required to compare $PM_{2.5}$ and UFP exposures in other locations in the world where the effects of different particle compositions on atherosclerosis, not only size, can be assessed. In addition, experimental designs need to include various doses and time points that allow assessment of toxicological parameters of importance such as threshold, latency of effects or potency. In addition, the interaction of particulates with gaseous components need to be assessed as well as the effects of particulates from various sources such as gasoline exhausts or diesel emissions could be affected by the concomitant presence of gas species present in those emissions, reported to lead to changes in the composition of atherosclerotic plaques[57] (Table 8.1).

8.5 How Do Pro-Oxidative UFP Enhance Atherosclerosis?

PM-induced atherosclerosis could likely be the result of systemic pro-oxidant and proinflammatory effects at vascular sites. Inhalation exposures to concentrated $PM_{2.5}$ have led to enhanced ROS generation in the aortic plaques, increased formation of 3-nitrotyrosine residues,[54,55] upregulation of the NADPH oxidase subunits p47phox and Rac1[55] as well as increased hepatic oxidative stress.[56] It is possible that UFP exposures could lead to greater systemic pro-oxidant and proinflammatory effects than does the larger $PM_{2.5}$, as judged by their greater ability to induce the upregulation of Nrf2-regulated antioxidant genes and UPR genes in the liver,[22,56] in parallel with their greater proatherogenic potential.

Five-week exposures to concentrated PM also led to the development of dysfunctional HDL,[56] a condition in which circulating HDL particles lose their antioxidant, anti-inflammatory and/or cholesterol-reverse transport capacity. The degree of HDL dysfunction can be induced and assessed as a continuous variable with a critical point at which, HDL promotes oxidation and/or inflammation rather than protecting against it. The UCLA study showed that UFP exposures can lead to a greater degree of HDL dysfunction than does $PM_{2.5}$.[56] While $PM_{2.5}$ exposures resulted in mere loss of HDL protective qualities, UFP exposures turned HDL into proinflammatory particles as assessed by a coculture monocyte chemotactic assay. This greater degree of HDL dysfunction correlated with the greater degree of atherosclerosis and a systemic Nrf2-regulated antioxidant response, supporting the idea that all these effects could be due to common mechanisms. Thus, dysfunctional HDL can be the result of systemic pro-oxidant and/or proinflammatory effects, it could mediate some of these effects *per se* or it could at least function as a marker of systemic inflammation. Regardless of its specific mode of generation, it may play a role in disease pathogenesis. It is well established that plasma HDL cholesterol and apoA1 levels are inversely correlated with the risk for coronary artery disease,[58–60] due to the well-characterized ability to promote reverse cholesterol transport[61] and protect against oxidation, inflammation and thrombotic activities.[62–64] However, high levels are not always protective in subjects, suggesting that not all HDLs prevent atherosclerosis.[65] Dysfunctional proinflammatory HDL may serve as a useful marker for predicting susceptibility for atherosclerosis in humans[66] and in rabbits,[37] where it has been found to be a better predictor than total or LDL cholesterol levels. It will be important to characterize the type of alterations that UFP exposures induce in HDL particles and confirm whether those occur in human subjects also. This can help to elucidate pathogenesis and may even lead to the identification of a potential biomarker.

Systemic proinflammatory effects have also been demonstrated in C57BL6 mice exposed to 5-month $PM_{2.5}$ inhalations, which exhibited increased levels of inflammatory markers such as tumor necrosis factor-α (TNF-α), interleukin-6 (IL-6), E-selectin and ICAM-1,[67] in association with increased visceral adiposity and insulin resistance, supporting a possible link between air pollutants and type-2 diabetes mellitus.[67] These findings are in agreement with evidence obtained with intrapharingeal administration of other types of PM in other animal models. Thus, the administration of I.T. PM_{10} to Watanabe rabbits led to the elevation of polymorphonuclear cells and circulating band cell counts 2 weeks after starting the exposures, with an increase in the size of the bone marrow mitotic pool as assessed by BrdU-labeling.[50] This is in support of the notion that PM can induce systemic inflammation. Likewise, I.T. instillation of residual oil fly ash (ROFA) in rats led to a greater degree of vascular ROS generation, resulting in a dose-dependent impairment of systemic endothelium-dependent arterial dilation, increased leukocyte rolling and adhesion in the spinotrapezius muscle microcirculation.[68,69]

Despite a large degree of uncertainty on the toxicological principle components of the particles, several factors could help to explain the greater cardiovascular toxicity portrayed by a small particle size,[8] such as the following.

8.5.1 Larger Particle Number

Particles in the ultrafine size range account for > 85–90% of the total $PM_{2.5}$ particle number[70] despite contributing very little to overall $PM_{2.5}$ mass. It is conceivable that larger particle numbers of a small particle size would result in better fractional deposition in the lungs and greater cardiovascular effects. However, little evidence has been reported in support of this since most studies have used the more traditional measures of PM exposure based on assessment of mass per space (*e.g.* daily, annual, average or peak µg/m^3) that might not reflect UFP exposures. In the UCLA study, the only animal study where the proatherogenic potential of UFP has been assessed to date, development of larger atherosclerotic lesions in the UFP exposures correlated with increased particle numbers rather than with overall PM mass.[56]

8.5.2 Greater Lung Retention

Small particle size allows better penetrability and diffusion into the lungs according to the IRCP 1994 model for particle deposition in the respiratory tract[71] that defines three anatomical regions for fractional deposition of particles during nose breathing (nasopharyngeal, tracheobronchial and alveolar regions). Indeed, UFP exhibit a greater fractional deposition in the tracheobronchial and alveolar regions than bigger particles. The greater retention of UFP could be partly due to increased Van der Waals forces, electrostatic or steric interactions, all of which contributes to "adhesive interactions".[72,73]

8.5.3 Larger Content of Redox Active Compounds

PM is constituted of hundreds of organic chemicals and transition metals that could trigger or exacerbate free-radical reactions that may be responsible for the systemic proatherogenic effects. In particular, UFP are enriched in organic carbon content as well as pro-oxidative PAHs that could lead to oxidative stress, inflammation and increase in atherosclerotic lesion development. Whether or not these compounds are the actual toxicants contributing to atherogenesis, they may serve as a proxy for the substances.

8.5.4 Greater Bioavailability

UFP exhibit a larger surface-to-mass ratio than the larger particles that could lead to a sizable increase in bioavailable surface.[74] The number of atoms or molecules that are displayed and packaged on the surface of particles increase exponentially as the size shrinks below 100 nm.[75] Thus, the presence of bioreactive chemicals (*e.g.* PAHs, transition metals) on the large surface area of UFP could make these chemicals more bioavailable at the sites of contact of the particles with cells and tissues where subsequent ROS generation takes place. This greater bioavailability could make UFP chemically more active than larger particles.

Despite the demonstrated pro-oxidative potential of PM in the *in vitro* studies, the generation of systemic pro-oxidative and proinflammatory effects *in vivo*, and the putative reasons that could make UFP more proatherogenic than larger particles, it is unclear how the exposure to PM, either by inhalation or by I.T. instillations, could result in those effects. Various possibilities to be considered include (1) systemic translocation of UFP and/or their chemical constituents that could synergize with ox-OPAPC generated within ox-LDL in the activation of proatherogenic pathways, (2) development of pulmonary inflammation with the release of inflammatory mediators into the systemic circulation and promotion of monocyte migration into the vasculature, (3) effects of circulating lipoproteins and development of HDL dysfunction with the loss of its antioxidant and anti-inflammatory properties. These possibilities are schematized in Figure 8.1 and belong to "general mediating" pathways shown in Figure 2.1 of Chapter 2. Various lines of experimental data suggest, although not conclusively yet, the plausibility for each one of these mechanisms as recently reviewed by us.[8]

8.6 Do UFP Enhance Atherosclerosis in Humans?

Exposure to PM has been associated with various measures of human atherosclerosis. Thus, Kunzli *et al.*[76] found a 5.9% increase in carotid intima-medial thickness for every 10 µg/m^3 rise in PM$_{2.5}$ levels[76] in a cross-sectional study that included 798 individuals where the degree of carotid intima-media thickness (CIMT) was correlated with the levels of annual mean concentrations of ambient PM$_{2.5}$ (Table 8.2). A recent study on a related population also showed that the annual rate of progression of CIMT among individuals living within 100 m of a highway was accelerated and more than twice the population mean progression.[77] Likewise, Hoffman *et al.* reported an association between

Table 8.2 Human studies linking air pollution exposure with atherosclerosis.

Study	Air pollutant	Evaluation of atherosclerosis	Major Findings	Reference
Kunzli et al. 2004	$PM_{2.5}$ Ozone	CIMT	5.9% increased in CIMT per 10 µg $PM_{2.5}/m^3$	76
Hoffman et al. 2007	$PM_{2.5}$ Distance to major road	CACS	Increased CAC scores with shorter distances to a major road	78
Diez Roux et al. 2008	PM_{10} $PM_{2.5}$	CIMT CACS BAI	1–3% increase in CIMT per 21 and 12.5 µg/m³ in PM_{10} and $PM_{2.5}$ respectively	79
Allen et al. 2009	$PM_{2.5}$ Distance to major road	Aortic calcification	6% increase in the risk of aortic calcification with a 10-µg/m³ contrast in $PM_{2.5}$	90
Kunzli et al. 2010	$PM_{2.5}$ Distance to highway or major road	CIMT	Greater annual progression of CIMT among individuals living <100 m from a highway	77

Studies are listed in chronological order based on the year of publication. CIMT = Carotid intima-media thickness, CACS = coronary artery calcium score, BAI = Brachial artery index.

long-term residential exposure to high traffic and coronary atherosclerosis, as assessed by coronary artery calcification scores.[78] They found in a German cohort study including 4494 participants, that as compared with subjects living >200 meters away from a major road, subjects living within 101 to 200 m, 51 to 100 or less than 50 m showed 8%, 34% and 63% increase in the probability of having a high coronary artery calcification (CAC) score, respectively. Data from the Multi-Ethnic Study of Atherosclerosis[79] also support the association of PM with atherosclerosis. Diez Roux et al. reported that PM_{10} exposures assessed over long-term (20-year means and 2001 mean) and 20-year $PM_{2.5}$ exposures did correlate with a 1–3% increase in CIMT per 21 µg/m³ increase in PM_{10} or 12.5 µg/m³ increase in $PM_{2.5}$, respectively. Likewise, Allen et al. reported that $PM_{2.5}$ exposures correlated with increased risk for aortic calcification in a related study (Table 8.2).

While epidemiological studies with PM_{10} and $PM_{2.5}$ data support that a smaller particle size correlates with larger cardiovascular effects, there are only few reports supporting the association of UFP with increased total or cardiorespiratory mortality[80,81] and there are no reports that support an association between UFP and atherosclerosis yet. This lack of conclusive evidence is partly due to the difficulty that represents the reliable measurement of UFP particle number and mass concentrations since they are very dependent on the proximity to the source of generation. In addition, routine air-pollution monitoring does not include measurement of UFP parameters, which has led to the unavailability of exposure metrics data and therefore, the difficulty to conduct studies on health effects that could prompt the delineation of standards for the regulation of UFPs.

Despite the lack of direct evidence, it has been suggested that associations between PM and CV mortality, CV morbidity and atherosclerosis could be even stronger with the UFP fractions for reasons that were discussed in the previous sections. Delfino *et al.* studied residents in independent living facilities of four large retirement communities in the Los Angeles air basin,[82] 60 elderly subjects with a history of CAD were followed up over 7-month periods with a very detailed pollutant-exposure characterization and blood collection for the determination of systemic inflammatory, antioxidant and coagulation markers. They found positive associations of particle number (dominated by UFPs) and outdoor quasi-ultrafine $PM_{0.25}$ ($<0.25\,\mu m$) with biomarkers of systemic inflammation such as interleukin (IL-6), soluble tumor necrosis factor receptor II (sTNF-RII) and C-reactive protein (CRP).[82,83] This study is in agreement with a previous report where exposure to UFP correlated with increased plasma levels of soluble CD40ligand (sCD40L), a marker for platelet activation that can cause increased coagulation and inflammation[84] and other repeated-measure studies showing associations between ambient air pollution and biomarkers of systemic inflammation in healthy young adults[85] and susceptible subjects with CAD.[86,87] Thus, there is need for human studies that evaluate the potential links between UFP exposure parameters and clinical measures of atherosclerosis, especially given the *in vitro* cellular and *in vivo* animal experimental work that support the greater cardiovascular toxicity of UFP.

8.7 Conclusions

Cumulative epidemiological data supports the association of exposure to air pollution with cardiovascular morbidity and mortality, mostly in relation to its particulate matter components. Experimental animal work using hypercholesterolemic rabbits, apoE null and LDLR null mice shows that ambient PM exposure promotes atherosclerotic lesion formation and it appears that the smaller the particles, the greater the proatherogenic effects. Enhancement of atherosclerosis correlates with the induction of systemic pro-oxidant and proinflammatory effects although the mechanisms for transduction of these effects are not clear. UFP particles may be more toxic based on their greater number, larger content of redox active compounds such as PAHs, greater surface-to-mass ratio and larger bioavailability of chemically active constituents. Although there is no direct epidemiological or clinical data supporting the association of UFP with atherosclerosis yet, it is likely that exposure to UFP could result in greater associations than with larger particles. Much work is needed to better characterize the main toxic compounds, mechanism(s) of pathogenesis, types of genetic susceptibility that exposed individuals may exhibit and degree of associations of UFP with cardiovascular mortality, cardiovascular morbidity and human atherosclerosis.

Acknowledgments

Writing of this article was supported by the National Institute of Environmental Health Sciences, NIH (RO1 ES016959 to Jesus A. Araujo).

References

1. C. A. Pope 3rd, M. J. Thun, M. M. Namboodiri, D. W. Dockery, J. S. Evans, F. E. Speizer and C. W. Heath Jr, Particulate air pollution as a predictor of mortality in a prospective study of US adults, *Am. J. Respir. Crit. Care Med.*, 1995, **151**, 669–674.
2. J. M. Samet, F. Dominici, F. C. Curriero, I. Coursac and S. L. Zeger, Fine particulate air pollution and mortality in 20 US cities, 1987–1994, *N. Engl. J. Med.*, 2000, **343**, 1742–1749.
3. D. W. Dockery, C. A. Pope 3rd, X. Xu, J. D. Spengler, J. H. Ware, M. E. Fay, B. G. Ferris Jr and F. E. Speizer, An association between air pollution and mortality in six US cities, *N. Engl. J. Med.*, 1993, **329**, 1753–1759.
4. C. A. Pope, R. T. Burnett, G. D. Thurston, M. J. Thun, E. E. Calle, D. Krewski and J. J. Godleski, Cardiovascular mortality and long-term exposure to particulate air pollution: epidemiological evidence of general pathophysiological pathways of disease, *Circulation*, 2004, **109**, 71–77.
5. K. A. Miller, D. S. Siscovick, L. Sheppard, K. Shepherd, J. H. Sullivan, G. L. Anderson and J. D. Kaufman, Long-term exposure to air pollution and incidence of cardiovascular events in women, *N. Engl. J. Med.*, 2007, **356**, 447–458.
6. A. Bhatnagar, Environmental cardiology: studying mechanistic links between pollution and heart disease, *Circ. Res.*, 2006, **99**, 692–705.
7. R. D. Brook, Cardiovascular effects of air pollution, *Clin. Sci. (London)*, 2008, **115**, 175–187.
8. J. A. Araujo and A. E. Nel, Particulate matter and atherosclerosis: role of particle size, composition and oxidative stress, *Part. Fibre Toxicol.*, 2009, **6**, 24.
9. R. D. Brook, B. Franklin, W. Cascio, Y. Hong, G. Howard, M. Lipsett, R. Luepker, M. Mittleman, J. Samet, S. C. Smith Jr and I. Tager, Air pollution and cardiovascular disease: a statement for healthcare professionals from the Expert Panel on Population and Prevention Science of the American Heart Association, *Circulation*, 2004, **109**, 2655–2671.
10. USEPA, Air quality criteria for particulate matter (EPA/600/P-99/002aF), ed. W. US Environmental Protection Agengy, DC, 2004.
11. C. A. Pope 3rd, R. L. Verrier, E. G. Lovett, A. C. Larson, M. E. Raizenne, R. E. Kanner, J. Schwartz, G. M. Villegas, D. R. Gold and D. W. Dockery,

Heart-rate variability associated with particulate air pollution, *Am. Heart J.*, 1999, **138**, 890–899.
12. D. R. Gold, A. Litonjua, J. Schwartz, E. Lovett, A. Larson, B. Nearing, G. Allen, M. Verrier, R. Cherry and R. Verrier, Ambient pollution and heart-rate variability, *Circulation*, 2000, **101**, 1267–1273.
13. A. Peters, S. Perz, A. Doring, J. Stieber, W. Koenig and H. E. Wichmann, Increases in heart rate during an air pollution episode, *Am. J. Epidemiol.*, 1999, **150**, 1094–1098.
14. A. Ibald-Mulli, J. Stieber, H. E. Wichmann, W. Koenig and A. Peters, Effects of air pollution on blood pressure: a population-based approach, *Am. J. Public Health*, 2001, **91**, 571–577.
15. R. D. Brook, J. R. Brook, B. Urch, R. Vincent, S. Rajagopalan and F. Silverman, Inhalation of fine particulate air pollution and ozone causes acute arterial vasoconstriction in healthy adults, *Circulation*, 2002, **105**, 1534–1536.
16. N. L. Mills, K. Donaldson, P. W. Hadoke, N. A. Boon, W. MacNee, F. R. Cassee, T. Sandstrom, A. Blomberg and D. E. Newby, Adverse cardiovascular effects of air pollution, *Nat. Clin. Pract. Cardiovasc. Med.*, 2009, **6**, 36–44.
17. A. E. Nel, D. Diaz-Sanchez, D. Ng, T. Hiura and A. Saxon, Enhancement of allergic inflammation by the interaction between diesel-exhaust particles and the immune system, *J. Allergy Clin. Immunol.*, 1998, **102**, 539–554.
18. T. S. Hiura, M. P. Kaszubowski, N. Li and A. E. Nel, Chemicals in diesel-exhaust particles generate reactive oxygen radicals and induce apoptosis in macrophages, *J. Immunol.*, 1999, **163**, 5582–5591.
19. N. Li, M. Wang, T. D. Oberley, J. M. Sempf and A. E. Nel, Comparison of the pro-oxidative and proinflammatory effects of organic diesel-exhaust particle chemicals in bronchial epithelial cells and macrophages, *J. Immunol.*, 2002, **169**, 4531–4541.
20. Y. wBai, A. K. Suzuki and M. Sagai, The cytotoxic effects of diesel-exhaust particles on human pulmonary artery endothelial cells *in vitro*: role of active oxygen species, *Free Radic. Biol. Med.*, 2001, **30**, 555–562.
21. S. Hirano, A. Furuyama, E. Koike and T. Kobayashi, Oxidative-stress potency of organic extracts of diesel exhaust and urban fine particles in rat heart microvessel endothelial cells, *Toxicology*, 2003, **187**, 161–170.
22. K. W. Gong, W. Zhao, N. Li, B. Barajas, M. Kleinman, C. Sioutas, S. Horvath, A. J. Lusis, A. Nel and J. A. Araujo, Air-pollutant chemicals and oxidized lipids exhibit genome-wide synergistic effects on endothelial cells, *Genome Biol.*, 2007, **8**, R149.
23. R. Li, Z. Ning, J. Cui, B. Khalsa, L. Ai, W. Takabe, T. Beebe, R. Majumdar, C. Sioutas and T. Hsiai, Ultrafine particles from diesel engines induce vascular oxidative stress *via* JNK activation, *Free Radic. Biol. Med.*, 2009, **46**, 775–782.
24. R. Li, Z. Ning, R. Majumdar, J. Cui, W. Takabe, N. Jen, C. Sioutas and T. Hsiai, Ultrafine particles from diesel vehicle emissions at different

driving cycles induce differential vascular proinflammatory responses: implication of chemical components and NF-kappaB signaling, *Part. Fibre Toxicol.*, **7**, 6.
25. N. Li, J. Alam, M. I. Venkatesan, A. Eiguren-Fernandez, D. Schmitz, E. Di Stefano, N. Slaughter, E. Killeen, X. Wang, A. Huang, M. Wang, A. H. Miguel, A. Cho, C. Sioutas and A. E. Nel, Nrf2 is a key transcription factor that regulates antioxidant defense in macrophages and epithelial cells: protecting against the proinflammatory and oxidizing effects of diesel exhaust chemicals, *J. Immunol.*, 2004, **173**, 3467–3481.
26. N. Li, C. Sioutas, A. Cho, D. Schmitz, C. Misra, J. Sempf, M. Wang, T. Oberley, J. Froines and A. Nel, Ultrafine particulate pollutants induce oxidative stress and mitochondrial damage, *Environ. Health Perspect.*, 2003, **111**, 455–460.
27. L. Ntziachristos, J. Froines, A. Cho and C. Sioutas, Relationship between redox activity and chemical speciation of size-fractionated particulate matter, *Part. Fibre Toxicol.*, 2007, **4**, 5.
28. J. G. Ayres, P. Borm, F. R. Cassee, V. Castranova, K. Donaldson, A. Ghio, R. M. Harrison, R. Hider, F. Kelly, I. M. Kooter, F. Marano, R. L. Maynard, I. Mudway, A. Nel, C. Sioutas, S. Smith, A. Baeza-Squiban, A. Cho, S. Duggan and J. Froines, Evaluating the toxicity of airborne particulate matter and nanoparticles by measuring oxidative stress potential" a workshop report and consensus statement, *Inhal. Toxicol.*, 2008, **20**, 75–99.
29. R. Ross, Atherosclerosis – an inflammatory disease, *N. Engl. J. Med.*, 1999, **340**, 115–126.
30. C. K. Glass and J. L. Witztum, Atherosclerosis. the road ahead, *Cell*, 2001, **104**, 503–516.
31. A. D. Watson, J. A. Berliner, S. Y. Hama, B. N. La Du, K. F. Faull, A. M. Fogelman and M. Navab, Protective effect of high density lipoprotein associated paraoxonase. Inhibition of the biological activity of minimally oxidized low density lipoprotein, *J. Clin. Invest.*, 1995, **96**, 2882–2891.
32. H. C. Stary, A. B. Chandler, R. E. Dinsmore, V. Fuster, S. Glagov, W. Insull Jr, M. E. Rosenfeld, C. J. Schwartz, W. D. Wagner and R. W. Wissler, A definition of advanced types of atherosclerotic lesions and a histological classification of atherosclerosis. A report from the Committee on Vascular Lesions of the Council on Arteriosclerosis, *American Heart Association, Circulation*, 1995, **92**, 1355–1374.
33. H. C. Stary, A. B. Chandler, S. Glagov, J. R. Guyton, W. Insull Jr, M. E. Rosenfeld, S. A. Schaffer, C. J. Schwartz, W. D. Wagner and R. W. Wissler, A definition of initial, fatty streak, and intermediate lesions of atherosclerosis. A report from the Committee on Vascular Lesions of the Council on Arteriosclerosis, American Heart Association, *Circulation*, 1994, **89**, 2462–2478.
34. A. J. Lusis, Atherosclerosis, *Nature*, 2000, **407**, 233–241.
35. G. Camejo, S. O. Olofsson, F. Lopez, P. Carlsson and G. Bondjers, Identification of Apo B-100 segments mediating the interaction of low

density lipoproteins with arterial proteoglycans, *Arteriosclerosis*, 1988, **8**, 368–377.
36. J. S. Frank and A. M. Fogelman, Ultrastructure of the intima in WHHL and cholesterol-fed rabbit aortas prepared by ultra-rapid freezing and freeze-etching, *J. Lipid Res.*, 1989, **30**, 967–978.
37. M. Navab, J. A. Berliner, A. D. Watson, S. Y. Hama, M. C. Territo, A. J. Lusis, D. M. Shih, B. J. Van Lenten, J. S. Frank, L. L. Demer, P. A. Edwards and A. M. Fogelman, The Yin and Yang of oxidation in the development of the fatty streak. A review based on the 1994 George Lyman Duff Memorial Lecture, *Arterioscler. Thromb. Vasc. Biol.*, 1996, **16**, 831–842.
38. D. Steinberg, Low density lipoprotein oxidation and its pathobiological significance, *J. Biol. Chem.*, 1997, **272**, 20963–20966.
39. J. C. Khoo, E. Miller, P. McLoughlin and D. Steinberg, Enhanced macrophage uptake of low density lipoprotein after self-aggregation, *Arteriosclerosis*, 1988, **8**, 348–358.
40. M. T. Quinn, S. Parthasarathy, L. G. Fong and D. Steinberg, Oxidatively modified low density lipoproteins: a potential role in recruitment and retention of monocyte/macrophages during atherogenesis, *Proc. Natl. Acad. Sci. USA*, 1987, **84**, 2995–2998.
41. T. B. Rajavashisth, A. Andalibi, M. C. Territo, J. A. Berliner, M. Navab, A. M. Fogelman and A. J. Lusis, Induction of endothelial cell expression of granulocyte and macrophage colony-stimulating factors by modified low-density lipoproteins, *Nature*, 1990, **344**, 254–257.
42. L. Gu, Y. Okada, S. K. Clinton, C. Gerard, G. K. Sukhova, P. Libby and B. J. Rollins, Absence of monocyte chemoattractant protein-1 reduces atherosclerosis in low density lipoprotein receptor-deficient mice, *Mol. Cell*, 1998, **2**, 275–281.
43. M. Navab, S. S. Imes, S. Y. Hama, G. P. Hough, L. A. Ross, R. W. Bork, A. J. Valente, J. A. Berliner, D. C. Drinkwater and H. Laks *et al*, Monocyte transmigration induced by modification of low density lipoprotein in cocultures of human aortic wall cells is due to induction of monocyte chemotactic protein 1 synthesis and is abolished by high density lipoprotein, *J. Clin. Invest.*, 1991, **88**, 2039–2046.
44. J. A. Araujo, E. L. Romano, B. E. Brito, V. Parthe, M. Romano, M. Bracho, R. F. Montano and J. Cardier, Iron overload augments the development of atherosclerotic lesions in rabbits, *Arterioscler. Thromb. Vasc. Biol.*, 1995, **15**, 1172–1180.
45. L. D. Orozco, M. H. Kapturczak, B. Barajas, X. Wang, M. M. Weinstein, J. Wong, J. Deshane, S. Bolisetty, Z. Shaposhnik, D. M. Shih, A. Agarwal, A. J. Lusis and J. A. Araujo, Heme oxygenase-1 expression in macrophages plays a beneficial role in atherosclerosis, *Circ. Res.*, 2007, **100**, 1703–1711.
46. J. Sullivan, N. Ishikawa, L. Sheppard, D. Siscovick, H. Checkoway and J. Kaufman, Exposure to ambient fine particulate matter and primary cardiac arrest among persons with and without clinically recognized heart disease, *Am. J. Epidemiol.*, 2003, **157**, 501–509.

47. Q. Sun, P. Yue, R. I. Kirk, A. Wang, D. Moatti, X. Jin, B. Lu, A. D. Schecter, M. Lippmann, T. Gordon, L. C. Chen and S. Rajagopalan, Ambient air particulate matter exposure and tissue factor expression in atherosclerosis, *Inhal. Toxicol.*, 2008, **20**, 127–137.
48. H. Yamawaki and N. Iwai, Mechanisms underlying nano-sized air-pollution-mediated progression of atherosclerosis: carbon black causes cytotoxic injury/inflammation and inhibits cell growth in vascular endothelial cells, *Circ. J.*, 2006, **70**, 129–140.
49. A. Montiel-Davalos, E. Alfaro-Moreno and R. Lopez-Marure, $PM_{2.5}$ and PM_{10} induce the expression of adhesion molecules and the adhesion of monocytic cells to human umbilical vein endothelial cells, *Inhal. Toxicol.*, 2007, **19**(Suppl 1), 91–98.
50. T. Suwa, J. C. Hogg, K. B. Quinlan, A. Ohgami, R. Vincent and S. F. van Eeden, Particulate air pollution induces progression of atherosclerosis, *J. Am. Coll. Cardiol.*, 2002, **39**, 935–942.
51. K. Yatera, J. Hsieh, J. C. Hogg, E. Tranfield, H. Suzuki, C. H. Shih, A. R. Behzad, R. Vincent and S. F. van Eeden, Particulate matter air-pollution exposure promotes recruitment of monocytes into atherosclerotic plaques, *Am. J. Physiol. Heart Circ. Physiol.*, 2008, **294**, H944–953.
52. L. C. Chen and C. Nadziejko, Effects of subchronic exposures to concentrated ambient particles (CAPs) in mice. V. CAPs exacerbate aortic plaque development in hyperlipidemic mice, *Inhal. Toxicol.*, 2005, **17**, 217–224.
53. L. C. Chen, C. Quan, J. S. Hwang, X. Jin, Q. Li, M. Zhong, S. Rajagopalan and Q. Sun, Atherosclerosis lesion progression during inhalation exposure to environmental tobacco smoke: a comparison to concentrated ambient air fine particles exposure, *Inhal. Toxicol.*, **22**, 449–459.
54. Q. Sun, A. Wang, X. Jin, A. Natanzon, D. Duquaine, R. D. Brook, J. G. Aguinaldo, Z. A. Fayad, V. Fuster, M. Lippmann, L. C. Chen and S. Rajagopalan, Long-term air-pollution exposure and acceleration of atherosclerosis and vascular inflammation in an animal model, *JAMA*, 2005, **294**, 3003–3010.
55. Z. Ying, T. Kampfrath, G. Thurston, B. Farrar, M. Lippmann, A. Wang, Q. Sun, L. C. Chen and S. Rajagopalan, Ambient particulates alter vascular function through induction of reactive oxygen and nitrogen species, *Toxicol. Sci.*, 2009, **111**, 80–88.
56. J. A. Araujo, B. Barajas, M. Kleinman, X. Wang, B. J. Bennett, K. W. Gong, M. Navab, J. Harkema, C. Sioutas, A. J. Lusis and A. E. Nel, Ambient particulate pollutants in the ultrafine range promote early atherosclerosis and systemic oxidative stress, *Circ. Res.*, 2008, **102**, 589–596.
57. M. J. Campen, A. K. Lund, T. L. Knuckles, D. J. Conklin, B. Bishop, D. Young, S. Seilkop, J. Seagrave, M. D. Reed and J. D. McDonald, Inhaled diesel emissions alter atherosclerotic plaque composition in ApoE(–/–) mice, *Toxicol. Appl. Pharmacol.*, **242**, 310–317.
58. G. Assmann and A. M. Gotto Jr, HDL cholesterol and protective factors in atherosclerosis, *Circulation*, 2004, **109**, III8–14.

59. P. J. Barter and K. A. Rye, Relationship between the concentration and antiatherogenic activity of high-density lipoproteins, *Curr. Opin. Lipidol.*, 2006, **17**, 399–403.
60. D. J. Rader, Mechanisms of disease: HDL metabolism as a target for novel therapies, *Nat. Clin. Pract. Cardiovasc. Med.*, 2007, **4**, 102–109.
61. C. J. Fielding and P. E. Fielding, Molecular physiology of reverse cholesterol transport, *J. Lipid Res.*, 1995, **36**, 211–228.
62. J. R. Nofer, B. Kehrel, M. Fobker, B. Levkau, G. Assmann and A. von Eckardstein, HDL and arteriosclerosis: beyond reverse cholesterol transport, *Atherosclerosis*, 2002, **161**, 1–16.
63. P. J. Barter, S. Nicholls, K. A. Rye, G. M. Anantharamaiah, M. Navab and A. M. Fogelman, Antiinflammatory properties of HDL, *Circ. Res.*, 2004, **95**, 764–772.
64. J. K. Bielicki and M. N. Oda, Apolipoprotein A-I(Milano) and apolipoprotein A-I(Paris) exhibit an antioxidant activity distinct from that of wild-type apolipoprotein A-I, *Biochemistry*, 2002, **41**, 2089–2096.
65. S. J. Nicholls, L. Zheng and S. L. Hazen, Formation of dysfunctional high-density lipoprotein by myeloperoxidase, *Trends Cardiovasc. Med.*, 2005, **15**, 212–219.
66. B. J. Ansell, M. Navab, S. Hama, N. Kamranpour, G. Fonarow, G. Hough, S. Rahmani, R. Mottahedeh, R. Dave, S. T. Reddy and A. M. Fogelman, Inflammatory/anti-inflammatory properties of high-density lipoprotein distinguish patients from control subjects better than high-density lipoprotein cholesterol levels and are favorably affected by simvastatin treatment, *Circulation*, 2003, **108**, 2751–2756.
67. Q. Sun, P. Yue, J. A. Deiuliis, C. N. Lumeng, T. Kampfrath, M. B. Mikolaj, Y. Cai, M. C. Ostrowski, B. Lu, S. Parthasarathy, R. D. Brook, S. D. Moffatt-Bruce, L. C. Chen and S. Rajagopalan, Ambient air pollution exaggerates adipose inflammation and insulin resistance in a mouse model of diet-induced obesity, *Circulation*, 2009, **119**, 538–546.
68. T. R. Nurkiewicz, D. W. Porter, M. Barger, V. Castranova and M. A. Boegehold, Particulate matter exposure impairs systemic microvascular endothelium-dependent dilation, *Environ. Health Perspect.*, 2004, **112**, 1299–1306.
69. T. R. Nurkiewicz, D. W. Porter, M. Barger, L. Millecchia, K. M. Rao, P. J. Marvar, A. F. Hubbs, V. Castranova and M. A. Boegehold, Systemic microvascular dysfunction and inflammation after pulmonary particulate matter exposure, *Environ. Health Perspect.*, 2006, **114**, 412–419.
70. C. Sioutas, R. J. Delfino and M. Singh, Exposure assessment for atmospheric ultrafine particles (UFPs) and implications in epidemiologic research, *Environ. Health Perspect.*, 2005, **113**, 947–955.
71. International Commission on Radiological Protection, Human respiratory model for radiological protection, *Ann. IRCP*, 1994, **24**, 1–300.
72. A. Peters, B. Veronesi, L. Calderon-Garciduenas, P. Gehr, L. Chen, M. Geiser, W. Reed, B. Rothen-Rutishauser, S. Schurch and H. Schulz,

Translocation and potential neurological effects of fine and ultrafine particles a critical update, *Part. Fibre Toxicol.*, 2006, **3**, 13.
73. M. Geiser, B. Rothen-Rutishauser, N. Kapp, S. Schurch, W. Kreyling, H. Schulz, M. Semmler, V. Im Hof, J. Heyder and P. Gehr, Ultrafine particles cross cellular membranes by nonphagocytic mechanisms in lungs and in cultured cells, *Environ. Health Perspect.*, 2005, **113**, 1555–1560.
74. G. Oberdorster, Pulmonary effects of inhaled ultrafine particles, *Int. Arch. Occup. Environ. Health*, 2001, **74**, 1–8.
75. G. Oberdorster, E. Oberdorster and J. Oberdorster, Nanotoxicology: an emerging discipline evolving from studies of ultrafine particles, *Environ. Health Perspect.*, 2005, **113**, 823–839.
76. N. Kungli, M. Jerrett, W. D. Mack, B. Beckerman, L. Labree, F. Gilliland, D. Thomas, J. Peters and H. N. Hodis, Ambient air pollution and atherosclerosis in Los Angeles, *Environ. Health Pers.*, 2005, **113**, 201–206.
77. N. Kunzli, M. Jerrett, R. Garcia-Esteban, X. Basagana, B. Beckermann, F. Gilliland, M. Medina, J. Peters, H. N. Hodis and W. J. Mack, Ambient air pollution and the progression of atherosclerosis in adults, *PLoS One*, **5**, e9096.
78. B. Hoffmann, S. Moebus, S. Mohlenkamp, A. Stang, N. Lehmann, N. Dragano, A. Schmermund, M. Memmesheimer, K. Mann, R. Erbel and K. H. Jockel, Residential exposure to traffic is associated with coronary atherosclerosis, *Circulation*, 2007, **116**, 489–496.
79. A. V. Diez Roux, A. H. Auchincloss, T. G. Franklin, T. Raghunathan, R. G. Barr, J. Kaufman, B. Astor and J. Keeler, Long-term exposure to ambient particulate matter and prevalence of subclinical atherosclerosis in the Multi-Ethnic Study of Atherosclerosis, *Am. J. Epidemiol.*, 2008, **167**, 667–675.
80. M. Stolzel, S. Breitner, J. Cyrys, M. Pitz, G. Wolke, W. Kreyling, J. Heinrich, H. E. Wichmann and A. Peters, Daily mortality and particulate matter in different size classes in Erfurt, Germany, *J. Expo. Sci. Environ. Epidemiol.*, 2007, **17**, 458–467.
81. H. E. Wichmann, C. Spix, T. Tuch, G. Wolke, A. Peters, J. Heinrich, W. G. Kreyling and J. Heyder, Daily mortality and fine and ultrafine particles in Erfurt, Germany part I: role of particle number and particle mass, *Res. Rep. Health Eff. Inst.*, 2000, **5–86**; discussion 87–94..
82. R. J. Delfino, N. Staimer, T. Tjoa, D. L. Gillen, A. Polidori, M. Arhami, M. T. Kleinman, N. D. Vaziri, J. Longhurst and C. Sioutas, Air-pollution exposures and circulating biomarkers of effect in a susceptible population: clues to potential causal component mixtures and mechanisms, *Environ. Health Perspect.*, 2009, **117**, 1232–1238.
83. R. J. Delfino, N. Staimer, T. Tjoa, A. Polidori, M. Arhami, D. L. Gillen, M. T. Kleinman, N. D. Vaziri, J. Longhurst, F. Zaldivar and C. Sioutas, Circulating biomarkers of inflammation, antioxidant activity, and platelet activation are associated with primary combustion aerosols in subjects with coronary artery disease, *Environ. Health Perspect.*, 2008, **116**, 898–906.

84. R. Ruckerl, R. P. Phipps, A. Schneider, M. Frampton, J. Cyrys, G. Oberdorster, H. E. Wichmann and A. Peters, Ultrafine particles and platelet activation in patients with coronary heart disease – results from a prospective panel study, *Part. Fibre Toxicol.*, 2007, **4**, 1.
85. K. J. Chuang, C. C. Chan, T. C. Su, C. T. Lee and C. S. Tang, The effect of urban air pollution on inflammation, oxidative stress, coagulation, and autonomic dysfunction in young adults, *Am. J. Respir. Crit. Care Med.*, 2007, **176**, 370–376.
86. S. D. Dubowsky, H. Suh, J. Schwartz, B. A. Coull and D. R. Gold, Diabetes, obesity, and hypertension may enhance associations between air pollution and markers of systemic inflammation, *Environ. Health Perspect.*, 2006, **114**, 992–998.
87. W. Yue, A. Schneider, M. Stolzel, R. Ruckerl, J. Cyrys, X. Pan, W. Zareba, W. Koenig, H. E. Wichmann and A. Peters, Ambient source-specific particles are associated with prolonged repolarization and increased levels of inflammation in male coronary artery disease patients, *Mutat. Res.*, 2007, **621**, 50–60.
88. Y. Niwa, Y. Hiura, T. Murayama, M. Yokode and N. Iwai, Nano-sized carbon black exposure exacerbates atherosclerosis in LDL-receptor knockout mice, *Circ. J.*, 2007, **71**, 1157–1161.
89. S. R. Soares, R. Carvalho-Oliveira, E. Ramos-Sanchez, S. Catanozi, L. F. da Silva, T. Mauad, M. Gidlund, H. Goto and M. L. Garcia, Air pollution and antibodies against modified lipoproteins are associated with atherosclerosis and vascular remodeling in hyperlipemic mice, *Atherosclerosis*, 2009, **207**, 368–373.
90. R. W. Allen, M. H. Criqui, A. V. Diez Roux, M. Allison, S. Shea, R. Detrano, L. Sheppard, N. D. Wong, K. H. Stukovsky and J. D. Kaufman, Fine particulate-matter air pollution, proximity to traffic, and aortic atherosclerosis, *Epidemiology*, 2009, **20**, 254–264.

CHAPTER 9

Air Pollution and Ischemic Heart Disease

A. PETERS

Helmholtz Zentrum München – German Research Center for Environmental Health, Ingolstädter Landstr. 1, D-85764 Neuherberg, Germany

9.1 Introduction

Analyses of time-series data suggest an association between ambient particulate matter and cardiovascular disease hospital admissions[1,2] indicating that concentrations of particulate matter could potentially exacerbate cardiovascular disease.

To explain this link, it has been suggested that inflammation in the lung induced by deposited ambient particles could lead to systemic responses in the entire body promoting atherosclerosis.[3,4] Figure 9.1 lists the steps in the development of atherosclerosis and plaque rupture that are likely to be influenced by external stimuli such as ambient air pollution.[4] Inflammation is believed to be a key feature in the initiation and progression of atherosclerosis,[5] and therefore this is currently considered to be a key pathophysiological mechanism linking ambient air pollution with ischemic heart disease[4,6] (Figure 9.2). In addition, activation of the endothelium and prothrombotic states have been associated with exposure to ambient particulate matter.[6] Through these mechanisms, ambient air pollution can increase the probability of plaque rupture and clot formation resulting in sustained obstruction of the coronary vessels and reduced oxygen supply of heart tissue, thereby inducing myocardial ischemia.

The evidence for these links will be reviewed in this chapter by highlighting experimental and epidemiological studies that contributed to the evolving

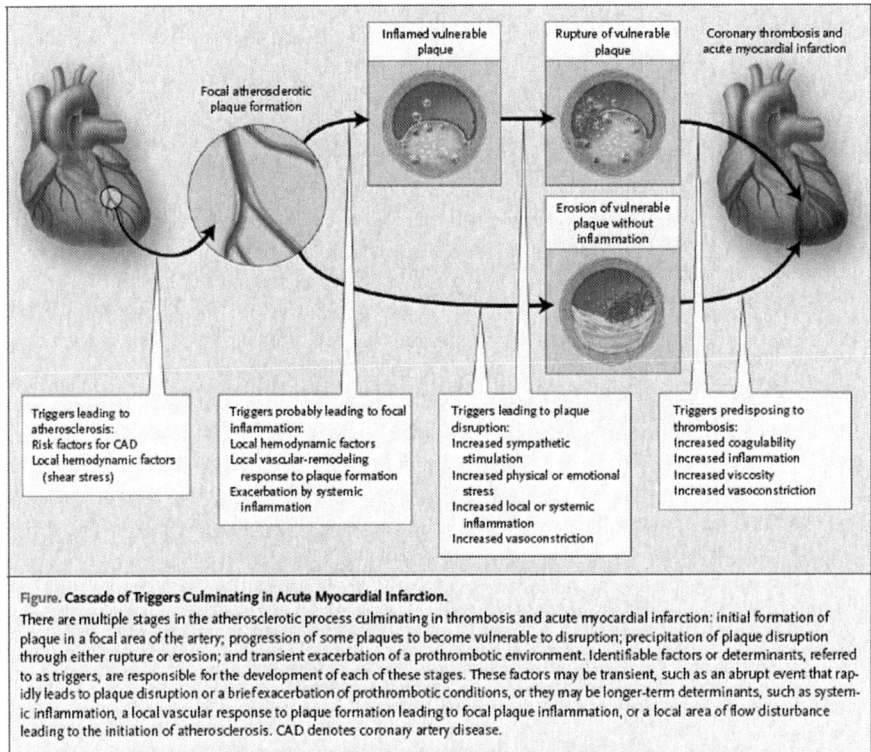

Figure 9.1 Development of acute coronary artery disease (Stone 2004). © 2004 Massachusetts Medical Society. All rights reserved

knowledge in this area. The evidence provided by these studies ranges from the involvement of ambient particulate matter in the promotion of atherosclerosis to the contribution of traffic exposure to the sudden onset of myocardial infarction.

9.2 Chronic Exposure to Particulate Matter and the Risk of Ischemic Heart Disease

Long-term health effects of ambient particulate matter smaller than 2.5 μm ($PM_{2.5}$) on cardiopulmonary disease were consistently reported in the Harvard Six Cities Study[7] and the American Cancer Society Study.[8,9] Cause-specific analyses within the American Cancer Society study based on more than 500 000 subjects showed for ischemic heart disease a hazard ratio of 1.18 (95% confidence interval (CI): 1.14 to 1.23) for a 10-μg/m^3 increase in long-term exposure to $PM_{2.5}$[10] (Figure 9.3). These were the largest estimates obtained for any of the cardiovascular disease categories in this analysis. The differences in $PM_{2.5}$ concentrations reflected differences in regional pollution concentrations.

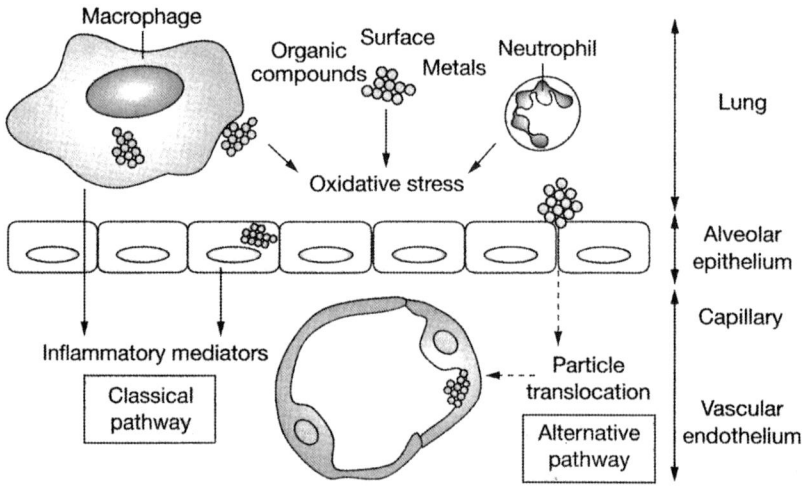

Figure 9.2 Summary of particle action in the cardiovascular bed (Mills *et al.* 2009).

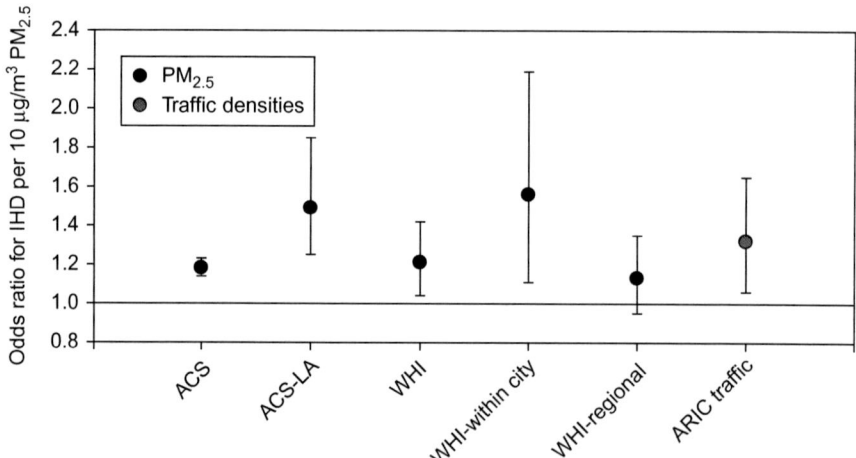

Figure 9.3 Summary of long-term effects of chronic particle exposure.

An analysis of the American Cancer Society Study data for the Los Angeles region in which exposure at the place of residence was evaluated for more than 22 000 subjects found a hazard ratio of 1.49 (95% CI: 1.25 to 1.85) per 10 μg/m³ increase in $PM_{2.5}$.[11] This data indicated that potentially spatial differences within urban areas contribute substantially to the risk associated with ambient PM exposure.

While these data had to rely on death certificates specifying the cause of death, Miller and colleagues were able to estimate the association between $PM_{2.5}$ and incident coronary artery disease in more than 65 000 women.[12]

They found that for coronary heart disease the hazard ratio was 1.21 (95% CI: 1.04 to 1.42) per $10\,\mu g/m^3$ $PM_{2.5}$, which was slightly smaller than the overall estimate for cardiovascular disease (1.24 (95% CI: 1.09 to 1.41)). Exposure estimation in this study was based on measurements from the nearest monitor, which was required to be within a 30 mile or 48 km radius of the women's homes. Comparing within city gradients for $PM_{2.5}$ resulted in larger risk estimates (1.56 (95% CI: 1.11 to 2.19) per $10\,\mu g/m^3$ $PM_{2.5}$) compared with between city gradients (1.13 (95% CI: 0.95 to 1.35) per $10\,\mu g/m^3$ $PM_{2.5}$). Incident coronary artery disease was also linked to traffic density in the Atherosclerosis Risk in Communities study, where an increased risk for coronary disease of 1.32 (95% CI: 1.06 to 1.65) was observed comparing the highest quartile of traffic density to the lowest.[13]

Collectively, these studies indicate that long-term exposure to ambient $PM_{2.5}$ is associated with an increased risk for ischemic heart disease as indicated by cohort studies with a follow-up between 6 and 16 years. The risk estimates were larger for within urban area contrasts than for regional contrasts.

9.3 Chronic Exposure to Particulate Matter and Atherosclerosis

Cross-sectional studies also suggest that cumulative exposure to PM is associated with an increase in clinical measures of atherosclerosis. An analysis of data obtained from 798 participants in two clinical trials conducted in the Los Angeles metro area showed that exposure to $PM_{2.5}$ was associated with increased carotid intima-media thickness (IMT), a measure of subclinical atherosclerosis.[14] A cross-sectional contrast in exposure of $10\,\mu g/m^3$ of $PM_{2.5}$ was associated with an approximately 4% increase in IMT. In a population-based sample from Germany,[15] it was found that traffic-related exposure was associated with coronary artery calcification (CAC), an alternative noninvasive measure of coronary atherosclerosis. Additional, evidence for an association with peripheral arterial disease was also found as indicated by an association between residence close to a major road and the ankle-brachial index.[16] In an analysis of three measures of subclinical disease (IMT, CAC, and ankle-brachial index) from the Multi-Ethnic Study of Atherosclerosis (MESA) only CAC was associated with long-term exposure to particulate air pollution.[17] In a related MESA study, another indicator of systemic atherosclerosis (abdominal aortic calcium, AAC), was associated with long-term $PM_{2.5}$ exposure, especially for residentially stable participants residing near a $PM_{2.5}$ monitor.[18]

These epidemiological data are supported by evidence from animal models indicating that instillation of ambient particles induces the progression of atherosclerotic plaque in mice on a high-fat diet.[19] Recent data has further substantiated these findings by extending it to inhalation of ambient concentrated particles.[20,21] In addition, experiments conducted in the Los Angeles area indicated that PM smaller than 100 nm, also called ultrafine particles, which are high in areas close to traffic, might be especially potent in inducing atherosclerosis in mice.[22]

Taken together, these findings suggest that long-term exposure to ambient $PM_{2.5}$ is associated with an increased risk for atherosclerosis and that traffic-related pollution such as ultrafine particles might be responsible for this association.

9.4 Inflammation as a Marker for Increased Cardiovascular Risk

Studies of day-to-day variation in ambient particles show that in random samples of the population, elevated levels of biomarkers indicative of an acute phase response, such as plasma viscosity, C-reactive protein, or fibrinogen[23–25] are associated with ambient PM levels. A number of recent studies have evaluated the effects of exposure on day-to-day variation in acute phase proteins in children and young adults[26,27] elderly populations[26–30] and in patients with coronary artery disease,[26–32] diabetes,[33] or chronic obstructive pulmonary disease.[34] They found associations between PM exposure and acute phase proteins; however, there was considerable variation in the strength of the associations. Evidence was found that subjects with a number of risk factors for coronary artery disease and metabolic syndrome exhibited stronger associations than healthier elderly study participants (for example).[30] Inflammatory responses due to ambient particles might be modified by genetic susceptibility as indicated for the PM_{10} – fibrinogen response in myocardial-infarction survivors, which was modified by a common single nucleotide polymorphism in the promoter of the fibrinogen B-chain gene.[35] On the other side, studies in patient populations suggested that treatment by lipid reducing medication such as statins may counteract the inflammatory effects of ambient PM.[32]

Some studies have also assessed the potential for increases in inflammatory markers due to cumulative or chronic air-pollution exposures. The effects of cumulative PM exposure on fibrinogen levels, as well as the counts of platelets and white blood cells have been reported.[36,37] C-reactive protein and fibrinogen levels were shown to increase in association with estimated residential $PM_{2.5}$ concentrations in the Ruhr area,[38] but the same study did not find an association with day-to-day variations in $PM_{2.5}$ levels or traffic exposure at the place of residence. Thus, inflammatory responses as a result of day-to-day variation in ambient particulate matter as well as long-term exposure to ambient particles may provide a link between air pollution and ischemic disease.

9.5 Evidence for Endothelial Cell Activation and Changes in Coagulation Markers

Systemic oxidative stress not only induces inflammation, but it could also activate endothelial cells, as well as circulating leukocytes and platelets.

Therefore, prothrombotic states may be induced, which could contribute to cardiovascular risk independent of inflammation. The potential consequences span from the induction and progression of atherosclerosis to the aggravation of clotting during acute ischemic events.

Current evidence shows that soluble adhesion molecules are upregulated in subjects with coronary artery disease and diabetes[31,33] and the von Willebrand Factor (vWF) is increased.[31] These changes may be indicative of endothelial dysfunction. Long-term exposure to elevated $PM_{2.5}$ was associated with increased levels of circulating endothelin-1 and elevated mean pulmonary arterial pressure in children living in Mexico City.[39]

In addition, evidence for platelet activation indicated by an increase in the levels of soluble CD40-ligand[40] was found recently in patients with coronary artery disease. Furthermore, increases in plasminogen activator inhibitor 1 (PAI1) were noted in healthy subjects[27] and in patients with coronary artery disease.[41] In a cross-sectional study, increased clotting times in blood samples were recorded in subjects previously exposed to elevated PM levels,[42] suggesting that exposure to ambient PM could transiently induce prothrombotic states.

9.6 ECG Recorded Ischemia

It has been reported that in patients with coronary artery disease, mild exercise in association with exposure to $PM_{2.5}$ and ultrafine particles induces myocardial ischemia.[43] Additional investigations by Pekkanen et al. 2002[43] using personal exposure data indicated that the onset of ischemia may be associated with personal exposures in the hours before the recording was made.[44] In an experimental setting, these findings were extended to exposure to diesel particle exposure in patients with coronary artery disease while exercising. Changes in ST-segment depressions were observed in patients exposed to diesel emissions, but not in those breathing clean air.[45] These changes were present within the hour of exposure. Newer evidence from epidemiological studies supports these findings in elderly subjects,[46] as well as in patients with coronary artery disease.[47] Thus, evidence for induction of ischemia based on day-to-day fluctuation in PM levels was observed in patients with coronary artery disease. There seem to be immediate responses to particulate matter when assessed by personal measurements or within experimental settings.

9.7 Acute Exposure to Particulate Matter and the Risk of Ischemic Heart Disease

The effects of day-to-day variation in ambient PM on ischemic heart have been studied by assessing emergency-room visits or hospital-admission records[1,2,48–56] or myocardial infarction registries.[57–59] Most of the studies identified an association between ambient particulate matter and ischemic heart disease. However, these studies also report that the effect of ambient PM is mostly seen in subjects that had underlying severe coronary artery disease and is likely to occur only in subjects with subclinical atherosclerosis. In addition, the composition of the air-pollution mixture seemed to be important and exposures to local traffic-related pollution or other regionally distributed air pollution from combustion sources seemed to be a prerequisite for the manifestation of these effects. A striking observation was the onset of myocardial infarction one hour after time spent in traffic, as obtained by interview.[60] This study, although not directly providing measures of traffic-related particle exposure, suggests an immediate impact of traffic-related PM and is consistent with the evidence from an experimental study in men with coronary artery disease.[45] These studies, which provide evidence for an increase in the risk of ischemic heart disease morbidity and mortality is observed based on time-series data, support the idea that acute exposure to PM could trigger adverse cardiovascular events, especially in subjects with underlying coronary disease.

9.8 Components of the Ambient Air-Pollution Mixture Associated with Ischemic Heart Disease

Multiple studies indicate that acute and chronic exposures to traffic-related particles, particularly diesel emissions, are associated with ischemic heart disease. Many studies, however, rely on measurements of particle mass ($PM_{2.5}$ or PM_{10}) or indirect measures of traffic-related exposures by distance of the residence to major roads or time spent in traffic. It is important to note that ambient air pollution in urban areas is a mixture of gaseous and particle-phase pollutants. Some of these could induce oxidative stress through different chemical mechanisms and, based on their size and solubility, have different fates once deposited in the lung. For example, data from 5 European cities suggest that gaseous pollutants as well as particle mass and particle number may be associated with hospital readmission for cardiac causes in myocardial-infarction survivors.[61] An issue of fierce debate for standard setting is, whether these pollutants *per se* are associated with ischemia, or whether they are markers of sources and therefore show associations with ischemia. At present, the issue remains particularly for pollutants such as NO_2, which is considered both as a criteria pollutant as well as an indicator of traffic in urban areas.[62]

Soot particles are suspected to be responsible at least in part for the observed associations as indicated by experimental studies.[45] A defining feature of these particles is their high surface are, which could support high oxidative reactivity. Thus, these particles can induce high oxidative stress locally, as well as systemically.[63] Because of their high surface area, these particles are able to transport biologically active components into the lung, from where they might be transferred into the blood stream either bound to the particles, or to proteins or in solution.[64] Thus, particles could plausibly induce systemic effects and exacerbate the progression of atherogenesis.[65]

Air Pollution and Ischemic Heart Disease 227

Table 9.1 Schematic summary of the existing evidence.

Outcomes	Acute health effects	Chronic health effect
Ischemic heart disease events	High confidence, underlying cardiovascular disease seems to be a prerequisite	High confidence
Signs of ischemia	Possible associations, only few studies	No studies
Atherosclerosis	Not applicable	Possible associations, only few studies
Inflammation	Possible associations, some diverging results	Possible associations, only few studies
Coagulation	Possible associations, few studies	Possible associations, only few studies

9.9 Overall Summary and Outlook

Evidence for acute as well as chronic exacerbation of ischemic heart disease has been observed upon exposure to ambient PM and is summarized in Table 9.1. The link between acute changes in intermediate phenotypes of cardiovascular disease and chronic elevated risks is also a subject of current research. The exact particle properties responsible for these associations are currently being investigated, but traffic-related particles seem to be among the likely culprits. Links between other forms of cardiovascular diseases exist and also susceptible subgroups based on either genetic susceptibility or other comorbidities are being investigated. Nevertheless, the existing evidence supporting a link between ambient PM and cardiovascular disease has been one of the driving forces supporting the rationale for further mitigation of particulate air pollution.[62] Recent evidence showing the salutary effects of antitobacco smoke legislation[66,67] further fuels the need for enforcing existing ambient air-quality standards.

References

1. F. Dominici, R. D. Peng, M. L. Bell, L. Pham, A. McDermott, S. L. Zeger and J. M. Samet, Fine particulate air pollution and hospital admission for cardiovascular and respiratory diseases, *JAMA*, 2006, **295**, 1127–1134.
2. J. Schwartz and R. Morris, Air pollution and hospital admissions for cardiovascular disease in Detroit, Michigan, *Am. J. Epidemiol.*, 1995, **142**, 23–35.
3. A. Seaton, W. MacNee, K. Donaldson and D. Godden, Particulate air pollution and acute health effects, *Lancet*, 1995, **345**, 176–178.
4. P. H. Stone, Triggering myocardial infarction, *N. Engl. J. Med.*, 2004, **351**, 1716–1718.
5. P. Libby, Inflammation in atherosclerosis, *Nature*, 2002, **420**, 868–874.
6. N. L. Mills, K. Donaldson, P. W. Hadoke, N. A. Boon, W. MacNee, F. R. Cassee, T. Sandstrom, A. Blomberg and D. E. Newby, Adverse

cardiovascular effects of air pollution, *Nat. Clin. Pract. Cardiovasc. Med.*, 2009, **6**, 36–44.
7. D. W. Dockery, C. A. Pope, X. Xu, J. D. Spengler, J. H. Ware, M. E. Fay, B. G. Ferris Jr and F. E. Speizer, Adverse cardiovascular effects of air pollution, *N. Engl. J. Med.*, 1993, **329**, 1753–1759.
8. C. A. Pope 3rd, R. T. Burnett, M. J. Thun, E. E. Calle, D. Krewski, K. Ito and G. D. Thurston, Lung cancer, cardiopulmonary mortality, and long-term exposure to fine particulate air pollution, *JAMA*, 2002, **287**, 1132–1141.
9. C. A. Pope 3rd, M. J. Thun, M. M. Namboodiri, D. W. Dockery, J. S. Evans, F. E. Speizer and C. W. Heath Jr, Particulate air pollution as predictor of mortality in a prospective study of US adults, *Am. J. Respir. Crit. Care Med.*, 1995, **151**, 669–674.
10. C. A. Pope 3rd, R. T. Burnett, G. D. Thurston, M. J. Thun, E. E. Calle, D. Krewski and J. J. Godleski, Cardiovascular mortality and long-term exposure to particulate air pollution: epidemiological evidence of general pathophysiological pathways of disease, *Circulation*, 2004, **109**, 71–77.
11. M. Jerrett, R. T. Burnett, R. Ma, C. A. Pope 3rd, D. Krewski, K. B. Newbold, G. Thurston, Y. Shi, N. Finkelstein, E. E. Calle and M. J. Thun, Spatial analysis of air pollution and mortality in Los Angeles, *Epidemiology*, 2005, **16**, 727–736.
12. K. A. Miller, D. S. Siscovick, L. Sheppard, K. Shepherd, J. H. Sullivan, G. L. Anderson and J. D. Kaufman, Long-term exposure to air pollution and incidence of cardiovascular events in women, *N. Engl. J. Med.*, 2007, **356**, 447–458.
13. H. Kan, G. Heiss, K. M. Rose, E. A. Whitsel, F. Lurmann and S. J. London, Prospective analysis of traffic exposure as a risk factor for incident coronary heart disease: the Atherosclerosis Risk in Communities (ARIC) study, *Environ. Health Perspect.*, 2008, **116**, 1463–1468.
14. N. Kunzli, M. Jerrett, W. J. Mack, B. Beckerman, L. LaBree, F. Gilliland, D. Thomas, J. Peters and H. N. Hodis, Ambient air pollution and atherosclerosis in Los Angeles, *Environ. Health Perspect.*, 2005, **113**, 201–206.
15. B. Hoffmann, S. Moebus, S. Mohlenkamp, A. Stang, N. Lehmann, N. Dragano, A. Schmermund, M. Memmesheimer, K. Mann, R. Erbel and K. H. Jockel, Residential exposure to traffic is associated with coronary atherosclerosis, *Circulation*, 2007, **116**, 489–496.
16. B. Hoffmann, S. Moebus, K. Kroger, A. Stang, S. Mohlenkamp, N. Dragano, A. Schmermund, M. Memmesheimer, R. Erbel and K. H. Jockel, Residential exposure to urban air pollution, ankle-brachial index, and peripheral arterial disease, *Epidemiology*, 2009, **20**, 280–288.
17. A. V. Diez Roux, A. H. Auchincloss, T. G. Franklin, T. Raghunathan, R. G. Barr, J. Kaufman, B. Astor and J. Keeler, Long-term exposure to ambient particulate matter and prevalence of subclinical atherosclerosis in the Multi-Ethnic Study of Atherosclerosis, *Am. J. Epidemiol.*, 2008, **167**, 667–675.

18. R. W. Allen, M. H. Criqui, A. V. Diez Roux, M. Allison, S. Shea, R. Detrano, L. Sheppard, N. D. Wong, K. H. Stukovsky and J. D. Kaufman, Fine particulate-matter air pollution, proximity to traffic, and aortic atherosclerosis, *Epidemiology*, 2009, **20**, 254–264.
19. T. Suwa, J. C. Hogg, K. B. Quinlan, A. Ohgami, R. Vincent and S. F. van Eeden, Particulate air pollution induces progression of atherosclerosis, *J. Am. Coll. Cardiol.*, 2002, **39**, 935–942.
20. Q. Sun, A. Wang, X. Jin, A. Natanzon, D. Duquaine, R. D. Brook, J. G. Aguinaldo, Z. A. Fayad, V. Fuster, M. Lippmann, L. C. Chen and S. Rajagopalan, Long-term air pollution exposure and acceleration of atherosclerosis and vascular inflammation in an animal model, *JAMA*, 2005, **294**, 3003–3010.
21. Q. Sun, P. Yue, R. I. Kirk, A. Wang, D. Moatti, X. Jin, B. Lu, A. D. Schecter, M. Lippmann, T. Gordon, L. C. Chen and S. Rajagopalan, Ambient air particulate matter exposure and tissue factor expression in atherosclerosis, *Inhal. Toxicol.*, 2008, **20**, 127–137.
22. J. A. Araujo, B. Barajas, M. Kleinman, X. Wang, B. J. Bennett, K. W. Gong, M. Navab, J. Harkema, C. Sioutas, A. J. Lusis and A. E. Nel, Ambient particulate pollutants in the ultrafine range promote early atherosclerosis and systemic oxidative stress, *Circ. Res.*, 2008, **102**, 589–596.
23. J. Pekkanen, E. J. Brunner, H. R. Anderson, P. Tiittanen and R. W. Atkinson, Daily concentrations of air pollution and plasma fibrinogen in London, *J. Occup. Environ. Med.*, 2000, **57**, 818–822.
24. A. Peters, A. Doring, H. E. Wichmann and W. Koenig, Increased plasma viscosity during air pollution episode: A link to mortality?, *Lancet*, 1997, **349**, 1582–1587.
25. A. Peters, M. Frohlich, A. Doring, T. Immervoll, H. E. Wichmann, W. L. Hutchinson, M. B. Pepys and W. Koenig, Particulate air pollution is associated with an acute phase response in men, *Eur. Heart J..*, 2001, **22**, 1198–1204.
26. L. Calderon-Garciduenas, R. Villarreal-Calderon, G. Valencia-Salazar, C. Henriquez-Roldan, P. Gutierrez-Castrellon, R. Torres-Jardon, N. Osnaya-Brizuela, L. Romero, A. Solt and W. Reed, Systemic inflammation, endothelial dysfunction, and activation in clinically healthy children exposed to air pollutants, *Inhal. Toxicol.*, 2008, **20**, 499–506.
27. K. J. Chuang, C. C. Chan, T. C. Su, C. T. Lee and C. S. Tang, The effect of urban air pollution on inflammation, oxidative stress, coagulation, and autonomic dysfunction in young adults, *Am. J. Respir. Crit. Care Med.*, 2007, **176**, 370–376.
28. C. A. Pope 3rd, M. L. Hansen, R. W. Long, K. R. Nielsen, N. L. Eatough, W. E. Wilson and D. J. Eatough, Ambient particulate air pollution, heart-rate variability, and blood markers of inflammation in a panel of elderly subjects, *Environ. Health Perspect.*, 2004, **112**, 339–345.
29. J. H. Sullivan, R. Hubbard, S. L. Liu, K. Shepherd, C. A. Trenga, J. Q. Koenig, W. L. Chandler and J. D. Kaufman, A community study of

the effect of particulate matter on blood measures of inflammation and thrombosis in an elderly population, *Environ. Health*, 2007, **6**, 3.
30. A. Zeka, J. R. Sullivan, P. S. Vokonas, D. Sparrow and J. Schwartz, Inflammatory markers and particulate air pollution: characterizing the pathway to disease, *Int. J. Epidemiol.*, 2006, **35**, 1347–1354.
31. R. Ruckerl, A. Ibald-Mulli, W. Koenig, A. Schneider, G. Woelke, J. Cyrys, J. Heinrich, V. Marder, M. Frampton, H. E. Wichmann and A. Peters, Air pollution and markers of inflammation and coagulation in patients with coronary heart disease, *Am. J. Respir. Crit. Care Med.*, 2006, **173**, 432–441.
32. R. Ruckerl, S. Greven, P. Ljungman, P. Aalto, C. Antoniades, T. Bellander, N. Berglind, C. Chrysohoou, F. Forastiere, B. Jacquemin, S. von Klot, W. Koenig, H. Kuchenhoff, T. Lanki, J. Pekkanen, C. A. Perucci, A. Schneider, J. Sunyer and A. Peters, Air pollution and inflammation (interleukin-6, C-reactive protein, fibrinogen) in myocardial-infarction survivors, *Environ. Health Perspect.*, 2007, **115**, 1072–1080.
33. M. S. O'Neill, A. Veves, J. A. Sarnat, A. Zanobetti, D. R. Gold, P. A. Economides, E. S. Horton and J. Schwartz, Air pollution and inflammation in type-2 diabetes: a mechanism for susceptibility, *J. Occup. Environ. Med.*, 2007, **64**, 373–379.
34. A. Seaton, A. Soutar, V. Crawford, R. Elton, S. McNerlan, J. Cherrie, M. Watt, R. Agius and R. Stout, Particulate air pollution and the blood, *Thorax*, 1999, **54**, 1027–1032.
35. A. Peters, S. Greven, I. M. Heid, F. Baldari, S. Breitner, T. Bellander, C. Chrysohoou, T. Illig, B. Jacquemin, W. Koenig, T. Lanki, F. Nyberg, J. Pekkanen, R. Pistelli, R. Ruckerl, C. Stefanadis, A. Schneider, J. Sunyer and H. E. Wichmann, Fibrinogen genes modify the fibrinogen response to ambient particulate matter, *Am. J. Respir. Crit. Care Med.*, 2009, **179**, 484–491.
36. J. C. Chen and J. Schwartz, Metabolic syndrome and inflammatory responses to long-term particulate air pollutants, *Environ. Health Perspect.*, 2008, **116**, 612–617.
37. J. Schwartz, Air pollution and blood markers of cardiovascular risk, *Environ. Health Perspect.*, 2001, **109**(Suppl 3), 405–409.
38. B. Hoffmann, S. Moebus, N. Dragano, A. Stang, S. Mohlenkamp, A. Schmermund, M. Memmesheimer, M. Brocker-Preuss, K. Mann, R. Erbel and K. H. Jockel, Chronic residential exposure to particulate-matter air pollution and systemic inflammatory markers, *Environ. Health Perspect.*, 2009, **117**, 1302–1308.
39. L. Calderon-Garciduenas, R. Vincent, A. Mora-Tiscareno, M. Franco-Lira, C. Henriquez-Roldan, G. Barragan-Mejia, L. Garrido-Garcia, L. Camacho-Reyes, G. Valencia-Salazar, R. Paredes, L. Romero, H. Osnaya, R. Villarreal-Calderon, R. Torres-Jardon, M. J. Hazucha and W. Reed, Elevated plasma endothelin-1 and pulmonary arterial pressure in children exposed to air pollution, *Environ. Health Perspect.*, 2007, **115**, 1248–1253.

40. R. Ruckerl, R. P. Phipps, A. Schneider, M. Frampton, J. Cyrys, G. Oberdorster, H. E. Wichmann and A. Peters, Ultrafine particles and platelet activation in patients with coronary heart disease – results from a prospective panel study, *Part. Fibre Toxicol.*, 2007, **4**, 1.
41. T. C. Su, C. C. Chan, C. S. Liau, L. Y. Lin, H. L. Kao and K. J. Chuang, Urban air pollution increases plasma fibrinogen and plasminogen activator inhibitor-1 levels in susceptible patients, *Eur. J. Cardiovasc. Prev. Rehabil.*, 2006, **13**, 849–852.
42. A. Baccarelli, A. Zanobetti, I. Martinelli, P. Grillo, L. Hou, S. Giacomini, M. Bonzini, G. Lanzani, P. M. Mannucci, P. A. Bertazzi and J. Schwartz, Effects of exposure to air pollution on blood coagulation, *J. Thromb. Haemost.*, 2007, **5**, 252–260.
43. J. Pekkanen, A. Peters, G. Hoek, P. Tiittanen, B. Brunekreef, J. de Hartog, J. Heinrich, A. Ibald-Mulli, W. G. Kreyling, T. Lanki, K. L. Timonen and E. Vanninen, Particulate air pollution and risk of ST-segment depression during repeated submaximal exercise tests among subjects with coronary heart disease: the Exposure and Risk Assessment for Fine and Ultrafine Particles in Ambient Air (ULTRA) study, *Circulation*, 2002, **106**, 933–938.
44. T. Lanki, G. Hoek, K. L. Timonen, A. Peters, P. Tiittanen, E. Vanninen and J. Pekkanen, Hourly variation in fine particle exposure is associated with transiently increased risk of ST segment depression, *J. Occup. Environ. Med.*, 2008, **65**, 782–786.
45. N. L. Mills, H. Tornqvist, M. C. Gonzalez, E. Vink, S. D. Robinson, S. Soderberg, N. A. Boon, K. Donaldson, T. Sandstrom, A. Blomberg and D. E. Newby, Ischemic and thrombotic effects of dilute diesel-exhaust inhalation in men with coronary heart disease, *N. Engl. J. Med.*, 2007, **357**, 1075–1082.
46. D. R. Gold, A. A. Litonjua, A. Zanobetti, B. A. Coull, J. Schwartz, G. MacCallum, R. L. Verrier, B. D. Nearing, M. J. Canner, H. Suh and P. H. Stone, Air pollution and ST-segment depression in elderly subjects, *Environ. Health Perspect.*, 2005, **113**, 883–887.
47. K. J. Chuang, B. A. Coull, A. Zanobetti, H. Suh, J. Schwartz, P. H. Stone, A. Litonjua, F. E. Speizer and D. R. Gold, Particulate air pollution as a risk factor for ST-segment depression in patients with coronary artery disease, *Circulation*, 2008, **118**, 1314–1320.
48. D. D'Ippoliti, F. Forastiere, C. Ancona, N. Agabiti, D. Fusco, P. Michelozzi and C. A. Perucci, Air pollution and myocardial infarction in Rome: a case-crossover analysis, *Epidemiology*, 2003, **14**, 528–535.
49. F. Forastiere, M. Stafoggia, S. Picciotto, T. Bellander, D. D'Ippoliti, T. Lanki, S. von Klot, F. Nyberg, P. Paatero, A. Peters, J. Pekkanen, J. Sunyer and C. A. Perucci, A case-crossover analysis of out-of-hospital coronary deaths and air pollution in Rome, Italy, *Am. J. Respir. Crit. Care Med.*, 2005, **172**, 1549–1555.
50. T. Lanki, J. Pekkanen, P. Aalto, R. Elosua, N. Berglind, D. D'Ippoliti, M. Kulmala, F. Nyberg, A. Peters, S. Picciotto, V. Salomaa, J. Sunyer,

P. Tiittanen, S. von Klot and F. Forastiere, Associations of traffic related air pollutants with hospitalisation for first acute myocardial infarction: the HEAPSS study, *J. Occup. Environ. Med.*, 2006, **63**, 844–851.
51. K. B. Metzger, P. E. Tolbert, M. Klein, J. L. Peel, W. D. Flanders, K. Todd, J. A. Mulholland, P. B. Ryan and H. Frumkin, Ambient air pollution and cardiovascular emergency department visits, *Epidemiology*, 2004, **15**, 46–56.
52. A. Peters, D. W. Dockery, J. E. Muller and M. A. Mittleman, Increased particulate air pollution and the triggering of myocardial infarction, *Circulation*, 2001, **103**, 2810–2815.
53. J. Schwartz, Air pollution and hospital admissions for cardiovascular disease in Tucson, *Epidemiology*, 1997, **8**, 371–377.
54. J. Schwartz, Air pollution and hospital admissions for heart disease in eight US counties, *Epidemiology*, 1999, **10**, 17–22.
55. A. Zanobetti, J. Schwartz and D. W. Dockery, T Airborne particles are a risk factor for hospital admissions for heart and lung disease, *Environ. Health Perspect.*, 2000, **108**, 1071–1077.
56. A. Zanobetti and J. Schwartz, The Effect of Particulate Air Pollution On Emergency Admissions for Myocardial Infarction: a Multi-city Case-crossover Analysis, *Environ. Health Perspect.*, 2005, **113**, 978–982.
57. A. Peters, S. von Klot, M. Heier, I. Trentinaglia, J. Cyrys, A. Hormann, M. Hauptmann, H. E. Wichmann and H. Lowel, Particulate air pollution and nonfatal cardiac events. Part I. Air pollution, personal activities, and onset of myocardial infarction in a case-crossover study, *Res. Rep. Health Eff. Inst.*, 2005, **20**, 1–66discussion 67–82, 141–148.
58. C. A. Pope 3rd, J. B. Muhlestein, H. T. May, D. G. Renlund, J. L. Anderson and B. D. Horne, Cardiovascular mortality and long-term exposure to particulate air pollution: epidemiological evidence of general pathophysiological pathways of disease, *Circulation*, 2006, **114**, 2443–2448.
59. J. Sullivan, L. Sheppard, A. Schreuder, N. Ishikawa, D. Siscovick and J. Kaufman, Relation between short-term fine-particulate matter exposure and onset of myocardial infarction, *Epidemiology*, 2005, **16**, 41–48.
60. A. Peters, S. von Klot, M. Heier, I. Trentinaglia, A. Hormann, H. E. Wichmann and H. Lowel, Exposure to traffic and the onset of myocardial infarction, *N. Engl. J. Med.*, 2004, **351**, 1721–1730.
61. S. von Klot, A. Peters, P. Aalto, T. Bellander, N. Berglind, D. D'Ippoliti, R. Elosua, A. Hormann, M. Kulmala, T. Lanki, H. Lowel, J. Pekkanen, S. Picciotto, J. Sunyer and F. Forastiere, Ambient air pollution is associated with increased risk of hospital cardiac readmissions of myocardial-infarction survivors in five European cities, *Circulation*, 2005, **112**, 3073–3079.
62. World-Health-Organization, Air Quality Guidelines, Global Update 2005, Particulate matter, ozone, nitrogen dioxide and sulfur dioxide, Scherfigsvej 8, DK-2100 O, Denmark, Copenhagen 2006, 1–496.
63. T. Stoeger, C. Reinhard, S. Takenaka, A. Schroeppel, E. Karg, B. Ritter, J. Heyder and H. Schulz, Instillation of six different ultrafine carbon

particles indicates a surface area threshold dose for acute lung inflammation in mice, *Environ. Health Perspect.*, 2006, **114**, 328–333.
64. A. Peters, B. Veronesi, L. Calderon-Garciduenas, P. Gehr, L. C. Chen, M. Geiser, W. Reed, B. Rothen-Rutishauser, S. Schurch and H. Schulz, Translocation and potential neurological effects of fine and ultrafine particles a critical update, *Part. Fibre Toxicol.*, 2006, **3**, 13.
65. A. B. Knol, J. J. de Hartog, H. Boogaard, P. Slottje, J. P. van der Sluijs, E. Lebret, F. R. Cassee, J. A. Wardekker, J. G. Ayres, P. J. Borm, B. Brunekreef, K. Donaldson, F. Forastiere, S. T. Holgate, W. G. Kreyling, B. Nemery, J. Pekkanen, V. Stone, H. E. Wichmann and G. Hoek, Expert elicitation on ultrafine particles: likelihood of health effects and causal pathways, *Part. Fibre Toxicol.*, 2009, **6**, 19.
66. G. Cesaroni, F. Forastiere, N. Agabiti, P. Valente, P. Zuccaro and C. A. Perucci, Effect of the Italian smoking ban on population rates of acute coronary events, *Circulation*, 2008, **117**, 1183–1188.
67. J. P. Pell, S. Haw, S. Cobbe, D. E. Newby, A. C. Pell, C. Fischbacher, A. McConnachie, S. Pringle, D. Murdoch, F. Dunn, K. Oldroyd, P. Macintyre, B. O'Rourke and W. Borland, Smoke-free legislation and hospitalizations for acute coronary syndrome, *N. Engl. J. Med.*, 2008, **359**, 482–491.

CHAPTER 10
Vehicular Emissions and Cardiovascular Disease

M. CAMPEN[1,2] AND A. LUND[2]

[1] College of Pharmacy, MSC09 5360, 1 University of New Mexico, Albuquerque, NM, 87131, USA; [2] Lovelace Respiratory Research Institute, Toxicology Division, 2425 Ridgecrest Dr. SE, Albuquerque, NM, 87108, USA

10.1 Introduction

Exposure to traffic-related air pollution is a ubiquitous, daily occurrence throughout the world. Evidence is emerging that implicates a causal role for traffic-related contaminants in promoting the incidence of cardiovascular morbidity and mortality. Our current understanding of the relationships between vehicular emissions and public health is muddled by the complexities of the combustion mixture chemistry, pathological mechanisms, and exposure dynamics. Vehicles generate air pollutants by three primary mechanisms: (1) direct generation of combustion-derived engine exhausts, (2) resuspension of road dust by mechanical agitation, and (3) off-gassing of petroleum-derived volatile organics from the gas tank and engine block. Furthermore, emissions from vehicles may have direct effects on human health, as well as indirectly leading to the generation of photochemical smog components (*e.g.*, ozone). Despite the complexity, a strong signal linking vehicular emissions and adverse cardiovascular health outcomes is emerging from both epidemiological and toxicological research. The present review describes the history of vehicular air pollution, current emission trends, and human health effects, including both epidemiological and toxicological research findings.

Issues in Toxicology No. 8
Environmental Cardiology: Pollution and Heart Disease
Edited by Aruni Bhatnagar
© The Royal Society of Chemistry 2011
Published by the Royal Society of Chemistry, www.rsc.org

10.1.1 Vehicle Emissions in the United States: Trends and Policy

It must be recognized from the outset that today's motor vehicles release far less pollution on a per-vehicle basis than in the early 1970s. The Department of Transportation's National Highway Traffic Safety Administration initiated the Corporate Average Fuel Economy (CAFE) program, which made fuel efficiency a legal requirement. While gas mileage does not perfectly correlate with emissions, the reduction in fuel combusted per mile driven was a major step in overall reductions. In 1972 the EPA passed regulations that removed lead from fuel, dramatically decreasing not only ambient air levels of lead, but also blood lead levels in the US population. With the passage of the Clean Air Act, regulations were established for numerous criteria pollutants.

Meanwhile, to keep up with regulations, the petroleum and motor vehicle industries have made significant engineering advances, from the catalytic converter to the new self-regenerating particle filter on contemporary diesel engines. Together, the regulations and engineering advancements have led to multiple, incremental decreases in the toxic substances released from the combustion of vehicle fuels. It is interesting to note how the emission trends in the United States seem to follow policy initiatives. Figure 10.1 shows how airborne levels of lead have fallen exponentially and how carbon monoxide from automobiles is exhibiting a gradual, but consistent decline, while total nitrogen oxides (NO_x) have been stable over approximately the past 20 years.[1,2] While some policies have led to more successful interventions than others, it is clear that allowing the free market to regulate environmental contamination is a strategy detrimental to both the environment and, as detailed below, public health. However, due to the number of vehicles on the road, the increased traffic congestion (which leads to prolonged duration of engine operation), and the growing and aging population, urban air-pollution levels remain a public-health concern despite the success of numerous environmental regulations.

10.1.2 Findings from Population Health Studies

Epidemiological associations between traffic exposure and cardiovascular sequelae have been described repeatedly over the past decade.[3] Among the earliest findings of an association between exposure to traffic and cardiovascular health outcomes was a study from southern Germany, wherein activity diaries were obtained from 691 patients with nonfatal myocardial infarction admitted to a hospital.[4] Among these patients, subgroups were established for those exposed to traffic-related air pollution while in cars, in public transportation (buses), or riding bicycles. Interestingly, the risk for myocardial infarction was highest among the latter group (odds ratio [OR] = 3.94), potentially owing to the direct proximity to fresh emissions or possibly an interaction with physical exercise, another common risk factor for acute myocardial infarction (AMI). Long-term exposure to traffic was similarly associated with an elevated risk of coronary heart disease events in a prospective population-based cohort

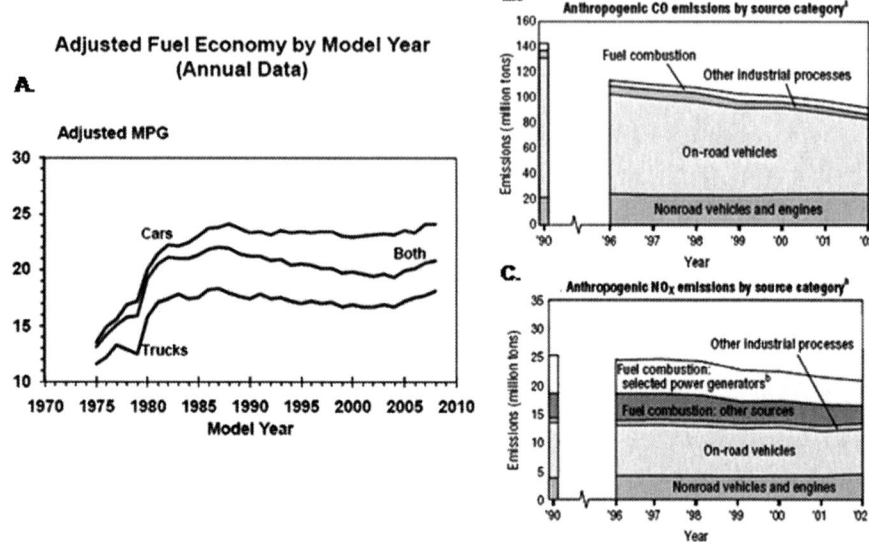

Figure 10.1 Recent trends in emissions from the United States automobile fleet. (A) Efficiency of vehicles in the Unites States has been stable since the mid-1980s as a result of CAFE standards implemented in the 1970s. (B) Carbon-monoxide emissions from on-road vehicles have consistently declined over the past decade, while other minor sources remain stable. (C) Oxides of nitrogen from on-road vehicles have shown relatively little change over the past decade. Figures reprinted from publicly available reports from the United Stated Environmental Protection Agency.[1,2]

from the Atherosclerosis Risk in Communities (ARIC) database.[5] In this study, traffic density was a better predictor of coronary events than was proximity to roadways, even after controlling for background PM_{10}, NO_2, and ozone levels. In another large cohort study (Multi-Ethnic Study of Atherosclerosis; MESA), Van Hee and colleagues[6] reported an association between roadway proximity to residence and indicators of left ventricular mass and blood pressure that could not be recapitulated by comparing levels of particulate matter (PM).

Using geographic information system (GIS) methods to compare relative distance to major roadways, Hoffmann et al.[7] examined the relationship between traffic-related air pollution and carotid artery calcification. In a cohort of >4500 subjects, a significant increase in the OR (1.6) for having high coronary artery calcification (above the 75th percentile) was noted for those living closer to major roadways, while similar comparisons with $PM_{2.5}$ revealed no significant association. A follow-up study revealed a similar positive association between distance-to-roadways and ankle-brachial index and peripheral arterial disease, further evidence of its role on vascular dysfunction and disease.[8] Similar geocoding techniques were used in a recent study that identified a potential relationship between proximity to traffic and venous thrombosis,[9] providing an interesting link to effects on coagulation dysregulation by vehicle-derived pollution. A number of studies have linked atherosclerosis and PM both epidemiologically[10,11] and toxicologically,[12,13] while others have failed to reproduce these findings of roadway proximity in other regions.[11] Thus, overall, the evidence remains somewhat controversial.

Approaching the question in another way, Sarnat and colleagues[14] used a source-apportionment method to ascertain the origins of the most biologically potent PM. Using chemical "fingerprints" of source pollutants, they compared concentrations of PM from specific sources with daily hospital admissions for respiratory and cardiovascular diseases. Interestingly, only markers for sulfate PM were associated with respiratory admissions. Cardiovascular admissions, on the other hand, were closely associated with the chemical markers for gasoline, diesel, and wood-smoke combustion. However, several other source apportionment studies have demonstrated conflicting results, with positive associations reported for secondary sulfates and coal-derived PM.[15] Thus, it is not clear if differences in regional pollution or source apportionment techniques may alter the estimated relationships between pollution and mortality.

10.1.3 Exposure Assessment

Although several population studies provide clear-cut findings regarding the potential public-health impact of traffic-related air pollution, these studies generally draw upon data sets with a substantial degree of uncertainty in terms of exposure variation.[16,17] People live primarily indoors and exposure to traffic-related air pollution can vary depending on personal behavior (outdoor activity, driving habits, *etc.*) and the quality of housing. While more elaborate exposure models have been developed in recent years, there is significant uncertainty in the levels of pollution within a given region. As different chemical components of a complex mixture have differing fates in the environment, the task of assessing the contribution of roadway emissions is far from trivial. PM will typically grow in size by agglomeration mechanisms as it leaves the tailpipe, while carbon monoxide (CO) is generally stable and follows a more predictable pattern of diffusion/dilution. Oxides of nitrogen will rapidly react such that measurements at the tailpipe may be many times greater than those obtained near a sidewalk.[18,19] Furthermore, because the contribution of environmental factors to cardiovascular disease is difficult to quantify, the complexity of susceptibility within a given population is equally difficult to characterize in epidemiological studies. For these and other reasons, there is a need to establish the plausibility of a causal relationship between air pollution and adverse health outcomes in controlled exposure studies.

10.1.4 Chemistry of Vehicular Emissions

Combustion of fossil fuels generates hundreds, if not thousands, of different chemical species. From a mass-concentration perspective, the bulk of emissions are in gaseous form, with only a small fraction of mass in solid or particulate form. Typically, diesel-engine emissions contain more particles per mass than gasoline-engine emissions; however, with contemporary diesel after-treatments, the particulate levels are dramatically reduced. Numerous reports detail the outputs of engines under specific conditions and with specific pollution-control devices.[20–22] Engine and vehicle manufacturers continue to implement new measures to reduce emissions, leading to the recent introduction of low-emission vehicles and zero-emission vehicles. Thus, while the current fleet is a mix of high and low polluters, the addition of new cars in the coming years will lead to considerable reductions in criteria pollutant emissions. Current regulations for vehicles in the United States can be found at http://www.epa.gov/otaq/standards.htm.

In a series of studies, researchers from the National Environmental Respiratory Center (NERC) conducted a thorough characterization of two pollutant atmospheres generated by both diesel- and gasoline-engine laboratory systems.[23–25] As the purpose of these studies was to create a real-world atmosphere that was consistent for toxicological studies, the engines were connected to a throttle-controlling dynamometer to alter the engine load in a manner resembling traffic conditions. This is crucial to understanding the chemistry, as constant engine operation at a low load will have a dramatically different chemistry profile from an engine operated at a high load. Furthermore, engine conditions (age, oil specifications, temperature) and environmental conditions (air temperature and humidity) can also affect the combustion output.

In general, the NERC diesel- and gasoline-engine systems were noted as having quite different emissions profiles. The gasoline-engine system (obtained from Chevy S10 trucks) generated a greater proportion of CO and hydrocarbons, while the diesel engine (Cummins 5.9 L) contained a greater contribution from NO_x, especially NO. Interestingly, and counter intuitively, the total mass concentration (gases and PM) of the gasoline-engine emissions (142 mg/m^3) was substantially greater than that of the diesel-engine emissions (92 mg/m^3) at identical dilutions. To a certain extent, these trends can be generalized to the overall gasoline- and diesel-engine fleets, although short-term driving with a gasoline engine will generate a greater amount of PM at cold start and long-term idling of diesel engines (such as at truck stops) will similarly produce a different composition. Additionally, owing to standard after-treatment devices, the levels of many vapor- and particulate-phase components are greatly reduced in magnitude in the contemporary fleet.

Thus, it must be kept in mind that the emissions from vehicles are comprised of both particulate and gaseous species. These constituents may have individual toxic effects, and they may have interactive effects. Carbon monoxide, for instance, is a prominent component of emissions, and recent findings in other organ systems have revealed adverse effects, primarily developmental, at environmentally relevant levels (12.5–50 ppm).[26,27] Using higher levels, *in utero* exposure to CO (150 ppm) led to persistent alterations in cardiomyocyte action potentials due to both increased calcium channel (L-type) currents and decreased potassium channel (transient outward rectifier) currents.[28] Other individual components may also cause cardiovascular effects or modify the effects of other toxicants, but limited evidence exists due primarily to an inadequate research portfolio. Future toxicological and epidemiological research must find novel and informative ways of assessing the cumulative impact of such complex pollutant atmospheres.

10.2 Toxicological Research Findings

10.2.1 Human Studies

Although only a few animal toxicological studies have examined the cardiovascular effects of fresh diesel emissions in animals, several studies have reported results from controlled human subjects. Mills et al.[29] first investigated the impact of diesel emissions exposure on forearm blood flow in 30 healthy subjects. On separate occasions, subjects were exposed to either filtered air or diesel emissions for 1 h. Characteristics of the emissions were reported elsewhere, but described as 300 μg PM/m^3, along with concentrations of major copollutants at 1.6 ppm NO_2, 4.5 ppm NO, and 7.5 ppm CO.[30] Forearm blood-flow responses to intravenously infused bradykinin, acetylcholine, and sodium nitroprusside were significantly reduced by exposure to diesel emissions. Diesel did not impact the vascular response to blockade of calcium channels with verapamil, suggesting that the effect related to the availability of NO in vascular smooth muscle cells. These effects persisted for up to 8 h after exposure. Furthermore, the report noted a reduced release of tissue plasminogen activator during bradykinin infusion. In a follow-up study, Törnqvist et al.[31] exposed healthy volunteers ($n = 15$) to filtered air or diesel exhaust and observed that vascular endothelial cell impairments persisted for 24 h post-exposure, while nitroprusside effects returned to normal. Furthermore, several plasma indicators of systemic inflammation (tumor necrosis factor-α, interleukin-6, etc.) were significantly elevated after exposure. More recently, it has been reported that arterial stiffness, another determinant of cardiovascular disease, was elevated acutely after inhaling fresh diesel emissions.[32]

In several follow-up studies, the individual roles of PM (as concentrated ambient PM) and NO_2 have been examined as drivers of the vascular effects of diesel exhaust.[33,34] While not representative of diesel PM specifically, exposure to ambient PM did not recapitulate the effects of the whole diesel mixture.[33] The authors speculated that dilution with nonvehicular PM, such as that originating from sea spray, may diminish the biological effect of diesel PM. Similarly, NO_2 caused no significant vascular effect.[34] The findings do not rule out that fresh diesel PM or some other gaseous component of the exhaust (for instance, CO) may be driving the effects; however, a more reasonable conclusion may be that the mixture of gaseous and particle components may be important in inducing endothelial dysfunction in humans.

Studying vulnerable populations has been key to identifying the ischemic effects of PM and copollutants. Mills et al.[35] studied a cohort with diagnosed stable coronary artery disease, defined stringently as patients with a recent history of myocardial infarction but no other major comorbidity, such as angina, arrhythmia, or diabetes. Eligible subjects underwent exposure to filtered air or whole diesel exhaust ($300\,\mu g/m^3$), concomitant with a stationary cycle exercise protocol. During the exercise period, no clear differences in heart rate were noted relative to the diesel exposure, but a significant and dramatic deviation in the ST-segment was observed during exposure to diesel exhaust compared with filtered air. The authors did not detect diesel-induced differences in forearm blood flow with dilatory agent infusion (both endothelium-dependent acetylcholine and endothelium-independent sodium nitroprusside), but they did observe significant increases in tissue plasminogen activator during infusions of increasing concentrations of infused bradykinin, which was reduced after diesel exposure. The absence of an effect of diesel exhaust on blood flow may be simply a reflection of the profound endothelial dysfunction already present in this population, although it suggests that coronary dilatation may not be the only significant factor in driving the ST-segment changes.

Lund et al.[36] examined circulating biomarkers in healthy humans exposed to diesel-engine emissions or filtered air on separate occasions, enabling paired comparisons (Figure 10.2). Volunteers ($N = 10$; 4 male/6 female; 18 to 40 years old) inhaled $100\,\mu g$ PM/m^3 whole diesel emissions for 2 h and serum was obtained before, immediately after, as well as 24 h after exposure. Matrix metalloproteinase-9 (MMP-9) concentration and activity were both significantly elevated in the plasma following diesel compared with filtered air.

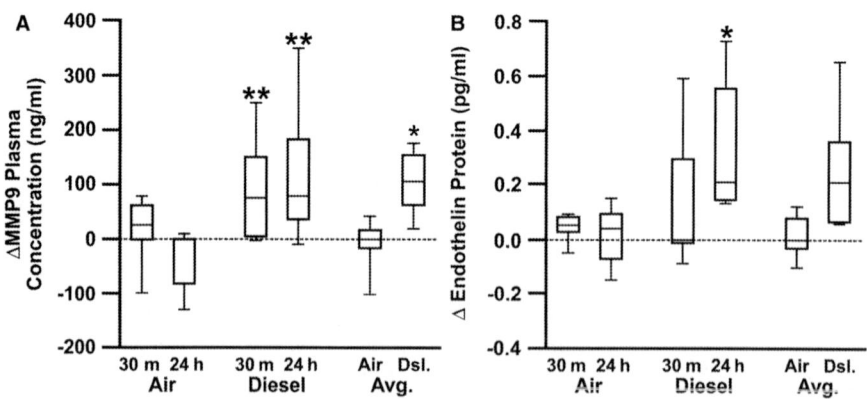

Figure 10.2 Diesel exposure-induced changes in circulating MMP-9 (A) and ET-1 (B). Healthy human volunteers ($N = 10$) were exposed to diesel-engine combustion emissions ($100\,\mu g$ PM/m^3) or filtered air on separate occasions for 2 h. Data shown represent differences between baseline values and values obtained within 30 min after exposures and 24 h after exposures. Additionally, averaged responses for both time points are shown. Asterisks denote significant differences between response to diesel exposures compared to response to air exposures by pairwise analysis of variance ($P < 0.05$). Reprinted from Lund et al.[36] with permission.

Additionally, endothelin-1 (ET-1) was significantly elevated 24 h after exposure to diesel. Plasma NO_x was acutely elevated, presumably due to the inhalation of NO (3.5 ppm) from the fresh emissions. These responses correlated well with findings in mice (described below), suggesting that these endpoints may have translational value as biomarkers. Moreover, while no major gender-related effects were observed, a subanalysis of the responses based on glutathione-S-transferase (GST) polymorphisms yielded compelling, albeit preliminary findings. The absence of GSTM1 was associated with a strong acute ET-1 response, while subjects with the gene showed no immediate effect at all (Figure 10.3A). Moreover, GSTP1, which has a polymorphism at Ile105Val, could determine ET-1 responses at 24-h postexposure, with individuals possessing the A/A allele being unresponsive, while those with G/A or G/G alleles showed a strong response (Figure 10.3B). Given the limited number of individuals in this cohort, segregational analysis provided only limited information into the potential vulnerability conferred by specific genotypes; however, these observations indicate that further inquiry into genetic as well as dietary and other lifestyle determinants of susceptibility is clearly warranted.

In other human exposures to diesel emissions researchers have noted changes in coagulation parameters, including increased rate of thrombus formation and platelet-neutrophil aggregates.[37] However, in a comparable study but with a population of individuals with metabolic syndrome, no alterations were noted in markers of coagulation, including D-dimer and plasminogen activator

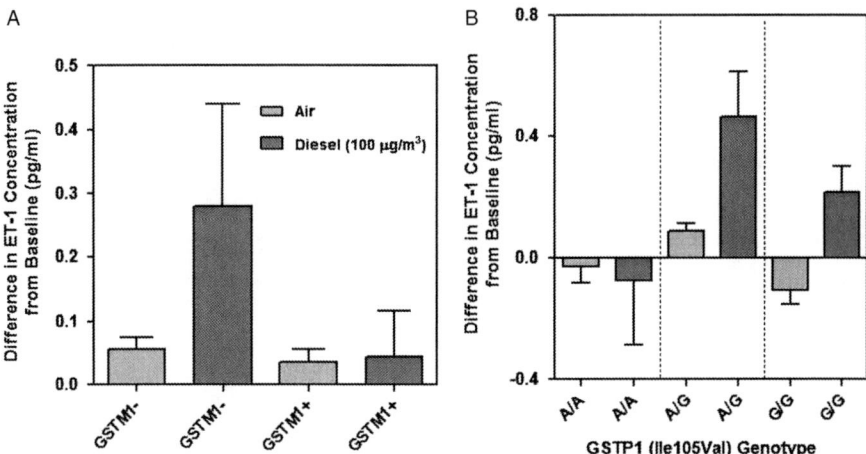

Figure 10.3 Segregational analysis of human data from Figure 2B based on polymorphisms for GSTM1 (A) and GSTP1 (B). Those individuals shown to be GSTM1-null responded acutely to diesel compared to air in terms of circulating ET-1, while subjects possessing GSTM1 showed no immediate effect at all (A). The GSTP1 Ile105Val polymorphism was also associated with responses to diesel wherein those possessing the A/A allele ($N=3$) were unresponsive, while those with G/A ($N=4$) or G/G ($N=2$) alleles showed a strong response (B).

inhibitor-1 (PAI-1); Von Willebrand Factor (VWF), a procoagulant molecule, actually decreased at 7-h postexposure.[38]

10.2.2 Animal Studies

Vascular effects of diesel-engine emissions also have been studied in numerous laboratory models. Diesel exhaust at concentrations of 0.5 and 3.5 mg/m^3 induced dramatic bradycardia and T-wave depression in apolipoprotein E-null (apoE$^{-/-}$) mice, but not their genetic background control, C57BL/6J mice.[39] These effects remained after filtration of the particles, suggesting that the gaseous components of the whole exhaust might be responsible for the cardiovascular findings. Interestingly, pulmonary inflammation in this model was apparent only when the atmosphere contained PM at the highest concentration and was not different between the apoE$^{-/-}$ and C57BL/6J mice, although the assessment of inflammation was limited to differential cell counts in the lavage and a few cytokines. This suggested that the cardiovascular effects were not dependent on pulmonary inflammation. However, in subsequent work with whole gasoline-engine emissions, the particulate-containing atmosphere (at a concentration of 60 μg/m^3) induced T-wave morphological alterations, while the filtered atmosphere did not.[40] Moreover, resuspended road dust, up to 3500 μg/m^3, had no impact on the electrocardiogram. Pulmonary inflammation was noted only in the road-dust exposures but not the gasoline-exhaust exposures, suggesting again that high PM is required for pulmonary, but not cardiovascular, effects. The effects from gasoline-engine PM were more subtle and required a more sophisticated analysis of averaged ECGs than did the effects of having a greater concentration of diesel-engine emissions. Circulating levels of ET-1 were elevated by gasoline emissions, but filtration of particles did not reduce this effect.

Subsequent research on the coronary vascular effects of whole diesel emissions demonstrated an enhanced vasoconstriction to ET-1 that was principally driven by an endothelial dysfunction.[41] In these studies, enhanced constriction by diesel exhaust was abolished by both denudation of the endothelium and coadministration of a NO synthase (NOS) inhibitor (Nω-Nitro-L-arginine). Diesel exhaust did not alter constriction to depolarizing potassium chloride, indicating a specific role for the endothelin receptors. Finally, it was shown that signaling through the endothelin B receptor on endothelial cells, which activates endothelial NOS, was diminished after diesel emissions. These mechanistic studies implicate NOS as a focal point for diesel-induced vascular dysfunction.

Beyond the coronary arteries, several other vascular beds have been examined for potential impacts on other syndromes. The venous circulation plays a prominent role in heart failure exacerbation.[42] One of the main roles of the major veins is to reserve blood volume for times of need, such as exercise, when additional volume can be shunted to enhance perfusion to muscles without compromising perfusion to the heart and brain. During heart failure, patients are often volume overloaded and therefore are placed on diuretics to alleviate symptoms of pulmonary congestion and chest pain. With regard to air pollution, heart-failure-linked hospital admissions are associated with both PM and other copollutants.[43] Knuckles et al.[44] hypothesized that if veins constrict in a manner similar to arteries following pollution exposure, then patients with severe heart failure may have temporary shunting of fluid to the pulmonary circulation that may elicit symptoms of heart failure, such as chest pain or dyspnea. Using a mouse model with both *in vivo* and *ex vivo* exposures to whole diesel emissions ($300\,\mu g/m^3 \times 4\,h$), the authors demonstrated a significant enhancement in vasoconstriction to ET-1 in veins with much weaker responses in arteries using an isolated, perfused vessel microscopy preparation. Similar to the results of Cherng et al.,[41] the effects were mimicked by the addition of the arginine analog Nω-Nitro-L-arginine methyl ester (a NOS inhibitor), suggesting that NOS, which is activated by endothelial endothelin B receptors, may be a target of diesel emissions. In the *ex vivo* studies, the authors removed most PM from the perfusate by filtration, implicating nonparticulate components of the whole emissions. The authors also hypothesized that representative volatile organic compounds might drive these effects, but no significant effects were noted for acetaldehyde, formaldehyde, or hexadecane among others.

Given the associations between traffic proximity and markers of atherosclerosis,[7,8] as well as recent studies showing a role for concentrated ambient PM in promoting atheromatous lesions in mouse models, Campen et al.[45] examined the impact of subchronic diesel emissions on atherosclerotic plaque morphology and markers of vascular remodeling in the $apoE^{-/-}$ mouse. Diesel exposures ranged from 100 to $1000\,\mu g\,PM/m^3$, with proportionate levels of gaseous copollutants as characterized by McDonald et al.[23] Following 50 days of exposure, the aortas of exposed mice showed concentration-related elevations in mRNA for ET-1, MMP-9, and tissue inhibitor of metalloproteinase-2 (TIMP-2). While exposure to diesel emissions did not appear to advance the size of the lesions, significant increases in macrophage infiltration and collagen deposition within the plaques were noted. Furthermore, a substantial induction of lipid peroxidation within the aortas of exposed mice was observed. While collagen deposition may reflect a stability factor, the presence of macrophages and lipid peroxides are suggestive of a more advanced plaque. Additional translational studies are urgently needed to fully assess the clinical relevance of these findings.

An additional arm of the diesel studies involved the removal of the particulate phase of emissions by filtration to compare toxicity. Vascular responses to PM-filtered diesel atmosphere were dynamic, with no obvious effects on the vascular mRNA changes, although there was a noticeable reduction in vascular lipid peroxides, collagen, and macrophage content, but these effects did not always reach statistical significance.[45] Overall, these studies suggest that both the particulate and gaseous phases are important and that studying individual components may underestimate the overall health effects of diesel emissions.

A more recent study of changes in atherosclerotic plaque by diesel-exhaust exposure found a similar lack of size progression in the $apoE^{-/-}$ mouse.[46]

This study compared the contributions of concentrated ambient PM and both whole diesel exhaust and PM-filtered diesel exhaust, along with combined concentrated PM and gaseous diesel emissions. The apoE$^{-/-}$ mice were exposed daily for 3 or 5 months, and vascular lesions were assessed in the aorta (*en face* preparation) and brachiocephalic artery (both by ultrasound and by histochemistry). While limited sample size precluded definite conclusions regarding the potency of atmospheres, it appeared that exposure to diesel emissions could augment the atherogenic effects of concentrated PM in specific vascular regions. When examining the contraction of aortas to phenylephrine or serotonin from pollutant-exposed mice it appeared that all permutations of pollutants could induce an enhanced constriction, and there was no obvious distinction in terms of relative potency.

Despite the ubiquitous nature of gasoline-powered motor vehicles, remarkably few studies have described the effects of fresh gasoline-engine emissions on cardiovascular health. Lund and colleagues[47] used whole emissions from gasoline exhaust to investigate changes in the transcriptional regulation of several gene products with known roles in both the chronic promotion and acute degradation/destabilization of atheromatous plaques. In these 50-day exposures (6 h/d, 7 d/wk) using apoE$^{-/-}$ mice on a high-fat chow, dilutions of gasoline-engine emissions induced a concentration-dependent increase in the mRNA of MMP isoforms (–3, –7, and –9), ET-1, and heme oxygenase-1. They also noted a marked increase in the level of lipid peroxides in the aortas. Using a high-efficiency particle trap, however, they established that the bulk, if not all of the effects, were caused by the gaseous portion of the emissions, not the particles. Indeed, the overall concentration of PM at the highest level was only 61 μg/m^3, while the concentrations of other gases such as CO and NO$_x$ were above the national ambient air-quality standards.

In a subsequent report, Lund *et al.*[36] described a potential role for endothelin receptors in driving responses to gasoline emissions. Tempol, a cell-permeable antioxidant, was capable of reversing only a portion of the assays investigated, while blockade of endothelin A receptors could fully abolish most of the vascular effects. More interestingly perhaps, it was reported that gasoline exhaust was able to induce a significant increase in aortic MMP activity after a single exposure, with further increases caused by additional exposures (Figure 10.4).[36] The authors further noted that in humans exposed to diesel emissions, temporally dependent upregulation of similar biomarkers (MMP-9, ET-1) was observed. These studies did not directly address lesion area in the mice, and therefore the results of the study cannot be directly compared with the concentrated PM studies by Sun *et al.*,[12] Chen and Nadziejko,[48] and Araujo *et al.*,[49] although the data suggest that the overall toxicity of the ambient airshed may be underestimated by modeling solely on PM levels.

Recently, Campen *et al.*[50] conducted a head-to-head comparison of several pollutant atmospheres and individual pollutant components, examining several of the sensitive biomarkers characterized previously. Using the apoE$^{-/-}$ mouse model and a standard 6 h/day×7-day exposure protocol, comparisons of vascular remodeling endpoints were conducted between gasoline emissions, diesel

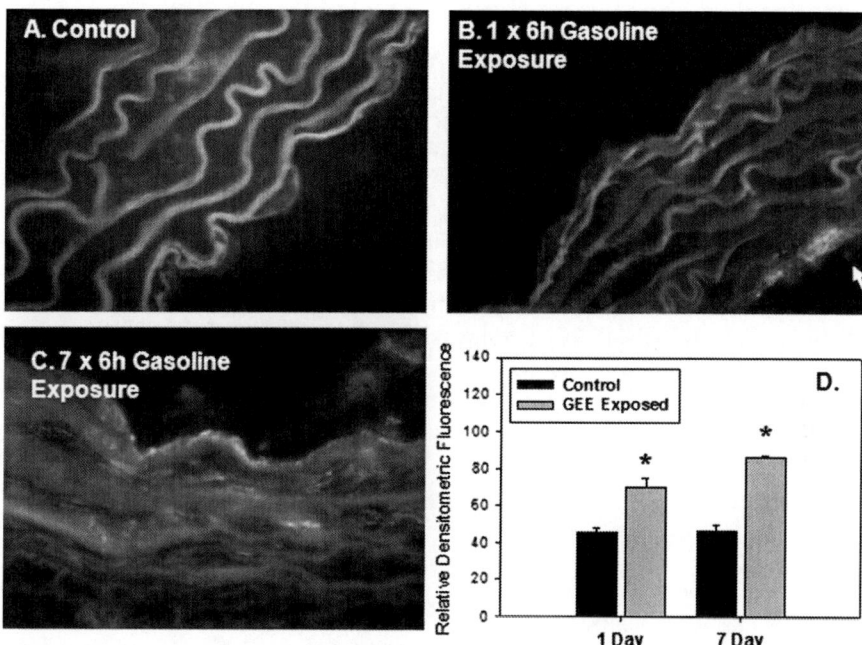

Figure 10.4 Zymographic analysis of aortic cryosections from – ApoE$^{-/-}$ mice exposed to (A) filtered air, (B) gasoline emissions for 1 day, and (C) gasoline emissions for 7 days. Briefly, following exposure and harvesting, cryosections of aorta were incubated with a dye-quenched gelatin, which when cleaved by active MMP-9 fluoresces. Quantification of fluorescence revealed significant ($P<0.05$) enhancement of aortic matrix MMP-9 activity after only a single exposure to gasoline exhaust (D). Reprinted with permission from Lund et al.[36]

emissions, hardwood smoke, and a simulated downwind coal-combustion atmosphere (SDCCA), along with exposures to two different types of secondary organic aerosol (SOA), CO, NO, and NO_2. The gasoline- and diesel-engine exhausts were clearly more toxic in several assays, including aortic lipid peroxidation and aortic mRNA for ET-1 and MMP-9. Hardwood smoke, SDCCA, and SOAs were unable to drive many of the observed effects accounted for from the vehicular emissions. However, CO and NO were able to recapitulate most of the aortic transcriptional effects, altering ET-1 and MMP-9 in a concentration-dependent manner. These monoxides also significantly upregulated aortic gelatinase activity. Aortic lipid peroxidation was not induced by any individual gas or SOA, suggesting that there is either something unique in the vehicular emissions, such as potentially volatile organic compounds (VOCs), or that there is a gas–particle interaction that was uncharacterized in this study. The potency of gasoline- and diesel-engine emissions relative to the other combustion atmospheres and individual components further underscores the importance of vehicle-derived pollution in driving cardiovascular effects.

In gasoline-engine emissions studies, the levels of monoxide gases are significantly above typical ambient levels, as measured at US Environmental Protection Agency (EPA) monitoring stations. However, as the source of these emissions is vehicles and EPA monitoring stations are purposefully positioned to minimize the roadway contribution, there is certainly room to consider "hotspot" exposures near roadways. Restrepo and colleagues[19] compared the levels of criteria pollutant recorded in official monitoring stations (15 m above ground) with a station located more proximal to the roadways (4 m above ground). While PM and a few other criteria pollutants were fairly consistent between monitors, the levels of CO and NO_2 were roughly 75% and 140% greater at "ground-level" monitoring station. A study of tunnel pollution noted extremely high levels of NO with mean values of approximately 1.3 ppm.[51] Similar hotspots for CO in tunnels[52] and tollbooths have been reported.[53]

10.3 Research Needs: Mechanisms, Interactions, and Sensitivities

It is becoming increasingly clear that specific pollutant classes, especially PM and CO, may have causal roles in promoting cardiovascular disease. Future research is required to delineate whether PM from specific sources can have greater cardiotoxic effects than others. Furthermore, given the complex physicochemical interactions between particulate, semivolatile, and VOCs, especially in the cloud of fresh engine emissions, a greater understanding of the modifying effects of gaseous copollutants on PM toxicity is needed. Gases may also prime biological pathways that lead to vulnerability to PM and *vice versa*. It is difficult to fully appreciate the health impact of exposure to traffic-related air pollutants until more sophisticated approaches are developed to examine the components and the biological effects of such complex mixtures.

Another issue that remains unresolved is the degree to which aging affects the relative toxicity of traffic-related pollutants. Certainly, reactive components such as NO and many VOCs will be reduced dramatically over time and distance, beyond the effects of simple dilution. However, these precursors can generate reactive products such as ozone or secondary PM. The PM derived directly from engines will also transform, with unknown consequences in terms of biological activity. Chronic exposure to concentrated ambient particles has been shown to be quite potent in terms of advancing atherosclerosis, hypertension, and metabolic syndrome.[49,54,55] The material in these studies varies chemically due to alterations in source contributions and meteorological conditions. The degree to which vehicular emissions drive the findings of these studies remains uncertain, although Lippmann *et al.*[56] found a contribution from regional nickel levels. Application of source apportionment models similar to that used by Sarnat *et al.*[14] might permit better estimates of the relative contribution of traffic-related pollutants to cardiovascular risk and disease. Traffic contributions to ambient PM levels are naturally reduced at increasing distance from the origin, and the relative pulmonary toxicity and allergenicity of ambient PM has been shown to be higher closer to roadways.[57] No comparable studies, however, have been conducted for cardiovascular outcomes.

Lastly, the issue of susceptibility will continue to be of high interest and importance. As individuals have wide variations in their genetics, diet, lifestyle, and overall health, it is imperative to consider the factors that confer vulnerability to better understand the subpopulations at risk. As described above, segregational analysis of genetic polymorphisms for GST subtypes identified groups that appeared protected (GSTP1 A/A genotype) or vulnerable (GSTM1-null). Such findings, while limited, are consistent with reports on allergic and pulmonary responses to inhaled pollutants, specifically diesel PM, which also demonstrate a role for GST polymorphisms.[58,59] Numerous other protective genes are likely to modify responses to vehicular emissions as well. Future studies are required to identify specific susceptibility factors and populations vulnerable to the cardiovascular toxicity of automobile emissions.

References

1. United States Environmental Protection Agency Report on the Environment, 2008. Available at: http://cfpub.epa.gov/ncea/cfm/recordisplay.cfm?deid = 190806.
2. United States Environmental Protection Agency Technical Report. 2008. Light-Duty Automotive Technology and Fuel Economy Trends: 1975 Through 2008. EPA420-R-08-015, available at http://www.epa.gov/otaq/cert/mpg/fetrends/420r08015.pdf.
3. S. D. Adar and J. D. Kaufman, Cardiovascular disease and air pollutants: evaluating and improving epidemiological data implicating traffic exposure, *Inhal. Toxicol.*, 2007, **19**, 135–149.
4. Peters, S. von Klot, M. Heier, I. Trentinaglia, A. Hörmann, H. E. Wichmann and H. Löwel, Cooperative Health Research in the Region of Augsburg Study Group, Exposure to traffic and the onset of myocardial infarction, *N. Engl. J. Med.*, 2004, **351**, 1721–1730.
5. H. Kan, G. Heiss, K. M. Rose, E. A. Whitsel, F. Lurmann and S. J. London, Prospective analysis of traffic exposure as a risk factor for incident coronary heart disease: the Atherosclerosis Risk in Communities (ARIC) study, *Environ. Health Perspect.*, 2008, **116**, 1463–1468.
6. V. C. Van Hee, S. D. Adar, A. A. Szpiro, R. G. Barr, D. A. Bluemke, A. V. Diez Roux, E. A. Gill, L. Sheppard and J. D. Kaufman, Exposure to traffic and left ventricular mass and function: the Multi-Ethnic Study of Atherosclerosis, *Am. J. Respir. Crit. Care Med.*, 2009, **179**, 827–834.
7. B. Hoffmann, S. Moebus, S. Möhlenkamp, A. Stang, N. Lehmann, N. Dragano, A. Schmermund, M. Memmesheimer, K. Mann, R. Erbel and K. H. Jöckel, Residential exposure to traffic is associated with coronary atherosclerosis, *Circulation*, 2007, **116**, 489–496.

8. B. Hoffmann, S. Moebus, K. Kröger, A. Stang, S. Möhlenkamp, N. Dragano, A. Schmermund, M. Memmesheimer, R. Erbel and K. H. Jöckel, Residential exposure to urban air pollution, ankle-brachial index, and peripheral arterial disease, *Epidemiology*, 2009, **20**, 280–288.
9. A. Baccarelli, I. Martinelli, V. Pegoraro, S. Melly, P. Grillo, A. Zanobetti, L. Hou, P. A. Bertazzi, P. M. Mannucci and J. Schwartz, Living near major traffic roads and risk of deep vein thrombosis, *Circ.*, 2009, **119**, 3118–3124.
10. N. Künzli, M. Jerrett, W. J. Mack, B. Beckermann, L. LaBree, F. Gilliland, D. Thomas, J. Peters and H. N. Hodis, Ambient air pollution and atherosclerosis in Los Angeles, *Environ. Health Perspect.*, 2005, **113**, 201–206.
11. R. W. Allen, M. H. Criqui, A. V. Diez Roux, M. Allison, S. Shea, R. Detrano, L. Sheppard, N. D. Wong, K. H. Stukovsky and J. D. Kaufman, Fine particulate matter air pollution, proximity to traffic, and aortic atherosclerosis, *Epidemiology.*, 2009, **20**, 254–264.
12. Q. Sun, A. Wang, X. Jin, A. Natanzon, D. Duquaine, R. D. Brook, J. G. Aguinaldo, Z. A. Fayad, V. Fuster, M. Lippmann, L. C. Chen and S. Rajagopalan, Long-term air-pollution exposure and acceleration of atherosclerosis and vascular inflammation in an animal model, *J. Am. Med. Assoc.*, 2005, **294**, 3003–3010.
13. K. Yatera, J. Hsieh, J. C. Hogg, E. Tranfield, H. Suzuki, C. H. Shih, A. R. Behzad, R. Vincent and S. F. van Eeden, Particulate matter air-pollution exposure promotes recruitment of monocytes into atherosclerotic plaques, *Am. J. Physiol. Heart Circ. Physiol.*, 2008, **294**, 944–953.
14. J. A. Sarnat, A. Marmur, M. Klein, E. Kim, A. G. Russell, S. E. Sarnat, J. A. Mulholland, P. K. Hopke and P. E. Tolbert, Fine particle sources and cardiorespiratory morbidity: an application of chemical mass balance and factor analytical source-apportionment methods, *Environ. Health Perspect.*, 2008, **116**, 459–466.
15. K. Ito, W. F. Christensen, D. J. Eatough, R. C. Henry, E. Kim, F. Laden, R. Lall, T. V. Larson, L. Neas, P. K. Hopke and G. D. Thurston, PM source apportionment and health effects: 2. An investigation of intermethod variability in associations between source-apportioned fine particle mass and daily mortality in Washington, DC, *J. Expo. Sci. Environ. Epidemiol.*, 2006, **16**, 300–310.
16. J. A. Sarnat, W. E. Wilson, M. Strand, J. Brook, R. Wyzga and T. Lumley, Panel discussion review: session 1-exposure assessment and related errors in air pollution epidemiologic studies, *J. Expo. Sci. Environ. Epidemiol.*, 2007, **17**(Suppl 2), 75–82.
17. A. J. Wheeler, M. Smith-Doiron, X. Xu, N. L. Gilbert and J. R. Brook, Intra-urban variability of air pollution in Windsor, Ontario – measurement and modeling for human exposure assessment, *Environ. Res.*, 2008, **106**, 7–16.
18. E. M. Fujita, D. E. Campbell, B. Zielinska, J. C. Sagebiel, J. L. Bowen, W. S. Goliff, W. R. Stockwell and D. R. Lawson, Diurnal and weekday variations in the source contributions of ozone precursors in California's South Coast Air Basin, *J. Air Waste Manag. Assoc.*, 2003, **53**, 844–863.

19. C. Restrepo, R. Zimmerman, G. Thurston, J. Clemente, J. Gorczynski, M. Zhong, M. Blaustein and L. C. Chen, A comparison of ground-level air quality data with New York State Department of Environmental Conservation monitoring stations data in South Bronx, New York, *Atmospher. Environ.*, 2004, **38**, 5295–5304.
20. D. Schuetzle, W. O. Siegl, T. E. Jensen, M. A. Dearth, E. W. Kaiser, R. Gorse, W. Kreucher and E. Kulik, The relationship between gasoline composition and vehicle hydrocarbon emissions: a review of current studies and future research needs, *Environ. Health Perspect.*, 1994, **102**(Suppl 4), 3–12.
21. A. B. Khan, N. N. Clark, G. J. Thompson, W. S. Wayne, M. Gautam, D. W. Lyons and D. Hawelti, Idle emissions from heavy-duty diesel vehicles: review and recent data, *J. Air Waste Manag. Assoc.*, 2006, **56**, 1404–1419.
22. G. A. Ban-Weiss, J. P. McLaughlin, R. A. Harley, K. J. Kean, E. Grosjean and D. Grosjean, Carbonyl and nitrogen dioxide emissions from gasoline- and diesel-powered motor vehicles, *Environ. Sci. Technol.*, 2008, **42**, 3944–3950.
23. J. D. McDonald, E. B. Barr, R. K. White, J. C. Chow, J. J. Schauer, B. Zielinska and E. Grosjean, Generation and characterization of four dilutions of diesel-engine exhaust for a subchronic inhalation study, *Environ. Sci. Technol.*, 2004, **38**, 2513–2522.
24. J. D. McDonald, M. D. Reed, M. J. Campen, E. G. Barrett, J. Seagrave and J. L. Mauderly, Health effects of inhaled gasoline-engine emissions, *Inhal. Toxicol.*, 2007, **19**(Suppl 1), 107–116.
25. J. D. McDonald, R. K. White, E. B. Barr, B. Zielinska, J. C. Chow and E. Grosjean, Generation and characterization of gasoline-engine exhaust inhalation exposure atmospheres, *Inhal. Toxicol.*, 2008, **20**, 1157–1168.
26. J. E. Stockard-Sullivan, R. A. Korsak, D. S. Webber and J. Edmond, Mild carbon monoxide exposure and auditory function in the developing rat, *J. Neurosci. Res.*, 2003, **74**, 644–654.
27. I. A. Lopez, D. Acuna, L. Beltran-Parrazal, A. Espinosa-Jeffrey and J. Edmond, Oxidative stress and the deleterious consequences to the rat cochlea after prenatal chronic mild exposure to carbon monoxide in air, *Neuroscience*, 2008, **151**, 854–867.
28. L. Sartiani, E. Cerbai, G. Lonardo, P. DePaoli, M. Tattoli, R. Cagiano, M. R. Carratù, V. Cuomo and A. Mugelli, Prenatal exposure to carbon monoxide affects postnatal cellular electrophysiological maturation of the rat heart: a potential substrate for arrhythmogenesis in infancy, *Circulation*, 2004, **109**, 419–423.
29. N. L. Mills, H. Törnqvist, S. D. Robinson, M. Gonzalez, K. Darnley, W. MacNee, N. A. Boon, K. Donaldson, A. Blomberg, T. Sandstrom and D. E. Newby, Diesel exhaust inhalation causes vascular dysfunction and impaired endogenous fibrinolysis, *Circulation*, 2005, **112**, 3930–3936.
30. S. Salvi, A. Blomberg, B. Rudell, F. Kelly, T. Sandstrom, S. Holgate and A. Frew, Acute inflammatory responses in the airways and peripheral

blood after short-term exposure to diesel exhaust in healthy human volunteers, *Am. J. Respir. Crit. Care Med.*, 1999, **159**, 702–709.
31. H. Törnqvist, N. L. Mills, M. Gonzalez, M. R. Miller, S. D. Robinson, I. L. Megson, W. Macnee, K. Donaldson, S. Söderberg, D. E. Newby, T. Sandström and A. Blomberg, Persistent endothelial dysfunction in humans after diesel exhaust inhalation, *Am. J. Respir. Crit. Care Med.*, 2007, **176**, 395–400.
32. M. Lundbäck, N. L. Mills, A. Lucking, S. Barath, K. Donaldson, D. E. Newby, T. Sandström and A. Blomberg, Experimental exposure to diesel exhaust increases arterial stiffness in man, *Part. Fibre Toxicol.*, 2009, **6**, 7.
33. N. L. Mills, S. D. Robinson, P. H. Fokkens, D. L. Leseman, M. R. Miller, D. Anderson, E. J. Freney, M. R. Heal, R. J. Donovan, A. Blomberg, T. Sandström, W. MacNee, N. A. Boon, K. Donaldson, D. E. Newby and F. R. Cassee, Exposure to concentrated ambient particles does not affect vascular function in patients with coronary heart disease, *Environ. Health Perspect.*, 2008, **116**, 709–715.
34. J. P. Langrish, M. Lundbäck, S. Barath, S. Söderberg, N. L. Mills, D. E. Newby, T. Sandström and A. Blomberg, Exposure to nitrogen dioxide is not associated with vascular dysfunction in man, *Inhal. Toxicol.*, 2010, **22**, 192–198.
35. N. L. Mills, H. Törnqvist, M. C. Gonzalez, E. Vink, S. D. Robinson, S. Söderberg, N. A. Boon, K. Donaldson, T. Sandström, A. Blomberg and D. E. Newby, Ischemic and thrombotic effects of dilute diesel-exhaust inhalation in men with coronary heart disease, *N. Engl. J. Med.*, 2007, **357**, 1075–1082.
36. A. K. Lund, J. A. Lucero, S. Lucas, M. C. Madden, J. D. McDonald, J. C. Seagrave, T. L. Knuckles and M. J. Campen, Vehicular Emissions Induce Vascular MMP-9 Expression and Activity *via* Endothelin-1 Mediated Pathways, *Arterioscler. Thromb. Vasc. Biol.*, 2009, **29**, 511–517.
37. A. J. Lucking, M. Lundback, N. L. Mills, D. Faratian, S. L. Barath, J. Pourazar, F. R. Cassee, K. Donaldson, N. A. Boon, J. J. Badimon, T. Sandstrom, A. Blomberg and D. E. Newby, Diesel exhaust inhalation increases thrombus formation in man, *Eur. Heart J..*, 2008, **29**, 3043–3051.
38. C. Carlsten, J. D. Kaufman, C. A. Trenga, J. Allen, A. Peretz and J. H. Sullivan, Thrombotic markers in metabolic syndrome subjects exposed to diesel exhaust, *Inhal. Toxicol.*, 2008, **20**, 917–921.
39. M. J. Campen, N. S. Babu, G. A. Helms, S. Pett, J. Wernly, R. Mehran and J. D. McDonald, Nonparticulate components of diesel exhaust promote constriction in coronary arteries from ApoE−/− mice, *Toxicol. Sci.*, 2005, **88**, 95–102.
40. M. J. Campen, J. D. McDonald, M. D. Reed and J. Seagrave, Fresh gasoline emissions, not paved road dust, trigger alterations in cardiac repolarization in ApoE−/− mice, *Cardiovasc. Toxicol.*, 2006, **6**, 199–210.
41. T. W. Cherng, M. J. Campen, T. L. Knuckles, L. Gonzalez-Bosc and N. L. Kanagy, Impairment of coronary endothelial cell ET_B receptor function

following short-term inhalation exposure to whole diesel emissions, *Am. J. Physiol. Regul. Integr. Comp. Physiol.*, 2009, **297**, 640–647.
42. B. K. Gehlbach and E. Geppert, The pulmonary manifestations of left heart failure, *Chest*, 2004, **125**, 669–682.
43. G. A. Wellenius, T. F. Bateson, M. A. Mittleman and J. Schwartz, Particulate air pollution and the rate of hospitalization for congestive heart failure among medicare beneficiaries in Pittsburgh, Pennsylvania, *Am. J. Epidemiology*, 2005, **161**, 1030–1036.
44. T. L. Knuckles, A. K. Lund, S. N. Lucas and M. J. Campen, Diesel exhaust exposure enhances venoconstriction through uncoupling of eNOS, *Toxicol. Appl. Pharmacol.*, 2008, **230**, 346–351.
45. M. J. Campen, A. K. Lund, T. L. Knuckles, D. J. Conklin, B. Bishop, D. Young, S. K. Sielkop, J. C. Seagrave, M. D. Reed and J. D. McDonald, Inhaled diesel emissions alter atherosclerotic plaque composition in ApoE$^{-/-}$ mice, *Toxicol. Appl. Pharmacol.*, 2010, **242**, 310–317.
46. C. Quan, Q. Sun, M. Lippmann and L. C. Chen, Comparative effects of inhaled diesel exhaust and ambient fine particles on inflammation, atherosclerosis, and vascular dysfunction, *Inhal. Toxicol.*, 2010, in press.
47. A. K. Lund, T. L. Knuckles, C. Obot Akata, R. Shohet, J. D. McDonald, A. Gigliotti, J. Seagrave and M. J. Campen, Gasoline exhaust emissions induce vascular remodeling pathways involved in atherosclerosis, *Toxicol. Sci.*, 2007, **95**, 485–494.
48. J. C. Chen and C. Nadziejko, Effects of subchronic exposures to concentrated ambient particles (CAPs) in mice. V. CAPs exacerbate aortic plaque development in hyperlipidemic mice, *Inhal. Toxicol.*, 2005, **17**, 217–224.
49. J. A. Araujo, B. Barajas, M. Kleinman, X. Wang, B. J. Bennett, K. W. Gong, M. Navab, J. Harkema, C. Sioutas, A. J. Lusis and A. E. Nel, Ambient particulate pollutants in the ultrafine range promote early atherosclerosis and systemic oxidative stress, *Circ. Res.*, 2008, **102**, 589–596.
50. M. J. Campen, A. K. Lund, M. Doyle-Eisele, J. D. McDonald, T. L. Knuckles, A. Rohr, E. Knipping and J. L. Mauderly, A Comparison of Vascular Effects from Complex and Individual Air Pollutants Indicates a Toxic Role for Monoxide Gases, *Environ. Health Perspect.*, 2010, in press. PMID: 20197249.
51. R. De Fré, P. Bruynseraede and J. G. Kretzschmar, Air pollution measurements in traffic tunnels, *Environ. Health Perspect.*, 1994, **102**(Suppl 4), 31–37.
52. M. Kamei and Y. Yanagisawa, Estimation of CO exposure of road construction workers in tunnel, *Ind. Health*, 1997, **35**, 119–125.
53. S. Niza and H. H. Jamal, Carbon monoxide exposure assessment among toll operators in Klang Valley, Kuala Lumpur, Malaysia, *Int. J. Environ. Health Res.*, 2007, **17**, 95–103.
54. Q. Sun, P. Yue, Z. Ying, A. J. Cardounel, R. D. Brook, R. Devlin, J. S. Hwang, J. L. Zweier, L. C. Chen and S. Rajagopalan, Air-pollution exposure potentiates hypertension through reactive oxygen species-

mediated activation of Rho/ROCK, *Arterioscler. Thromb. Vasc. Biol.*, 2008, **28**, 1760–1766.
55. Q. Sun, P. Yue, J. A. Deiuliis, C. N. Lumeng, T. Kampfrath, M. B. Mikolaj, Y. Cai, M. C. Ostrowski, B. Lu, S. Parthasarathy, R. D. Brook, S. D. Moffatt-Bruce, L. C. Chen and S. Rajagopalan, Ambient air pollution exaggerates adipose inflammation and insulin resistance in a mouse model of diet-induced obesity, *Circulation*, 2009, **119**, 538–546.
56. J. Lippmann, K. Ito, J. S. Hwang, P. Maciejczyk and L. C. Chen, Cardiovascular effects of nickel in ambient air, *Environ. Health Perspect.*, 2006, **114**, 1662–1669.
57. M. T. Kleinman, C. Sioutas, J. R. Froines, E. Fanning, A. Hamade, L. Mendez, D. Meacher and M. Oldham, Inhalation of concentrated ambient particulate matter near a heavily trafficked road stimulates antigen-induced airway responses in mice, *Inhal. Toxicol.*, 2007, **19**(Suppl 1), 117–126.
58. F. D. Gilliland, W. J. Gauderman, H. Vora, E. Rappaport and L. Dubeau, Effects of glutathione-S-transferase M1, T1, and P1 on childhood lung function growth, *Am. J. Respir. Crit. Care Med.*, 2002, **166**, 710–716.
59. F. D. Gilliland, Y. F. Li, A. Saxon and D. Diaz-Sanchez, Effect of glutathione-S-transferase M1 and P1 genotypes on xenobiotic enhancement of allergic responses: randomised, placebo-controlled crossover study, *Lancet*, 2004, **363**, 119–125.

CHAPTER 11
Manufactured Nanoparticles

G. S. KANG, P. A. GILLESPIE AND L. C. CHEN

Department of Environmental Medicine, New York University School of Medicine, 57 Old Forge Road, Tuxedo, New York 10987, USA

11.1 Nanoparticles and Nanotoxicology

Nanotechnology is defined as "a wide range of technologies that measure, manipulate, or incorporate materials and/or physical features with at least one dimension between approximately 1 and 100 nanometers (nm)".[1] There is rapidly growing interest in nanotechnology for various medical and technological applications, such as imaging, drug-delivery devices, cosmetics, sensors, and electronics, and it is estimated to become a $1 trillion market by 2015.[2] Although humans have been exposed to natural (*e.g.* viruses and condensation aerosols from volcanic eruptions and forest fires) or anthropogenic (*e.g.*, condensation aerosol effluents from power plants, jet engines, welding fumes) nano-sized particles (NSPs, particles < 100 nm) throughout history, recent industrial developments in nanotechnology are likely to add yet other sources for human exposure to NSPs: *i.e.* manufactured nanoparticles (NPs, also called engineered nanoparticles).[3] Considering the rapid growth of nanotechnology, it is essential to identify, quantify, and manage potential health risks related to NPs exposure.[4]

NPs have unique characteristics due to their small size, large surface area, and high reactivity. These characteristics contribute to their merits in application, but they also could be major factors contributing to their toxic potential. Indeed, various *in vitro* and *in vivo* studies have shown that, on an equal-mass basis, NPs induce stronger inflammatory or toxic responses than larger-sized counterparts of the same chemical composition.[3,5–7] This seemingly increased and/or different toxic potential of NPs may be related to the following three factors: (1) the unique physicochemical properties of NPs themselves; (2) different and/or more efficient deposition or uptake; and/or (3) transmembrane translocation.

NPs have much higher surface-to-volume ratios than larger particles, so a higher proportion of their atoms are on the surface, allowing them to more readily react with adjacent atoms and substances[8] including other particles or surrounding media. For example, it is often expected that nanoscale particulate materials dissolve more quickly and to a greater extent than macroscopic particles of the same material,[9] and this property may play an important role in inducing biological responses. Also, reduction in size may alter electrostatic properties of the material by increasing the number of structural defects and disrupting the well-structured electronic configuration, and this may lead to formation of superoxide radicals.[2] Recently, one study tested different types of titanium dioxide (TiO_2) particles varying in size and crystal structure for their potential to generate reactive oxygen species (ROS) in a cell-free system, and suggested that the inherent oxidant capacity of NPs was related to available surface defect sites.[10]

Particle size may affect patterns of their deposition, clearance and cellular uptake. For instance, it is well known that the deposition efficiency of inhaled particles in different anatomic regions of the respiratory tract (*i.e.* nasopharyngeal, tracheobronchial, and gas-exchange airways) is highly dependent on particle size.[3] Since the lung has extensive, location-specific defense systems such as mucociliary clearance in the conductive airways and macrophage clearance in the gas-exchange airways, different patterns of deposition, depending on the particle size, may lead to different mechanisms and efficiencies of clearance, and ultimately differences in potential for toxic effects.[9] Furthermore, it has been suggested that the size of particles may affect their uptake into different cell types.[11,12] For example, it has been shown that inhaled NPs are not recognized as efficiently as bigger particles by alveolar macrophages (AM), leading to increased contact with epithelial cells, and/or penetration into deeper regions of the lung,[13–15] potentially causing direct particle-induced effects.[3]

Another unique mechanism regarding NPs-induced toxicity is transmembrane translocation. It has been indicated that inhaled NPs, after deposition in the respiratory tract, can escape pulmonary clearance systems and translocate into other tissues including the blood stream, liver, heart and brain.[13,16,17] This phenomenon may provide an explanation for epidemiological findings of cardiovascular effects associated with inhaled ambient UFPs as a consequence of direct particle effect on the cardiovascular system.[3] There is also substantial evidence that systemic or oral administered NPs in rodents are captured by phagocytic cells in the spleen, liver, kidney and bone marrow.[18–20] One recent study has revealed that the biodistribution of intravenously administered gold particles is size dependent, showing NPs had greater accumulation in various organs.[21] Although it is more likely that the extent of translocation is highly dependent on physicochemical properties of particles,[3] these examples suggest that translocation may be a key factor in understanding the systemic toxicity induced by NPs.

In spite of a growing body of evidence indicating the unique toxic potentials of NPs, the environmental and health effects of NPs remain largely unknown.

Moreover, the present occupational and environmental exposure limit is based on mass, does not discriminate among particles of different size and takes no account of the likely enhanced toxicity of NPs in exposed individuals. Therefore, there is an urgent need to evaluate the potential health effects of NPs in occupational and environmental settings.

11.2 NP Exposure and Cardiac Toxicity

Human exposure to NPs is possible *via* six principle routes: intravenous, dermal, subcutaneous, inhalation, intraperitoneal, and oral,[22] and numerous studies have shown potential toxicity on various organs and/or tissues of human and animal followed by exposures to certain NPs.[3,5,6,23,24] Cardiac toxicity is one of the major concerns regarding NPs-induced toxicity, because: (1) some NPs are developed to be administered directly for targeting the cardiovascular system for diagnostic or therapeutic purposes; and (2) the cardiovascular system can be affected by secondary effects from NP exposure occurring *via* other routes.

11.2.1 Direct Cardiac Exposure to NPs

Over the last few decades, there has been growing interest in various applications of NPs to improve diagnosis and therapy for many diseases, and nanotechnology is widely used in practically every branch of medicine such as imaging, biosensing, drug delivery, tissue engineering, implants, and microsurgery.[20,24,25] The relatively large surface area of NPs is highly amenable for functionalization to attach peptides or antibodies that can precisely target cell types or tissues.[24] Based on this property, NPs have the potential to revolutionize; (1) biological imaging at the cellular level; (2) cancer detection and treatment; (3) radio- and chemo-sensitizing agents; and (4) targeted drug delivery.[26] In addition, protein nanochips are being developed to detect traces of proteins in biological fluids with much higher sensitivity than conventional bioassays and nanobiosensors can provide platforms to develop portable or implantable detection and monitoring devices.[25]

Cardiovascular research is one of the major fields for the aforementioned applications. For example, noninvasive assessment of atherosclerotic plaques is a highly desirable clinical goal; it has been a topic of active research using ligand-conjugated NPs to recognize the stage-specific markers of atherosclerosis.[27] More than ten types of NPs have been used in clinical and preclinical studies for stage-specific visualization of plaque progression, including quantum dots (QD) and iron oxide NPs.[27] In addition, one recent study reported the development of polymer-based nanomaterial for a synthetic heart valve based on its superior biocompatibility and *in vivo* biostability.[28] Another study found polyelectrolyte-coated gold nanorods had the potential to modulate cell-mediated matrix remodeling in cardiac fibroblasts, and suggested that these NPs could be applied for antifibrotic therapies.[29]

Most of these applications require direct administration of nanomaterials into the target cardiovascular tissues to be effective, and that means exposure to relatively high concentrations of NPs is likely to occur in those target tissues. This poses an emerging health concern, since there is accumulating evidence to suggest that NPs may exert adverse effects on the cardiovascular system.

11.2.2 Cardiovascular Effects by Pulmonary NP Exposure

In addition to cardiac effects resulting from direct exposure to NPs, it is possible that the cardiovascular system may experience secondary effects from NP exposure at distant organs. Among the six principle routes, inhalation is considered the major route of exposure for NPs,[3,30] especially in occupational settings. Moreover, several studies report associations between inhaled ambient NSPs (variably called ultrafine particles – UFPs, as well) and increased risk of cardiopulmonary disease.[31] Therefore, this chapter will focus on the cardiovascular effects of pulmonary exposure to NPs.

Epidemiological studies suggest that respiratory exposure to ambient ultrafine particles (UFPs) is associated with an increased risk of cardiopulmonary disease and mortality.[31] Recent studies show that exposure to fine ambient particles (including ambient NPs) for six months could enhance development of atherosclerosis in hyperlipidemic mice.[32,33] Since UFPs and NPs have several physical characteristics and toxicological properties in common,[34] it has been suggested that inhaled NPs might have the potential to induce systemic cardiovascular toxicity, as well.

To explain how pulmonary NPs exposure can elicit cardiovascular responses, several mechanisms have been proposed. The first hypothesis is that NPs deposited in the lung act through neural mechanisms to alter cardiac autonomic function.[35] The second hypothesis proposes that NPs deposited in the lung initiate local inflammatory responses *via* oxidative stress that further develop into systemic oxidative stress/inflammation.[36] Pulmonary oxidative stress and inflammation after NPs or PM exposure are well documented[3,37] and systemic pro-oxidative and proinflammatory reactions after PM exposure are also widely studied.[37] More recently, a third hypothesis was proposed whereby NPs deposited in the lung could translocate into the systemic circulation and directly interact with target tissues, including vascular endothelial cells (ECs), to induce injury, inflammation, and destabilization of plaques.[36] It should be noted that there is evidence to support all three mechanisms, and that they are not mutually exclusive. Some of the recent studies supporting these mechanisms are discussed below.

11.3 Study Review – Cardiac Toxicity by NP Exposure

There are reportedly 1300 nanomaterials that are either currently used or are being considered for use in industrial or commercial applications.[38]

Considering the rapid increase of publications in nanotechnology and nanotoxicology, reviewing every publication related to NP toxicity is not feasible, and is beyond the scope of this chapter. Therefore, in this section, studies investigating the most widely used nanomaterials are reviewed by categories; the four key NP classes, which are used in many commercial applications. These are: (1) fullerenes; (2) carbon nanotubes; (3) quantum dots; and (4) metallic and metal oxide-based nanoparticles.[38]

Because particle characterization is a crucial element in nanotoxicology, only studies that report detailed characterization are considered here. We have covered studies investigating the biological effects of NP exposure on the cardiovascular system, and therefore, studies mainly focusing on toxicokinetics or biodistribution of NPs are not discussed, unless they were relevant to cardiac toxicity.

11.3.1 Fullerenes

Fullerenes (C_x) are novel allotropes of carbon consisting of carbon atoms joined to form hollow spheres with a cage-like structure,[39] and the most popular form of fullerene is composed of 60 carbon atoms (C_{60}), which resembles a soccer-ball (also called Buckminsterfullerene or "Bucky balls"[40]).

Fullerenes are characterized by having numerous points of attachment, which provide an ample platform for functionalization.[30] In addition to their industrial usage as components in a variety of plastics, including filtration membranes,[41] they are envisioned as promising new platforms for drug delivery and other medical applications, due to the facts that: (1) they have hollow structures; and (2) the diameter of C_{60} is about 0.7 nm, which is roughly the size of many small pharmaceutical molecules.[39] Fullerenes and their hydroxylated derivatives (*i.e.* $C_{60}(OH)_{24}$) have been extensively studied, and they have been reported to have potent antioxidant and antibacterial properties.[41] Several studies have, however, raised safety concerns by showing potential cytotoxic effects of fullerenes and their derivatives.[41–43] Despite their popularity as candidates for drug-delivery devices, which often requires intravenous administration, little is known with regards to their potential cardiovascular toxicity. Some of the representative studies are summarized below.

Bosi *et al.* investigated the hemolytic effect of various water-soluble C_{60} derivatives on human red blood cells *in vitro*.[44] After 30-min incubation, C_{60} fullerenes with two cationic chains resulted in hemolysis of 40 to 50% of the cells at concentrations ranging from 20 to 60 µM, whereas compounds with bis-functionalized chains bearing carboxylic functions or only one cationic chain did not show any hemolytic effects up to 80 µM. This result suggests that the hemolytic potentials of fullerenes depend upon numbers and positions of cationic chains on the fullerene.

Another *in vitro* study regarding the acute toxicity of a polyhydroxylated fullerene derivative was conducted by Yamawaki and Iwai using human umbilical vein endothelial cells (HUVEC).[45] In this study, a 24-h exposure to 1–100 mg/l of fullerenol ($C_{60}(OH)_{24}$) induced cytotoxic morphological changes in HUVECs, such as vacuole formation in the cytosol and decreased cell density in a dose-dependent manner. Fullerene aggregates were accumulated in numerous autophagosomes in the cells after 24 h exposure and after 10 d exposure to 100 mg/l, vacuoles and spindles were observed along with decreased growth rates.

One recent study by Gelderman et al. also used HUVEC to determine the adverse effects of C_{60} and $C_{60}(OH)_{24}$ on endothelial cells.[41] After 24-h exposure to C_{60} water suspension (nC_{60}; 4 µg/ml) or $C_{60}(OH)_{24}$ (10, 50, and 100 µg/ml), both materials caused cell cycle arrests (G1) and acute Ca^{2+} influx of HUVECs. At a high concentration (100 µg/ml) of $C_{60}(OH)_{24}$, cell surface expression of intercellular adhesion molecule 1 (ICAM-1, also called CD54) and tissue factor (CD142) on HUVECs were significantly elevated. In addition, increased apoptosis was observed in the cells exposed to 100 µg/ml of $C_{60}(OH)_{24}$. These results indicate that exposure to fullerenes or their derivatives may induce proinflammatory and proapoptotic effects on the endothelium.

Satoh et al. investigated the effects of water-soluble C_{60} derivatives (mono- or di-malonic acid C_{60}) on vasomotor dysfunction, specifically on endothelium-dependent relaxation, using isolated rabbit aorta.[46,47] After incubation with the C_{60} derivatives (10 µM), potent and selective inhibitory effects were observed on the endothelium-dependent relaxation induced either by agonists (i.e. acetylcholine) or by endogenous nitric oxide (NO). Since these inhibitory effects of the malonic-acid C_{60} derivatives were masked in the presence of superoxide dismutase, it was suggested that these C_{60} derivatives inhibit endothelium-dependent agonist-induced relaxation of the aorta through the production of superoxide.

One recent study investigated the association between intraperitoneal injection of pristine C_{60} fullerenes and vasomotor dysfunction in the aorta of apoE$^{-/-}$ mice of different ages (11–13 vs. 40–42 weeks old). The vasomotor dysfunction was determined one hour after an intraperitoneal injection of 0.05 or 0.5 mg/kg of pristine C_{60} using the aorta segments mounted in myographs. In general, intraperitoneal injection of pristine C_{60} affected mainly the response to vasorelaxation in young apoE$^{-/-}$ mice, whereas the vasomotor dysfunction in old apoE$^{-/-}$ mice was less affected. Both endothelium-dependent and endothelium-independent vasorelaxation responses were slightly decreased in young apoE$^{-/-}$ mice after the pristine C_{60} treatments. These findings indicate that intraperitoneal administration is associated with a moderate decrease in the vascular function of mice with atherosclerosis.[48]

When interpreting the fullerene toxicity data, especially some of the earlier studies, it is important to consider the purity of the test materials. It has been demonstrated that residual tetrahydrofuran (THF) used for C_{60} solubilization substantially influenced the observed toxicity[41] and according to Andrievsky et al.,[49] the toxic effects observed by one earlier study[50] are associated with the presence of a large quantity of impurities ($\sim$10% of organic impurities).

11.3.2 Carbon Nanotubes

Carbon nanotubes (CNTs) are one of the most extensively used nanoparticles. Their global production is hundreds of metric tons per year and is expanding rapidly.[40] CNTs are allotropes of carbon with nanostructures that resemble a sheet of graphite (a hexagonal network of carbon atom) rolled into a hollow cylinder several millimeters in length, with diameters as small as 0.7 nm. There are two main forms of manufactured CNTs; single-walled carbon nanotubes (SWCNTs) composed of a single layer of cylinder, and multiwalled carbon nanotubes (MWCNTs) with several layers of concentric SWCNTs.[51]

Due to their unique physical and chemical properties (*i.e.* ultralight weight, high mechanical strength, high electrical conductivity), a large amount of research has been dedicated to their novel applications including electronic devices, polymer composites, and biochemical application such as enzymatic films, nanostructured medical devices such as tissue-engineered scaffolds, and constructs for intracellular drug/gene delivery.[30,40] However, it has been suggested that some of the unique properties that make CNTs attractive for various engineering applications may contribute to their potential for biological toxicity. Their length/width ratio, reactive surface chemistry and poor solubility have provided reasons for safety concerns, especially considering past experience with hazardous fibers, such as asbestos.[52–54]

The cellular uptake, cytotoxicity, and stimulatory effects of CNTs have been examined in several *in vitro* studies (reviewed by Shvedova *et al.*, 2009) and several dozen animal studies have shown that pulmonary exposure to SWCNTs or MWCNTs causes acute pulmonary inflammation as well as chronic responses such as fibrosis.[53,54] However, studies on the cardiovascular effects of CNTs are still scarce. At the time of this writing, only a few reports on the cardiovascular toxicity of CNTs have been published.

In one such study, Raja *et al.* exposed rat aortic smooth muscle cells (SMC) to SWCNTs and monitored the cell growth rates for 3.5 days.[55] To determine the effects of SWCNT aggregates, the authors filtered the SWCNT-suspended media and compared the toxicity before and after the filtration. Both types of exposure media resulted in significant inhibition in cell growth between day 1 and day 2.5, in a SWCNT dose-dependent manner (0.01–0.1 mg/ml). A comparison of filtered and unfiltered media showed that the finely dispersed NPs were significantly associated with the cell-growth inhibition, despite their low (<1%, v/v) concentration in the media.

In addition, Walker *et al.* investigated whether exposure to purified CNTs could alter cell function and induce cytotoxicity in human aortic endothelial cells (HAEC).[54] At 24 h postexposure, SWCNT or MWCNT at 50 and 150 μg/10^6 cells (1.5 and 4.5 μg/ml) caused a marked disruption of actin filaments and vascular endothelial (VE) cadherin cytotoxicity was observed with reduced tubule formation. These effects were not observed with carbon black (a control material) exposure or lower concentrations of CNTs. These findings suggest that CNT exposure induces specific effects on endothelial cells in a dose-dependent manner.

In another recent *in vitro* study, neonatal rat ventricular cardiomyocytes were exposed to SWCNTs, and changes in impulse conduction microfibrillar structure and the formation of reactive oxygen species were examined.[56] After exposure to SWCNTs at 0.25–25 μg/ml for 24 h, slight stimulating effects were observed in regard to heart cell function, *i.e.* impulse conduction velocity and action potential. These results indicate that SWCNTs can directly affect the function of the cardiac cells, although the effects of SWCNTs were milder when compared with other types of particles that were studied – diesel-exhaust particles or titanium dioxide.

In one of the few *in vivo* studies published so far, Li *et al.* demonstrated that pulmonary exposure to SWCNTs induced oxidative stress in the aorta and exacerbated plaque formation in hyperlipidemic mice (apoE$^{-/-}$).[57] First, they exposed wild-type C57BL/6 mice to SWCNTs by a single intrapharyngeal instillation (10 and 40 μg/mouse) and found that pulmonary SWCNT exposure increased mitochondrial DNA damage and protein carbonyl formation, and decreased mitochondrial glutathione in the aorta, at 7, 28, 60 days post-exposure. In the second part of their study, they repeatedly exposed apoE$^{-/-}$ mice to 20 μg of SWCNTs (every 2 w for 8 w) and quantified the plaque formation in the aorta. A significant increase in the relative plaque area based on surface area of the whole aorta (*en face*) was observed along with a significant increase in plaque lesions in the brachiocephalic arteries. These results clearly suggest that pulmonary exposure to SWCNTs can exacerbate cardiovascular disease in mice.

In another study, Nemmar *et al.* investigated the thrombogenic effects of pulmonary exposure to MWCNTs.[58] The authors exposed Swiss mice to 200 and 400 μg of MWCNTs by intratracheal instillation. At 24 h post-exposure, MWCNT-exposed mice showed substantial neutrophil influx in the lung, elevated plasma coagulant microvesicular tissue factor activity, and enhanced peripheral thrombogenicity. Circulating platelet–leukocyte conjugates were detected exclusively at 6 h post exposure. This indicates that there was early but transient activation of platelets induced by pulmonary MWCNT exposure. When P-selectin was neutralized *in vivo* using blocking antibodies, the MWCNT-induced thrombogenesis was mostly abrogated. Based on this result, the authors suggest that the mild lung inflammation induced by MWCNT exposure can translate into P-selectin-mediated systemic inflammation *via* platelet activation, which may lead to inflammation-induced procoagulant activity and prothrombotic risk.

It is important to consider that, due to the manufacturing process, CNTs may contain high levels of several impurities, including toxic metals (*i.e.* Co, Fe, Ni). These impurities can significantly affect CNT toxicity.[53,59,60] For example, Shvedova *et al.* reported that SWCNTs synthesized by certain methods could have significant levels of metal impurities (*i.e.* 30% of Fe or 20% of Ni), which could induce marked oxidative stress and decrease in viability in BEAS-2B cells.[61] It has also been suggested that the agglomeration status of CNTs in dispersion media may affect the observed toxicity.[53,60] Despite controversy regarding artifacts related to CNT toxicity, it is generally accepted that well-dispersed, purified CNTs exhibit relatively low cytotoxicity *in vitro*. Further studies with thoroughly characterized particles will help in understanding better the toxic potential of CNTs.

11.3.3 Quantum Dots

Quantum dots (QDs) are heterogeneous NPs consisting of a colloidal nucleus surrounded by one or more surface coatings, with sizes ranging from 2–100 nm.[22,26] The nuclei of quantum dots are made of either metals (*i.e.* Cd, Co, and Fe) or semiconductors. Due to wide diversity in their chemical composition, it has been suggested that quantum dots should not be considered a single uniform group of materials.[62] QDs have unique optical and electronic properties. They are highly stable, with "size-tunable" fluorescence,[26] which makes them particularly attractive for imaging and diagnostic purposes.[24] As the application of QDs to biomedical devices is advancing,[27] there has been rising concern over the fact that certain types of QDs have inherently toxic elements (*i.e.* Cd) in their core. Since cardiomyocytes have been reported to be more sensitive to cadmium-induced toxicity than other cell types,[63] there is an even stronger rationale for investigating the potential cardiovascular toxicity following QD exposure.

Unfortunately, little attention has been paid to the cardiovascular toxicity of QDs. Geys *et al.* have published the only *in vivo* toxicity study with QDs. These investigators delivered amine- and carboxyl-coated commercially available CdSe/ZnS QDs to mice by intravenous injections.[64] At high doses (144, 720, 3600 pmol/mouse), both types of QDs caused pulmonary vascular thrombosis, but the carboxyl-QDs were more potent in inducing this effect than amine-QDs. They also found that pretreatment with heparin mitigated pulmonary thrombosis induced by carboxyl-QDs, indicating that negatively charged QDs activate the coagulation process *via* contact activation rather than platelet activation.

11.3.4 Metallic and Metal-Oxide-Based NPs

Many metallic and metal-oxide-based NPs have been developed and are in use for various applications. For example, gold NPs are used extensively in electron microscopy and they are also approved for treatment of rheumatoid arthritis.[27] Active research is in progress on applications of superparamagnetic iron oxide NPs for cardiac magnetic resonance imaging (MRI).[65] However, the potential health effects of this growing class of new material are largely unknown. One of the most studied metal-based NPs for their toxicity is titanium dioxide (TiO_2).

TiO_2 is widely used as a white pigment for a wide range of products including paints, food colorants, cosmetics, and disinfectants[66] and it was once considered to be physiologically inert, posing little risk to humans.[67] However, a number of recent studies have shown that TiO_2 NPs (or ultrafine TiO_2) can cause significant cytotoxicity, genotoxicity, and trigger inflammatory responses.[3,67]

Although, TiO_2 NPs have been studied extensively, studies on the cardiovascular toxicity by TiO_2 NPs are still scarce.

Nurkiewicz and his colleagues recently published two articles regarding microvascular dysfunction induced by inhaled TiO_2. In this study, Nurkiewicz et al. exposed rats to fine or ultrafine TiO_2 aerosols (primary particle diameters: ∼1 μm vs. ∼21 nm) in a whole-body inhalation chamber at concentrations that did not induce pulmonary inflammation or lung damage determined by bronchoalveolar lavage markers.[6] Then, they infused Ca^{2+} ionophore A23187 intraluminally to evaluate endothelium-dependent arteriolar dilation in the spinotrapezius muscle. In rats exposed to ultrafine TiO_2, A23187 infusion produced arteriolar constrictions or significantly impaired vasodilator responses compared with responses observed in control rats or in those exposed to a similar pulmonary load of fine TiO_2. These results suggest that on an equivalent-pulmonary-deposition basis, ultrafine TiO_2 inhalation induces greater remote microvascular dysfunction compared to fine TiO_2.

The other study by Nurkiewicz et al.[68] was designed to identify potential mechanisms for the remote miscovascular dysfunction induced by inhaled ultrafine TiO_2 in the previous study.[6] They used the same exposure protocol, and measured the pulmonary deposition of two types of TiO_2 particles. After ultrafine TiO_2 exposure, the endothelium-dependent dilation in response to A23187 was significantly attenuated. Ultrafine TiO_2 exposure significantly increased microvascular oxidative stress by approximately 60%, and also elevated nitrosative stress dramatically (∼4-fold). In parallel, NO production was decreased in a particle deposition dose-dependent manner. They also found that radical scavenging or inhibition of certain enzymes (myeloperoxidase or NADPH) could partially restore NO production and microvascular function. These results indicate that inhaled ultrafine TiO_2 can affect the remote vascular system by not only inducing microvascular dysfunction but also by decreasing NO bioavailability.

11.4 A Case Study: Subchronic Effects of Inhaled Nickel Nanoparticles on the Progression of Atherosclerosis in a Hyperlipidemic Mouse Model

Since most nanotoxicology studies have investigated adverse effects by short-term exposures, our group designed the following study to determine whether long-term exposure to inhaled NPs induces adverse cardiovascular effects. Our hypothesis was that inhaled NPs could induce oxidative stress and trigger inflammatory responses, not only in the lung, but also in the cardiovascular system, and in the long term whether these effects exacerbate atherogenesis in $apoE^{-/-}$ mice.

Nickel was selected as the test material due to its popular use in industry and its potential cardiovascular toxicity.[69,70] Nickel-based NPs were generated in the laboratory using a Palas spark generator, and chemical composition of the generated particles was confirmed as nickel hydroxide, Ni(OH)$_2$ (nano-NH) *via* X-ray photoelectron spectroscopy (XPS).

For subacute and subchronic exposure, 5-month-old male apoE$^{-/-}$ mice were exposed either to filtered air or nano-NH (diameter of primary particle: 5 nm, count median diameter of agglomerates: ∼40 nm) at ∼80 µg Ni/m^3, less than 10% of the current permissible exposure limit (PEL) for nickel hydroxide set by the Occupational Safety and Health Administration (OSHA), for 5 h/d, 5 d/w, for either 1 week or 5 months. Various indicators of oxidative stress and inflammation were measured in the lung and cardiovascular tissue, and plaque formation on the ascending aorta were determined after 5 months of exposure.

The analyses using bronchoalveolar lavage fluid (BALF) revealed significant oxidative stress and pulmonary inflammation at both time points. These data were consistent with analyses of gene expression in the lung showing upregulation of genes like heme-oxygenase-1 (HO-1), interleukin-6 (IL-6) and monocyte chemotactic protein-1 (MCP-1). These 3 genes were also upregulated in the heart tissue after 5 months of exposure, indicating that inhaled nano-NH induces systemic effects. Furthermore, MCP-1 was overexpressed in the aorta tissue, along with CD68 and vascular cell-adhesion molecule-1 (VCAM-1), after the 5-month exposure. This phenomenon coincides with increased plaque formation in the ascending aorta, providing a molecular mechanism for the exacerbation of atherogenesis.

These results suggest that at occupationally relevant levels, chronic exposure to inhaled nano-NH significantly exacerbates atherosclerosis. These findings contribute to the further understanding of the potential risks and mechanisms of NPs-induced toxicity, and will help establish a database required for the development of nanoparticle-specific regulations in occupational and environmental settings.

11.5 Future Studies

In the last few years, significant progress has been made in nanotoxicology, both quantitatively and qualitatively. However, several important research topics need to be addressed to better understand the potential health risk of NPs, particularly their cardiovascular toxicity.

11.5.1 Human Data

First, because no human toxicity data are available, the potential adverse effects of NPs in humans remain unknown.[24] Although based on various animal studies and past experience with ambient particles or asbestos, it is reasonable to suspect that NP exposure may be deleterious to humans, and the lack of reliable predictive human data is one of the major limitations in establishing appropriate guidelines for occupational and environmental exposures.

11.5.2 Thorough Particle Characterization

Because numerous physicochemical parameters have been suggested to influence the toxicity of NPs, such as chemical composition, dissolution, particle size, size distribution, shape, and surface area,[71] it has been recommended that for toxicological studies NPs should be characterized as completely as practically feasible.[72] It has even been recommended that purchased NPs, which usually come with listed physicochemical properties provided by the manufacturer, should be independently characterized because of batch-to-batch variation, difficulties with quality control for the large production scale,[73] and the possibility of change during transport or storage. A thorough characterization of NPs will help not only in establishing a relationship between important physicochemical properties and toxic potential, but also in ensuring that the toxicological studies are reproducible and reliable.

Without sufficient characterization, it is extremely challenging to interpret the results of individual studies, and virtually impossible to compare the results of different studies. Although there are no current universally accepted set of parameters that are deemed necessary for NP characterization, recent reports have highlighted several key physicochemical elements that are strongly recommended to be reported. This includes methods of synthesis, size, size distribution, shape, composition, crystal structure, aggregation and agglomeration status, dissolution, purity, surface area, and other surface characteristics. Characterization of NPs in the context of the experimental exposure media (cell culture media, dosing solution, aerosol, *etc.*) is also of considerable importance as some physicochemical parameters are likely to differ depending on whether they are determined in the experimental media or in the bulk (*i.e.* "as received") state.[24]

11.5.3 Relevant Exposure Scenario

Although there is growing interest in evaluating the toxicity of NPs, most studies use either an *in vitro* system or nonphysiological *in vivo* exposure routes (*i.e.* intratracheal instillation [IT] or oropharngeal aspiration [OPA]). While these types of studies are useful for providing certain toxicological information, they have important limitations that must be taken into account. They often use extraordinarily high doses or exposure concentrations, which may deliver the NPs as a bolus (extremely high dose rate), so there is no or limited relevancy to real or anticipated human exposure scenarios.[34] They also lack consideration for clearance mechanisms or uptake/deposition pattern, which may play key roles in inducing NP-specific toxicity. In addition, studies utilizing *in vitro*, IT or OPA as their exposure methods are only able to measure short-term effects.

The aforementioned are well-known inherent limitations of *in vitro* tests and IT/OPA studies regardless of the test material. However, when applied to test NPs, additional concerns must arise because these methods usually administer NPs in the form of an aqueous suspension. The problem is NPs tend to agglomerate and clump in suspensions and agglomerates (a group of particles held together by relatively weak forces, *e.g.*, van der Waals or capillary that may break apart into smaller particles[1]) have greater diameter than a single particle. Since many aspects of NPs-induced toxicity are considered size dependent, delivering heavily agglomerated NPs could lead to inaccurate assessment of toxicity.[74] The strong tendency of NPs to agglomerate may also make it difficult to perform reproducible and reliable toxicological studies with them. Due to lack of standardization, many researchers have used a variety of different approaches to suspend NPs, which might be one of the reasons for contradictory results.[75] Recent studies have actually shown that dispersion pattern and state of agglomeration of the particles can vary significantly, depending upon the dispersing media[74,75] and these changes may greatly affect biological reactivity of the same material.[75–77]

11.6 Summary

In the last decade, engineered nanoparticles have become an important and unique class of new materials.[30] With a growing interest in the research and development of new nanotechnology, it is important to address concerns about the potential adverse health effects of NPs.

The cardiovascular system is one of the major targets for NP. Cardiovascular tissues can come into direct contact with NPs especially those that are developed to diagnose or treat cardiovascular diseases. In addition, the cardiovascular function could be affected by secondary effects from NP at the site of exposure. As has been summarized here, extensive data from *in vitro* and *in vivo* studies illustrate that various types of engineered NPs have the capacity to exert adverse cardiovascular effects. Additional research in this field will lead expectedly to a better understanding of the potential hazards associated with NP exposure and ultimately, perhaps, to the development of safe and effective production and applications of NPs.

Acknowledgment

The authors would like to thank Dr. Morton Lippmann at Dept. of Environmental Medicine, New York University School of Medicine, for his critical review of this chapter.

References

1. ASTM International, Standard terminology relating to nanotechnology, *ASTM Int.*, 2006, **E2456–06**.
2. A. Nel, Atmosphere. Air pollution-related illness: effects of particles, *Science*, 2005, **308**, 804–806.
3. G. Oberdorster, E. Oberdorster and J. Oberdorster, Nanotoxicology: an emerging discipline evolving from studies of ultrafine particles, *Environ. Health Perspect.*, 2005, **113**, 823.

4. B. T. Mossman, P. J. Borm, V. Castranova, D. L. Costa, K. Donaldson and S. R. Kleeberger, Mechanisms of action of inhaled fibers, particles and nanoparticles in lung and cardiovascular diseases, *Part. Fibre Toxicol.*, 2007, **4**, 4.
5. C. de Haar, I. Hassing, M. Bol, R. Bleumink and R. Pieters, Ultrafine but not fine particulate matter causes airway inflammation and allergic airway sensitization to co-administered antigen in mice, *Clin. Exp. Allergy*, 2006, **36**, 1469.
6. T. R. Nurkiewicz, D. W. Porter, A. F. Hubbs, J. L. Cumpston, B. T. Chen, D. G. Frazer and V. Castranova, Nanoparticle inhalation augments particle-dependent systemic microvascular dysfunction, *Part. Fibre Toxicol.*, 2008, **5**, 1.
7. Q. Zhang, Y. Kusaka, X. Zhu, K. Sato, Y. Mo, T. Kluz and K. Donaldson, Comparative toxicity of standard nickel and ultrafine nickel in lung after intratracheal instillation, *J. Occup. Health.*, 2003, **45**, 23.
8. M. C. Powell and M. S. Kanarek, Nanomaterial health effects – part 1: background and current knowledge, *WMJ*, 2006, **105**, 16.
9. P. Borm, F. C. Klaessig, T. D. Landry, B. Moudgil, J. Pauluhn, K. Thomas, R. Trottier and S. Wood, Research strategies for safety evaluation of nanomaterials, part V: role of dissolution in biological fate and effects of nanoscale particles, *Toxicol. Sci.*, 2006, **90**, 23.
10. J. Jiang, G. Oberdorster, A. Elder, R. Gelein, P. Mercer and P. Biswas, Does nanoparticle activity depend upon size and crystal phase?, *Nanotoxicology*, 2008, **2**, 33.
11. L. K. Limbach, Y. Li, R. N. Grass, T. J. Brunner, M. A. Hintermann, M. Muller, D. Gunther and W. J. Stark, Oxide nanoparticle uptake in human lung fibroblasts: effects of particle size, agglomeration, and diffusion at low concentrations, *Environ. Sci. Technol.*, 2005, **39**, 9370.
12. B. Rothen-Rutishauser, C. Muhlfeld, F. Blank, C. Musso and P. Gehr, Translocation of particles and inflammatory responses after exposure to fine particles and nanoparticles in an epithelial airway model, *Part. Fibre Toxicol.*, 2007, **4**, 9.
13. M. Geiser, M. Casaulta, B. Kupferschmid, H. Schulz, M. Semmler-Behnke and W. Kreyling, The role of macrophages in the clearance of inhaled ultrafine titanium dioxide particles, *Am. J. Respir. Cell Mol. Biol.*, 2008, **38**, 371–376.
14. M. Semmler-Behnke, S. Takenaka, S. Fertsch, A. Wenk, J. Seitz, P. Mayer, G. Oberdorster and W. G. Kreyling, Efficient elimination of inhaled nanoparticles from the alveolar region: evidence for interstitial uptake and subsequent re-entrainment onto airways epithelium, *Environ. Health Perspect.*, 2007, **115**, 728.
15. K. Unfried, C. Albrecht, L. Klotz, A. V. Mikecz, S. Grether-Beck and R. P. Schins, Cellular responses to nanoparticles: Target structures and mechanisms, *Nanotoxicology*, 2007, **1**, 52.
16. A. Elder, R. Gelein, V. Silva, T. Feikert, L. Opanashuk, J. Carter, R. Potter, A. Maynard, Y. Ito, J. Finkelstein and G. Oberdorster,

Translocation of inhaled ultrafine manganese oxide particles to the central nervous system, *Environ. Health Perspect.*, 2006, **114**, 1172.
17. L. E. Yu, L. L. Yung, C. N. Ong, Y. L. Tan, K. S. Balasubramaniam, D. Hartono, G. Shui, M. R. Wenk and W. Ong, Translocation and effects of gold nanoparticles after inhalation exposure in rats, *Nanotoxicology*, 2007, **1**, 235.
18. E. Fabian, R. Landsiedel, L. Ma-Hock, K. Wiench, W. Wohlleben and B. van Ravenzwaay, Tissue distribution and toxicity of intravenously administered titanium dioxide nanoparticles in rats, *Arch. Toxicol.*, 2008, **82**, 151.
19. E. Sadauskas, H. Wallin, M. Stoltenberg, U. Vogel, P. Doering, A. Larsen and G. Danscher, Kupffer cells are central in the removal of nanoparticles from the organism, *Part. Fibre Toxicol.*, 2007, **4**, 10.
20. J. X. Wang, Y. B. Fan, Y. Gao, Q. H. Hu and T. C. Wang, TiO(2) nanoparticles translocation and potential toxicological effect in rats after intra-articular injection, *Biomaterials*, 2009, **30**, 4590.
21. G. Sonavane, K. Tomoda and K. Makino, Biodistribution of colloidal gold nanoparticles after intravenous administration: effect of particle size, *Colloids Surf. B Biointerf.*, 2008, **66**, 274.
22. J. P. Ryman-Rasmussen, J. E. Riviere and N. A. Monteiro-Riviere, Variables influencing interactions of untargeted quantum dot nanoparticles with skin cells and identification of biochemical modulators, *Nano Lett.*, 2007, **7**, 1344.
23. A. R. Murray, E. Kisin, S. S. Leonard, S. H. Young, C. Kommineni, V. E. Kagan, V. Castranova and A. A. Shvedova, Oxidative stress and inflammatory response in dermal toxicity of single-walled carbon nanotubes, *Toxicology*, 2009, **257**, 161.
24. J. W. Card, D. C. Zeldin, J. C. Bonner and E. R. Nestmann, Pulmonary applications and toxicity of engineered nanoparticles, *Am. J. Physiol. Lung Cell. Mol. Physiol.*, 2008, **295**, L400.
25. K. K. Jain, Nanomedicine: application of nanobiotechnology in medical practice, *Med. Princ. Pract.*, 2008, **17**, 89.
26. B. A. Rzigalinski and J. S. Strobl, Cadmium-containing nanoparticles: perspectives on pharmacology and toxicology of quantum dots, *Toxicol. Appl. Pharmacol.*, 2009, **238**, 280.
27. K. Douma, L. Prinzen, D. W. Slaaf, C. P. Reutelingsperger, E. A. Biessen, T. M. Hackeng, M. J. Post and M. A. van Zandvoort, Nanoparticles for optical molecular imaging of atherosclerosis, *Small*, 2009, **5**, 544.
28. A. G. Kidane, G. Burriesci, M. Edirisinghe, H. Ghanbari, P. Bonhoeffer and A. M. Seifalian, A novel nanocomposite polymer for development of synthetic heart valve leaflets, *Acta Biomater.*, 2009, **5**(7), 2409–2417.
29. P. N. Sisco, C. G. Wilson, E. Mironova, S. C. Baxter, C. J. Murphy and E. C. Goldsmith, The effect of gold nanorods on cell-mediated collagen remodeling, *Nano Lett.*, 2008, **8**, 3409.
30. C. Medina, M. J. Santos-Martinez, A. Radomski, O. I. Corrigan and M. W. Radomski, Nanoparticles: pharmacological and toxicological significance, *Br. J. Pharmacol.*, 2007, **150**, 552.

31. C. A. Pope 3rd, R. T. Burnett, G. D. Thurston, M. J. Thun, E. E. Calle, D. Krewski and J. J. Godleski, Cardiovascular mortality and long-term exposure to particulate air pollution: epidemiological evidence of general pathophysiological pathways of disease, *Circulation*, 2004, **109**, 71.
32. L. C. Chen and C. Nadziejko, Effects of subchronic exposures to concentrated ambient particles (CAPs) in mice. V. CAPs exacerbate aortic plaque development in hyperlipidemic mice, *Inhal. Toxicol.*, 2005, **17**, 217.
33. Q. Sun, A. Wang, X. Jin, A. Natanzon, D. Duquaine, R. D. Brook, J. G. Aguinaldo, Z. A. Fayad, V. Fuster, M. Lippmann, L. C. Chen and S. Rajagopalan, Long-term air-pollution exposure and acceleration of atherosclerosis and vascular inflammation in an animal model, *JAMA*, 2005, **294**, 3003.
34. G. Oberdorster, V. Stone and K. Donaldson, Toxicology of nanoparticles: A historical perspective, *Nanotoxicology*, 2007, **1**, 2.
35. T. R. Nurkiewicz, D. W. Porter, M. Barger, L. Millecchia, K. M. Rao, P. J. Marvar, A. F. Hubbs, V. Castranova and M. A. Boegehold, Systemic microvascular dysfunction and inflammation after pulmonary particulate matter exposure, *Environ. Health Perspect.*, 2006, **114**, 412.
36. H. Yamawaki and N. Iwai, Mechanisms underlying nano-sized air-pollution-mediated progression of atherosclerosis: carbon black causes cytotoxic injury/inflammation and inhibits cell growth in vascular endothelial cells, *Circ. J.*, 2006, **70**, 129.
37. R. D. Brook, Cardiovascular effects of air pollution, *Clin. Sci. (London)*, 2008, **115**, 175.
38. T. Papp, D. Schiffmann, D. Weiss, V. Castranova, V. Vallyathan and Q. Rahman, Human health implications of nanomaterial exposure, *Nanotoxicology*, 2008, **2**, 9.
39. G. D. Nielsen, M. Roursgaard, K. A. Jensen, S. S. Poulsen and S. T. Larsen, In vivo biology and toxicology of fullerenes and their derivatives, *Basic Clin. Pharmacol. Toxicol.*, 2008, **103**, 197.
40. A. Poma and M. L. Di Giorgio, Toxicogenomics to improve comprehension of the mechanisms underlying responses of *in vitro* and *in vivo* systems to nanomaterials: a review, *Curr. Genomics*, 2008, **9**, 571.
41. M. P. Gelderman, O. Simakova, J. D. Clogston, A. K. Patri, S. F. Siddiqui, A. C. Vostal and J. Simak, Adverse effects of fullerenes on endothelial cells: fullerenol $C_{60}(OH)_{24}$ induced tissue factor and ICAM-I membrane expression and apoptosis *in vitro*, *Int. J. Nanomedicine*, 2008, **3**, 59.
42. C. M. Sayes, A. M. Gobin, K. D. Ausman, J. Mendez, J. L. West and V. L. Colvin, Nano-C_{60} cytotoxicity is due to lipid peroxidation, *Biomaterials*, 2005, **26**, 7587.
43. S. Zhu, E. Oberdorster and M. L. Haasch, Toxicity of an engineered nanoparticle (fullerene, C60) in two aquatic species, Daphnia and fathead minnow, *Mar. Environ. Res.*, 2006, **62**(Suppl), S5.
44. S. Bosi, L. Feruglio, T. Da Ros, G. Spalluto, B. Gregoretti, M. Terdoslavich, G. Decorti, S. Passamonti, S. Moro and M. Prato, Hemolytic effects of water-soluble fullerene derivatives, *J. Med. Chem.*, 2004, **47**, 6711.

45. H. Yamawaki and N. Iwai, Cytotoxicity of water-soluble fullerene in vascular endothelial cells, *Am. J. Physiol. Cell. Physiol.*, 2006, **290**, C1495.
46. M. Satoh, K. Matsuo, H. Kiriya, T. Mashino, M. Hirobe and I. Takayanagi, Inhibitory effect of a fullerene derivative, monomalonic acid C60, on nitric oxide-dependent relaxation of aortic smooth muscle, *Gen. Pharmacol.*, 1997, **29**, 345.
47. M. Satoh, K. Matsuo, H. Kiriya, T. Mashino, T. Nagano, M. Hirobe and I. Takayanagi, Inhibitory effects of a fullerene derivative, dimalonic acid C_{60}, on nitric oxide-induced relaxation of rabbit aorta, *Eur. J. Pharmacol.*, 1997, **327**, 175.
48. L. K. Vesterdal, J. K. Folkmann, N. R. Jacobsen, M. Sheykhzade, H. Wallin, S. Loft and P. Moller, Modest vasomotor dysfunction induced by low doses of C_{60} fullerenes in apolipoprotein E knockout mice with different degree of atherosclerosis, *Part. Fibre Toxicol.*, 2009, **6**, 5.
49. G. Andrievsky, V. Klochkov and L. Derevyanchenko, Is the C_{60} fullerene molecule toxic?, *Fullerenes, Nanotubes, Carbon Nanostruct.*, 2005, **13**, 363.
50. E. Oberdorster, Manufactured nanomaterials (fullerenes, C_{60}) induce oxidative stress in the brain of juvenile largemouth bass, *Environ. Health Perspect.*, 2004, **112**, 1058.
51. Y. Ju-Nam and J. R. Lead, Manufactured nanoparticles: an overview of their chemistry, interactions and potential environmental implications, *Sci. Total Environ.*, 2008, **400**, 396.
52. L. A. Mitchell, J. Gao, R. V. Wal, A. Gigliotti, S. W. Burchiel and J. D. McDonald, Pulmonary and systemic immune response to inhaled multi-walled carbon nanotubes, *Toxicol. Sci.*, 2007, **100**, 203.
53. A. A. Shvedova, E. R. Kisin, D. Porter, P. Schulte, V. E. Kagan, B. Fadeel and V. Castranova, Mechanisms of pulmonary toxicity and medical applications of carbon nanotubes: Two faces of Janus?, *Pharmacol. Ther.*, 2009, **121**, 192.
54. V. G. Walker, Z. Li, T. Hulderman, D. Schwegler-Berry, M. L. Kashon and P. P. Simeonova, Potential *in vitro* effects of carbon nanotubes on human aortic endothelial cells, *Toxicol. Appl. Pharmacol.*, 2009, **236**, 319.
55. P. M. Raja, J. Connolley, G. P. Ganesan, L. Ci, P. M. Ajayan, O. Nalamasu and D. M. Thompson, Impact of carbon nanotube exposure, dosage and aggregation on smooth muscle cells, *Toxicol. Lett.*, 2007, **169**, 51.
56. M. Helfenstein, M. Miragoli, S. Rohr, L. Muller, P. Wick, M. Mohr, P. Gehr and B. Rothen-Rutishauser, Effects of combustion-derived ultra-fine particles and manufactured nanoparticles on heart cells *in vitro*, *Toxicology*, 2008, **253**, 70.
57. Z. Li, T. Hulderman, R. Salmen, R. Chapman, S. S. Leonard, S. H. Young, A. Shvedova, M. I. Luster and P. P. Simeonova, Cardiovascular effects of pulmonary exposure to single-wall carbon nanotubes, *Environ. Health Perspect.*, 2007, **115**, 377.
58. A. Nemmar, P. H. Hoet, P. Vandervoort, D. Dinsdale, B. Nemery and M. F. Hoylaerts, Enhanced peripheral thrombogenicity after lung inflammation is

mediated by platelet-leukocyte activation: role of P-selectin, *J. Thromb. Haemost.*, 2007, **5**, 1217.
59. K. Donaldson, R. Aitken, L. Tran, V. Stone, R. Duffin, G. Forrest and A. Alexander, Carbon nanotubes: a review of their properties in relation to pulmonary toxicology and workplace safety, *Toxicol. Sci.*, 2006, **92**, 5.
60. C. W. Lam, J. T. James, R. McCluskey, S. Arepalli and R. L. Hunter, A review of carbon nanotube toxicity and assessment of potential occupational and environmental health risks, *Crit. Rev. Toxicol.*, 2006, **36**, 189.
61. A. A. Shvedova, T. M. Sager, A. R. Murray, E. Kisin, D. W. Porter and S. S. Leonard, Critical issues in the evaluation of possible effects resulting from airborne nanoparticles in *Nanotechnology: Characterization, Dosing and Health Effects*, ed. N. Monteiro-Riviere and L. Tran, Informa Healthcare, Philadelpia, PA, 2007, p. 221.
62. R. Hardman, A toxicologic review of quantum dots: toxicity depends on physicochemical and environmental factors, *Environ. Health Perspect.*, 2006, **114**, 165.
63. D. A. Limaye and Z. A. Shaikh, Cytotoxicity of cadmium and characteristics of its transport in cardiomyocytes, *Toxicol. Appl. Pharmacol.*, 1999, **154**, 59.
64. J. Geys, A. Nemmar, E. Verbeken, E. Smolders, M. Ratoi, M. F. Hoylaerts, B. Nemery and P. H. Hoet, Acute toxicity and prothrombotic effects of quantum dots: impact of surface charge, *Environ. Health Perspect.*, 2008, **116**, 1607.
65. K. W. Au, S. Y. Liao, Y. K. Lee, W. H. Lai, K. M. Ng, Y. C. Chan, M. C. Yip, C. Y. Ho, E. X. Wu, R. A. Li, C. W. Siu and H. F. Tse, Effects of iron oxide nanoparticles on cardiac differentiation of embryonic stem cells, *Biochem. Biophys. Res. Commun.*, 2009, **379**, 898.
66. E. J. Park, J. Yoon, K. Choi, J. Yi and K. Park, Induction of chronic inflammation in mice treated with titanium dioxide nanoparticles by intratracheal instillation, *Toxicology*, 2009, **260**, 37.
67. J. Chen, X. Dong, J. Zhao and G. Tang, In vivo acute toxicity of titanium dioxide nanoparticles to mice after intraperitioneal injection, *J. Appl. Toxicol.*, 2009, **29**, 330.
68. T. R. Nurkiewicz, D. W. Porter, A. F. Hubbs, S. Stone, B. T. Chen, D. G. Frazer, M. A. Boegehold and V. Castranova, Pulmonary nanoparticle exposure disrupts systemic microvascular nitric oxide signaling, *Toxicol. Sci.*, 2009, **110**, 191.
69. M. J. Campen, J. P. Nolan, M. C. Schladweiler, U. P. Kodavanti, P. A. Evansky, D. L. Costa and W. P. Watkinson, Cardiovascular and thermoregulatory effects of inhaled PM-associated transition metals: a potential interaction between nickel and vanadium sulfate, *Toxicol. Sci.*, 2001, **64**, 243.
70. M. Lippmann, K. Ito, J. S. Hwang, P. Maciejczyk and L. C. Chen, Cardiovascular effects of nickel in ambient air, *Environ. Health Perspect.*, 2006, **114**, 1662.
71. D. B. Warheit, P. J. Borm, C. Hennes and J. Lademann, Testing strategies to establish the safety of nanomaterials: conclusions of an ECETOC workshop, *Inhal. Toxicol.*, 2007, **19**, 631.

72. K. W. Powers, M. Palazuelos, B. M. Moudgil and S. M. Roberts, Characterization of the size, shape, and state of dispersion of nanoparticles for toxicological studies, *Nanotoxicology*, 2007, **1**, 42.
73. J. M. Pettibone, A. Adamcakova-Dodd, P. S. Thorne, P. T. O'Shaughnessy, J. A. Wedert and V. H. Grassian, Inflammatory response of mice following inhalation exposure to iron and copper nanoparticles, *Nanotoxicology*, 2008, **2**, 189.
74. T. M. Sager, D. W. Porter, V. A. Robinson, W. G. Lindsley, D. E. Schwegler-Berry and V. Castranova, Improved method to disperse nanoparticles for *in vitro* and *in vivo* investigation of toxicity, *Nanotoxicology*, 2007, **1**, 118.
75. M. C. Buford, R. F. Hamilton Jr and A. Holian, A comparison of dispersing media for various engineered carbon nanoparticles, *Part. Fibre Toxicol.*, 2007, **4**, 6.
76. A. A. Shvedova, E. Kisin, A. R. Murray, V. J. Johnson, O. Gorelik, S. Arepalli, A. F. Hubbs, R. R. Mercer, P. Keohavong, N. Sussman, J. Jin, J. Yin, S. Stone, B. T. Chen, G. Deye, A. Maynard, V. Castranova, P. A. Baron and V. E. Kagan, Inhalation *vs.* aspiration of single-walled carbon nanotubes in C57BL/6 mice: inflammation, fibrosis, oxidative stress, and mutagenesis, *Am. J. Physiol. Lung Cell. Mol. Physiol.*, 2008, **295**, L552.
77. R. R. Mercer, J. Scabilloni, L. Wang, E. Kisin, A. R. Murray, D. Schwegler-Berry, A. A. Shvedova and V. Castranova, Alteration of deposition pattern and pulmonary response as a result of improved dispersion of aspirated single-walled carbon nanotubes in a mouse model, *Am. J. Physiol. Lung Cell. Mol. Physiol.*, 2008, **294**, L87.

CHAPTER 12
Metals in Environmental Cardiovascular Diseases

A. BARCHOWSKY

University of Pittsburgh, Department of Environmental and Occupational Health, Graduate School of Public Health Street, Bridgeside Point, 100 Technology Drive, Suite 328, Pittsburgh, PA 15219-3130, USA

12.1 Introduction

Metals are both essential for general health and among the oldest toxicants known. The impact of many metals on human cardiovascular diseases is often unrealized and mechanistic understanding of their cellular and molecular actions or pathogenic effects is limited or controversial. Controversy often results from interpretation of high-dose toxicity studies designed to identify lethal dosages or acute toxicities in animals and attempts to translate these studies to identify causal disease etiologies in humans. Generally the etiology of chronic disease promotion for environmental or even most occupational metal exposures differs greatly from acute lethal mechanisms. Environmental metal exposures are widespread and may produce greater effects in susceptible populations or when exposures occur during development. Mechanisms of metals or metalloid action are often thought to be mediated by nonspecific interactions with peptide or protein sulfhydryls, produced by random oxidant injury, or caused by disrupting the normal signaling of selective ion channels. However, recent epidemiological studies and studies in genetic rodent models indicate that pathogenic effects of metals on cell signaling are not random and cannot be accounted for by direct competition with endogenous ions or excessive oxidant-mediated injury. Resolution of the molecular understanding of these pathogenic effects and toxicities remains a challenge. This chapter will

---

Issues in Toxicology No. 8
Environmental Cardiology: Pollution and Heart Disease
Edited by Aruni Bhatnagar
© The Royal Society of Chemistry 2011
Published by the Royal Society of Chemistry, www.rsc.org

present pathogenic effects of metals in the heart and blood vessels with an emphasis on the cellular and molecular actions underlying disease or clinically significant toxicities that are directly linked to human exposures to metals.

12.2 Overview of Metal Exposures

The majority of the elements in the periodic table are metals or metalloids. However, most of these are not toxic and many are essential to health. As elements, unlike organic toxicants, they can be redistributed but not destroyed in the environment. Ingestion of metals in food and water or inhalation of particles and fumes are the main routes of environmental or occupational exposure. In addition, metals are injected as therapeutics and these therapies have associated toxicities. In general, exposures to high, acutely toxic levels of many metals have been reduced by present-day occupational and environmental standards. In contrast, exposures to the more rare metals such as indium, gallium, yttrium, strontium, or gadolinium are increasing as they are used in manufacture of electronic components, advanced imagining technologies and therapeutic medical implants. Many of the toxicities of these rare metals and their formulations in novel applications (*e.g.* gallium arsenide semiconductors or nanoparticulates) are yet to be discovered. More importantly, there is growing interest in the mechanistic understanding of clinical and subclinical pathologies produced by chronic, lower dose exposures to metals and metalloids. These chronic exposures are highly relevant to the health of millions of humans worldwide and the cardiovascular toxicities of these exposures, as well as their role in what is often termed idiopathic disease, seems to be underappreciated.

Most cell types in the cardiovascular system are primary targets of metal toxicities. Since the endothelial cells see the highest circulating concentration of metals, endothelial dysfunction often precedes parenchymal cell toxicity. This was elegantly demonstrated for cadmium, which when infused into rats caused liver endothelial cells to lose junctional integrity and slough before hepatocyte injury was apparent.[1,2] These studies also demonstrate that toxicity depends upon increased uptake of the metal by the endothelial cells relative to hepatocytes and genetic differences in the uptake mechanism.[1,2] In addition to direct effects on smooth muscle contractility, metal-induced endothelial cell dysfunction often limits release of nitric oxide (NO) required for agonist stimulated vasodilation. Finally, the cardiac myocytes are well recognized targets of metals, which block ion channels and induce cardiomyopathy.

The concentration–response relationships for metal toxicity are complicated by both the concentration of the metal in a given tissue and the duration of exposure. Determining the dose or concentration of the active form of a metal at the tissue or cellular level remains difficult. This inability further complicates establishing accurate dose–response relationships when determining disease risks. Tissue uptake and the biological half-lives of metals in tissues vary greatly. For example, cadmium is actively transported into cells and may have a biological half-life in the kidney of 20–30 years, whereas arsenic freely follows water and is readily excreted with a biological half-life of hours to days. Dietary factors may also dictate the bioavailability or action of the metal making it difficult to demonstrate causal relationships for the metals in disease etiology.

Preferential uptake of metals into the different types of cardiovascular cells or their organelles significantly influences the pathophysiologic effects of the different metals. Uptake of the metals, their intracellular actions, and their cellular efflux depend on both the oxidation state of the metal and a number of selective transport and binding proteins. For example, trivalent arsenic (arsenite) is generally several orders of magnitude more toxic than pentavalent arsenic (arsenate) due to rapid uptake of arsenite through aquaporins relative to poor uptake of arsenate, which competes with phosphate for cell entry. However, once inside the cells, both species of arsenic exert actions on signaling events or metabolic functions with arsenite binding to critical sulfhydryls in proteins or arsenate mimicking phosphate in phosphotransfer reactions.[3] The oxidation states of iron and copper are also critical for their exchange between binding and transport proteins, as well as their cellular actions and catalysis of redox reactions.

Generally, it is the free or labile pool of the metal that participates in toxic actions, but fortunately metals are rarely free in the circulation or inside of cells. Instead, a number of selective metal transporters and chaperone proteins tightly bind the metals to facilitate tissue distribution and storage. Primary examples of the importance of sequestering proteins in preventing cardiovascular toxicity are seen with copper- or iron-binding proteins and transporters. Copper release from gastrointestinal enterocytes into the blood and its elimination from hepatocytes into bile are regulated by the Cu^+-transporting P-type ATPases, ATP7A and ATP7B respectively. Mutations in these two proteins cause either Menkes or Wilson's diseases that both present with cardiovascular sequelae. Menkes disease (ATP7A mutation) is an X-linked lethal disorder of intestinal copper hyperaccumulation with severe copper deficiency in peripheral tissues. This causes deficits in copper-dependent enzymes that lead to the clinical hallmarks of the disease including abnormal vascular development.[4,5] While free or labile copper participates in oxidant generation, copper deficiency may make the vasculature more susceptible to oxidative stress due to reduced copper in circulating, extracellular superoxide (SOD3).[5] Wilson's disease (ATP7B mutation) is an autosomal recessive disease characterized by striking hepatic and neuronal copper overload with oxidant stress,[4] as well as left ventricular remodeling and a relatively high frequency of benign supraventricular tachycardias and extra systolic beats.[6] The majority of copper is bound to ceruloplasmin and expression of this transporter in the liver is tightly linked to circulating copper levels.

Likewise, circulating iron is primarily bound to transferrin and uptake of transferrin by its cell-surface receptor and intracellular iron transfer for storage in ferritin are tightly regulated by expression of the respective proteins. Labile iron binds to iron regulatory proteins (IRPs/aconitase) that bind and regulate iron response elements (IREs) on the 3′- and 5′-untranslated regions of their respective transcripts.[7] In iron deficiency, IRP bound to DNA increases transferrin receptor mRNA stability and blocks translation of ferritin mRNA. In overload, iron-bound IRP is released from the mRNAs to destabilize the transcripts or enhance translation, respectively. Again, the purpose is to tightly regulate the labile iron pool to provide enough essential metal for enzyme reactions without providing excess iron that catalyzes ROS generation and enhances damage to the vessel walls.[8] Another example of the importance of metal-transport proteins is the increased endothelial surface expression of a specific cadmium-transport protein. This protein, ZIP8, is highly expressed on testicular endothelial cells and promotes hyperaccumulation of the metal that confers sensitivity to cadmium-induced testicular ischemia following endothelial cell injury and vascular leak.[9]

12.3 Mechanisms of Metal Action

There are multiple mechanisms through which metals exert their effects on cardiovascular functions. These could be direct effects caused by binding of metal to a critical protein or enzyme substrate and indirect effects mediated by increased ROS formation. Ligand binding is the most fundamental chemical process in metal-stimulated cell signaling, toxicity, and cellular defenses against metal toxicity. Depending on their oxidation state, metals readily react with sulfhydryls, carbonyl, or phosphate groups. For example, the coordination chemistry of copper ions varies with charge, since Cu^{1+} prefers sulfur donor ligands, such as cysteine or methionine. In contrast, Cu^{2+} prefers nitrogen donors such as histidine or oxygen donors such as glutamate or aspartate.[4] When bound as the divalent cation, copper is capable of enzyme inactivation through redox cycling and ROS generation.[10] The strong interaction of many metals with sulfur in cysteine and nitrogen in histidine underlies the coordination of the metals in the catalytic centers of enzymes and provides tertiary structure for ligand recognition. This is evident in the Cys/Cys or Cys/His coordination of zinc in protein kinase C family members that provides lipid cofactor binding and in the zinc fingers transcription factors that provide specificity for binding DNA sequences. Displacement of essential metals from these coordination complexes is an important mechanism of toxicity. Displacement of metals from storage proteins, such as cadmium or zinc release from the multiple cysteine binding sites in metallothionein or release of iron from ferritin, results in indirect cellular actions of some metals. A number of metals at high concentrations interfere with calcium channels (nickel, cobalt, magnesium, manganese, cadmium, and lead) or potassium (thallium) channels. However, this inference may be relevant only to acute toxicity caused by very high occupational or accidental exposures (*e.g.* arrhythmias in hypermagnesemia resulting from errors in therapeutic ion replacement). More commonly, metals produce secondary effects by increasing free concentrations of endogenous metals to produce adverse signaling, such as mobilization of intracellular calcium and or mitochondrial calcium leak, which are secondary to increased ROS generation.

ROS-mediated cell signaling and increased oxidative stress are independent indirect mechanisms of metal actions and toxicities. The two are defined by the level ROS generated and the mechanism of their generation. Signal-generated low levels of ROS, such H_2O_2, are second messengers for many receptor-mediated vasoactive and mitogen responses in endothelial and smooth muscle cells. High ROS levels from respiratory bursts or inflammatory cells or mitochondrial leak following injury overwhelm cellular antioxidant defenses, promote mitochondrial calcium leak, and damage or kill cells. The ability of metals to generate ROS differs since not all metals affect the same signaling mechanisms and not all metals coordinate electron transfer for redox reactions. Several metals, such as free iron and copper, are catalysts that propagate formation of cell-membrane-damaging oxygen-centered lipid radicals and peroxides, as well as incomplete reduction of molecular oxygen through Haber–Weiss and Fenton reactions with superoxide and H_2O_2.

$$Fe^{2+} + O_2 \rightarrow (FeO_2)^{2+}(\text{oxidant}) \rightarrow \text{Lipid}\cdot$$

$$\text{Lipid}\cdot + O_2 \rightarrow \text{LipidOO}\cdot$$

$$Fe^{2+} + H_2O_2 \rightarrow Fe^{3+} + OH + OH^-$$

The end result is generation of lipid or hydroxyl radicals that react indiscriminately with and damage proteins and DNA.[4,11] This damage is only relevant to increased macromolecule degradation and cell death, since reactivity of the radicals is diffusion limited and too random to coherently affect cell signaling. In contrast, most cardiovascular toxic metals are not capable of catalyzing oxygen-centered radical or peroxide generation at concentrations relevant to human exposures and under biological or physiological conditions. Instead, they indirectly act stimulate signaling pathways that increase NADPH oxidase-generated ROS or disrupt mitochondrial enzymes to collapse respiration and initiate apoptotic cascades.

Mechanistic understanding of toxicity to cardiac or vascular cells for many metals is confused by biphasic or multiphasic concentration response relationships. Functional or phenotypic changes usually occur at low metal concentrations. In contrast, apoptosis or necrosis occurs at metal levels that promote the generation of overwhelming levels of ROS. For example, arsenic has both positive and toxic effects on vascular cells that can be prevented by increasing antioxidant capacity. Low, environmentally relevant levels of arsenic stimulate both endothelial and smooth muscle cell NAD(P)H oxidases[12,13] to rapidly increase superoxide generation. This rapid increase may result from stimulation of G-protein coupled receptors that initiate signal amplification to activate the membrane-bound NADPH oxidase complex.[14] The levels of superoxide and hydrogen peroxide generated by NAD(P)H oxidase in response to arsenic or other metals are in keeping with the amount of ROS generated by the enzyme when cells are stimulated by endogenous peptides or lipids, such as angiotensin II, platelet-derived growth factor, or sphingosine-1-phosphate. These endogenous peptides and lipids are known to be smooth muscle cell mitogens, proangiogenic, and hypertensive and excess signaling in response to these ligands is a longitudinal CVD risk factor.[15,16] Thus, arsenic and some other metals may act through the receptors or signaling pathways associated with the receptors for these endogenous risk factors to elicit disease. This contrasts with the generation of lethal oxidative stress. In this case, higher concentrations of arsenic promote lethal oxidant stress in endothelial cells by targeting mitochondrial enzymes and interfering with phosphotransfer reactions. These levels are relevant to the vascular toxicity promoted in therapies designed to destroy tumor blood supplies.[3,17,18]

Copper has several unique mechanisms for promoting disease through oxidant stress. As indicated, copper directly generates ROS by redox cycling.[4,10] However, copper deficiency increases oxidant stress and vascular toxicity by decreasing the activity of copper-containing antioxidant enzymes, such as Cu/Zn SOD. Oxidative stress from copper deficiency also compromises the immune system, resulting in increased virulence of cardiac pathology from amyocarditic and myocarditic viruses.[4,19] Copper is not alone in this potentiation, nickel, cadmium and mercury induced inflammatory cardiac disease is also affected by viral infections.[20,21] Increased copper levels in plasma correlate with increased levels of oxidized lipoproteins; although the mechanisms for this increase are unclear and not simply defined by direct redox-catalyzed reactions.[22] Similarly, copper increases the levels of the known endogenous cardiovascular toxicant, homocysteine, which may contribute to its association with oxidative vascular dysfunction and increased peripheral vascular disease.[23–25] Copper, as well as iron, combines with homocysteine to promote proatherosclerotic LDL oxidation and enhanced oxidative vessel injury in aging.[8,26]

Metals influence gene transcription by directly binding to transcription factors. In addition to iron-responsive IRP/aconitase, there are several important regulators of transcription that are metal responsive and induce genes encoding proteins for adaptation, defense through metal sequestering, and stress toxicity. The metal-responsive-element-binding transcription factor-1 (MTF-1) contains 6 zinc finger motifs that are critical to its ability to transactivate important protective genes, such as the metallothioneins.[27,28] The protein responds only to zinc and stimulation of MTF-1 transactivation by other metals results from displacing zinc from intracellular pools, such as metallothionein itself, or indirect signaling effects on MTF-1 post-translational modifications. Deficiency of MTF-1 greatly contributes to the general toxicity of metals in many tissues, but also specifically to toxicity in vascular endothelial cells.[29] Nuclear factor-E2 related factor 2 (Nrf2) is another metal-responsive factor that regulates the transcription of protective or stress-responsive genes. Metals either directly or indirectly promote oxidation of key thiols in the cytoplasmic anchor protein KEAP1 and its posttranslational phosphorylation. Modified KEAP1 releases Nrf2 allowing Nrf2 nuclear transport and transactivation of antioxidant response elements (ARE) or stress-response elements (SRE) in the promoters of genes, such as hemooxygenase-1,[30–33] glutamate-cysteine ligase,[34] and thioredoxin.[35]

Another significant target for metal-induced gene expression is the hypoxia-inducible factor-1α (HIF-1α). In normoxia, HIF-1α protein is rapidly degraded following post-translational hydroxylation of conserved proline residues in a region of the peptide referred to as the oxygen-dependent degradation (ODD) domain.[36–39] This hydroxylation is catalyzed by 2-oxoglutarate-dependent prolyl-4-hydroxylases (PHD), which contain iron, and the hydroxylated proline residues in the ODD facilitate proteosome degradation following recognition by the von Hippel–Lindau.[36–39] HIF-1α is a an important mediator of induced cardiovascular protection, especially during late preconditioning, which protects against myocardial ischemic injury.[38] However, HIF-1α participates in pathological vessel remodeling as well, such as in tumor angiogenesis and pulmonary arterial remodeling that leads to pulmonary hypertension.[36,37] The PHDs are targets of many metals, although there appears to be little commonality amongst the metals (e.g. nickel, cobalt, or arsenic) in the mechanism by which they affect PHD-induced destabilization of HIF-1α. Low concentrations of cobalt mimic hypoxia to cause HIF-1α-induced preconditioning and improve cardiac contractility.[40] Similar preconditioning by cobalt infusions protects against ischemic injury in a variety of tissues.[41,42] In contrast, cadmium inhibits myocardial hypoxic preconditioning by preventing HIF-1α stabilization and gene transactivation.[43] It is important to remember that transcription factors are highly interactive in complexes that transactivate or repress promoters genes involved in both protection and cardiovascular disease etiology.[44] In fact, metal induction of metallothioneins may result from interactions between activated MTF-1 and HIF-1α.[45]

12.4 Pathogenic Actions of Metals in the Heart

12.4.1 Metal-Induced Cardiomyopathies

Cardiomyopathies, especially dilated cardiomyopathy, are significant diseases of the heart muscle characterized by progressive cardiac dilation and contractile dysfunction, often with signs of hypertrophy. Most cases of dilated cardiomyopathy are idiopathic and the importance of metal-related cardiomyopathies is underscored by reports of elevated trace metals in cardiac tissues of patients with idiopathic dilated cardiomyopathies.[46] In general, metal-related myopathies are chronic diseases rather than acute toxic manifestations of effects on cardiac myocytes or the blockage of myocyte ion channels. Trace elements claimed to account for idiopathic disease include cadmium,[29] mercury,[46] antimony,[46] and arsenic.[47,48] Iron overload also produces cardiomyopathy; however, these overloads are often associated with genetic diseases, such as thalassemia.[49] Metals may enhance the efficacy of other toxicants in causing myopathy, such as reports of patients presenting with often fatal cardiomyopathies after consuming alcoholic beverages contaminated with either cobalt[50] or arsenic.[51] In contrast, deficiencies in essential metals, especially copper, zinc, and selenium, enhance myopathies due to increased oxidative stress or enhanced virulence of myocarditic strains of viruses.[19,52] The latter may be mediated by increased inflammatory responses and decreased innate immune responses due to lack of copper- or selenium-containing antioxidant and protective enzymes.[19]

Metals alter myocardial cell function to produce dilated myopathy *via* multiple mechanisms. Disrupting energy production by direct metal binding to cysteines in critical mitochondrial enzymes is a primary mechanism for cadmium-, cobalt-, mercury-, and arsenic-induced myopathy. Cadmium accumulates in the heart and binds to complex III in the mitochondrial respiratory chain to uncouple oxidative phosphorylation and increase mitochondrial release of oxidants. Higher levels of cadmium interfere with intracellular calcium mobilization, but it is not clear whether this effect is primary or secondary to the increased oxidant stress.[29] Recent evidence indicates that chronic low-level cadmium exposures target vascular endothelial cells in the heart microvessels, instead of myocytes, to cause myopathies.[29] Cadmium decreases vascularization of the heart by promoting endothelial cell dysfunction and loss of junctional integrity[29] and increased endothelial cell permeability has been a common finding for cadmium effects outside of the heart.[9,53] It is not clear whether this effect is linked to mitochondrial enzyme inhibition or to changes in adhesion molecules and proteins in VE-cadherin-dependent junctions that maintain endothelial cell integrity.

Chronic cobalt exposure targets mitochondrial enzymes in complex II and complex III, with modest effects on respiration and a minimal lowering of ATP levels.[54] Chronic exposures cause pronounced decreases in mitochondrial manganese superoxide dismutase levels in rat hearts, which may contribute to oxidative myocardial injury.[54] In addition, cobalt-stabilized HIF-1α increases expression of endothelin-1,[55,56] which could trigger cardiac hypertrophy and remodeling.

12.4.2 Cardiac Arrhythmias

Metals disrupt electrical conduction in the heart through several mechanisms. The simplest mechanism is interference with or blockage of calcium, potassium, or sodium channels. High, mM concentrations of nickel are commonly used in research to block cardiac calcium channels. Clinically, magnesium toxicities in hypermagnesiumemia also relate to calcium ion channel blockade, but this occurs only at very high, accidental exposures to the metal. These acute accidental or adverse clinical exposures result in metal concentrations that exceed blood calcium concentrations. In contrast, acquired long QT syndrome (acLQTS) is a significant dose-limiting side effect of antileukemic arsenic therapies that occurs when the levels of circulating arsenic approach 5 μM.[57–59] Prolonged QT intervals have also been associated with chronic environmental exposures to arsenic in drinking water when circulating arsenic levels are often submicromolar.[60] LQTS reflects slowed ventricular repolarization at the cellular level and is characterized by a prolongation of the QT interval on the electrocardiogram. As a direct consequence of abnormal repolarization in LQTS, arsenic-sensitive patients may present with syncope, torsades de pointes, arrhythmias, or sudden cardiac death.[57,58] The cardiac hERG (human ether-a-go-go related gene) potassium channel is the α-subunit of the rapid delayed rectifier current IKr in ventricular myocytes. This channel contributes prominently to terminal repolarization and is the target of many chemicals and therapeutics that cause acLQTS.[58,61] However, in contrast to most of these agents that bind and block the channel, arsenic is devoid of direct acute effects on cardiac repolarization, and it does not inhibit hERG/IKr.[58,61] Instead, arsenic reduces trafficking of hERG/IKr to myocyte membranes by affecting molecular chaperone proteins, such as Hsp90.[58,59]

Trivalent antimony, which is cardiotoxic and similar to trivalent arsenic, also inhibits hERG/IKr trafficking.[58] Pentavalent antimony is a mainstay in treating humans for leshmaniasis and schistosomiasis. However, its use is limited by severe cardiac side effects after conversion to the trivalent form. Trivalent antimony is more acutely cardiotoxic and causes pronounced EEG alteration with QT prolongation and ST wave flattening.[62] In addition to effects on the ion channels, antimony impairs ventricular muscle contractility. These effects are related to mitochondrial stress that can be prevented by feeding L-carnitine to preserve β-oxidation of fatty acids.[62] Thus, antimony, like arsenic, could produce dysrrhythmias both by affecting ion-channel expression and by causing myocyte injury.

Increased exposure to ambient airborne particulate matter (PM) has been correlated with increased emergency-department admissions for cardiac arrhythmias[63,64] and $PM_{2.5}$ exposure increases the risk of negative cardiovascular events, especially in women.[65] Epidemiological evidence suggests that these adverse health effects depend on both the amount and duration of exposure with long-term exposures having larger and more persistent cumulative effects than short-term exposures.[66] Human and experimental animal studies demonstrate that both acute neural and protracted inflammatory mechanisms account for the pathogenic cardiovascular responses to PM.[63] This makes identifying the dominance of inflammatory or accelerated atherosclerosis or altered cardiac autonomic function mechanisms difficult. In addition, it is important to keep in perspective the much lower risk of cardiovascular effects from ambient exposures relative to much greater risk from direct or indirect exposure to tobacco smoke. However, it is clear that the effects of ambient fine particle are additive, if not synergistic, with tobacco smoke in increasing the risk of cardiac ischemic disease and dysrrhythmias.[63]

The composition of PM can vary greatly and a number of metal constituents within PM, especially nickel, vanadium, and zinc, have been implicated in promoting adverse cardiac events and heart-rhythm variability.[67] Acute changes in heart rate and its variability occur in mice exposed to concentrated fine ambient particles containing Ni, relative to control or vanadium-containing particles.[68] Daily cardiovascular mortality rates analyzed in the 60-city

National Mortality and Morbidity Air pollution Study were significantly associated with average nickel and vanadium, but not with other measured metal species.[68] In contrast, a recent epidemiological study demonstrated that communities with higher $PM_{2.5}$ content of nickel, vanadium, and elemental carbon have higher risks of hospitalizations for cardiac and pulmonary events associated with short-term $PM_{2.5}$ exposures.[67] As mentioned above, these effects cannot be ascribed to the well-known effect of blocking cardiac calcium channels, since the amounts of Ni in the particulate matter in these studies and in most ambient exposures are negligible compared to circulating calcium levels. Known biological mechanisms cannot account for the significant associations between Ni and the acute cardiac functional changes in mice or cardiovascular mortalities in people.[68] It is also not clear whether the effects are at the level of an autonomic reaction to the inhalation of the nickel or chronic systemic inflammatory injury.

Zinc represents a significant soluble component of some PM that has been shown to contribute to cardiac injury both *in vivo* and in isolated cardiac myocytes.[69–71] Acute and prolonged exposure of cultured rat ventricular myocytes to zinc reduced spontaneous beat rates and differentially induced expression of a number of potassium channel proteins.[70] *In vivo*, intratracheal or nose-only exposure of WYK rats to environmentally relevant zinc-containing PM for 8–16 weeks decreased the number of myocardial granulated mast cells, and produced multifocal myocardial degeneration, chronic-active inflammation, and fibrosis, relative to rats exposed to clean air or nonzinc-containing PM.[71,72] At the cellular level, these zinc PM exposures caused ventricular cell mitochondrial enzyme inhibition.[71] In addition to confirming that genes involved in ion channel function were induced, this study found zinc caused modest changes in gene transcripts encoding for proteins involved in signaling, oxidative stress, mitochondrial fatty-acid metabolism, and cell-cycle regulation.

12.4.3 Ischemic Diseases and Atherosclerosis

Arteriosclerosis, and especially atherosclerosis, or occlusive disease is the most common pathologic process underlying cardiovascular diseases. Atherogenesis can be systemic or confined to individual organs. Environmental exposures to metals, such as cadmium and arsenic, as well as seemingly normal accumulation of iron or copper may play a role in atherogenesis. However, many large epidemiological and clinical studies provide conflicting results that preclude drawing conclusions about direct causal relationships between metals and atherogenesis and their mechanisms.

The roles of iron and copper in the etiology of age-related atherogenesis have been extensively studied. In a recent review, Brewer concluded that there is stronger evidence that levels of iron and copper that are normal and adequate in reproductive years become clear risks for age-related atherosclerosis.[8] Many of the studies that failed to find associations of iron or ferritin levels with disease failed to accurately measure the labile pool of iron that would participate in oxidative injury to the cardiovascular tissues.[8] In addition to citing support from definitive, well-controlled animal studies, recent review of the literature indicates that there is high-level iron deposition in human atherosclerotic lesions, that H- and L ferritin mRNAs are higher in human and rabbit atherosclerotic vessels than in normal ones, that iron colocalizes with ceroid (an insoluble polymer of oxidized lipid and protein) in human atherosclerotic tissue, and that iron chelators inhibit low-density lipoprotein (LDL) oxidation.[8,73] In addition, as discussed for copper, labile iron and homocysteine cooperate to enhance oxidant generation and atherogenic lipid oxidation that contributes to disease etiology.[74] Many mechanistic studies, however, fail to make causal links between critical, rate-limiting iron effects in atherosclerotic pathogenesis that are distinguished from effects that result from the general atherogenic progression.

12.4.4 Copper

Elevated labile copper is a well-established risk for atherogenesis, as well as coronary artery disease. Molecular studies are generally supportive of a relationship of copper to the atherogenic process and mechanistic studies find elevated levels of copper in human atherosclerotic plaques. These studies also find that copper oxidizes LDL and that apolipoprotein E may owe its antioxidant effects to inhibiting this oxidation.[8] A significant portion of copper-related ischemic disease results from effects on and interactions with homocysteine. Risk of both ischemic disease and impaired endothelial-dependent vasorelaxation is reduced by decreasing copper to lower circulating homocysteine levels.[8,23,24] There is a correlation between total homocysteine and copper levels in patients with peripheral vascular disease[23] and copper chelation with penicillamine prevents high homocysteine levels from inhibiting vasodilation.[24] Mechanistically, copper could increase homocysteine levels by decreasing its degradation in a transulferation metabolic pathway.[25] Finally, the homocysteine at high levels contributes to the increase of copper toxicity by binding the metal and facilitating catalysis of LDL oxidation.[24]

12.4.5 Arsenic

Environmental exposures to arsenic in drinking water[75-79] or occupational exposures through inhalation[80] have been linked to increased cardiac ischemic diseases and infarctions. However, many epidemiological studies have technical flaws or lack direct arsenic exposure measurements. This confounds conclusive interpretations and prevents clear identification of the mechanism for the atherogenic potential of arsenic.[70] Nonetheless, there is substantial evidence supporting the view that arsenic promotes ischemic heart disease in humans and mice. Arsenic-exposed individuals in Taiwan present with thickened coronary and carotid arteries, even years after the exposures cease.[76,81,82] Recent evidence from carefully controlled studies of widespread exposures in Bangladesh indicate that arsenic may contribute to disease by promoting long-term vascular inflammation, as indicated by persistent elevation of adhesion molecules that regulate pathogenic leukocyte–endothelial cell interactions.[83,84] In addition to not measuring individual exposures, a potentially confounding factor that weakens earlier epidemiological studies is failure to determine an individual's age at the onset of exposure. This is important since *in utero* exposure or exposure to arsenic early in childhood has been shown to greatly increase the risk of acute myocardial infarction, presumably related to ischemic disease later in life.[79] The atherogenic potential of arsenic may also be enhanced by *in utero* exposures, since mice exposed to arsenic *in utero*, but not after birth, retained increased sensitivity to atheroma formation when placed on high-fat diets.[84,85] Other studies have indicated that direct damage to the endothelial cell monolayer, altered nitric-oxide metabolism, and possibly loss of barrier function may contribute to arsenic-induced atherogenesis in adult rodent models.[86–88] There is also strong evidence that, as with endogenous risk factors for atherosclerosis, arsenic signaling in both endothelial and smooth muscle cells stimulates NADPH oxidase generation of atherogenic ROS. Arsenic promotes pathogenic differetiation and capillarization of liver sinusoidal endothelial cells in mice that reduces lipid and protein scavenging across the liver.[89] Liver sinusoids in $p47^{phox}$ $-/-$ mice lacking NOX2-based NADPH oxidase were protected from arsenic exposures and hydrogen peroxide, but not arsenic, promoted capillarization of liver sinusoidal endothelial cells isolated from the knockout mice.[89] These studies support a critical role for the NADPH oxidase-derived peroxide in signaling for functional changes in the endothelium that may impact lipid deposition in systemic atherogenesis.

Blackfoot disease is a unique peripheral vascular disease found in certain populations exposed to high levels of arsenic (>500 ppb) in their drinking water. First described in endemic regions of southwest Taiwan, this disease is a form of arteriosclerosis obliterans that promotes systemic ischemic disease, dry gangrene, and spontaneous amputation of the affected extremity. Blackfoot disease has also been described in mining areas of Central and South America. An epidemic of a similar peripheral vascular disease was seen in German vintners; although this form may have resulted from combined effects of arsenic and alcohol exposure.[51,75,77] In addition, individuals with reduced capacity to methylate arsenic appear to be more at risk for Blackfoot or other peripheral ischemic disease.[90,91] In support of this mechanism, it has been reported that in comparison with females, males are more prone to arsenic-related peripheral vascular disease and have reduced methylating capacity.[90]

12.4.6 Cadmium

Cadmium exposure has been positively linked to atherosclerosis and peripheral vascular disease.[29,53,92,93] A recent report from the Austrian Atherosclerosis Risk Factors in Female Youngsters (ARFY) study indicated that in otherwise healthy young women, cadmium (Cd) level was independently associated with early atherosclerotic vessel wall thickening (intima-media thickness exceeding the 90th percentile of the distribution).[53] Through a series of mouse and *in vitro* studies, these authors demonstrated that the effects of cadmium on aortic lipid deposition and vascular cell permeability could be reduced by increased zinc status.[53] Selenium and zinc have been shown by others to protect rodents from both the atherogenic and hypertensive effects of chronic low levels of cadmium exposures through increased expression of metallothionein and cadmium-sequestering proteins.[29,93] The mechanism for cadmium-promoted atherogenesis appears to be related to endothelial cell injury.[29,93] Cadmium induces the release of a several proinflammatory mediators from endothelial cells and it stimulates the release of antithrombolytic agents to facilitate adhesion of leukocytes and platelets to the vessel wall.[29,94] These factors combined with cadmium's effect of decreasing endothelial junctional integrity and barrier function would contribute to atheroma formation.[53] In addition, cadmium promotes smooth muscle cell proliferation and enhances the extracellular matrix production to increase vessel-wall stiffness. Finally, cadmium produces a unique and intense ischemic injury in rodent testes that is promoted by cadmium concentration due to highly expression levels of the ZIP8 metal transporter on the testicular endothelial cells.[9]

Several large-population studies have found associations between high levels of mercury exposure and increased incidence of coronary heart disease and acute myocardial infarction.[93] Most studies of environmental exposures to mercury and methylmercury focus on fish consumption as the primary route of mercury exposure and found only modest increases in disease at the very highest levels of consumption.[95] In Minamata, Japan where the worst environmental exposures have occurred, there is only a slight increase in ischemic heart disease and hypertension in males.[95,96] Likewise, studies in Finland show elevated risks of coronary heart disease and acute myocardial infarction that directly correlate with hair levels of methylmercury, but only in the upper quintile of exposures in males.[95,97] Mechanistically, there appears to be a strong correlation between the incidences of mercury-related coronary disease and elevated immune complexes with titers for oxidized LDL.[97]

12.4.7 Hypertension

Sustained hypertension or elevated blood pressure is the most important risk factor for enhanced ischemic disease, atherosclerosis, and cerebrovascular disease. It affects the function and structure of primarily small muscular arteries and arterioles. Despite extensive research, approximately 90–95% of hypertension remains characterized as idiopathic, meaning that it is derived from unknown causes and possibly from the background environment. A majority of idiopathic disease is primary or essential hypertension resulting from direct pathogenic events in vascular cells. Most of the remaining 5–10% of hypertensive disease is secondary to renal disease or renovascular hypertension and infrequently from abnormal adrenal glands (*e.g.* Cushing's syndrome, primary aldosteronism, or pheochormacytoma). Metal and metalloid exposures have been associated with modest elevation of both primary and secondary hypertension. Additionally, exposure to high levels of metals produces acute, pronounced increases in blood pressure. Very few metals in concentrations relevant to human exposures directly contract intact vascular smooth muscle,[98] suggesting that a majority of metal-induced hypertension result from primary effects on acquired changes in vascular cell phenotype or are secondary to systemic changes. Metals can also promote hypertensive disease etiologies that are common to other longitudinal risk factors, such as excess levels of endogenous vasoactive or mitogenic angiotensin II, endothelin-1, and catecholamines. These etiologies include: stimulation of NADPH oxidase to generate ROS; loss of endothelial nitric-oxide production needed for vasodilation; enhanced vascular leak resulting in platelet plugging and release of vasoconstrictors; and enhanced smooth muscle hypertrophy and proliferation leading to increased vessel-wall stiffening. Sustained hypertension can result in loss of capillaries and stimulation of inflammatory angiogenesis. Since the metals share common basic mechanisms for promoting hypertensive diseases with endogenous or other environmental diseases, it is often difficult to clearly link exposure to a given metal with disease incidence.

Individuals who are predisposed to develop hypertension, in general, may also be more prone to increased risk of hypertensive responses to metals. For example, polymorphisms in human NADPH oxidase gene may be linked to both angiotensin II[99] and arsenic-associated[100] hypertension. Overall, the role of environmental metal exposures in directly causing primary or essential hypertension remains controversial. Nonetheless, it is clear that the ability of these metals to stimulate vascular pathogenesis and contribute to hypertensive risk is an important public-health concern and may explain a significant portion of idiopathic hypertension.

The hypertensive effects of cadmium have been studied in greater detail than most other metals. Large epidemiological studies have produced conflicting evidence of an association between cadmium exposures and elevated diastolic or systolic blood pressures. In part, interstudy differences may arise from the exposures biomarkers measured in the studies, since there is a consensus that the effect of environmental cadmium exposures on increased pressures is only seen when blood cadmium levels are compared.[101] A better correlation with blood relative to urine levels may indicate that blood pressure is more sensitive to recent cadmium exposures or that cadmium is more biologically active in blood.[101] Low-dose chronic exposures of animals to cadmium have been linked to both hypertension and atherosclerosis.[9] As discussed above for atherosclerosis, vascular endothelial cell dysfunction appears to be the target of cadmium toxicity leading to essential hypertension.[9] As indicated above for atherogenesis, this toxicity can be modulated by other metals, such as zinc and selenium, which increase cadmium sequestering by inducing metallothionein.[23,93]

Studies of direct effects on vascular cells indicate that cadmium promotes oxidative stress that decreases nitric-oxide release, increases the expression of adhesion molecules and matrix proteins, disrupts cell junctions, and increases smooth muscle cell proliferation.[9] Although it is not clear that all of these mechanisms are relevant responses to the levels of cadmium in blood from environmentally or occupationally exposed humans, they would contribute to decreased vessel reactivity and increased vessel-wall stiffness underlying modest increases in systolic and diastolic pressures.[101] Cadmium accumulates in the kidney and the resulting chronic changes in renal metabolism and tubular injury it causes may contribute to secondary hypertension by elevating sodium retention and renin release.[102]

A number of human studies have found that occupational lead exposure is correlated with hypertension (reviewed in ref. 9). However, the exposures were very high and the hypertension is probably secondary to severe kidney damage. The case for a relationship between environmental lead exposure and either primary or secondary hypertension remains controversial.[9,103,104] Several studies that examined the effects of lower lead exposures and incidence of cardiovascular disease indicate that mechanisms other than hypertension, impaired renal function, and inflammation might underlie any observed associations.[92,104,105] In addition to studies finding only modest effects on blood pressure, some of the controversy regarding a clear causal relationship between lead exposures and primary hypertension may be due to hypertensive risk being relevant only to susceptible populations. Chronic occupational exposures produced hypertension in male, but not female, workers and only years after exposures.[106]

Analysis of the large NHANES database revealed that increased systolic and diastolic pressures correlated with lead exposures in black males and females, but blood pressure was not changed in exposed whites or Hispanics of either sex.[107] It is plausible that finding a role for lead in primary hypertension relates more to individuals predisposed to develop hypertension, in general, and their risk is further enhanced by responses to lead. The array of potential mechanisms revealed by animal and cell-based studies that underlie lead-induced hypertension is extensive.[9] Endothelial cell dysfunction and inflammatory cytokine release, as well as altered smooth muscle cell calcium signaling with enhanced contractility are central to these hypothetical mechanisms. However, without a clearly established relationship between blood lead levels and human hypertension, the relevance of these concentration- and time-dependent effects to clinically significant regulation of blood pressure is unclear.

Numerous large and small epidemiological studies from different exposed populations suggest that exposure to arsenic increases cardiovascular disease risk. The associations between ischemic and cardiac diseases and arsenic exposure mentioned above are stronger than for hypertension. However, recent studies demonstrate that arsenic increases systolic and pulse pressures, but not diastolic pressure, even at relatively low level chronic exposures of 50 µg/L.[108,109] These effects are more prevalent in exposed women and individuals with poor folate nutrition.[108,109] Others find that a reduced capacity to fully methylate arsenic to the excreted dimethyl metabolite increases risk of hypertension and peripheral vascular disease in males.[90,91] In addition to causing systemic hypertension, arsenic exposures have been linked to liver portal hypertension.[110] Arsenic targets both endothelial and smooth muscle to signal for hypertensive effects. NAD(P)H oxidases in both cell types are activated in response to arsenic and the ROS generated decrease endothelial cell-dependent vasodilation and increase smooth muscle cell growth. In isolated aortic rings, low levels of arsenic decrease endothelial nitric-oxide release and increase smooth muscle calcium sensitization to enhance stimulated contraction.[111,112]

Humans exposed to arsenic in their drinking water have reduced capacity to produce nitric oxide[113] and *in vivo* rat studies demonstrate that chronic arsenic exposures impair acetylcholine-stimulated dilation.[111] The mechanism for this reduced capacity was investigated in rabbits and found to result from uncoupling of endothelial nitric oxide synthase as arsenic decreased the levels of its cofactor tetrahydrobiopterin.[114] This uncoupling provides an additional source of oxidants that, together with NADPH oxidase-generated ROS, promote vascular dysfunction. Higher levels of arsenic impair endothelial cell-independent vasodilation by decreasing the capacity of smooth muscle cells to generate cGMP.[112] In liver, arsenic-induced vascular remodeling and channeling in humans and rodents has been associated with portal hypertension.[110] As discussed above, arsenic promotes NADPH-oxidase-dependent defenestration and capillarization of liver sinusoidal endothelial cells,[89,115,116] which may increase pressure in, and across, this vascular bed. In addition, prolonged exposures to arsenic in drinking water decreases luminal diameters in hepatic arterioles and promotes hyperplasia in bile ducts.[115,116] While these studies emphasize the central role of NADPH oxidase ROS generation in the vascular response to low-level arsenic exposures, they did not measure hepatic or blood pressure and a relationship to hypertension can only be inferred.

Hypertensive effects of mercury appear to be much stronger from prenatal exposures and exposures in early childhood. While some studies found weak associations between mercury exposure and adult hypertension,[93] the majority of well-conducted epidemiological studies have been negative.[95] Cord blood mercury levels at birth were positively correlated with both increased systolic and diastolic blood pressure in seven-year olds from a large study in the Faroe Islands, but the association appears to disappear by young adulthood.[95,117,118] In another population study, a slight association between mercury and diastolic pressure remained in male children at age 15.[118] However, the consequence of this hypertension early in life on adult disease is unclear. Mechanistically, it is more probable that mercury causes secondary hypertension, since its effects on blood pressure mostly relate to increased local and circulating levels of catecholamines or angiotensin II. Mercury is only marginally constrictive in direct assays of intact vessels.[98] However, it may promote ROS and angiotensin II-mediated endothelial dysfunction to increase vasoconstrictor efficacy[119] at the vessel level. High-level mercury intoxication, especially caused by exposure to methylmercury from broken thermometers, causes acrodynia in infants and young adults that presents with profound arterial hypertension resembling pheochromocytoma.[120,121] The effect of elemental and methylmercury on catecholamine release and metabolism is well studied in neural tissues where the metal concentrates, but not in the cardiovascular system. Mercury disruption of complex III of mitochondrial respiration causes oxidative injury and the resulting calcium leak in nerve endings promotes vesicular catecholamine release.[121] High micromolar to low millimolar concentrations of mercury interfere with catecholamine catabolism.[122] At very high levels, it is possible that this interference is caused by mercury binding to and reducing levels of S-adenosylmethionine; the cosubstrate for catecholamine-O-methyltransferase catalyzed inactivation of norepinephrine, epinephrine, and dopamine.[120]

12.4.8 Angiogenesis

Angiogenesis, the growth of new vessels from pre-existing blood vessels, represents an important axis for metals to promote aberrant vascular remodeling and tumorigenesis. Closely related, neovascularization in development is also affected by different metal exposures and this can contribute to their teratogenic effects.

Copper is essential for adequate angiogenic responses. The mechanisms through which copper contributes to angiogenesis are multifaceted and complex, since they include: induction and adequate expression of a number of proangiogenic cytokines;[123] supporting the activity of CuZn SOD1;[124,125] and possibly promoting circulating levels of endothelial cell progenitor cells.[124] Chelating copper with tetrathiomolybdate is an effective means of reducing tumor size and burden in many animal models and in phase 1 and 2 human clinical trials. Tumor angiogenesis requires inflammation and cytokine recruitment of leukocytes and progenitor cells.[126] Copper promotes transactivation of NF-κB driven expression of several important cytokines that produce an oxidative state for cell proliferation and migration. Copper coordination in SOD1 plays an essential role in protecting those cells from excess oxidative injury that allows tumors to grow. Thus, producing a copper-deficient state with chelation both reduces the drive for angiogenesis and promotes oxidative injury and apoptosis in the angiogenic endothelium and tumor cells.[124]

Arsenic causes concentration-dependent biphasic effects on angiogenesis. At low to moderate environmental and therapeutic levels, arsenic increases angiogenesis in a number of developmental and tumorigenesis animal models.[127–130] Arsenic is not genotoxic, and thus the tumor- and metastatic-enhancing effects of arsenic-induced angiogenesis are significant in that they may provide a mechanism for arsenic to promote tumor growth. Higher therapeutic arsenic levels are used to kill angiogenic endothelial cells in tumors.[17,128] However, as indicated above, this use is limited by the narrow therapeutic window between levels required to kill endothelial cells and those that increase cardiac arrhythmias. Arsenic induces angiogenic genes in both endothelial and smooth muscle cells. It increases the recruitment of inflammatory CD45$^+$ cells to support full development of patent angiogenic vessels. Arsenic induction of angiogenic genes depends on stimulating cell-signaling cascades through activation of Rac1-GTPase and NADPH oxidases.[14,131] The low levels of oxidants generated in turn activates multiple transcription factors, including NF-κB and HIF-1α.[132,133] In contrast, higher arsenic levels activate apoptosis through oxidative stress and inhibit NF-κB activation and transactivation.[3,134,135] Toxic levels of arsenic target cysteine-containing mitochondrial enzymes and interfere with phosphotransfer to decrease ATP levels and collapse the mitochondrial electron gradient. The net result is endothelial apoptosis.[17,128] It is important to note that inorganic arsenic ingested in drinking water or injected therapeutically is metabolized to methylated species that are even more active on endothelial cells than the parent compound.[136] As indicated above, sex differences in methylation capacity may account for differences for cardiovascular disease incidence or severity between males and females.

Cadmium inhibits angiogenesis by disrupting intercellular contacts and preventing endothelial cell migration.[9,29,53] Endothelial cells are relatively resistant to the cytotoxic effects of cadmium. However, noncytotoxic levels of cadmium inhibit migration and tube formation by disrupting angiogenesis by redistributing VE-cadherin from the endothelial cell surface.[9,29] Inhibitory effects on cell-adhesion molecules may also limit endothelial cell interactions with circulating cells that provide for stable vessel formation. In addition, inhibition of angiogenesis by cadmium may be associated with a decreased production of nitric oxide,[9] possibly by inhibiting HIF-1α-driven gene expression.[43]

12.5 Conclusions

Environmental, occupational, and accidental metal exposures produce a range of adverse cardiovascular effects. These effects may be through direct actions of the metal or secondary to tissue injury caused by selective metal accumulation and oxidative stress. It is clear that certain metals are acutely toxic to both the heart and blood vessels following high levels of exposure. There appears to be a hierarchy of cell sensitivities to acute metal injury with endothelial cells that receive the highest dose of metal being injured before underlying tissues. However, the primary and secondary effects of lower levels of metals that interact and contribute to the etiology of chronic cardiovascular diseases are complex and mechanisms for these effects often remain unresolved. In general, the subtle nature of environmental metal exposures on cardiovascular function and injuries has a complicated linkage of exposure to disease etiology or modification. Moreover, several epidemiological studies revealed that certain populations or subgroups of individuals are more susceptible to the cardiovascular effects of metals and that nutrition modifies many of these effects. Without appropriate stratification of epidemiological data to account for these individuals or factors, the true impacts of metal exposures on cardiovascular disease are often lost. It is evident from the large numbers of epidemiological studies, clinical case reports, animal exposure studies, and cell-based studies that metals are significant public-health concerns that cause unique cardiovascular toxicities and enhance disease risks.

The study of the molecular pathogenesis of the different metals is complicated by the bioavailability of the different species of the metals in the environment, dynamic speciation of the metals in the circulation or in the cells, uptake of the different species into cells, and differential interactions of the species with cellular macromolecules. These interactions include both activating cell-signaling cascades at lower levels of exposure and promoting macromolecule damage or degradation at higher levels. The inability to separate the direct effects of the metals on these macromolecules from secondary effects on cell signaling and ROS production hinders attempts to define the initial sites and rate-limiting steps of their cellular actions and toxicities. Despite these limitations, much is known about the molecular actions of metals on the individual cardiovascular cells. It is evident that the cardiovascular toxicity of metals cannot be attributed to single mechanisms or simple modes of action, such as random interactions with thiols or oxidative stress. Instead, individual metals and metal species have unique properties that provide selectivity in reacting with cellular targets and in the mechanisms that initiate or suppress cell functions. Understanding of the molecular pathogenesis of metals in the heart and blood vessels continues to improve as mechanistic investigation of disease etiologies has shifted to investigating endpoints *in vivo* that are linked to real-world metal exposures.

References

1. J. Liu, W. C. Kershaw, Y. P. Liu and C. D. Klaassen, Cadmium-induced hepatic endothelial cell injury in inbred strains of mice, *Toxicology*, 1992, **75**, 51–62.
2. J. M. McKim Jr, J. Liu, Y. P. Liu and C. D. Klaassen, Distribution of cadmium chloride and cadmium-metallothionein to liver parenchymal, Kupffer, and endothelial cells: their relative ability to express metallothionein, *Toxicol. Appl. Pharmacol.*, 1992, **112**, 324–330.
3. S. J. Ralph, Arsenic-based antineoplastic drugs and their mechanisms of action, *Met. Based Drugs*, 2008, **2008**, 260146.
4. B. E. Kim, T. Nevitt and D. J. Thiele, Mechanisms for copper acquisition, distribution and regulation, *Nat. Chem. Biol.*, 2008, **4**, 176–185.
5. Z. Qin, S. Itoh, V. Jeney, M. Ushio-Fukai and T. Fukai, Essential role for the Menkes ATPase in activation of extracellular superoxide dismutase: implication for vascular oxidative stress, *FASEB J*, 2006, **20**, 334–336.
6. Z. Hlubocka, Z. Marecek, A. Linhart, E. Kejkova, L. Pospisilova, P. Martasek and M. Aschermann, Cardiac involvement in Wilson disease, *J. Inherit. Metab. Dis.*, 2002, **25**, 269–277.
7. L. Zecca, M. B. Youdim, P. Riederer, J. R. Connor and R. R. Crichton, Iron, brain ageing and neurodegenerative disorders, *Nat. Rev. Neurosci.*, 2004, **5**, 863–873.

8. G. J. Brewer, Iron and copper toxicity in diseases of aging, particularly atherosclerosis and Alzheimer's disease, *Exp. Biol. Med. (Maywood)*, 2007, **232**, 323–335.
9. W. C. Prozialeck, J. R. Edwards, D. W. Nebert, J. M. Woods, A. Barchowsky and W. D. Atchison, The vascular system as a target of metal toxicity, *Toxicol. Sci.*, 2008, **102**, 207–218.
10. N. H. Gokhale, S. Bradford and J. A. Cowan, Stimulation and oxidative catalytic inactivation of thermolysin by copper. Cys-Gly His-Lys, *J. Biol. Inorg. Chem.*, 2007, **12**, 981–987.
11. D. Galaris and K. Pantopoulos, Oxidative stress and iron homeostasis: mechanistic and health aspects, *Crit. Rev. Clin. Lab. Sci.*, 2008, **45**, 1–23.
12. S. Lynn, J. R. Gurr, H. T. Lai and K. Y. Jan, NADH oxidase activation is involved in arsenite-induced oxidative DNA damage in human vascular smooth muscle cells, *Circ. Res.*, 2000, **86**, 514–519.
13. K. R. Smith, L. R. Klei and A. Barchowsky, Arsenite stimulates plasma membrane NADPH oxidase in vascular endothelial cells, *Am. J. Physiol.*, 2001, **280**, L442–L449.
14. A. C. Straub, L. R. Klei, D. B. Stolz and A. Barchowsky, Arsenic requires sphingosine-1-phosphate type 1 receptors to induce angiogenic genes and endothelial cell remodeling, *Am. J. Pathol.*, 2009, **174**, 1949–1958.
15. M. Y. Lee and K. K. Griendling, Redox signaling, vascular function, and hypertension, *Antioxid. Redox. Signal.*, 2008, **10**, 1045–1059.
16. S. L. Peters and A. E. Alewijnse, Sphingosine-1-phosphate signaling in the cardiovascular system, *Curr. Opin. Pharmacol.*, 2007, **7**, 186–192.
17. G. J. Roboz, S. Dias, G. Lam, W. J. Lane, S. L. Soignet, R. P. Warrell Jr. and S. Rafii, Arsenic trioxide induces dose- and time-dependent apoptosis of endothelium and may exert an antileukemic effect *via* inhibition of angiogenesis, *Blood*, 2000, **96**, 1525–1530.
18. S. Waxman and K. C. Anderson, History of the development of arsenic derivatives in cancer therapy, *Oncologist*, 2001, **6**(Suppl 2), 3–10.
19. A. D. Smith, S. Botero and O. A. Levander, Copper deficiency increases the virulence of amyocarditic and myocarditic strains of coxsackievirus B3 in mice, *J. Nutr.*, 2008, **138**, 849–855.
20. N. G. Ilback, L. Wesslen, J. Fohlman and G. Friman, Effects of methyl mercury on cytokines, inflammation and virus clearance in a common infection (coxsackie B3 myocarditis), *Toxicol. Lett.*, 1996, **89**, 19–28.
21. N. G. Ilback, U. Lindh, J. Fohlman and G. Friman, New aspects of murine coxsackie B3 myocarditis – focus on heavy metals, *Eur. Heart J.*, 1995, **16**(Suppl O), 20–24.
22. M. J. Burkitt, A critical overview of the chemistry of copper-dependent low density lipoprotein oxidation: roles of lipid hydroperoxides, alpha-tocopherol, thiols, and ceruloplasmin, *Arch. Biochem. Biophys.*, 2001, **394**, 117–135.
23. M. A. Mansoor, C. Bergmark, S. J. Haswell, I. F. Savage, P. H. Evans, R. K. Berge, A. M. Svardal and O. Kristensen, Correlation between

plasma total homocysteine and copper in patients with peripheral vascular disease, *Clin. Chem.*, 2000, **46**, 385–391.
24. N. Shukla, A. Koupparis, R. A. Jones, G. D. Angelini, R. Persad and J. Y. Jeremy, Penicillamine administration reverses the inhibitory effect of hyperhomocysteinaemia on endothelium-dependent relaxation and superoxide formation in the aorta of the rabbit, *Eur. J. Pharmacol.*, 2006, **531**, 201–208.
25. E. O. Uthus, P. G. Reeves and J. T. Saari, Copper deficiency decreases plasma homocysteine in rats, *J. Nutr.*, 2007, **137**, 1370–1374.
26. E. Nakano, M. P. Williamson, N. H. Williams and H. J. Powers, Copper-mediated LDL oxidation by homocysteine and related compounds depends largely on copper ligation, *Biochim. Biophys. Acta.*, 2004, **1688**, 33–42.
27. R. Heuchel, F. Radtke, O. Georgiev, G. Stark, M. Aguet and W. Schaffner, The transcription factor MTF-1 is essential for basal and heavy metal-induced metallothionein gene expression, *EMBO J.*, 1994, **13**, 2870–2875.
28. J. H. Laity and G. K. Andrews, Understanding the mechanisms of zinc-sensing by metal-response element binding transcription factor-1 (MTF-1), *Arch. Biochem. Biophys.*, 2007, **463**, 201–210.
29. W. C. Prozialeck, J. R. Edwards and J. M. Woods, The vascular endothelium as a target of cadmium toxicity, *Life Sci.*, 2006, **79**, 1493–1506.
30. S. B. Cullinan, J. D. Gordan, J. Jin, J. W. Harper and J. A. Diehl, The Keap1-BTB protein is an adaptor that bridges Nrf2 to a Cul3-based E3 ligase: oxidative stress sensing by a Cul3-Keap1 ligase, *Mol. Cell Biol*, 2004, **24**, 8477–8486.
31. A. T. Dinkova-Kostova, W. D. Holtzclaw and N. Wakabayashi, Keap1, the sensor for electrophiles and oxidants that regulates the phase 2 response, is a zinc metalloprotein, *Biochemistry*, 2005, **44**, 6889–6899.
32. P. Gong, D. Stewart, B. Hu, C. Vinson and J. Alam, Multiple basic-leucine zipper proteins regulate induction of the mouse heme oxygenase-1 gene by arsenite, *Arch. Biochem. Biophys.*, 2002, **405**, 265–274.
33. X. M. Liu, K. J. Peyton, D. Ensenat, H. Wang, A. I. Schafer, J. Alam and W. Durante, Endoplasmic reticulum stress stimulates heme oxygenase-1 gene expression in vascular smooth muscle, role in cell survival, *J. Biol. Chem.*, 2005, **280**, 872–877.
34. M. M. Cortese, C. V. Suschek, W. Wetzel, K. D. Kroncke and V. Kolb-Bachofen, Zinc protects endothelial cells from hydrogen peroxide via Nrf2-dependent stimulation of glutathione biosynthesis, *Free Radic. Biol. Med.*, 2008, **44**, 2002–2012.
35. T. Sakurai, T. Ohta and K. Fujiwara, Inorganic arsenite alters macrophage generation from human peripheral blood monocytes, *Toxicol. Appl. Pharmacol.*, 2005, **203**, 145–153.
36. J. Hanze, N. Weissmann, F. Grimminger, W. Seeger and F. Rose, Cellular and molecular mechanisms of hypoxia-inducible factor driven vascular remodeling, *Thromb. Haemost.*, 2007, **97**, 774–787.

37. G. H. Fong, Regulation of angiogenesis by oxygen sensing mechanisms, *J. Mol. Med.*, 2009, **87**, 549–560.
38. G. Loor and P. T. Schumacker, Role of hypoxia-inducible factor in cell survival during myocardial ischemia-reperfusion, *Cell Death. Differ.*, 2008, **15**, 686–690.
39. R. V. Shohet and J. A. Garcia, Keeping the engine primed: HIF factors as key regulators of cardiac metabolism and angiogenesis during ischemia, *J. Mol. Med.*, 2007, **85**, 1309–1315.
40. H. Endoh, T. Kaneko, H. Nakamura, K. Doi and E. Takahashi, Improved cardiac contractile functions in hypoxia-reoxygenation in rats treated with low concentration Co^{2+}, *Am. J. Physiol. Heart Circ. Physiol.*, 2000, **279**, H2713–H2719.
41. M. Matsumoto, Y. Makino, T. Tanaka, H. Tanaka, N. Ishizaka, E. Noiri, T. Fujita and M. Nangaku, Induction of renoprotective gene expression by cobalt ameliorates ischemic injury of the kidney in rats, *J. Am. Soc. Nephrol.*, 2003, **14**, 1825–1832.
42. F. R. Sharp, M. Bergeron and M. Bernaudin, Hypoxia-inducible factor in brain, *Adv. Exp. Med. Biol.*, 2001, **502**, 273–291.
43. E. Belaidi, P. C. Beguin, P. Levy, C. Ribuot and D. Godin-Ribuot, Prevention of HIF-1 activation and iNOS gene targeting by low-dose cadmium results in loss of myocardial hypoxic preconditioning in the rat, *Am. J. Physiol. Heart Circ. Physiol.*, 2008, **294**, H901–H908.
44. N. Adhikari, N. Charles, U. Lehmann and J. L. Hall, Transcription factor and kinase-mediated signaling in atherosclerosis and vascular injury, *Curr. Atheroscler. Rep.*, 2006, **8**, 252–260.
45. B. J. Murphy, T. Kimura, B. G. Sato, Y. Shi and G. K. Andrews, Metallothionein induction by hypoxia involves cooperative interactions between metal-responsive transcription factor-1 and hypoxia-inducible transcription factor-1alpha, *Mol. Cancer Res.*, 2008, **6**, 483–490.
46. A. Frustaci, N. Magnavita, C. Chimenti, M. Caldarulo, E. Sabbioni, R. Pietra, C. Cellini, G. F. Possati and A. Maseri, Marked elevation of myocardial trace elements in idiopathic dilated cardiomyopathy compared with secondary cardiac dysfunction, *J. Am. Coll. Cardiol.*, 1999, **33**, 1578–1583.
47. Y. J. Kang, Molecular and cellular mechanisms of cardiotoxicity, *Environ. Health Perspect.*, 2001, **109**(Suppl 1), 27–34.
48. Y. Li, X. Sun, L. Wang, Z. Zhou and Y. J. Kang, Myocardial toxicity of arsenic trioxide in a mouse model, *Cardiovasc. Toxicol.*, 2002, **2**, 63–73.
49. J. C. Wood, C. Enriquez, N. Ghugre, M. Otto-Duessel, M. Aguilar, M. D. Nelson, R. Moats and T. D. Coates, Physiology and pathophysiology of iron cardiomyopathy in thalassemia, *Ann. N. Y. Acad. Sci.*, 2005, **1054**, 386–395.
50. K. T. Weber, A Quebec quencher, *Cardiovasc. Res.*, 1998, **40**, 423–425.
51. A. L. Klatsky, Alcohol and cardiovascular diseases: a historical overview, *Ann. N. Y. Acad. Sci.*, 2002, **957**, 7–15.

52. Y. Jiang, C. Reynolds, C. Xiao, W. Feng, Z. Zhou, W. Rodriguez, S. C. Tyagi, J. W. Eaton, J. T. Saari and Y. J. Kang, Dietary copper supplementation reverses hypertrophic cardiomyopathy induced by chronic pressure overload in mice, *J. Exp. Med.*, 2007, **204**, 657–666.
53. B. Messner, M. Knoflach, A. Seubert, A. Ritsch, K. Pfaller, B. Henderson, Y. H. Shen, I. Zeller, J. Willeit, G. Laufer, G. Wick, S. Kiechl and D. Bernhard, Cadmium is a novel and independent risk factor for early atherosclerosis mechanisms and *in vivo* relevance, *Arterioscler. Thromb. Vasc. Biol.*, 2009, **29**, 1392–1398.
54. N. Clyne, C. Hofman-Bang, Y. Haga, N. Hatori, S. L. Marklund, S. K. Pehrsson and R. Wibom, Chronic cobalt exposure affects antioxidants and ATP production in rat myocardium, *Scand. J. Clin. Lab. Invest.*, 2001, **61**, 609–614.
55. Y. Kakinuma, T. Miyauchi, T. Suzuki, K. Yuki, N. Murakoshi, K. Goto and I. Yamaguchi, Enhancement of glycolysis in cardiomyocytes elevates endothelin-1 expression through the transcriptional factor hypoxia-inducible factor-1 alpha, *Clin. Sci. (London)*, 2002, **103**(Suppl 48), 210S–214S.
56. Y. Kakinuma, T. Miyauchi, K. Yuki, N. Murakoshi, K. Goto and I. Yamaguchi, Novel molecular mechanism of increased myocardial endothelin-1 expression in the failing heart involving the transcriptional factor hypoxia-inducible factor-1α induced for impaired myocardial energy metabolism, *Circulation*, 2001, **103**, 2387–2394.
57. J. T. Barbey, Cardiac toxicity of arsenic trioxide, *Blood*, 2001, **98**, 1632–1634.
58. A. Dennis, L. Wang, X. Wan and E. Ficker, hERG channel trafficking: novel targets in drug-induced long QT syndrome, *Biochem. Soc. Trans.*, 2007, **35**, 1060–1063.
59. E. Ficker, Y. A. Kuryshev, A. T. Dennis, C. Obejero-Paz, L. Wang, P. Hawryluk, B. A. Wible and A. M. Brown, Mechanisms of arsenic-induced prolongation of cardiac repolarization, *Mol. Pharmacol.*, 2004, **66**, 33–44.
60. J. L. Mumford, K. Wu, Y. Xia, R. Kwok, Z. Yang, J. Foster and W. E. Sanders, Chronic arsenic exposure and cardiac repolarization abnormalities with QT interval prolongation in a population-based study, *Environ. Health Perspect.*, 2007, **115**, 690–694.
61. A. N. Katchman, J. Koerner, T. Tosaka, R. L. Woosley and S. N. Ebert, Comparative evaluation of HERG currents and QT intervals following challenge with suspected torsadogenic and nontorsadogenic drugs, *J. Pharmacol. Exp. Ther.*, 2006, **316**, 1098–1106.
62. M. Alvarez, C. O. Malecot, F. Gannier and J. M. Lignon, Antimony-induced cardiomyopathy in guinea-pig and protection by L-carnitine, *Br. J. Pharmacol.*, 2005, **144**, 17–27.
63. C. A. Pope III, R. T. Burnett, G. D. Thurston, M. J. Thun, E. E. Calle, D. Krewski and J. J. Godleski, Cardiovascular mortality and long-term exposure to particulate air pollution: epidemiological evidence of general pathophysiological pathways of disease, *Circulation*, 2004, **109**, 71–77.

64. N. L. Mills, K. Donaldson, P. W. Hadoke, N. A. Boon, W. MacNee, F. R. Cassee, T. Sandstrom, A. Blomberg and D. E. Newby, Adverse cardiovascular effects of air pollution, *Nat. Clin. Pract. Cardiovasc. Med.*, 2009, **6**, 36–44.
65. K. A. Miller, D. S. Siscovick, L. Sheppard, K. Shepherd, J. H. Sullivan, G. L. Anderson and J. D. Kaufman, Long-term exposure to air pollution and incidence of cardiovascular events in women, *N. Engl. J. Med.*, 2007, **356**, 447–458.
66. C. A. Pope III, Mortality effects of longer term exposures to fine particulate air pollution: review of recent epidemiological evidence, *Inhal. Toxicol.*, 2007, **19**(Suppl 1), 33–38.
67. M. L. Bell, K. Ebisu, R. D. Peng, J. M. Samet and F. Dominici, Hospital admissions and chemical composition of fine particle air pollution, *Am. J. Respir. Crit. Care Med.*, 2009, **179**, 1115–1120.
68. M. Lippmann, K. Ito, J. S. Hwang, P. Maciejczyk and L. C. Chen, Cardiovascular effects of nickel in ambient air, *Environ. Health Perspect.*, 2006, **114**, 1662–1669.
69. M. J. Campen, J. P. Nolan, M. C. Schladweiler, U. P. Kodavanti, P. A. Evansky, D. L. Costa and W. P. Watkinson, Cardiovascular and thermoregulatory effects of inhaled PM-associated transition metals: a potential interaction between nickel and vanadium sulfate, *Toxicol. Sci.*, 2001, **64**, 243–252.
70. D. W. Graff, W. E. Cascio, J. A. Brackhan and R. B. Devlin, Metal particulate matter components affect gene expression and beat frequency of neonatal rat ventricular myocytes, *Environ. Health Perspect.*, 2004, **112**, 792–798.
71. U. P. Kodavanti, M. C. Schladweiler, P. S. Gilmour, J. G. Wallenborn, B. S. Mandavilli, A. D. Ledbetter, D. C. Christiani, M. S. Runge, E. D. Karoly, D. L. Costa, S. Peddada, R. Jaskot, J. H. Richards, R. Thomas, N. R. Madamanchi and A. Nyska, The role of particulate matter-associated zinc in cardiac injury in rats, *Environ. Health Perspect.*, 2008, **116**, 13–20.
72. U. P. Kodavanti, C. F. Moyer, A. D. Ledbetter, M. C. Schladweiler, D. L. Costa, R. Hauser, D. C. Christiani and A. Nyska, Inhaled environmental combustion particles cause myocardial injury in the Wistar Kyoto rat, *Toxicol. Sci.*, 2003, **71**, 237–245.
73. X. M. Yuan and W. Li, Iron involvement in multiple signaling pathways of atherosclerosis: a revisited hypothesis, *Curr. Med. Chem.*, 2008, **15**, 2157–2172.
74. B. Pfanzagl, F. Tribl, E. Koller and T. Moslinger, Homocysteine strongly enhances metal-catalyzed LDL oxidation in the presence of cystine and cysteine, *Atherosclerosis*, 2003, **168**, 39–48.
75. R. R. Engel, C. Hopenhayn-Rich, O. Receveur and A. H. Smith, Vascular effects of chronic arsenic exposure: a review, *Epidemiol. Rev.*, 1994, **16**, 184–208.
76. C. H. Wang, C. L. Chen, C. K. Hsiao, F. T. Chiang, L. I. Hsu, H. Y. Chiou, Y. M. Hsueh, M. M. Wu and C. J. Chen, Increased risk of QT

prolongation associated with atherosclerotic diseases in arseniasis-endemic area in southwestern coast of Taiwan, *Toxicol. Appl. Pharmacol.*, 2009, **239**, 320–324.
77. A. Navas-Acien, A. R. Sharrett, E. K. Silbergeld, B. S. Schwartz, K. E. Nachman, T. A. Burke and E. Guallar, Arsenic Exposure and Cardiovascular Disease: A Systematic Review of the Epidemiologic Evidence, *Am. J. Epidemiol.*, 2005, **162**, 1037–1049.
78. C. H. Wang, C. K. Hsiao, C. L. Chen, L. I. Hsu, H. Y. Chiou, S. Y. Chen, Y. M. Hsueh, M. M. Wu and C. J. Chen, A review of the epidemiologic literature on the role of environmental arsenic exposure and cardiovascular diseases, *Toxicol. Appl. Pharmacol.*, 2007, **222**, 315–326.
79. Y. Yuan, G. Marshall, C. Ferreccio, C. Steinmaus, S. Selvin, J. Liaw, M. N. Bates and A. H. Smith, Acute myocardial infarction mortality in comparison with lung and bladder cancer mortality in arsenic-exposed region II of Chile from 1950 to 2000, *Am. J. Epidemiol.*, 2007, **166**, 1381–1391.
80. I. Hertz-Picciotto, H. M. Arrighi and S. W. Hu, Does arsenic exposure increase the risk for circulatory disease?, *Am. J. Epidemiol.*, 2000, **151**, 174–181.
81. C. H. Tseng, C. K. Chong, C. P. Tseng, Y. M. Hsueh, H. Y. Chiou, C. C. Tseng and C. J. Chen, Long-term arsenic exposure and ischemic heart disease in arseniasis-hyperendemic villages in Taiwan, *Toxicol. Lett.*, 2003, **137**, 15–21.
82. C. H. Wang, J. S. Jeng, P. K. Yip, C. L. Chen, L. I. Hsu, Y. M. Hsueh, H. Y. Chiou, M. M. Wu and C. J. Chen, Biological gradient between long-term arsenic exposure and carotid atherosclerosis, *Circulation*, 2002, **105**, 1804–1809.
83. Y. Chen, R. M. Santella, M. G. Kibriya, Q. Wang, M. Kappil, W. J. Verret, J. H. Graziano and H. Ahsan, Association between Arsenic exposure from drinking water and plasma levels of soluble cell adhesion molecules, *Environ. Health Perspect.*, 2007, **115**, 1415–1420.
84. J. C. States, S. Srivastava, Y. Chen and A. Barchowsky, Arsenic and cardiovascular disease, *Toxicol. Sci.*, 2009, **107**, 312–323.
85. S. Srivastava, S. E. D'Souza, U. Sen and J. C. States, In utero arsenic exposure induces early onset of atherosclerosis in ApoE−/− mice, *Reprod. Toxicol.*, 2007, **23**, 449–456.
86. M. Bunderson, D. M. Brooks, D. L. Walker, M. E. Rosenfeld, J. D. Coffin and H. D. Beall, Arsenic exposure exacerbates atherosclerotic plaque formation and increases nitrotyrosine and leukotriene biosynthesis, *Toxicol. Appl. Pharmacol.*, 2004, **201**, 32–39.
87. F. E. Pereira, J. D. Coffin and H. D. Beall, Activation of protein kinase C and disruption of endothelial monolayer integrity by sodium arsenite – Potential mechanism in the development of atherosclerosis, *Toxicol. Appl. Pharmacol.*, 2007, **220**, 164–177.
88. P. P. Simeonova, T. Hulderman, D. Harki and M. I. Luster, Arsenic exposure accelerates atherogenesis in apolipoprotein E(−/−) mice, *Environ. Health Perspect.*, 2003, **111**, 1744–1748.

89. A. C. Straub, K. A. Clark, M. A. Ross, A. G. Chandra, S. Li, X. Gao, P. J. Pagano, D. B. Stolz and A. Barchowsky, Arsenic-stimulated liver sinusoidal capillarization in mice requires NADPH oxidase-generated superoxide, *J. Clin. Invest.*, 2008, **118**, 3980–3989.
90. C. H. Tseng, Y. K. Huang, Y. L. Huang, C. J. Chung, M. H. Yang, C. J. Chen and Y. M. Hsueh, Arsenic exposure, urinary arsenic speciation, and peripheral vascular disease in blackfoot disease-hyperendemic villages in Taiwan, *Toxicol. Appl. Pharmacol.*, 2005, **206**, 299–308.
91. Y. K. Huang, C. H. Tseng, Y. L. Huang, M. H. Yang, C. J. Chen and Y. M. Hsueh, Arsenic methylation capability and hypertension risk in subjects living in arseniasis-hyperendemic areas in southwestern Taiwan. Toxicol, *Appl. Pharmacol.*, 2007, **218**, 135–142.
92. A. Navas-Acien, E. K. Silbergeld, R. Sharrett, E. Calderon-Aranda, E. Selvin and E. Guallar, Metals in urine and peripheral arterial disease, *Environ. Health Perspect.*, 2005, **113**, 164–169.
93. M. C. Houston, The role of mercury and cadmium heavy metals in vascular disease, hypertension, coronary heart disease, and myocardial infarction, *Altern. Ther. Health Med.*, 2007, **13**, S128–S133.
94. E. M. Jeong, C. H. Moon, C. S. Kim, S. H. Lee, E. J. Baik, C. K. Moon and Y. S. Jung, Cadmium stimulates the expression of ICAM-1 *via* NF-kappaB activation in cerebrovascular endothelial cells, *Biochem. Biophys. Res. Commun.*, 2004, **320**, 887–892.
95. A. H. Stern, A review of the studies of the cardiovascular health effects of methylmercury with consideration of their suitability for risk assessment, *Environ. Res.*, 2005, **98**, 133–142.
96. H. Tamashiro, H. Akagi, M. Arakaki, M. Futatsuka and L. H. Roht, Causes of death in Minamata disease: analysis of death certificates, *Int. Arch. Occup. Environ. Health*, 1984, **54**, 135–146.
97. J. T. Salonen, K. Seppanen, K. Nyyssonen, H. Korpela, J. Kauhanen, M. Kantola, J. Tuomilehto, H. Esterbauer, F. Tatzber and R. Salonen, Intake of mercury from fish, lipid peroxidation, and the risk of myocardial infarction and coronary, cardiovascular, and any death in eastern Finnish men, *Circulation*, 1995, **91**, 645–655.
98. D. H. Evans and K. Weingarten, The effect of cadmium and other metals on vascular smooth muscle of the dogfish shark, *Squalus acanthias*, *Toxicology*, 1990, **61**, 275–281.
99. G. Zalba, G. San Jose, M. U. Moreno, A. Fortuno and J. Diez, NADPH oxidase-mediated oxidative stress: genetic studies of the p22(phox) gene in hypertension, *Antioxid. Redox. Signal.*, 2005, **7**, 1327–1336.
100. Y. M. Hsueh, P. Lin, H. W. Chen, H. S. Shiue, C. J. Chung, C. T. Tsai, Y. K. Huang, H. Y. Chiou and C. J. Chen, Genetic polymorphisms of oxidative and antioxidant enzymes and arsenic-related hypertension, *J. Toxicol. Environ. Health A*, 2005, **68**, 1471–1484.
101. M. Tellez-Plaza, A. Navas-Acien, C. M. Crainiceanu and E. Guallar, Cadmium exposure and hypertension in the 1999–2004 National Health

and Nutrition Examination Survey (NHANES), *Environ. Health Perspect.*, 2008, **116**, 51–56.
102. S. Satarug, S. M. Nishijo, J. M. Lasker, R. J. Edwards and M. R. Moore, Kidney dysfunction and hypertension: role for cadmium, p450 and heme oxygenases?, *Tohoku J. Exp. Med.*, 2006, **208**, 179–202.
103. Y. Cheng, J. Schwartz, D. Sparrow, A. Aro, S. T. Weiss and H. Hu, Bone lead and blood lead levels in relation to baseline blood pressure and the prospective development of hypertension: the Normative Aging Study, *Am. J. Epidemiol.*, 2001, **153**, 164–171.
104. T. S. Nawrot and J. A. Staessen, Low-level environmental exposure to lead unmasked as silent killer, *Circulation*, 2006, **114**, 1347–1349.
105. A. Menke, P. Muntner, V. Batuman, E. K. Silbergeld and E. Guallar, Blood lead below 0.48 micromol/L (10 microg/dL) and mortality among US adults, *Circulation*, 2006, **114**, 1388–1394.
106. B. S. Glenn, W. F. Stewart, J. M. Links, A. C. Todd and B. S. Schwartz, The longitudinal association of lead with blood pressure, *Epidemiology*, 2003, **14**, 30–36.
107. S. Vupputuri, J. He, P. Muntner, L. A. Bazzano, P. K. Whelton and V. Batuman, Blood lead level is associated with elevated blood pressure in blacks, *Hypertension*, 2003, **41**, 463–468.
108. Y. Chen, P. Factor-Litvak, G. R. Howe, J. H. Graziano, P. Brandt-Rauf, F. Parvez, A. van Geen and H. Ahsan, Arsenic exposure from drinking water, dietary intakes of B vitamins and folate, and risk of high blood pressure in Bangladesh: A population-based, cross-sectional study, *Am. J. Epidemiol.*, 2007, **165**, 541–552.
109. R. K. Kwok, P. Mendola, Z. Y. Liu, D. A. Savitz, G. Heiss, H. L. Ling, Y. Xia, D. Lobdell, D. Zeng, J. M. Thorp Jr, J. P. Creason and J. L. Mumford, Drinking water arsenic exposure and blood pressure in healthy women of reproductive age in Inner Mongolia, China, *Toxicol. Appl. Pharmacol.*, 2007, **222**, 337–343.
110. D. N. Mazumder, Effect of chronic intake of arsenic-contaminated water on liver, *Toxicol. Appl. Pharmacol.*, 2005, **206**, 169–175.
111. M. Y. Lee, B. I. Jung, S. M. Chung, O. N. Bae, J. Y. Lee, J. D. Park, J. S. Yang, H. Lee and J. H. Chung, Arsenic-induced dysfunction in relaxation of blood vessels, *Environ. Health Perspect.*, 2003, **111**, 513–517.
112. M. Y. Lee, Y. H. Lee, K. M. Lim, S. M. Chung, O. N. Bae, H. Kim, C. R. Lee, J. D. Park and J. H. Chung, Inorganic arsenite potentiates vasoconstriction through calcium sensitization in vascular smooth muscle, *Environ. Health Perspect.*, 2005, **113**, 1330–1335.
113. J. Pi, Y. Kumagai, G. Sun, H. Yamauchi, T. Yoshida, H. Iso, A. Endo, L. Yu, K. Yuki, T. Miyauchi and N. Shimojo, Decreased serum concentrations of nitric oxide metabolites among Chinese in an endemic area of chronic arsenic poisoning in inner Mongolia, *Free Radic. Biol. Med.*, 2000, **28**, 1137–1142.
114. J. Pi, S. Horiguchi, Y. Sun, M. Nikaido, N. Shimojo, T. Hayashi, H. Yamauchi, K. Itoh, M. Yamamoto, G. Sun, M. P. Waalkes and

Y. Kumagai, A potential mechanism for the impairment of nitric oxide formation caused by prolonged oral exposure to arsenate in rabbits, *Free Radic. Biol. Med.*, 2003, **35**, 102–113.
115. A. C. Straub, D. B. Stolz, M. A. Ross, A. Hernandez-Zavala, N. V. Soucy, L. R. Klei and A. Barchowsky, Arsenic stimulates sinusoidal endothelial cell capillarization and vessel remodeling in mouse liver, *Hepatology*, 2007, **45**, 205–212.
116. A. C. Straub, D. B. Stolz, H. Vin, M. A. Ross, N. V. Soucy, L. R. Klei and A. Barchowsky, Low level arsenic promotes progressive inflammatory angiogenesis and liver blood vessel remodeling in mice, *Toxicol. Appl. Pharmacol.*, 2007, **222**, 327–336.
117. N. Sorensen, K. Murata, E. Budtz-Jorgensen, P. Weihe and P. Grandjean, Prenatal methylmercury exposure as a cardiovascular risk factor at seven years of age, *Epidemiology*, 1999, **10**, 370–375.
118. S. W. Thurston, P. Bovet, G. J. Myers, P. W. Davidson, L. A. Georger, C. Shamlaye and T. W. Clarkson, Does prenatal methylmercury exposure from fish consumption affect blood pressure in childhood?, *Neurotoxicology*, 2007, **28**, 924–930.
119. G. A. Wiggers, I. Stefanon, A. S. Padilha, F. M. Pecanha, D. V. Vassallo and E. M. Oliveira, Low nanomolar concentration of mercury chloride increases vascular reactivity to phenylephrine and local angiotensin production in rats, *Comp Biochem. Physiol. C. Toxicol. Pharmacol.*, 2008, **147**, 252–260.
120. A. D. Torres, A. N. Rai and M. L. Hardiek, Mercury intoxication and arterial hypertension: report of two patients and review of the literature, *Pediatrics*, 2000, **105**, E34.
121. C. Beck, B. Krafchik, J. Traubici and S. Jacobson, Mercury intoxication: it still exists, *Pediatr. Dermatol.*, 2004, **21**, 254–259.
122. L. R. Faro, J. L. do Nascimento, J. M. San Jose, M. Alfonso and R. Duran, Intrastriatal administration of methylmercury increases *in vivo* dopamine release, *Neurochem. Res.*, 2000, **25**, 225–229.
123. Q. Pan, C. G. Kleer, K. L. van Golen, J. Irani, K. M. Bottema, C. Bias, M. De Carvalho, E. A. Mesri, D. M. Robins, R. D. Dick, G. J. Brewer and S. D. Merajver, Copper deficiency induced by tetrathiomolybdate suppresses tumor growth and angiogenesis, *Cancer Res.*, 2002, **62**, 4854–4859.
124. F. Donate, J. C. Juarez, M. E. Burnett, M. M. Manuia, X. Guan, D. E. Shaw, E. L. Smith, C. Timucin, M. J. Braunstein, O. A. Batuman and A. P. Mazar, Identification of biomarkers for the antiangiogenic and antitumour activity of the superoxide dismutase 1 (SOD1) inhibitor tetrathiomolybdate (ATN-224), *Br. J. Cancer*, 2008, **98**, 776–783.
125. J. C. Juarez, O. Betancourt Jr, S. R. Pirie-Shepherd, X. Guan, M. L. Price, D. E. Shaw, A. P. Mazar and F. Donate, Copper binding by tetrathiomolybdate attenuates angiogenesis and tumor cell proliferation through the inhibition of superoxide dismutase 1, *Clin. Cancer Res.*, 2006, **12**, 4974–4982.

126. A. Albini, F. Tosetti, R. Benelli and D. M. Noonan, Tumor inflammatory angiogenesis and its chemoprevention, *Cancer Res.*, 2005, **65**, 10637–10641.
127. C. D. Kamat, D. E. Green, S. Curilla, L. Warnke, J. W. Hamilton, S. Sturup, C. Clark and M. A. Ihnat, Role of HIF signaling on tumorigenesis in response to chronic low-dose arsenic administration, *Toxicol. Sci.*, 2005, **86**, 248–257.
128. B. Liu, S. G. Pan, X. S. Dong, H. Q. Qiao, H. C. Jiang, G. W. Krissansen and X. Y. Sun, Opposing effects of arsenic trioxide on hepatocellular carcinomas in mice, *Cancer Science*, 2006, **97**, 675–681.
129. N. V. Soucy, M. A. Ihnat, C. D. Kamat, L. Hess, M. J. Post, L. R. Klei, C. Clark and A. Barchowsky, Arsenic stimulates angiogenesis and tumorigenesis *in vivo*, *Toxicol. Sci.*, 2003, **76**, 271–279.
130. N. V. Soucy, D. Mayka, L. R. Klei, A. A. Nemec, J. A. Bauer and A. Barchowsky, Neovascularization and angiogenic gene expression following chronic arsenic exposure in mice, *Cardiovasc. Toxicol.*, 2005, **5**, 29–42.
131. L. R. Klei and A. Barchowsky, Positive Signaling Interactions between Arsenic and Ethanol for Angiogenic Gene Induction in Human Microvascular Endothelial Cells, *Toxicol. Sci.*, 2008, **102**, 319–327.
132. A. Barchowsky, E. J. Dudek, M. D. Treadwell and K. E. Wetterhahn, Arsenic Induces Oxidant Stress And NF-KappaB Activation In Cultured Aortic Endothelial Cells, *Free Radic. Biol. Med.*, 1996, **21**, 783–790.
133. N. V. Soucy, L. R. Klei, D. D. Mayka and A. Barchowsky, Signaling pathways for arsenic-stimulated vascular endothelial growth factor-a expression in primary vascular smooth muscle cells, *Chem. Res. Toxicol*, 2004, **17**, 555–563.
134. A. Barchowsky, R. R. Roussel, L. R. Klei, P. E. James, N. Ganju, K. R. Smith and E. J. Dudek, Low Levels of Arsenic Trioxide Stimulate Proliferative Signals in Primary Vascular Cells without Activating Stress Effector Pathways, *Toxicol. Appl. Pharmacol.*, 1999, **159**, 65–75.
135. R. R. Roussel and A. Barchowsky, Arsenic Inhibits NF-κB-mediated gene transcription by blocking IκB kinase activity and IκBα phosphorylation and degradation, *Arch. Biochem. Biophys.*, 2000, **377**, 204–212.
136. S. Hirano, Y. Kobayashi, X. Cui, S. Kanno, T. Hayakawa and A. Shraim, The accumulation and toxicity of methylated arsenicals in endothelial cells: important roles of thiol compounds, *Toxicol. Appl. Pharmacol.*, 2004, **198**, 458–467.

CHAPTER 13
Environmental Aldehydes and Cardiovascular Disease

D. J. CONKLIN, P. HABERZETTL, J. LEE AND
S. SRIVASTAVA

Diabetes and Obesity Center, Division of Cardiovascular Medicine, Department of Medicine, University of Louisville, 580 S. Preston Street, Louisville, KY 40292, USA

13.1 Introduction

Aldehydes are compounds with a terminal carbonyl (HC=O) moiety, and can be saturated or unsaturated, *i.e.* containing one or more carbon–carbon double bonds. This primary structural difference increases the chemical reactivity of unsaturated aldehydes by two to three orders of magnitude more than saturated aldehydes of similar carbon length. Nonetheless, aldehydes, in general, are ubiquitous components of the environment and their increased abundance is associated with increased risk of cardiovascular disease in humans. However, increased aldehyde abundance in the environment is not always a negative for human health because several familiar foods contain a variety of nontoxic aldehydes that, in addition to providing flavor to foods (*e.g.*, anisealdehyde, benzaldehyde, cinnamaldehyde, citralaldehyde) and beverages (*e.g.*, anisealdehyde), likely stimulate beneficial actions systemically in humans, such as anti-inflammatory action as ascribed to cinnamaldehyde. Yet these naturally derived aldehydes are to some extent in "competition" with aldehydes present in or generated in foods during the cooking/heating process, and some of which, in addition to their odor/flavor-enhancing qualities (*e.g.*, acrolein, formaldehyde), could very well be detrimental to cardiovascular health of the consumer. Similarly, respirable air is teeming with aldehydes generated during organic combustion, *e.g.*, airplane fuel, automobile exhaust, cigarette smoke, forest fires, power plants, *etc.*, resulting in quantitative enhancement of saturated and unsaturated aldehyde levels in the air, including acrolein, α-ethylacrolein, formaldehyde, crotonaldehyde among many others.

Although environmental aldehydes are inhaled by large numbers of the population, a shorter list of aldehydes through their direct use in industrial processes can lead to potentially dangerous occupational exposures associated with the generation/use of these aldehydes within these industries, such as in embalming and perfumery practices (*e.g.*, formaldehyde). To exacerbate the ever present "environmental aldehyde load," many industrial occupational exposures occur with compounds that are metabolized in the body to reactive aldehydes, which are implicated in downstream organ toxicity. This scenario has led to more than one illustrative example of the potentially harmful effects of aldehyde exposure. For example, plastics factory workers exposed to high levels of vinyl chloride over several decades can develop hepatic hemangiosarcoma, a rare liver endothelial cell tumor, at rates significantly higher than in nonexposed workers. Although the precise mechanism is not known, the aldehyde metabolite 2-chloroacetaldehyde is implicated. Similarly, the detrimental effects of 1,3-butadiene exposure in humans (and in experimental models) are attributed to its conversion to crotonaldehyde, a four-carbon, α,β-unsaturated aldehyde. Even better documented are the toxic side effects (*e.g.*, cardiotoxicity, urinary bladder toxicity) of the anticancer drug, cyclophosphamide (Cytoxan®), which are largely attributed to the generation of the three-carbon α,β-unsaturated aldehyde, acrolein, and these effects are recapitulated in experimental models using cyclophosphamide, acrolein alone, or with another acrolein precursor, allylamine. To complicate the situation, many aldehydes present in the environment or generated by environmental exposures to precursors are also produced as byproducts of metabolism in the human body. For example, acrolein production is enhanced during conditions of oxidative stress, diabetes and inflammation, and thus, exposure to environmental aldehydes would likely promote an "aldehyde-induced aldehyde release", which invariably could exacerbate pathogenesis.

Despite a wealth of associative evidence for aldehyde-induced cardiovascular effects and toxicity, the basic underlying mechanism(s) of aldehyde action remains elusive although several recent studies have revealed potentially important endogenous targets of environmental aldehydes. For example, recent work has highlighted aldehyde interaction with the transient receptor potential (TRP) class of nonselective cationic (calcium) membrane channels, including TRPA1, a receptor for 4-hydroxy-trans-2-nonenal (HNE) and acrolein that mediates pain sensation *via* peripheral sensory C-fibers located throughout the body. In addition, endogenous and exogenous aldehydes can stimulate a generalized mechanism of cellular stress whereby increased accumulation of protein–aldehyde adducts in the endoplasmic reticulum (ER) triggers ER stress and a multifaceted, complex unfolded protein response (UPR), which has been implicated in type-II diabetes and cardiovascular disease associated with diabetic condition. Moreover, protein–aldehyde adducts are increased in a variety of disease states and abundant protein–acrolein and protein–HNE adducts are present in atherosclerotic plaques as well as being present in the CNS plaques of Alzheimer's disease patients, which provide additional support for a link between aldehyde-induced UPR and pathogenesis in this condition as well. Many other protein and nonprotein targets exist and aldehyde-induced adducts in DNA, lipids and carbohydrates are increased under a variety of oxidative conditions. The contribution or role of specific protein–aldehyde adducts localized to the ER, mitochondria, DNA, or various membrane channels under disease-promoting conditions is not clear but is becoming an important focus of future studies to elucidate the underlying "cause and effect" relationship between aldehyde exposure and cardiovascular toxicity. These new studies likely will identify novel therapeutic targets that could with treatment ameliorate some of the untoward effects of aldehyde exposure in cardiovascular disease.

Regardless of the source of aldehyde or its chemical structure, the cardiovascular system (heart and blood vessels) is exquisitely sensitive to the effects of aldehyde exposure (*e.g.*, vasodilation, vasopressor, myocardial stunning) but additionally is well equipped to defend itself from aldehyde-induced toxicity *via* a variety of metabolic and detoxification enzymes. In fact, induction of aldehyde-metabolizing enzymes using model dietary oxidants, such as dithiole-3-thione (D3T), protects cardiovascular cells against subsequent oxidant and aldehyde challenge providing evidence for the protective function of these enzyme systems. Gene polymorphisms in aldehyde-metabolizing genes, such as glutathione *S*-transferases (GSTs), represent alterations in aldehyde-metabolizing capacity and/or substrate specificity in ways that alter aldehyde metabolism, and thus, could explain variation in susceptibility to aldehyde-induced cardiovascular toxicity as well as toxicity in other organs. These natural loss-of-function "experiments" are highlighted by epidemiological studies. For example, the *GSTT1*-null genotype is associated with increased cardiovascular disease morbidity and mortality in diabetic smokers compared with matched reference population with *GSTT1* gene. Similar studies using genetically engineered animal models have begun to interrogate the specific deficiency of the null genotypes, and thus, reveal the protective role of these enzyme systems in cardiovascular organs during aldehydic stress. Almost certainly, additional GWAS (genome-wide association studies) will be useful in providing a more complete and panoramic view of the genetic relationships between suspected gene deficiencies and other polymorphisms (SNPs) that could alter aldehyde-metabolizing systems.

Some of these environmental aldehydes, especially those in low abundance including unsaturated aldehydes, are difficult to measure accurately in the environment. Moreover, their chemical composition can change dramatically over short time periods due to chemical interactions in the air (*e.g.*, ozone) through complex photo-dependent processes, and thus, result in new potentially dangerous compounds as well. Methods for measuring environmental aldehydes are particularly important given that government-based (*e.g.*, EPA) regulations will be dependent on the accuracy and precision of these methods and reported measurements. We have addressed the methodological approaches currently in practice, EPA-approved and otherwise, for measurement of various environmental aldehydes. A comparison of benefits and pitfalls of these methods is also discussed.

Overall, environmental aldehydes represent an important component of a growing world-wide catastrophe, air pollution, but the specific contribution of these reactive constituents to the burden of cardiovascular disease risk has not been calculated. The cardiovascular system is exquisitely sensitive to the effects of exogenous aldehyde exposure, especially α,β-unsaturated aldehydes such as acrolein and 4HNE. Yet, not all responses in the cardiovascular system, in and of themselves, are deleterious, and recent studies reveal that several plant-derived aromatic and unsaturated aldehydes stimulate sensitive and nontoxic endothelium-dependent vasorelaxation *via* release of NO and/or EDHF. Of course in toxicology, the dose makes the poison and aldehyde exposures are no exception to this rule. Induction of a variety of antioxidant enzymes by plant-derived allylic and thiol-containing substances, protects cardiovascular tissues from toxicity of subsequent aldehyde exposure implicating direct metabolism of aldehydes as a primary determinant of tissue sensitivity to aldehyde exposure. This conclusion is well supported by a plethora of epidemiological/single nucleotide polymorphism (SNP) studies that demonstrate increased risk of cardiovascular disease associated with elevated aldehydic stress and altered metabolic enzyme capacity (*e.g.*, GST-null genotypes) or profile. Additionally, transgenic and knockout animal models have provided a more focused view of the protective role of these enzymes in cardiovascular tissues against aldehyde exposure. Future studies will continue to home in on the fundamental mechanisms that drive aldehyde toxicity and the complex mechanisms that govern these processes at the cellular, tissue, organ and whole-animal level.

13.2 Epidemiology of Environmental Aldehydes and Cardiovascular Disease

13.2.1 Levels of Environmental Aldehydes

In keeping with the general nature of the *Environmental Cardiology* theme that each individual toxicant of the environment exerts a quantifiable level of risk for cardiovascular disease, we believe that environmental aldehydes are toxic individually or in concert with other pollutants, *e.g.*, particulate matter, one of the more nefarious of these agents.[1,2] This interpretation of aldehydes comes not solely because of their extreme reactivity as electrophiles, but also unlike other environmental pollutants, such as PM or arsenic (see other chapters in this book), aldehyde exposure can lead to additional endogenous aldehyde generation and exposure. For example, inhaled or ingested acrolein can induce endogenous acrolein generation by a variety of mechanisms. As such, it is devilish to distinguish the effects of exogenous acrolein exposure *per se* from the effects of endogenous acrolein generation. There does not appear to be another class of compounds where so many of the environmentally relevant constituents are also formed biologically *in vivo* (*e.g.*, acetaldehyde, formaldehyde, acrolein, propionaldehyde). Although it is true enough to say that "PM exposure does not generate endogenous PM", or that "arsenic exposure does not generate endogenous arsenic", we cannot ignore that exposure to either PM or arsenic also will result in quantitative aldehyde generation *in vivo*, perhaps *via* oxidative stress and lipid peroxidation or by stimulating inflammation. In either case, this argument highlights the beguiling nature of the interaction between aldehydes and the cardiovascular system because of their ubiquitous presence and also their role as perfunctory "middlemen" that "extract a heavy toll" for their involvement regardless of temporal appearance or initiation source.

Aldehydes are ubiquitous and plentiful compounds present in the environment – and relevant to this compendium – these are in high concentrations in automobile exhaust and smog. They are primary products of combustion of organic material of any form (*e.g.*, coal, gasoline, oil, paper, wood, *etc* . . .) but also are generated in secondary decomposition and photochemical processes. As such, they constitute 1–2% of the volatiles in automobile exhaust and are generated during the burning of fossil fuels (Feron *et al.* 1991). Moreover, alterations in gasoline composition to include greater methanol/ethanol fraction will increase aldehyde generation. As a prime exemplar, cigarette smoke contains 50–70 ppm (parts per million) acrolein, and 0.04–2.2 ppm acrolein has been detected near petrochemical plants.[3] In addition, acrolein, hexenal, and related aldehydes are also present in high abundance in several food substances, and their concentration is particularly high in fried foods and reheated oils.[3,4] With the exception of metals, aldehydes are the major toxicants in drinking water. In fact, 36 different aldehydes are found in drinking water, of which acrolein and endrin aldehyde are classified as the two highest priority pollutants.[3] Chemical structures of some of the more common aldehydes in foods/ beverages and air pollution are presented (Figure 13.1; also Conklin and Bhatnagar, 2010), and summaries of aldehyde levels in air, water, and foods are presented in Tables 13.1–13.3 (additionally see Wang *et al.* 2008 for levels of acrolein in many of these same sources).[5] Estimates of total aldehyde consumption *via* food and beverages range from $\sim 7\,\text{mg/kg/day}$ for total aldehydes and $\approx 5\,\text{mg/kg/day}$ for unsaturated aldehydes. In addition, an adult human breathes $\approx 8640\,\text{l}$ ($8.64\,\text{m}^3$) of air per day containing ≈ 0.6–$0.8\,\text{mg}$ total aldehydes (urban air) of which $<5\%$ are unsaturated aldehydes (see Table 13.2). Thus, the combined total aldehyde intake in humans is ≈ 7–$8\,\text{mg/kg/day}$, which is likely an underestimation of real-world exposure (*e.g.*, add dermal uptake to total burden). How to detect and measure environmental aldehydes is addressed in Section 13.5, Figure 13.7.

13.2.2 Epidemiology of Aldehyde Exposures

As indicated above, environmental aldehydes are ubiquitous, but it is not clear if any one aldehyde causes cardiovascular disease in humans. However, there are a number of epidemiological studies that highlight associations between a particular aldehyde exposure with the level of human cardiovascular disease

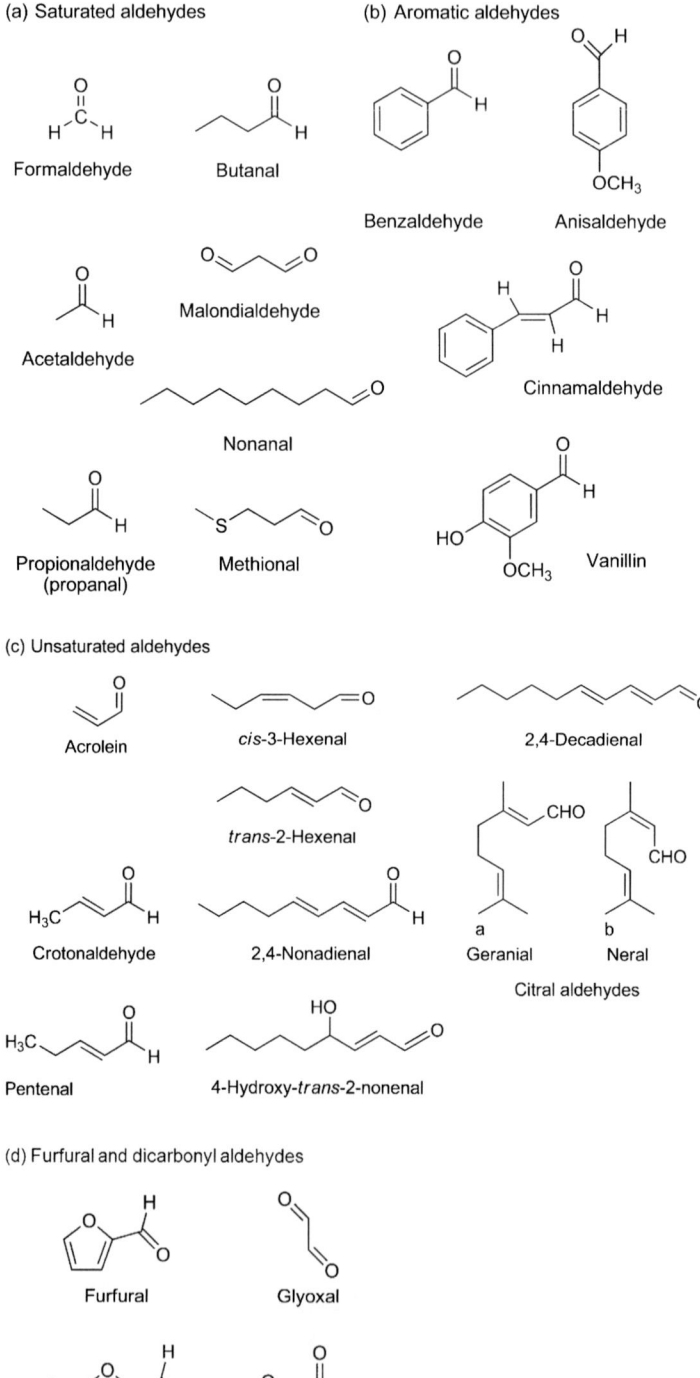

burden. For example, there is an increased risk of atherosclerotic heart disease in plant workers producing formaldehyde, and higher incidence of heart disease in embalmers, perfumery workers, and undertakers that is associated with increased exposure to formaldehyde.[6–9] Similarly, a higher incidence of heart disease in Swedish bus drivers may be dependent on aldehyde exposure.[10] Although numerous studies abound linking environmental tobacco smoke (ETS) or air pollution (in general defined as increased level of $PM_{2.5}$) exposure with cardiovascular disease risk in humans,[11–18] the direct role of aldehydes in these associations has not been defined. This is largely due to the lack of measurements of aldehyde exposures in those studies. There is no question that aldehydes, especially acrolein and crotonaldehyde, are at high levels in tobacco smoke (low ppm range) and could account for immediate respiratory and cardiovascular changes including increased morbidity/mortality but the question is whether they participate in precipitating these outcomes. Recently, we showed that cigarette-smoke-induced endothelial dysfunction is recapitulated by acrolein exposure (1 ppm) in glutathione S-transferase P-deficient mice, which provides some of the first data that a well-known cardiovascular effect of cigarette smoke (endothelial dysfunction) is dependent on an aldehyde and aldehyde metabolism.[19] Intriguingly, in a study of Devlin *et al.*, they also show that air-pollution exposure in healthy male state troopers induced cardiac changes that ape best associated with real-time aldehyde measurements made in the patrol car.[20] These are some of the first data to establish a potential direct and predominant "cause and effect" status for aldehydes in air-pollution-induced effects on the cardiovascular system.

Importantly, acrolein is reported to be present in ambient air at 10–100x the RfC level (reference concentration = no effect level) at numerous locations around the US, including in the air in Louisville, Kentucky and in air near several elementary schools in eastern Kentucky. In fact, a popular press story made headlines in a USA Today newspaper article in the fall 2009. However, data presented in the USA Today article were not direct measurements of acrolein levels but were estimates made from modeled air pollution data. Yet, a study of the US EPA indicated that acrolein levels in ambient air exceeds the chronic reference concentration (RfC; at which no health effect would occur) level (0.02 $\mu g/m^3$; 8.7 ppt) by 10–22 times.[21] Similarly, a study of 3 locations in northern California, including at a toll booth on the San Francisco Bay Bridge, measured elevated levels of acrolein.[22,23] Importantly, Woodruff *et al.*, attempt to estimate health risk (as decrement in lung function) due to these elevated levels of ambient acrolein *via* extrapolation using benchmark dose modeling from rat inhalation exposure data collected 20 years earlier.[24] These data highlight 2

Figure 13.1 Chemical structures of common environmental aldehydes. Although many of these aldehydes are present in air, food, and water, many are also generated endogenously as products of lipid peroxidation and enzymatic processes. Aldehydes are grouped according to structural features including: (A) saturated, (B) aromatic, (C) unsaturated, and (D) furfur aldehydes and dicarbonyl-containing aldehydes.

Table 13.1 Exposure limit values and reference concentrations (RfC) of ambient aldehydes.

Aldehyde	MW (g/mole)	Exposure limit value ([a]TWA)	[c]Reference concentration (RfC)
Formaldehyde	30.03	0.75 ppm (OSHA, 1992)	9.8 µg/m^3 (0.008 ppm) (ATSDR)
Acetaldehyde	44.06	25 ppm ([b]STEL) (ACGIH, 2004)	9.0 µg/m^3 (0.005 ppm) (IRIS)
Propanal (Propionaldehyde)	58.08	20 ppm (ACGIH, 2004)	8.0 µg/m^3 (0.004 ppm) (IRIS)
Pentanal (Valeraldehyde)	86.15	50 ppm (ACGIH, 2005)	420 µg/m^3 (0.119 ppm) (OEHHA)
Butanal (Butyraldehyde)	72.12	25 ppm (AIHA, 2003)	15.0 µg/m^3 (0.005 ppm) (NYSDEC)
Acrolein (Propenal)	56.06	0.1 ppm (ACGIH, 2004)	0.02 µg/m^3 (8.7×10^{-6} ppm) (IRIS)
Crotonaldehyde	70.09	2 ppm (OSHA, 1992)	Not Available
Benzaldehyde	106.12	Not Available	9.0 µg/m^3 (0.002 ppm) (NYSDEC)
Hexanal (hexanaldehyde)	100.18	Not Available	20.0 µg/m^3 (0.005 ppm) (NYSDEC)

[a]TWA: Time-Weighted Average, 8 h
[b]STEL: Short-Term Exposure Limit, 15 min
[c]RfC: non-cancer inhalation Reference Concentration. The ppm values were determined using the following formula:

$$\text{Gas concentration}(mg/m^3) = \frac{[\text{gas conc.}(ppm) \times \text{molecular weight}(g/mole)]}{\text{molar gas volume}}$$

where, molar gas volume = 24.4 L/mole at 25 °C. Abbreviations: OSHA, Occupational Safety and Health Administration; ACGIH, American Conference of Governmental Industrial Hygienists; AIHA, American Industrial Hygiene Association; ATSDR, Agency for Toxic Substance Disease Registry; IRIS, Integrated Risk Information System; OEHHA, Office of Environmental Health Hazard Assessment, California; NYSDEC, New York State Department of Environmental Conservation.

Table 13.2 Level of aldehydes in air.

Air source	Aldehyde levels	Measurement method	Reference
Research lab (indoor)	Formaldehyde: 2.4–2.9 ppb Acetaldehyde: 2.6–4.3 ppb	Cartridge/capillary electrophoresis-UV	Pereira et al., 2003
Urban air/industrial/incinerator emission	Formaldehyde: 1.6–8.5 ppb (urban) 20–6400 ppb (industrial) 260–680 ppb (incinerator) Acetaldehyde: 1.7–9.0 ppb (urban) 20–330 ppb (industrial) 880–1120 ppb (incinerator) Acrolein: 220–270 ppb (incinerator) Propionaldehyde: 0.8–5.9 ppb (urban)	Sep-PAK C_{18} cartridge/HPLC-UV	Kuwata et al., 1983
Coastal air (Savannah, Georgia)	Formaldehyde: 0.17–6.80 μg/m^3 (*0.1–5.5 ppb*) Acetaldehyde: 0.07–7.60 μg/m^3 (*0.04–4.2 ppb*) Propionaldehyde: 0.02–9.10 μg/m^3 (*0.01–4.1 ppb*)	Adsorbent tube/HPLC-UV	Macintosh et al., 2000
Urban ambient (Bangkok, Thailand) and Indoor (office building)	(Urban ambient) Formaldehyde: 10.1 μg/m^3 (*8.2 ppb*) Acetaldehyde: 7.9 μg/m^3 (*4.4 ppb*) (Indoor office) Formaldehyde: 35.5 μg/m^3 (*28.8 ppb*) Acetaldehyde: 17.1 μg/m^3 (*9.5 ppb*)	Cartridge/HPLC-UV	Ongwandee et al., 2009
Urban ambient air	Formaldehyde: 9–26 ng/L (*7.3–21 ppb*) Acetaldehyde: 10–34 ng/L (*6–19 ppb*) Acrolein: 0.1–1.0 ng/L (*0.04–0.4 ppb*) Propionaldehyde: 0.1–2.9 ng/L (*0.04–1.3 ppb*)	Impinger/HPLC-UV	Schlitt, 1997
Urban ambient air (traffic rush hour, San Francisco)	Acrolein: 0.031–0.103 μg/m^3 (0.013–0.045 ppb) Methacrolein: 0.061–0.188 μg/m^3 (0.021–0.066 ppb)	Impingers/high-resolution gas chromatography with ion trap mass spectrometry	Destaillats et al., 2002

Table 13.2 (*Continued*)

Air source	Aldehyde levels	Measurement method	Reference
	Crotonaldehyde: 0.052–0.149 µg/m^3 (0.018–0.052 ppb) p-Tolualdehyde: 0.006–0.128 µg/m^3 (0.001–0.026 ppb) Methylglyoxal: 0.039–0.095 µg/m^3 (0.013–0.032 ppb) Benzaldehyde: 0.090–0.162 µg/m^3 (0.021–0.037 ppb) Glycoaldehyde: 0.093–0.474 µg/m^3 (0.038–0.193 ppb)		
Urban ambient air (Rome)	Formaldehyde: 8.2–27.7 ppb Acetaldehyde: 2.9–17.4 ppb Acrolein: 0.3–1.0 ppb Propionaldehyde: 0.5–3.0 ppb Methylvinylketone/ Methacrolein: 0.2–1.0 ppb Butanal/2-butanone: 0.6–2.4 ppb Benzaldehyde: 0.3–1.0 ppb	DNPH cartridge/ HPLC-UV	Possanzini *et al.*, 1996

Note: the *ppb values in parentheses* are calculated using the following formula:
Gas conc. (mg/m^3) = [gas conc. (ppm) × molecular weight (g/mole)]/molar gas volume. Where, molar gas volume is 24.4 L/mole at 25 °C.
Estimated inhaled total aldehydes per day = Σ(saturated + unsaturated aldehyde content in urban air) in µg/L × 8,640 liters per day (TV × bpm × 1440 min/day).

shortcomings of current research in aldehyde-inhalation toxicology: (1) lack of knowledge regarding what levels of aldehydes constitute elevated risk, and (2) lack of aldehyde exposure data that demonstrate cardiovascular system alterations. It was concluded that human exposure to elevated air levels of acrolein (above the RfC) does increase risk of pulmonary injury (*i.e.*, decreased compliance and increased residual volume/total lung capacity ratio) by a median of 2.5 per 1000 cases of decreased compliance.[21] It should be noted that indoor levels of acrolein can be orders of magnitude greater (in the ppm range not the ppb range) than outdoor levels, especially where tobacco smoke is present, which implies that a much higher risk from aldehyde exposure occurs under these

Table 13.3 Levels of aldehydes in water.

Water source	Aldehyde levels	Measurement method	Reference
Tap water	Formaldehyde: 0.7–3.3 µg/L Acetaldehyde: ∼1.1 µg/L	Headspace-GC/MS	Sugaya et al., 2001
Commercial mineral water	Formaldehyde: ∼59 µg/L Aectaldehyde: ∼260 µg/L Propionaldehyde: ∼0.9 µg/L n-butylaldehyde: ∼0.3 µg/L		
Rain	Formaldehyde: 120 ppb	Capillary electrophoresis and UV detector	Asthana et al., 1998
Commercial drinking water	Formaldehyde: <129 ppb	HPLC/UV detector	Tsai et al., 2003
Mineral water in PET bottles	Japan Formaldehyde: 10.1–27.9 µg/L Acetaldehyde: 44.3–107.8 µg/L Europe Formaldehyde: 7.4–13.7 µg/L Acetaldehyde: 35.9–46.9 µg/L North America Formaldehyde: 13.6–19.5 µg/L Acetaldehyde: 41.4–44.8 µg/L	HPLC/UV detector	Mutsuga et al., 2006
Snow water	n-hexanal: 0.681–1.577 µg/L n-heptanal: 0.253–0.371 µg/L n-octanal: 0.324–0.594 µg/L n-nonanal: 0.837–1.891 µg/L n-decanal: 0.574–1.053 µg/L	Headspace-solid-phase dynamic extraction-GC/MS	Sieg et al., 2008

conditions. However, there are no estimates for cardiovascular disease risk for ambient acrolein (or other aldehydes) exposure (acute or chronic). Occupational exposure limits (time-weighted averages for an 8-h shift) for common aldehydes are typically in the 1–25 ppm range for saturated aldehydes and <2 ppm for unsaturated aldehydes, highlighting the basic difference between reactivity of

Table 13.4 Saturated aldehyde levels in foods.

	Measurements method: GC/flame photometric detector Units: μg/ml (liquid material) or μg/g (solid material)				
Data source 1: (Kataoka et al., 1995)	Propanal	Butanal	Isopentanal	Hexanal	Nonanal
Sesame oil	2.4 ± 0.1	6.9 ± 0.3	6.7 ± 0.1	4.0 ± 0.1	11.9 ± 0.7
Soybean oil	1.6 ± 0.1	5.4 ± 0.3	5.1 ± 0.3	1.8 ± 0.1	8.1 ± 0.5
Rapeseed oil	2.5 ± 0.1	5.3 ± 0.3	4.9 ± 0.3	3.6 ± 0.2	10.6 ± 0.6
Cottonseed oil	1.6 ± 0.1	5.3 ± 0.3	5.1 ± 0.3	2.3 ± 0.1	10.0 ± 0.6
Corn oil	1.3 ± 0.1	5.0 ± 0.2	4.0 ± 0.4	2.2 ± 0.1	9.5 ± 0.2
Olive oil	1.8 ± 0.1	5.5 ± 0.2	4.8 ± 0.3	3.5 ± 0.2	12.6 ± 1.0
Sardine oil	5.9 ± 0.1	6.2 ± 0.2	3.3 ± 0.2	6.5 ± 0.2	23.7 ± 1.2
Beef fat	2.7 ± 0.3	5.1 ± 0.3	5.4 ± 0.2	2.4 ± 0.2	12.6 ± 1.7
Lard	3.8 ± 0.2	5.5 ± 0.3	4.7 ± 0.3	2.6 ± 0.1	9.0 ± 0.9
Butter	2.1 ± 0.1	4.9 ± 0.2	5.1 ± 0.3	2.0 ± 0.2	13.0 ± 0.8
Margarine	1.9 ± 0.2	5.1 ± 0.4	4.8 ± 0.3	1.2 ± 0.1	8.6 ± 0.8
Mayonnaise	2.1 ± 0.3	3.8 ± 0.2	3.8 ± 0.4	2.2 ± 0.2	10.6 ± 0.5
Cheese	2.4 ± 0.1	4.6 ± 0.2	3.4 ± 0.1	3.5 ± 0.1	22.8 ± 1.3
Egg yolk	2.7 ± 0.3	4.0 ± 0.2	3.5 ± 0.2	1.5 ± 0.1	9.0 ± 0.2
Fresh milk	3.0 ± 0.5	3.7 ± 0.4	5.1 ± 0.1	1.5 ± 0.1	9.4 ± 0.5
Chocolate	1.9 ± 0.2	4.4 ± 0.3	3.5 ± 0.4	5.1 ± 0.2	19.1 ± 0.5
Potato chips	2.0 ± 0.1	3.2 ± 0.2	4.1 ± 0.2	1.8 ± 0.1	9.9 ± 0.9

	Measurements method: HPLC/cation micromembrane suppressor and enzyme reactor	
Data source 2: (Schultheiss et al., 2000)	Hydroxylmethylfuraldehyde (HMF)	Furaldehyde
Coffee powder	209–606 mg/kg	70–160 mg/kg
Instant coffee powder	959–6181 mg/kg	14–95 mg/kg
Honey	0.1–41 mg/kg	0–0.5 mg/kg
Dry fruits	114–1196 mg/kg	0–9 mg/kg
Breakfast cereal	26–44 mg/kg	Not detected
Fruit Juice	0.1–0.7 mg/L	0.3–5.9 mg/L

these compounds, although surprisingly the TWA for formaldehyde (0.75 ppm) is lower than the TWA for crotonaldehyde, an unsaturated aldehyde (2 ppm) (Table 13.1).

Because aldehydes are present in high levels in tobacco smoke and because smoking increases cardiovascular disease risk, it is hypothesized that aldehydes contribute to the increased cardiovascular disease risk in active and passive smokers. Certainly, exposure to unsaturated aldehydes, acrolein and crotonaldehyde, can recapitulate the effects of tobacco smoke on upper-airway inflammation and hyperreactivity[25–27] and endothelial dysfunction[19] in animal models. The former effects are dependent on the presence of TRPA1 receptor, which appears to act as an "unsaturated aldehyde receptor" as well as a receptor for other oxidants, which mediate pain and local inflammatory responses (*e.g.*, increased blood flow, edema, smooth muscle constriction; *vide infra*).[28–30] Because cardiovascular disease risk of cigarette smoking (or passive smoking) is significantly higher than lung cancer risk of smoking, it is thought that the cardiovascular system is a more sensitive target of these aldehydes in tobacco smoke. This will provide much fertile ground for future research in an attempt to "close the loop" by both identifying the agents present in cigarette smoke that are responsible for the increased threat to cardiovascular health and identify the mechanism by which tobacco smoke mediates these untoward effects (*e.g.*, TRPA1 receptor).[29,31,32] Although smoking cessation has dramatic beneficial effects on cardiovascular health, including decreasing the risk of MI, stroke, and endothelial dysfunction within weeks to months presumably *via* decreasing aldehyde exposure, it will require much tougher action to decrease urban-air aldehyde burden given the predominant use of fossil fuels in the generation of heat, power/electricity and for transportation.

13.2.3 Aldehyde Exposure as a Product of Xenobiotic Metabolism

Additional data implicating aldehyde exposure in burden of cardiovascular disease come from studies of aldehyde generation from exogenous compounds. We highlight 4 examples of metabolically-derived aldehydes in humans that lead to specific cardiovascular effects and for which substantial supporting data in animal models has also been gained.

13.2.3.1 1,3-Butadiene (Crotonaldehyde)

Exposure to 1,3-butadiene, which is metabolized to crotonaldehyde, is associated with increased incidence of atherosclerosis in male African-American workers (but not in matched Caucasians) in the styrene-butadiene industry for the years, 1943–1982.[33] These studies implicate a role for ethnically-related metabolism in outcome differences. Similarly, exposure of cockerels to 20 ppm 1,3-butadiene accelerates atherosclerosis in aorta in this model.[34,35] Additionally, cardiac angiosarcomas are increased in animals exposed to 1,3-butadiene. Although crotonaldehyde is a highly reactive aldehyde component of tobacco smoke and is implicated in the deleterious respiratory effects of ETS exposure,[29] its specific role in cardiovascular effects of 1,3-butadiene has not been rigorously established.[36–38]

13.2.3.2 Vinyl Chloride (Chloroacetaldehyde)

Exposure to the aldehyde-generating compound vinyl chloride, which forms chloroacetaldehyde, in workers in the plastics industry significantly increases cardiovascular disease, including myocardial infarction, hypertension, and other circulatory disorders.[39] Most interestingly, exposed workers have a higher incidence of the very rare hepatic hemangiosarcoma, which is thought to be due to hepatic endothelial cell-localized cytochromes P450-mediated metabolism of vinyl chloride to chloroacetaldehyde.[40–42]

13.2.3.3 Cyclophosphamide (Acrolein)

The anticancer drug cyclophosphamide (Cytoxan®) leads to formation of acrolein (2-propene), which is associated with deleterious side effects of urinary bladder hemorrhagic cystitis and cardiotoxicity.[43–48] Although the latter toxicity has been dramatically reduced in the acute clinical setting due to better understanding of maximal tolerated dosing, there are still too many cases of cardiac toxicity in a variety of settings.[48] Additionally, anticancer therapy in children may carry an additional burden of increased risk of heart failure but the role of cyclophosphamide/acrolein in these outcomes remains uncertain.[49]

13.2.3.4 Allylamine (Acrolein)

Along the same lines as cyclophosphamide, the metabolism of the xenobiotic allylamine by SSAO leads to formation of a toxic mixture of acrolein, H_2O_2, and NH_3 in equal molar amounts. This metabolism is highly localized to SSAO expression pattern in large arteries, such as aorta, in a variety of species.[50] Inhibition of SSAO with semicarbazide or more specific drugs prevents the untoward cardiovascular toxicity of allylamine, including atherosclerosis-like lesions, subendocardial myocardial infarction, and spontaneous hypercontracture (vasospasm) in isolated human coronary artery bypass graft (CABG) blood vessels.[51–53] These effects have been attributed largely to the generation of acrolein in the blood vessel wall, and in some models the effects of acrolein exposure alone have recapitulated the actions of allylamine.[53,54] In parallel studies, the role of SSAO in generation of formaldehyde from methylamine is implicated in vascular effects in experimental models of diabetes and in isolated human CABG blood vessels.[55,56] Thus, it remains an open question whether exogenous substrates (i.e., primary amines) or perhaps inhibitors, e.g., semicarbazide and β-aminopropionitrile, could interact with the highly vascular-specific SSAO enzyme to promote vascular dysfunction and structural derangements via aldehyde formation that ultimately increase cardiovascular disease risk.[57–59] For more detailed information regarding SSAO and vascular biology and pathology see Boor and colleagues.[60,61]

13.3 Cardiovascular Effects and Signaling Mechanisms of Aldehyde Exposure

The direct effects of aldehydes in the cardiovascular system of animals has been reviewed recently and will not be restated here. However, recent insight into mechanisms by which enal aldehydes stimulate cardiovascular responses and recently observed effects of systemic enal aldehyde treatments (mostly acrolein) will be a focus of this portion of the review.

13.3.1 Enals and Cell Signaling

13.3.1.1 Effect of Acrolein on the Activation of MAP Kinases

In vitro, enals, such as acrolein and HNE, affect multiple signaling pathways. Several studies suggest that MAP kinases (MAPK) play a pivotal role in acrolein-induced cell signaling. In endothelial cells, acrolein induces the expression of cyclooxygenase-2 (COX-2) in a p-38 and PKCδ-dependent manner and protein–acrolein adducts colocalize with COX-2 in human atherosclerotic lesions.[62] Acrolein also increases the expression of HSP72 in endothelial cells in PKCδ/JNK and calcium-dependent manner.[63] In rat vascular smooth muscle cells, acrolein activates all the three MAPK, i.e. ERK1/2, p38 and JNK, and causes protein tyrosine phosphorylation.[64] In human bronchial epithelial cells acrolein induces the expression of heme oxygenase-1 (HO-1),[65] which is prevented by the silencing of Nrf-2 gene. Inhibition of PKCδ diminishes acrolein-induced Nrf-2 nuclear translocation and HO-1 induction, whereas inhibition of PI3K attenuates acrolein-induced HO-1-expression, but does not prevent the nuclear translocation of Nrf-2, suggesting that PI3K is associated with the downstream signaling of Nrf-2. In Chinese hamster ovary (CHO) cells, acrolein-induced apoptosis is suggested to be mediated by ERK and p38 MAP kinases.[66] HNE activates JNK in several vascular cells,[67–69] and in stellate cells HNE covalently modifies the p46 and p54 isoforms of JNK.[70] Collectively, these studies suggest some of the signaling properties of enals could be mediated by MAPK. However, additional studies are required to examine the cause-and-effect relationship between MAPK (and/or other signaling molecules) in enal-induced cardiovascular dysfunction and inflammation.

13.3.1.2 Cell-Signaling Properties of Glutathionyl Conjugates of Enals

Although enals such as acrolein and HNE readily form glutathione conjugates, this is not sufficient for the inactivation of the aldehyde. We recently observed that in rat vascular smooth muscle cells, GS-HNE activates PKC and stimulates NF-κB and AP-1-dependent gene transcription and enhances smooth muscle cell growth.[71] We also observed that in polymorphonuclear leukocytes (PMNs) in vitro, GS-HNE stimulates superoxide radical formation and induces the expression of CD11b, and intraperitoneal injection with GS-HNE significantly augments peritoneal leukocyte infiltration and enhances the formation of proinflammatory lipid mediators.[72] Our results are in agreement with previous studies showing that the glutathionyl conjugate of acrolein increases superoxide production[73] and glutathione conjugates of enals are toxic and induce DNA damage or enhance free-radical production.[74,75] Recently, we also observed that even the reduced form of the glutathionyl conjugate of HNE (GS-DHN) has signaling properties. It stimulated protein kinase C, NF-κB, AP-1 and augmented smooth muscle cell proliferation.[76] Antibodies against the transporters of glutathione conjugates, RLIP (Ral-binding protein) or multidrug resistance protein-2 (MDR-2), or the knockdown of RLIP by siRNA increased cell proliferation, whereas overexpression of RLIP decreased the mitogenic signaling of the glutathionyl conjugates. Warnke et al., observed that RLIP-knockout mice accumulate GS-HNE in the liver.[77] Collectively, these recent studies suggest that glutathionyl conjugates of enals are involved in cell signaling, although the pathophysiological significance of this signaling remains to be examined.[78,79]

13.3.1.3 Role of TRPA1 Receptor in Enal Effects

How aldehydes or their glutathione conjugates activate signaling pathways is still a question, but some protein targets (receptors?) have been identified. For example, recent work has highlighted aldehyde interaction with the transient receptor potential (TRP) class of nonselective calcium membrane channels, including TRPA1 (where A = ankryin), an endogenous receptor for 4-hydroxy-*trans*-2-nonenal (HNE) and acrolein that mediates pain sensation *via* peripheral sensory C-fibers located throughout the body (see review of Jordt and Bessac, 2008).[28,31,80,81] Recently, the distribution of TRPA1 receptors has been expanded to include non-neuronal locations, such as the heart, endothelium, and urothelium.[82–84] Direct effects of unsaturated aldehydes on these endothelium-located receptors could result in stimulation of nonspecific increases in calcium, and thus, activation of eNOS and subsequent vasodilation *via* release of NO. Similarly, increased endothelial Ca^{++} could act as a trigger for release of endothelial-derived hyperpolarizing factor (EDHF),[82] and as described for acrolein-induced dilatation in the perfused rodent mesenteric bed.[85] Whether acrolein-induced vasodilatation is dependent on TRPA1 receptors located in the endothelium or elsewhere is not known, but the dilation mechanism so far investigated requires a "second look".

Exposure to environmental tobacco smoke (ETS) induces pulmonary inflammation and edema and airway hyperreactivity, effects that are mediated, in part, by TRPA1 receptors.[28,29,31,32] In fact, mice deficient in TRPA1 receptors have significantly less pulmonary injury in response to ETS than do similarly exposed wild-type mice.[29] Moreover, isolated airways exposed to acrolein respond with TRPA1-dependent edema and hyperreactivity, both demonstrating the direct effects of acrolein on these tissues and supporting the role of unsaturated aldehydes, such as acrolein and crotonaldehyde, as mediators of ETS-induced pulmonary inflammation.[29] Whether TRPA1 receptors mediate the deleterious effects of ETS or acrolein exposure in the cardiovascular system is unknown but provides a testable hypothesis that links well-described acute effects of ETS on cardiovascular physiology/pathology with a plausible biological mechanism.

Despite over two decades of observations linking second-hand tobacco-smoke exposure with endothelial dysfunction in animal models and humans,[11,12,86–93] the components in ETS responsible for this effect have not been discerned. Recently, we added support to the idea that unsaturated aldehydes, such as acrolein, in ETS contribute to endothelial dysfunction following subchronic exposure. In our model, glutathione S-transferase P (GSTP) deficient mice are susceptible to endothelial dysfunction following 3-day exposure to ETS or acrolein while wild-type mice are protected under these exposure conditions.[19] Moreover, we show that direct exposure of isolated aorta to acrolein resulted in a GSTP-dependent endothelial dysfunction, which supports a direct effect of acrolein on the endothelium and a role of acrolein in ETS-derived cardiovascular effects. Again, the role of TRPA1 in this effect was not explored but is a potentially attractive target of acrolein action in the vascular wall. Moreover, we provide evidence that implicates real-time metabolism of aldehydes in the vascular wall, specifically the endothelial cells, as an important determinant of cardiovascular protection and health.[19,94,95]

We recently reviewed the role of specific enzymes in the cardiovascular metabolism of aldehydes.[96] Importantly, and in addition to the metabolic function of these enzymes, other functions of GSTs such as protein–protein binding, could contribute significantly to the deleterious effects of aldehyde exposure. For example, GSTP-deficient mice are more sensitive to cyclophosphamide-induced urinary bladder cystitis than are wild-type mice – an effect paralleled by increased JNK and c-Jun phosphorylation and inflammatory infiltration.[97] It is difficult to separate the deficit in acrolein metabolism with the strict loss of GSTP-JNK binding that such a whole-body genetic knockout model presents, but it is well established that monomeric GSTP interacts with JNK and c-Jun proteins in a complex that prevents c-Jun activation by JNK.[98,99] Moreover, loss of total GSTP appears to result in constitutive activation of bone marrow, liver and lung JNK promoting constitutive c-Jun phosphorylation and presumably AP-1 activation. Interestingly, no constitutive JNK or c-Jun phosphorylation is detected in urinary bladder despite the absence of GSTP protein and an elevated level of basal protein–acrolein adducts.[97] Although increased protein–aldehyde adducts provide supportive evidence for a decreased metabolism of acrolein under basal conditions as well as during cyclophosphamide-induced cystitis, the overall metabolism of nontoxic doses of cyclophosphamide or acrolein to hydroxypropyl mercapturic acid (HPMA) in urine is similar between wild-type and GSTP-null mice.[97] These data indicate that additional mechanisms, such as JNK binding, could contribute in meaningful ways to subsequent alterations in tissue sensitivity to aldehyde exposure and aldehyde-induced injury.[19,97]

As indicated above, ingestion of aldehydes is at minimum equal if not a quantitatively greater route of aldehyde exposure than is inhalation exposure. Moreover, it is likely that some portion of inhaled aldehydes end up being processed by the gastrointestinal tract as constituents of swallowed pulmonary material provided by the mucociliary transport mechanism, which in addition to mucin and compounds trapped therein, will also contain alveolar macrophages (activated and apoptotic), particulate matter, and aldehyde-containing compounds (*e.g.*, GS-acrolein, protein–aldehyde adducts). These aldehydes and those in foods and beverages could also have a variety of effects (both direct and indirect) on the cardiovascular system. In recent work, the effects of ingested and inhaled enal aldehydes in acute settings in normal and dyslipidemic animal models were evaluated in order to assess the potential effects of aldehyde feeding on cardiovascular disease risk.

13.3.2 Cardiovascular Effects of Aldehydes

13.3.2.1 Effect of Enals on 5-Lipoxygenase Activation and Matrix Metalloproteinases

We recently showed that acrolein enhances the secretion of matrix metalloproteinase (MMP)-9 in differentiated THP-1 macrophages in a ROS and intracellular calcium-dependent manner.[100] These observations are in agreement with an earlier study showing that exposure to acrolein stimulates the cleavage of pro-hMMP9 in the lung and this was associated with mucin production.[101] We also observed that MMP activity was significantly increased in the atherosclerotic lesions of apoE-null mice following incubation with acrolein *ex vivo*, suggesting that exposure to acrolein can destabilize atherosclerotic lesions and could contribute to altered plaque stability and rupture.[100] In a recent study, Kim *et al.* reported that acrolein activates the 5-lipoxygenase (5-LO) pathway of leukotriene synthesis in an ERK-dependent manner.[102] Earlier studies from the same group showed that the structurally-related enal, HNE, enhances MMP-9 production in murine macrophages in an ERK and p38 MAPK-dependent manner.[103] The 5-LO pathway has been implicated in atherosclerosis.[104] However, additional studies are required to examine the effect of acrolein and HNE-derived activation of 5-LO and MMP-9 formation in the plaque and its potential effect on plaque destabilization. Contrary to the acrolein- and HNE-mediated activation of 5-LO pathway in macrophages, Berry *et al.* found that in GM-CSF/fMLP-stimulated human neutrophils acrolein inhibited the formation of 5-LO-dependent products in a concentration-dependent manner.[105] Further studies are required to examine the mechanisms of action of acrolein and structurally-related enals/compounds on the 5-LO pathway in various cell types and the overall significance of 5-LO on cell function under pathophysiological conditions.

13.3.2.2 Effect of Aldehydes and Cigarette Smoke Extract (CSE) on Cytokines

Smoking is a well-known risk factor for cardiovascular diseases. Several of the deleterious effects of cigarette smoke have been attributed to aldehydes, especially the α,β-unsaturated aldehydes such as acrolein and crotonaldehyde present in the vapor phase of cigarette-smoke extract (CSE). The concentration of these reactive aldehydes is significantly decreased in the cigarette smoke by using a charcoal filter. Scherer et al., reported that the use of charcoal filters significantly decreases the urinary excretion rates of mercapturic acids of acrolein, crotonaldehyde, 1,2 butadiene and benzene, but does not change the uptake of carbon monoxide and nicotine.[199]

Accumulating data suggest that atherogenic and inflammatory effects of CSE could be at least in part due to enals. Kirkham et al., reported that modification of collagen IV by CSE, acrolein and HNE enhances the adhesion of macrophages.[106] The increased macrophage adhesion is mediated by Type-A macrophage scavenger receptor (SRA), it was prevented by SRA ligand fucoidan and an anti-SRA antibody. CSE and acrolein also cause the necrosis of neonatal smooth muscle cells isolated from carotid arteries, splenic arteries and main and resistance pulmonary arteries. DTT and GSH diminished the CSE-induced cell death, suggesting that the toxicity of CSE can at least be attributed to the electrophilic nature of acrolein.[107] We have recently shown that acrolein causes endoplasmic reticulum stress in endothelial cells and increases the expression of the chemokine, IL-8, in endothelial cells (see below). Our observations were in agreement with a previous report showing that CSE increases the release of IL-8 from human macrophages.[108] The CSE-induced IL-8 release was attributed to its constituents, acrolein and crotonaldehyde, because these aldehydes by themselves increase the release of IL-8. Similarly, pre-incubation of macrophages with NAC or the cell permeable ester of glutathione also prevented the CSE-induced IL-8 release from macrophages, presumably by forming a Michael adduct between the -SH groups of thiols across the double bond of the aldehyde. CSE and acrolein have also been shown to augment the release of IL-8 from human bronchial epithelial cells.[109]

In contrast to the studies described above, multiple studies have reported that exposure to CSE and enals inhibits cytokine production. In bronchial epithelial cells acrolein inhibits IL-8 formation by inhibiting NF-κB.[110] Lambert et al., observed that CSE significantly inhibited the production of IL-1β, IL-2, GM-CSF, IL-6, IL-8, and TNF-α by human PBMC.[111] Among the volatile compounds tested in the CSE only acrolein and crotonaldehyde were effective in inhibiting cytokine production with IC_{50s} of 3 and 6 μM, respectively. Acrolein also inhibited the generation of IL-2, GM-CSF and TNF-α by T-lymphocytes. Moreover, pre-incubation of cells with NAC abolished the CSE-induced cytokine production, suggesting that the electophilic nature of enals is likely responsible for the inhibition by CSE of cytokine production in PBMC and T-lymphocytes. These investigators also observed that in human T-cells, acrolein and crotonaldehyde inhibited the formation of IL-2, IL-10, IFN-γ and GM-CSF but did not affect IL-8 formation.[112] Saturated aldehydes such as propanal and butanal had no effect on cytokine production in these cells. Acrolein, but not crotonaldehyde, also significantly inhibited IκBα activation and nuclear translocation of NF-κB. However, the concentrations of acrolein required for the inhibition of nuclear translocation of NF-κB is relatively higher than that required for cytokine production, suggesting that other transcription factors could be involved in acrolein-mediated cytokine production. Both acrolein and crotonaldehyde inhibited the p50 DNA binding. However, acrolein was 2000-fold more potent in inhibiting DNA binding than crotonaldhyde. Mass-spectroscopic analysis showed that acrolein alkylated residues Cys-61 and Arg-307 in the DNA binding domain, whereas crotonaldehyde only modified the residue Cys-61 in p50, which has been suggested to be a target of regulation of NF-κB activity.[113] Further studies are required to establish the significance of acrolein- or enal-induced inhibition of NF-κB.

As described, we showed that acrolein increases the expression of IL-8 in endothelial cells *in vitro*. However, our unpublished observations show that oral exposure to acrolein (2.5 mg/kg/day for 8 weeks) in 8-week old apoE-null mice maintained on normal chow accelerated atherosclerosis and enhanced macrophage accumulation in the lesions, but significantly decreased the expression of IL-1β and IL-6 in the spleen. Expression of TNF-α in the acrolein-exposed mice was similar with controls. Our data are in agreement with a recent report by Kasahara *et al.*, showing that exposure to acrolein (5 ppm; 6 h/day; 1–3 days) after intratracheal exposure to lipopolysaccharide (LPS; 300 μg/kg) in C57BL/6 mice significantly increases macrophage number and reduced lymphocyte recruitment in the lung.[114] Both basal and LPS-induced activation of NF-κB in the lungs is significantly inhibited in acrolein-exposed mice. This was accompanied by a significant decrease in the levels of IL-2, IL-10, IL-12 but a significant increase in the levels of GM-CSF in the lavage fluid in the lungs.[114]

Clearly, the conflicting reports on the effect of CSE and enals, such as acrolein, on cytokine production suggest that the enal role in and mechanisms of inflammation are quite complex. Because most of these experiments were done under diverse experimental conditions (*e.g.*, different concentration/doses; different incubation/exposure conditions; different cell types), collectively, these data should be interpreted with caution and in the context of the specific experimental conditions. Thus, further studies are required to resolve and understand the conflicting effects of CSE and enals on cytokine production.

13.3.2.3 Anti-Inflammatory Actions of α,β-Unsaturated Aromatic Carbonyls

While there is still debate on the pro- or anti-inflammatory properties of enals such as acrolein, most of the published reports show that α,β-unsaturated aromatic carbonyls display anti-inflammatory properties. Cinnamaldehyde, elicits antiatherogenic, anti-inflammatory and antioxidant activity in macrophages. Liao *et al.*, observed that cinnamaldehyde inhibits the expression of adhesion molecules ICAM-1 and VCAM-1 on endothelial cells and the adhesion of monocytic cells to endothelial cells.[115] Cinnamaldehyde diminished TNF-α-induced endothelial activation by two distinct signaling mechanisms. Acute effects of cinnamaldehyde on TNF-α activated endothelial cells were NF-κB dependent as it significantly prevented the degradation of IκBα. Prolonged pre-incubation with cinnamaldehyde exerted its anti-inflammatory effects in a Nrf-2-dependent manner.[115] Under these conditions cinnamaldehyde upregulated Nrf-2 in nuclear extract, increased ARE-luciferase activity, and upregulated the expression of Nrf-2-regulated proteins, thioredoxin reductase-1 and HO-1.[115] Pharmacological inhibition of HO-1 abolished the anti-inflammatory effects of cinnamaldehyde. In an independent study, Chao *et al.*, reported that in macrophages, cinnamaldehyde inhibits LPS- and lipotecichoic acid-induced formation of IL-1β and TNF-α formation as well as the generation of ROS and phosphorylation of ERK and JNK.[116]

Similar to effects of cinnamaldehyde, chalcone, an α,β-unsaturated flavonoid exhibited anti-inflammatory effects.[117] The authors observed that chalcone, inhibited IL-6- and LPS-induced expression of ICAM-1 and adhesion of THP-1 cells to endothelial cells. Pre-incubation of endothelial cells with chalcone, inhibited the IL-6-induced STAT3 tyrosine phosphorylation and LPS-induced NF-κB activation. Free-radical scavengers and pharmacological inhibitors of HO-1 and tyrosine phosphatase had no effect on chalcone-mediated inhibition of STAT3 phosphorylation. Together, these studies suggest that aromatic carbonyls, such as cinnamaldehyde and chalcone, elicit anti-inflammatory properties *in vitro*. However, the anti-inflammatory role of these carbonyls and structurally similar compounds under acute and chronic inflammatory conditions *in vivo* remains to be tested.

13.3.2.4 Dyslipidemia and Vascular Effects of Enals

Because of the need to explore the actions of enals in a systemic setting, we monitored systemic and cardiovascular effects of aldehydes *via* gavage exposure.

a. Acute dyslipidemia: Acute acrolein feeding induces dose- and time-dependent alteration in plasma lipoprotein electrophoretic mobility and significant changes in total lipids (cholesterol, triglycerides and phospholipids) within 12–24 h after feeding. Because altered lipoprotein mobility on Agarose gels precedes absolute changes in plasma lipid levels, the former could affect the latter. In the absence of an increase in systemic inflammatory cytokines, decreased hepatic lipase mRNA and activity corresponded with changes in plasma lipids and provides one potential mechanism by which ingested aldehydes could affect dyslipidemia.[118] Because changes in lipids induced by acute feeding abated over 72 h, the long-term effects will need to be studied after repeated dosing.

b. Acute effects on aorta and mesenteric vasculature: Although acute dyslipidemia is associated with endothelial dysfunction, the direct effects of aldehydes on the vasculature are equally dramatic. For example, acrolein induces endothelium-dependent vasorelaxation in rodent aorta and perfused mesenteric bed, as well as, inducing relaxation in isolated human coronary artery bypass graft (CABG) blood vessels (internal mammary artery, radial artery, saphenous vein). In human CABG blood vessels, acrolein (or allylamine) also results in spontaneous hypercontraction that is calcium dependent and irreversible.[52] Similarly, short-duration exposure (30 min) of mouse aorta with acrolein (10 µM) leads to endothelium injury, an effect that is exacerbated in the absence of GSTP,[19] a finding that highlights the potential role of aldehyde metabolism in protection of peripheral vasculature. In any case, the mechanism of acrolein-induced vasorelaxation or injury is not clear although dilation in mesenteric vasculature appears independent of NO and dependent on a hyperpolarizing factor (EDHF) in mesenteric vascular bed of mouse and rat.[85] Thus, the direct (dilatory/injury) and indirect (dyslipidemia) effects of acrolein feeding on the systemic vasculature requires further investigation. Subchronic feeding by definition comprises multiday feeding exposures of less than 10 days of continuous feeding (*i.e.* chronic). We evaluated the effects of subchronic acrolein feeding in both mice and rats to examine whether acute dyslipidemia was sustained over multiple feedings. Additionally, acute effects of acrolein perfusion in mesenteric bed included vasodilatation,[85] an effect that could be enhanced or diminished under subchronic aldehyde exposure – an effect that was evaluated in an experiment using rats.

c. Effects on lipids: Subchronic feeding of acrolein in mice or rats resulted in sustained elevation of cholesterol but not triglycerides, indicating that these two parameters are under the control of separate processes in rodents.[119] We concluded this from our initial investigation of single-dosing responses to acrolein in mice because the dose relationship between acrolein and increased cholesterol or increased triglycerides was different (*i.e.*, linear in the former; exponential in the latter).[119]

d. Effects on aorta and mesenteric vasculature: Treating rats for 5 consecutive days with acrolein at 2 mg/kg/d significantly depressed body weight gain (% change) $+4.6 \pm 2.2$ ($n=4$) but not at 0.5 or 1 mg/kg/d, respectively, versus $+11.4 \pm 3.0$ in controls ($n=6$) (Table 13.5). Independent of weight gain, mean arterial blood pressure was significantly elevated to 122.0 ± 9.4 and 121.1 ± 3.0 mm Hg in rats treated with either 0.5 or 1 mg/kg/d, respectively ($n=4,4$), versus 94.5 ± 4.5 mm Hg in water-fed controls ($n=6$) (Table 13.5). These data show that acrolein-induced hypertension at the lowest tested dose (0.5 mg/kg/d) – an effect that was not associated with changes in body weight. Similarly, plasma lipids were altered only at the highest acrolein dose (2 mg/kg/d; Table 13.5). Collectively, these data indicate that cardiovascular changes occur at doses that do not induce systemic toxicity (*i.e.*, a decrease in weight gain) or sustained dyslipidemia.

Acrolein feeding did not affect PE-induced constrictions in mesenteric bed in normal PSS, however, in the presence of 20 mM K^+ PSS beds of acrolein-fed

Table 13.5 Effects of 5-day water- (control) and acrolein (0.5, 1, and 2 mg/kg bwt wt/d, po) treatment in male Sprague–Dawley rats *in vivo*.

Acrolein (mg/kg/d)	Total body weight change (%)	MAP[1]	Cholesterol[2]	PPL[2]	Triglycerides[2]
0 ($n=5$)	11.4 ± 3.0	95.4 ± 4.5	65.0 ± 5.4	109.6 ± 8.4	32.7 ± 3.6
0.5 ($n=4$)	12.6 ± 2.8	122.0 ± 9.4[3]	49.0 ± 5.0	–	27.3 ± 4.1
1 ($n=4$)	7.3 ± 1.2	121.1 ± 3.0[3]	75.3 ± 3.3	118.3 ± 6.2	47.7 ± 9.4
2 ($n=4$)	4.6 ± 2.2	–	102.3 ± 1.5[3]	148.3 ± 5.4[3]	48.7 ± 7.2

Abbr.: MAP, mean arterial pressure; Units: [1], mm Hg; [2], mg/dl; –, not measured; Values are mean ± SE; [3], $P<0.05$ between value with asterisk and control value.

rats (0.5 and 1 mg/kg/d) showed enhanced PE-induced constrictions at the 50 and 100 nmol PE doses compared with control responses (Figure 13.2A and B). In contrast, the aorta of rats fed 2 mg/kg/d acrolein had a significantly reduced PE-induced contraction compared with controls. Specifically, as reported before (45), PE (1 μM) induced aortic contraction of 127.7 ± 6.4% (% of initial 80 mM K^+ contraction; $n=4$) in control rats versus 124.1 ± 6.0% ($n=4$), 117.8 ± 6.0% ($n=4$), and 88.4 ± 5.2% ($n=4$) in acrolein-treated rats (0.5, 1, and 2 mg/kg/d, respectively, Tables 13.6 and 13.7).

The ACh-induced (0.1–10 nmol) dilator response in PE-preconstricted mesenteric bed in normal PSS was attenuated in 1 mg/kg/d acrolein-fed rats only (Figure 13.3A). The ACh-induced dilator responses were significantly reduced in the control and acrolein-treated rat mesenteric beds perfused in the presence of 20 mM K^+PSS compared with ACh responses induced in normal PSS (Figure 13.3B). The ACh-induced dilatory response was significantly reduced in 20 mM K^+PSS in the mesenteric beds of rats exposed to 2 mg/kg/d acrolein compared with controls (Figure 13.3B). The maximal ACh-induced dilator response in normal PSS was attenuated in 20 mM K^+PSS from 92.5 ± 4.2% ($n=6$) to 59.0 ± 6.8% ($n=5$) in the control compared with the change from 85.5 ± 2.9% ($n=4$) to 19.8 ± 4.2% ($n=4$) in acrolein-treated rats (2 mg/kg/d; Figure 13.3B). Thus, although there was no change in absolute total ACh-induced dilation in normal PSS, there was an acrolein-induced, dose-dependent increase in the K^+-sensitive component (*i.e.*, EDHF) of the ACh-induced dilation in perfused mesenteric bed compared with controls (Figure 13.3C). The total dilation represents the sum of both K^+-insensitive (EDRF) and K^+-sensitive (EDHF) components, which are all depicted in Figure 13.3D. These results indicate that the mesenteric bed reorganizes its dilatory components to maintain homeostasis in the face of acrolein-induced downregulation of EDRF. Whether this also happens in other vascular beds remains to be tested.

To identify potential mechanisms by which acrolein feeding affects the hyperpolarizing pathways of smooth muscle relaxation, the 11,12-EET-induced

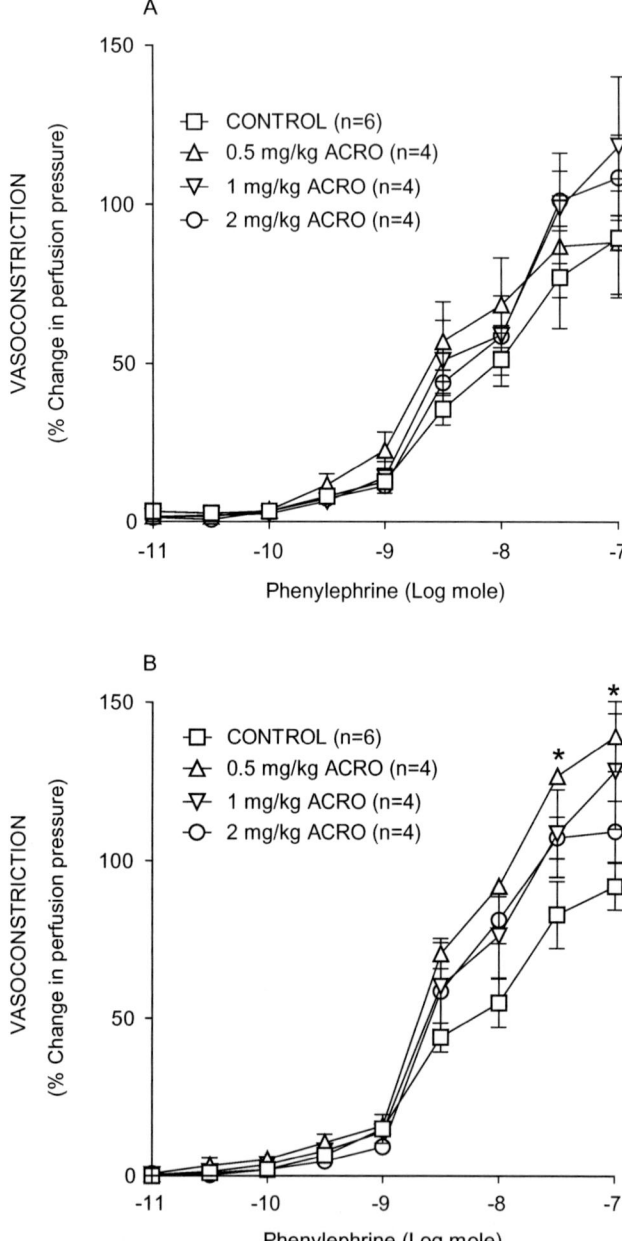

Figure 13.2 The effects of acrolein on constrictor responses to phenylephrine (PE). Mesenteric beds isolated from control (water) and acrolein-treated (0.5, 1, or 2 mg/kg/d; p.o. 5 consecutive days) rats were perfused with PSS (A), and then PSS plus 20 mM K^+ (B), and dose-dependent responses to increasing bolus doses of PE (0.01–100 nmol) were measured. Constrictor responses are expressed as a percentage of the maximum change in perfusion pressure produced by 100 nmol PE in control rat mesenteric bed. Values are mean ± SE. n = number of rats in each series. * $P < 0.05$ vs. control at the same PE dose.

Table 13.6 Effect of acrolein exposure *in vivo* and *in vitro* on aortic function *ex vivo*.

A In vivo dose (mg/kg/day)	PE response in vitro acrolein concentration [μM]		
Acrolein	0	10	40
0 (n=5)	98.4 ± 6.7	76.1 ± 4.5[2]	7.5 ± 7.5[1]
0.5 (n=4)	90.8 ± 6.9	70.4 ± 7.1[2]	1.0 ± 0.7[1]
1 (n=4)	75.8 ± 4.0	85.8 ± 6.3	14.7 ± 6.6[1]
2 (n=4)	81.0 ± 3.2	65.6 ± 7.3	6.7 ± 2.7[1]

B In vivo dose (mg/kg/day)	ACh response in vitro acrolein concentration [μM]		
Acrolein	0	10	40
0 (n=5)	70.4 ± 5.4	47.3 ± 4.7[1]	NA
0.5 (n=4)	66.8 ± 5.7	59.5 ± 8.2	NA
1 (n=4)	73.2 ± 5.5	71.2 ± 7.8	NA
2 (n=4)	83.6 ± 8.1	68.7 ± 4.3	NA

C In vivo dose (mg/kg/day)	SNP response in vitro acrolein concentration [μM]		
Acrolein	0	10	40
0 (n=5)	84.7 ± 4.7	92.6 ± 2.6	NA
0.5 (n=4)	96.1 ± 2.1	98.7 ± 0.9	NA
1 (n=4)	89.0 ± 3.6	89.4 ± 4.0	NA
2 (n=4)	90.9 ± 2.7	91.7 ± 4.0	NA

Values are mean ± SE. n = number of rats in each series., significant difference ($P > 0.05$ between acrolein-treated aorta (10 or 40 μM) and 0 acrolein control. NA = not available due to limited PE-induced tone (i.e., < 15% of pre-acrolein PE-induced tone). Phenylephrine-induced (PE; 1 μM; A) contractions and acetylcholine- (ACh; 1 μM; B) and sodium nitroprusside-induced (SNP; 100 μM; C) relaxations were measured in isolated aorta of water- (control) and acrolein-gavaged (0.5, 1 and 2 mg/kg bwt/d; p.o.) rats and the latter expressed as a percentage decrease in PE-induced tone.

(30 nmol) dilator response was measured in PE-preconstricted perfused mesenteric bed from control and acrolein-treated rats. The 11,12-EET-induced dilation was significantly reduced in acrolein-treated rats compared with the control rats at all doses (Figure 13.4A). As observed with ACh-induced dilation, 20 mM K^+ PSS further attenuated 11,12-EET dilator response in perfused mesenteric beds of control and acrolein-treated rats (Figure 13.4B). These data indicate that the K^+-sensitive component of ACh- and 11,12-EET-induced dilations (EDHF) was enhanced in acrolein-fed rats in a dose-dependent manner. This appeared to be an EDHF-specific response because no significant difference in dilator responses to SNP (10 nmol) in the PE-preconstricted perfused mesenteric bed was observed between control and acrolein-treated rats. SNP maximally dilated PE-induced preconstrictions by 78.6 ± 5.9% (n=4) in control rats versus 76.1 ± 3.8% (n=4), 81.5 ± 3.7% (n=5), and 73.8 ± 7.3% (n=4) in rats fed 0.5, 1, and 2 mg/kg/d acrolein, respectively. Thus, the increased EDHF component of 11,12-EET-stimulated dilation (~3×)

Chapter 13

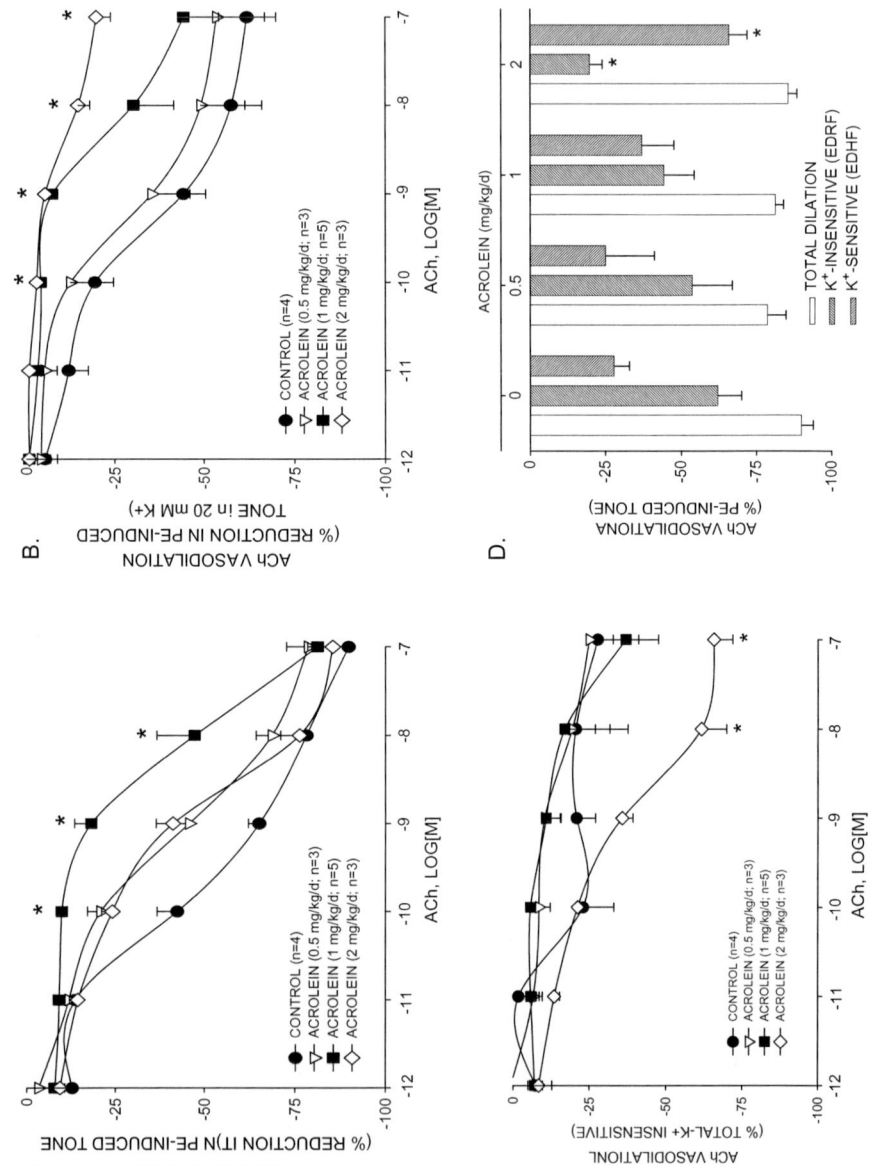

likely reflected an endothelium-specific, upregulated dilatory response induced by acrolein. We found that acrolein feeding reduces EDRF and increases EDHF dilatory components in perfused mesenteric bed with 2 independent dilatory agonists, ACh and 11,12-EET, which work *via* distinct pathways to elicit dilatation in the mesenteric bed. The acrolein-induced switch from agonist-stimulated EDRF-dominated dilatory response to a largely EDHF-mediated response appears as a coordinated and dose-dependent effect. Our data indicate that decreased EDRF component occurred at the lowest dose of acrolein (0.5 mg/kg/d) tested, whereas, a significant increase in EDHF component was observed only at the highest dose (2 mg/kg/d) tested. We infer that the loss of EDRF (endothelial dysfunction) likely contributes to the observed systemic hypertension, enhanced mesenteric bed contractility, and to the subsequent upregulation of EDHF component(s), although the specific relationship between these two dilatory pathways remains unclear.

13.3.2.5 Acrolein, Platelet Activation and Thrombosis

Our preliminary studies showed that acrolein causes platelet activation.[120] Blood platelets are an integral component of haemostasis and play a critical role in the coagulant activity of blood. Under normal physiological conditions, platelets are involved in wound healing and in arresting blood loss. However, hyper-platelet activation in the vasculature can cause thrombotic activity and lead to the occlusion of blood vessels. Thromboembolic events have been implicated in the acute cardiovascular effects of cigarette smoke.[121] Secondhand smoke increases platelet activation and aggregation in humans[122,123] and exposure of experimental animals to tobacco smoke decreases the clotting time.[92,93] The high acute cardiovascular toxicity of environmental pollutants and particulate matter has also been suggested to be, at least in part, due to the ability of environmental pollutants to promote a procoagulant state and to precipitate acute thrombotic events resulting in acute myocardial infarction and stroke.[1,2,18,124,125]

Our ongoing studies suggest that *in vitro* acrolein enhances the agonist (ADP and arachidonic acid)-induced platelet activation of human and murine platelets, both in the platelet-rich plasma and in washed platelets. Moreover,

---

Figure 13.3   Effects of acrolein feeding on endothelial-dependent dilation to acetylcholine (ACh). Dose-dependent dilator responses to ACh (0.001–100 nmol) in perfused, phenylephrine-preconstricted (PE; 5 μM) mesenteric bed were determined in control (water) and acrolein-treated rats (0.5, 1, or 2 mg/kg/d; p.o. 5 consecutive days) with PSS only (A; total dilation) or PSS plus 20 mM K^+ (B; K^+-insensitive dilation; EDRF). The K^+-sensitive dilatory component (C; *i.e.* EDHF) was calculated as the difference between total dilation (A) and K^+-insensitive dilation (B). The respective contributions of ACh-induced K^+-insensitive and K^+-sensitive dilatory responses to the maximal dilation (*i.e.*, total dilation) are expressed as a percentage reduction of PE-induced tone (D). Values are mean ± SE. $n =$ number of rats in each series.* $P < 0.05$ *vs.* control.

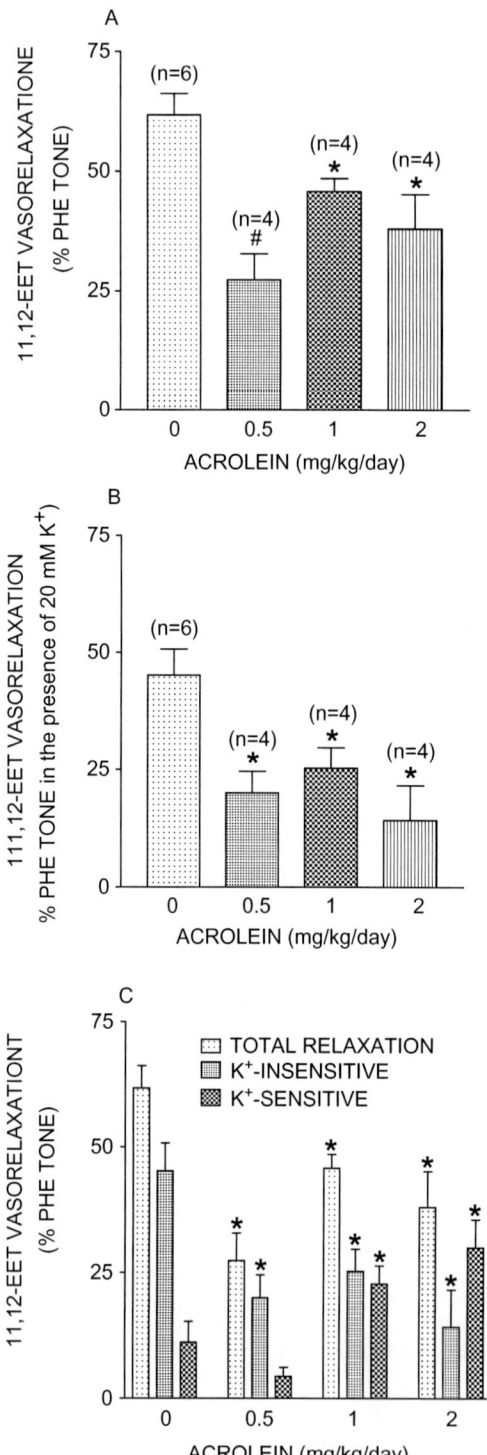

platelets obtained from C57BL/6 mice exposed to acrolein acutely or subchronically by inhalation or orally by gavage show a greater propensity to aggregate *ex vivo* in the presence of ADP compared with platelets obtained from respective air-exposed or water-fed control mice.[120] The effect of acrolein on platelet activation is similar to that of tobacco smoke. Cigarette smoke has been reported to enhance platelet activation and much of the ischemic heart-disease risk of smoking could be attributed to platelet hyperactivation.[87] We also observed that exposure of C57BL/6 mice to acrolein or tobacco smoke resulted in an increased plasma concentration of platelet factor 4 (PF4).[120] PF4 is mainly synthesized in megakaryocytes and then stored in the α-granules of platelets. Upon platelet activation, PF4 is released from the platelet granules and binds to heparin-like molecules leading to the activation of antithrombin,[126] thereby facilitating clotting at the site of vessel injury. Our data are in agreement with recent studies showing that acrolein *in vitro* inhibits plasma antithrombin activity by inhibiting heparin affinity.[127,128] Lower antithrombin activity levels could augment thrombin availability in the blood, and thus, add another prothrombotic component to the effects of acrolein exposure.

Activated platelets interact with leukocytes *in vivo* to form platelet–leukocyte aggregates in the blood.[129] The interaction between platelets and leukocytes is primarily through the mobilization of P-selectin, an adhesive protein, on the cell membrane of activated platelets.[129,130] Increased formation of platelet–leukocyte aggregates is an established marker of increased thrombosis *in vivo*.[131] Therefore, to examine whether platelets are activated *in situ* in animals exposed to acrolein, we measured the formation of platelet–leukocyte aggregation by flow cytometry in C57BL/6 mice following acute or subchronic exposure to acrolein. Significantly, we observed a 1.5–3-fold increase in the number of platelet–leukocyte aggregates in the blood of acrolein-exposed mice.[120] These data suggest exposure to acrolein causes hyperactivation of platelets *in situ* in the vasculature of the mice.

We also observed that exposure to acrolein by inhalation or gavage in C57BL/6 mice decreased the tail-bleeding time, further suggesting that the exposure to acrolein causes prothrombotic shift in the platelet hemostatic parameters, and thus, affecting the *in vivo* coagulation process.[120] Similar

---

Figure 13.4 Effects of acrolein feeding on the K^+-sensitive component of 11,12-epoxyeicosatrienoic acid (11,12-EET)-induced dilation. Dilator responses to 11,12-epoxyeicosatrienoic acid (30 nmol) in phenylephrine-preconstricted (PE; 5 µM) mesenteric beds from control (water) and acrolein-fed (0.5, 1, or 2 mg/kg/d, p.o. 5 consecutive days) rats were measured in PSS only (A; total) and then in PSS plus 20 mM K^+ (B; K^+-insensitive; EDRF). The K^+-sensitive dilatory component (C; EDHF) was calculated as the difference between total dilation (A) and K^+-insensitive dilation (B). The respective contributions of 11,12-EET-induced K^+-insensitive and K^+-sensitive dilatory responses to the maximal dilation (*i.e.* total dilation) are expressed as a percentage reduction of PE-induced tone (D). Values are mean ± SE. n = number of rats in each series. * $P < 0.05$ *vs.* control.

increase in platelet activation is reported in individuals exposed to ultrafine particles,[132] diesel exhaust,[133] or tobacco smoke.[134] Mutlu et al., have recently shown that exposure to PM accelerates blood coagulation,[135] and the effects of acrolein on blood coagulation in our ongoing studies are in agreement with those reported for cigarette smokers.[136] Collectively, these data suggest that environmental, occupational, or dietary exposure to acrolein (or related aldehydes) could potentiate thrombosis in humans and may thereby elevate the risk of developing cardiovascular disease.

13.3.2.6 Aldehydes and Atherosclerosis

Atherosclerosis is a complex disease. Interplay of multiple factors including hypercholesterolemia, endothelial cell, platelet and macrophage activation, smooth muscle cell proliferation and migration and inflammation, is well documented in the pathogenesis of atherosclerosis. Multiple studies have suggested that aldehydes generated from lipoprotein oxidation could be atherogenic. *In vitro* incubation of endothelial cells with oxidized PAPC or the phospholipid aldehyde, POVPC, enhances monocyte adhesion and increases cytokine production. Aldehyde-modified lipids are taken up by scavenger receptors on macrophages and this facilitates foam-cell formation. Aldehyde-modified proteins are present in the atherosclerotic lesions and immunization of atherogenic mice with oxidized LDL or lipid peroxidation-derived aldehydes, such as MDA, diminishes atherosclerosis. Recently, we showed that pharmacological inhibition or genetic ablation of aldose reductase (AR), an important enzyme in the metabolism and detoxification of aldehydes generated from the oxidation of lipids, exacerbates atherosclerotic lesion formation in apoE-null mice.[137] The AR/apoE double-knockout mice have increased accumulation of macrophages in the aortic lesions. A significant increase in the concentration of aldehydes, which are substrates for AR, is present in the plasma. Moreover, the accumulation of protein–HNE adducts in the lesions of AR/apoE double-knockout mice was also augmented as compared with the apoE-null mice control. When the AR/apoE double-knockout mice were made diabetic with streptozotocin (STZ) treatment, there was a significant increase in the lesion size as compared with the corresponding controls, the diabetic apoE-null mice.[137] *In vitro* the AR enzyme efficiently catalyzed the reduction of several carbonyls (*e.g.*, glyoxal, methyglyoxal, deoxyglucosone, *etc.*), which are precursors for advanced glycation end products (AGEs).[138] Accumulation of the AGEs was also significantly increased in lesions in the innominate artery of AR/apoE double-knockout mice. Together, these data suggest that aldehydes could be involved in the etiology of atherosclerosis. However, little is known about the effect of environmental aldehydes on atherosclerosis.

Ongoing studies in our laboratories show that 8 weeks of acrolein feeding (by gavage) in 8-week old apoE-null mice maintained on normal chow exacerbated atherosclerotic lesion formation (data not shown). These data are consistent with the report that exposure of apoE-null mice to sidestream cigarette smoke accelerates atherosclerosis.[139] Immunohistochemical analysis showed that lesions in acrolein-fed mice were rich in macrophages. In these early lesions, we did not observe any protein–acrolein adducts in the lesions of control or acrolein-fed mice. *In vitro*, Shao *et al.*, found that acrolein binds to the lysine residues of apolipoprotein A-1 (apoA-1), the most abundant protein in HDL, and inhibits reverse cholesterol transport.[140] In our studies, we observed that acrolein feeding in apoE-null mice significantly increased the concentration of total cholesterol in the plasma. Moreover, NMR analysis of plasma lipoproteins showed a significant increase in the levels of small and medium VLDL particles in acrolein-fed mice, and these changes are similar to those present in lipoproteins in acrolein-fed C57BL/6 mice.[119] These smaller lipoprotein particles can readily accumulate in the subendothelial space, activate the endothelium and enhance monocyte recruitment and transmigration. In the subendothelial space, these monocytes, upon differentiation into macrophages, can engulf more VLDL and LDL particles and augment foam-cell formation. This is supported by our recent *in vitro* study showing that incubation of endothelial cells with low concentrations of acrolein augments monocyte adhesion and transmigration. These observations are in agreement with our earlier studies showing that acrolein and structurally-related lipid peroxidation-derived aldehyde, 4-hydroxynonenal (HNE), enhances the TNF-α-induced expression of ICAM-1.[141] Together, these data suggest that acrolein-induced hypercholesterolemia and endothelial activation could augment atherosclerosis.

In keeping with our data showing that acrolein enhances platelet activation (*vide supra*), recent studies suggest that platelet activation could be involved in the development of atherogenesis.[142,143] In humans, adherence of platelets to the damaged endothelium increases thrombus formation, accelerates atherogenesis and increases the risk of ischemic heart disease.[87] Deficiency of P-selectin, which enhances platelet–endothelial interaction, attenuates atherosclerotic lesion formation in atherogenic mice.[144,145] Massberg *et al.*, observed that in apoE-null mice maintained on Western diet platelet adhesion in the carotid artery precedes leukocyte adhesion suggesting that platelets play a pivotal role in early atheosclerotic lesion formation.[142] As described above, our preliminary studies suggest that acute exposure to acrolein enhances platelet activation and thrombosis in C57BL/6 mice. The chemokine, PF4, released from the granules of activated platelets can augment the accumulation of lipoproteins in vascular cells and macrophages[146–148] and enhance the binding of oxLDL to endothelial cells, smooth muscle cells and macrophages.[149] Atherogenic mice lacking PF4 have diminished atherosclerotic lesion formation.[150] In the present study, we observed that chronic acrolein feeding in apoE-null mice increased the concentration of PF4 in the plasma and accumulation of PF4 in the lesions in the aortic valve. Increased PF4 in acrolein-fed mice could enhance the accumulation of lipoprotein particles in the subendothelial space and promote atherogenesis. Alternatively, the acrolein-induced hypercholesterolemia in these mice could enhance platelet activation, resulting in the release of proatherogenic PF4 from the platelets. We also observed that both acrolein and PF4 by themselves cause endothelial activation. Our observations are in agreement with previous studies showing that PF4 induces the expression of adhesion molecule E-selectin in endothelial cells,[148] and activated platelets enhance the recruitment of monocytes to the vessel wall by increasing the expression of MCP-1 and ICAM-1 in endothelial cells.[151] Significantly, we also observed that a subthreshold concentration of PF4 augments acrolein-induced transmigration of monocytic cells (data not shown). Concentration of PF4 in the plasma and accumulation of PF4 in the lesions were enhanced in acrolein-fed apoE-null mice *in vivo*. Also, we observed that incubation of endothelial cells with the plasma from acrolein-fed apoE-null mice significantly increased the transmigration of monocytic cells, and anti-PF4 antibody abolished the increase in transmigration of monocytes. Together, our preliminary observations suggest that acrolein could be responsible for the well-documented proatherogenic effects of direct or second-hand smoke exposure or exposure to polluted environments that are rich in combustion products.

13.3.3 Role of Protein–Aldehyde Adducts

Regardless of the specific effects of aldehyde exposure, the underlying mechanism by which aldehydes trigger these effects is not clear. Because all aldehydes possess the terminal carbonyl, a potential to form a protein–aldehyde adduct is a quality of aldehyde reactivity. Moreover, unsaturated aldehydes increase electron delocalization, enhancing both carbonyl reactivity for amines and Schiff base formation as well as providing a strong electrophilic site for thiol (cysteine) nucleophilic attack (*i.e.*, Michael adducts). The accumulation of such protein–aldehyde adducts could trigger a common mechanism that is shared across aldehydes.

13.3.3.1 Protein–Acrolein Adducts, Formation and Detection

Aldehydes are soft electrophiles, which can adduct by 1,4-Michael addition with amino acids of proteins, influencing protein function. The origin of the adducting aldehyde could be the environment, as aldehydes are found as ubiquitous pollutions or endogenous sources. Uchida and colleagues have demonstrated by utilizing the formation of aldehyde–protein adducts that acrolein is endogenously formed independent of exogenous sources. Using an acrolein-trapping ELISA in which acrolein reacts with the ε-amino group of lysine forming the N^{ε}-(3-formyl-3,4-dehydropiperidino)lysine (FDP-Lys) product, the group identified the FDP-lysine adduct and revealed that acrolein is endogenously formed as a lipid peroxidation product.[152,153] In further investigations the group showed protein-bound acrolein in atherosclerotic lesions[153] by an immunohistochemical approach using antisera against acrolein-modified proteins. The antiserum is raised by immunizing rabbits with acrolein-modified KLH (Keyhole Limpet Hemocyanin) as described in Figure 13.5A. KLH, a lysine-rich protein, is incubated with acrolein. During incubation, acrolein reacts with the ε-amino group of the lysine-residues of KLH to *bis*-Michael adducts, and thus, forms under dehydration the cyclic FDP-lysine adduct (Figure 13.5B). After semipurification the antibody can be used for various applications, *e.g.* Slot blot to show the total abundance of protein–acrolein adducts (Figure 13.5C), Western blot after 1- (Figures 13.5D and G) or 2- dimensional (Figure 13.5F) electrophoresis to identify the modified proteins, immunocytochemistry and immunohistochemistry to localize adducts in tissues, cells and subcellular organelles (Figure 13.5H).[19,97]

In other studies, protein–acrolein adducts formed by acrolein of endogenous origin were detected in photo-damaged skin,[154] in brain tissue of patients with Alzheimer's disease,[155,156] in plasma of patients after brain infarction,[157] and in saliva of Sjögren's syndrome patients.[158] In a recent study, FDP-Lys-adducts in albumin are shown in brain tissue after infarction using an antibody raised against bovine serum albumin conjugated acrolein.[235] Additionally, increased plasma levels of FDP-lysine are shown in patients with chronic renal failure.[159,160] Furthermore, it was demonstrated that protein–acrolein adducts accumulate in renal tissue of patients with diabetic-nephropathy,[161–164] and levels of adducts in serum and hemoglobin correlate with progression of diabetic retinopathy.[165]

In addition to endogenous acrolein formation *via* lipid peroxidation, acrolein is a common environmental pollutant of various sources. Acrolein is a combustion byproduct of cigarette smoke (CS). Exposing C57BL/6 mice for 5 h to CS, generated from Kentucky 2R4F reference cigarettes (Tobacco Research Institute of Kentucky, Lexington, KY; continuously 10 cigarettes were burned at one time over 5 h delivering a mixture of 89% sidestream, 11% main stream, standard puff of 35 cm^3, into a smoke chamber; Teague Enterprise-10) with a specifically high content of acrolein, resulted in a significant increase in the level of protein–acrolein adducts in plasma and lung as detected by Slot blot (Figure 13.5C), Western blot (Figures 13.5E and G), and by immunohistochemistry in lung (Figure 13.5H). Thereby, the specificity of the antibody was tested by Western blotting using heart homogenates incubated for 2 h in the absence or presence of acrolein (50 µM) (Figure 13.5D). After direct exposure to acrolein or indirect acrolein treatment using cigarette smoke (CS) by inhalation, levels of protein–acrolein adducts were increased in lung, plasma and aorta.[19] Injection of the chemotherapeutic agent cyclophosphamide (CY), which is metabolized to phosphoramide mustard and acrolein, leads to increased levels of protein–acrolein adducts in bladder tissue.[97] However, the protein–acrolein adducts appear to be rapidly removed from the plasma as observed in a investigation detecting adducts in plasma at 15, 30 and 60 min after gavage feeding of acrolein. Similarly, gavage feeding mice acrolein resulted in an increase of $\approx$ 1.45-fold of a protein band at a molecular weight of $\approx$ 250 kDa after 15 min and a peak at 30 min in a protein band at a molecular weight of $\approx$ 150 kDa (unpublished observations). In rats injected with radioactive-labeled albumin without or after incubation with acrolein, the albumin–acrolein adduct is removed faster from the circulation than the unmodified labeled albumin.[166] Analysis of plasma of mice exposed to acrolein (5 ppm, 6 h, inhalation) by two-dimensional (2D) gel electrophoresis and Western blot

(Figure 13.5F) followed by MALDI/TOF analysis revealed peptide matching with several common proteins that could be modified by acrolein (see table Figure 13.5F), *e.g.* albumin and hemopexin, although definitive identification of residue modification with acrolein was not obtained.

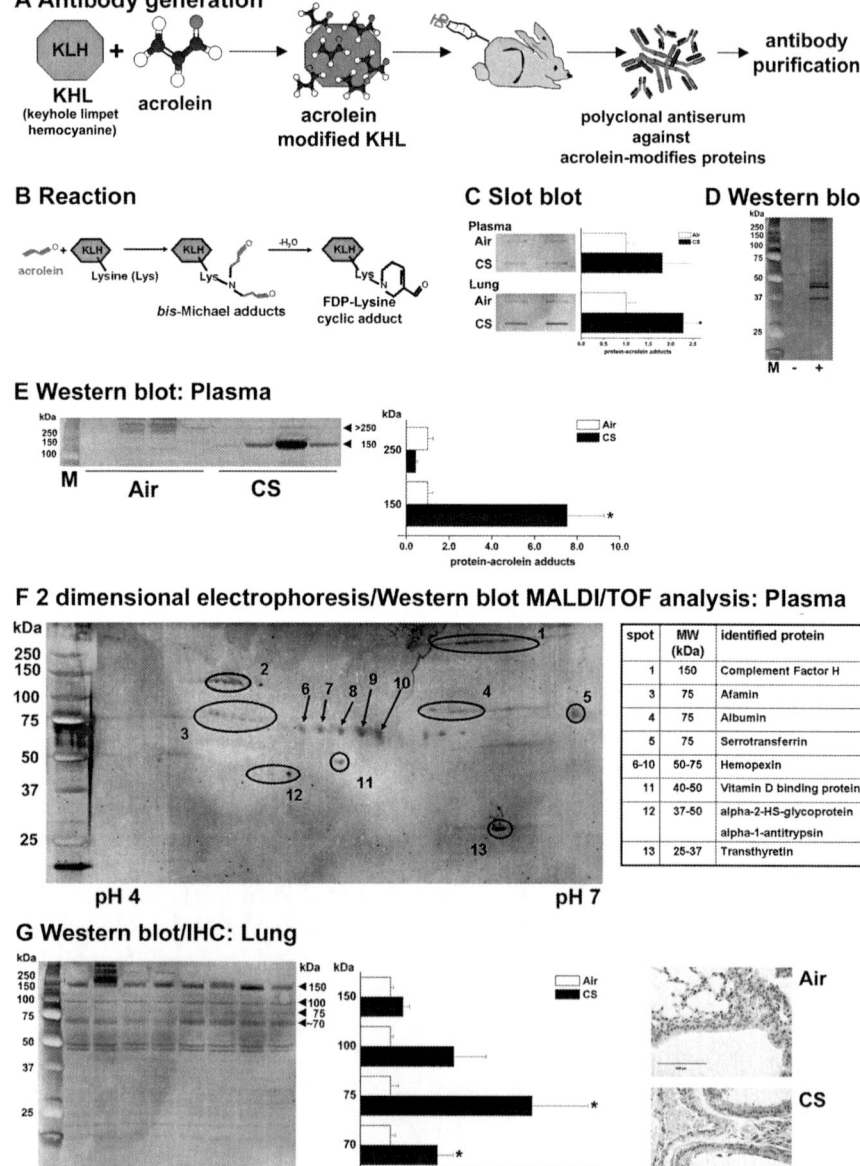

Due to its method of generation, the polyclonal antibody detects mainly protein adducts formed with lysine residues, but adduction with cysteine, histidine and arginine residues has been described *ex vivo* as well.[112,167–169] Hence, acrolein rapidly forms adducts with lysine amino group of proteins, compounds with free thiol or amino groups like Mesna or hydralazine, which have been suggested to "scavenge" acrolein and other aldehydes by forming stable Michael adducts. Hydralazine has been shown to react with acrolein or with the acrolein–Michael-adducted protein – scavenging acrolein or trapping the protein–acrolein adduct – protecting against acrolein toxicity directly or indirectly by preventing further crosslinking of protein–acrolein adducts.[170–173] Similarly, the "protein–acrolein trapping mechanisms" could be useful in future studies to capture protein–acrolein adducts from *in vivo* sources for further bioanalytical characterization to more readily identify acrolein-modified proteins by MALDI/TOF-MS. In conclusion, the nature and stability of protein–acrolein adducts make the identification of proteins modified *in vivo* especially difficult. Another challenge is the analysis of whether the protein-adducted acrolein actually derives from exogenous or endogenous sources.

Figure 13.5   Measurement of protein–acrolein adducts in tissues of mice exposed to acrolein. (A) Scheme: Antibody generation. Acrolein is incubated with Keyhole Limpet hemocyanin (KLH) a lysine-rich carrier glycoprotein leading to acrolein-modified KLH. For antibody generation the acrolein modified protein is injected into a rabbit and the isolated and purified polyclonal antiserum against acrolein-modified proteins can be used for investigations using Slot blot, 1- and 2-dimensional electrophoresis followed by Western blot and for immunostaining (ICC, IHC). (B) Proposed chemical reaction of KLH with acrolein. Acrolein binds to lysine on the KLH protein and forms the *bis*-Michael adducts, which forms the cyclic FDP-lysine adduct. (C) Slot blot. Slot blots preparations of plasma or lung lysates from mice exposed to air or cigarette smoke (CS) for 6 h detected with protein–acrolein antibody. Group data are presented as mean $\pm$ SEM ($n=4$, *$p<0.05$) normalized to air control. (D) Western blot. Western blot detection of protein–acrolein adducts in heart tissue incubated *ex vivo* with or without acrolein used as positive and negative control to test the purified antibody. (E) Western blot of protein–acrolein adducts in plasma. Western blot and densitometrical analysis of protein–acrolein adducts in plasma of mice exposed for 6 h to air or CS. Group data are presented as mean $\pm$ SE ($n=4$, *$p<0.05$) normalized to control (air). (F) Identification of protein–acrolein adducts in plasma. Analysis of protein–acrolein adducts in plasma of mice ($n=4$, plasma of 4 mice was combined) exposed to 5 ppm acrolein for 6 h by 2-dimensional electrophoresis/Western blot followed by MALDI/TOF. Identified spots are numbered and proteins are presented in the table. (G) Western blot and imunnohistochemistry of protein–acrolein adducts in lung tissue. Western blot and densitometrical analysis (left) of protein–acrolein adducts in lung homogenates of mice exposed to air or CS (6 h). Group data are presented as mean $\pm$ SE ($n=4$, *$p<0.05$) normalized to control. Representative images of lung sections of mice exposed to air or CS (6 h) immunolabeled with protein–acrolein antibody and stained with a DAB kit.

13.3.3.2 Contribution of Endoplasmic Reticulum Stress (ER-Stress) and Unfolded Protein Response (UPR)

The function of the secretory pathway of the endoplasmic reticulum (ER) is the synthesis, modification, folding and delivery of active proteins to their target location. Conditions under which the influx of nascent, unfolded proteins exceeds the folding capacity of the ER activate signaling pathways of the unfolded protein response (UPR) to restore physiological function. The UPR has adaptive and alarm phases directed by PERK (PKR-like endoplasmic reticulum kinase), IRE1 (inositol-requiring 1) and ATF6 (activating transcription factor 6) as summarized in the scheme in Figure 13.6.

Aldehydes, *e.g.* acrolein, form adducts with proteins as shown (Figure 13.6) for protein-lysine and acrolein resulting in the formation of the cyclic FDP-Lysine dimer adduct. Formation of such adducts with proteins within the folding process might lead to problems in protein folding in the endoplasmic reticulum (ER), resulting in the accumulation of unfolded or misfolded proteins in the ER triggering the UPR. Another possibility is that ER resident proteins are adducted by aldehydes leading to malfunctions of the folding machinery of the ER likewise resulting in the accumulation of un- or misfolded proteins. Modifications of ER-resident proteins by the lipid peroxidation product HNE are described *in vivo* after ethanol feeding for protein disulfide isomerase (PDI),

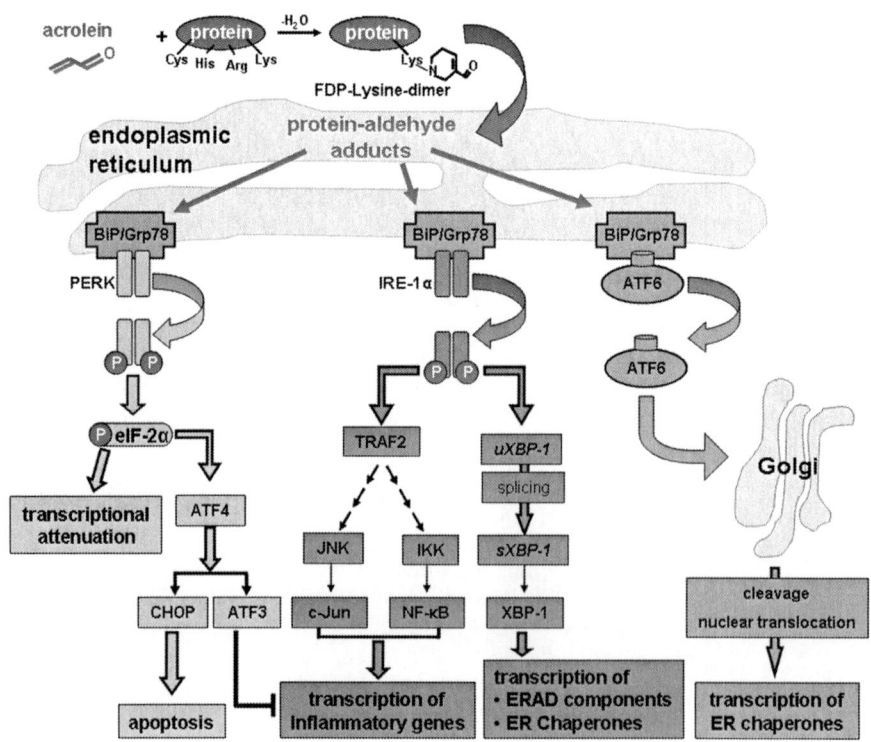

an ER chaperone supporting the formation of disulfide bridges. Incubation of PDI with HNE or acrolein resulted in an inhibition of the insulin disulfide reduction capacity of the protein.[174] Treatment of RKO human colorectal carcinoma cells with HNE alters the expression of ER-stress related targets (HERP, homocysteine), ER-stress inducible protein 1; ATF3, activating transcription factor 3) as analyzed by a microarray approach and confirmed by Western blot.[175]

Incubation of mouse fibroblasts (SA 3T3 cells) with aqueous CS extracts or the cigarette smoke-related aldehyde, acrolein, triggered ER stress and the UPR as shown by the phosphorylation of eIF2α. CS treatment further induced translational attenuation, phosphorylation of PERK and ATF4 induction, transcription of ATF4 target genes and XBP-1 splicing.[176] In human umbilical vein cells (HUVEC), treatment with acrolein induces ER-stress and activates several pathways of the UPR, shown in the phosphorylation of eIF2α, the induction of ATF3 and 4, activation of NF-κB, JNK phosphorylation and the splicing of XBP-1[177] (see Figure 13.6). However, induction of the proapoptotic protein CHOP is not found. Pre-incubation with the chemical chaperone

---

Figure 13.6 Potential role of protein–aldehyde adducts in stimulation of ER-stress. Aldehydes, e.g. acrolein, form adducts with proteins as shown for protein-lysine and acrolein resulting in the formation of the cyclic FDP-lysine dimer adduct. Formation of theses adducts could disturb protein folding or interfere with the protein-folding capacity of the endoplasmic reticulum (ER), resulting in the accumulation of unfolded or misfolded proteins in the ER. Accumulation of misfolded or unfolded proteins in the ER, changes in calcium homeostasis or oxidative stress trigger ER-stress resulting in the unfolded protein response (UPR). The UPR has distinct alarm and adaptive phases characterized by three distinct branches: the PERK-, IRE-1α- and ATF6-pathway. The PERK pathway is activated by the phosphorylation of PERK (PKR-like endoplasmic reticulum kinase), which dissociates from Grp78 and forms a dimer leading to the phosphorylation and deactivation of eIF-2α (eukaryotic translation initiation factor 2) causing transcriptional attenuation to block further protein production to resolve ER stress. Induction of the activating transcription factors ATF4 and ATF3 triggers CHOP (CCAAT/enhancer binding protein (C/EBP) homologous protein) induction leading to apoptosis or to the inhibition of the transcription of inflammatory genes. The phosphorylation and dissociation from Grp78 of IRE-1-α (inositol-requiring 1) activates the pathway inducing splicing of XBP-1 (X-box binding protein 1). The bZIP transcription factor XBP-1 is activated by unconventional splicing inducing the transcription of genes of ERAD (ER-associated degradation) components and ER chaperones to resolve ER stress. In contrast, the activation of the TRAF pathway results in the transcription of inflammatory genes by the activation of JNK and NFκB inducing an alarm response. The third arm of the UPR is the activation of the activating transcription factor 6 (ATF6). ATF6 dissociates from Grp78 and is translocated into the Golgi apparatus where the protein is cleaved. After nuclear translocation of the processed protein the transcription factor induces the transcription of ER chaperones again to resolve ER-stress.

phenylbutyric acid (PBA) inhibits the UPR response confirming the involvement of the UPR in acrolein-induced stress.[177] In conclusion, these studies indicate that aldehydes can induce ER-stress and activate the UPR (Figure 13.6), and the direct contribution of protein–acrolein adducts requires further investigation.

13.4 Aldehyde Metabolism

13.4.1 *In Vitro* Kinetic Studies

The α,β-unsaturation in aldehydes (enals) generates two reactive centers in the molecule: a polarized carbonyl that can form Schiff bases with lysine and histidine residues of proteins and an electrophilic γ-carbon that can avidly react with cellular nucleophiles such as protein sulfhydryl groups (-SH) of cellular antioxidant glutathione (GSH) or proteins by Michael-addition reaction.[178] We observed that incubation of GSH with equimolar concentrations of *trans*-2-alkenals at room temperature and neutral pH results in spontaneous conjugation of alkenals with the sulfhydryl group of GSH.[179] Our data showed that the shorter the chain length of *trans*-2-alkenals the faster the reaction with the cysteines of GSH. The rate of reaction of acrolein with GSH was >25-fold faster than that for crotonaldehyde and GSH, and the rate of GS–enal conjugation decreased further with increase in the hydrocarbon chain length of alkenals. Introduction of an additional double bond between C4 and C5 of the hydrocarbon chain completely abolished aldehyde reactivity with GSH. We also observed that acrolein reacted with lysine at neutral pH and at room temperature, however, the rate of reaction of acrolein–lysine adduct was >90-fold slower than that for the formation of the GS–acrolein conjugate.

In vitro, enals can be metabolized either by direct conjugation with glutathione or by oxidative or reductive transformations. It has also been suggested that alkenal/one oxidoreductase also known as leukotriene B4 12-hydroxydehydrogenase, 15-oxo-prostaglandin 13-reductase, and dithiole-3-thione-inducible (D3T) gene-1, can convert the α,β-unsaturated aldehyde to corresponding saturated aldehyde by reducing the α,β C=C to a single bond.[180] We observed that recombinant human aldose reductase efficiently catalyzed the reduction of medium- to long-chain unbranched saturated and unsaturated aldehydes.[179] The catalytic efficiency for the reduction of short-chain aldehydes such as propanal, acrolein and crotonaldehyde was much lower than the medium- or longer-chain aldehydes. Some of the other members of the aldo-ketoreductase superfamily, especially AKR1B10 (aldo-keto reductase 1B10), are expressed in several tissues in humans and FR-1 (fibroblast growth factor regulated protein-1) is expressed in mice, which catalyzes the reduction of several medium and longer-chain enals.[181,182] Moreover, we also observed that, human aldose reductase also catalyzes the reduction of the branched aliphatic (2-methyl/2-ethyl/2-butyl) or aromatic (phenyl/methyl phenyl) derivatives of acrolein with high efficiency.[179] Rikans reported that acrolein can be oxidized by two cytosolic ($K_m \approx 26$ or 700 µM; $V_{max} \approx 7$ or 15 nmoles NADH formed/min/mg protein), two mitochondrial ($K_m \approx 17$ or 430 µM; $V_{max} \approx 29$ or 40 nmoles NADH formed/min/mg protein) and one microsomal ($K_m \approx 1500$ µM; $V_{max} \approx 30$ nmoles NADH formed/min/mg protein) rat liver aldehyde dehydrogenases.[183] Contrary to the reports, Mitchell and Petersen did not detect acrolein oxidation by rat liver cytosolic, mitochondrial or microsomal aldehyde dehydrogenases.[184,185]

Conjugation with GSH is another route for the metabolism of enals. As described above, enals can spontaneously react with the -SH group of GSH. However, glutathione-S-transferase (GST) efficiently catalyzes the conjugation of enals with GSH. We observed that aldose reductase efficiently catalyzes the reduction of the glutathione conjugates of acrolein and HNE. The catalytic efficiency for the reduction of glutathionyl acrolein conjugate by AR was actually > 100-fold higher than acrolein or propanal.[179] The glutathionyl conjugate of acrolein can also be reduced by rat liver alcohol dehydrogenase;[186] however, the K_m for alcohol dehydrogenase ($K_m \approx 900$ µM) is much higher than that for aldose reductase ($K_m \approx 7$ µM). Mitchell and Petersen also observed that glutathionyl conjugate of acrolein can be oxidized by two cytosolic ($K_m \approx 200$ µM; $V_{max} \approx 3$ or 11 nmoles NADH formed/min/mg protein) and two microsomal ($K_m \approx 300$ or 7440 µM; $V_{max} \approx 3$ nmoles NADH formed/min/mg protein) aldehyde dehydrogenases.[186]

13.4.2 Regulation of the Enzyme Activity of Aldehyde-Metabolizing Enzymes by Enals

Enals can also regulate the activity of aldehyde-metabolizing enzymes. We observed that while no inactivation of the enzyme occurs during catalytic reduction of the enals, incubation of the apoenzyme with the enals causes the modification of the active site thiol residue of the protein-Cys-298.[179] Incubation of the protein with short-chain aldehydes, such as acrolein and crotonaldehyde, leads to a 4- to 8-fold increase in enzyme activity. Our data are consistent with an earlier report that acrolein activates AR.[187] We also observed a progressive decrease in the extent of activation with increasing hydrocarbon chain and, in fact, the C-8 and C-9 alkenals, *trans*-2-octenal and *trans*-2-nonenal inhibited the enzyme activity by approximately 30–50%. Similarly, HNE (C-9 hydroxy alkenal) inhibits enzyme activity, whereas incubation of the AR with *trans, trans*-2,4-alkadienals had no significant effect on the enzyme activity. Mitchell and Petersen observed that acrolein can inactivate rat liver mitochondrial and cytosolic high-affinity aldehyde dehydrogenases,[184] whereas crotonaldehyde is reported to be a competitive inhibitor of acetaldehyde dehydrogenase.[188] The GS–aldehyde adducts also inhibit GSTs.[189] Additional studies are required to examine whether the enal-mediated regulation of aldehyde-metabolizing enzyme activity occurs in intact cells/tissues and organs of the cardiovascular system.

13.4.3 Cardiovascular Metabolism of Enals

To the best of our knowledge,[96] no studies have been performed to examine the cardiovascular metabolism of short-chain environmental aldehydes such as acrolein and crotonaldehyde. Our studies in human and rat smooth muscle cells show that these cells metabolize HNE aldehyde either by direct conjugation with glutathione or by oxidative or reductive transformations. The glutathionyl conjugate of HNE, which constitutes up to 40–60% of HNE metabolites, is present predominantly in the reduced form (GS-DHN). Pharmacological inhibition of aldose reductase significantly prevented the reduction of GS-HNE, resulting in greater abundance of GS-HNE, suggesting that aldose reductase plays a significant role in the reduction of the glutathionyl conjugates of enals. Direct reduction of HNE to its corresponding alcohol, 1,4-dihydroxynonene (DHN), was a minor pathway and accounted for only 2–4% of total metabolism. On the contrary, oxidation of HNE to 4-hydroxynonanoic acid (HNA) was another major route ($\approx$40% of total metabolism) of aldehyde metabolism. Aldehyde dehydrogenase inhibitors significantly prevented the oxidation of HNE.[94] We also observed that aldose reductase inhibitors enhanced the accumulation of protein–HNE adducts in the neointima of balloon-injured rat carotid arteries and significantly prevented smooth muscle cell proliferation.[190] Spycher et al., showed that in cultured smooth muscle cells inhibition of aldose reductase enhances HNE-induced cell death.[191] Similarly, our unpublished data shows that inhibition of aldose reductase accelerated HNE-induced cell death of rat and mice cardiac myocytes (data not shown). Collectively, these studies suggest that aldose reductase plays a pivotal role in the detoxification of enals in cardiovascular tissues. Consistent with these observations, we also observed that in aortic endothelial cells, aldehyde-dehydrogenase-catalyzed oxidation of HNE and AR-catalyzed reduction of glutathionyl conjugate of HNE, are the major routes for the metabolism of HNE.[95] Likewise, we observed that perfusion of aerobic rat hearts with HNE, formation of glutathione conjugates (30%) and oxidation to HNA (60%) are the major routes for aldehyde metabolism.[192] Direct reduction of HNE to DHN was a minor pathway and accounted for <4% of total metabolites. Pharmacological inhibition of aldose reductase and aldehyde dehydrogenase significantly prevented the reduction of GS-HNE and oxidation of HNE to HNA, respectively. However, in ischemic rat hearts, HNA formation was significantly inhibited and elevated levels of DHN and glutathionyl conjugates of HNE were observed.[193] Reperfusion of rat hearts restored the capacity of HNE oxidation to HNA. Ischemia increased the myocardial levels of endogenous HNE and 1,4-dihydroxynonene, but not of 4-hydroxynonenoic acid (HNA). In isolated cardiac mitochondria, oxidation of HNE to HNA was the major route of the metabolism (>90%). Mitochondria isolated from ischemic hearts had a 2-fold increase in protein-bound HNE than the cytosolic fraction. Mitochondrial HNE oxidation was inhibited at $NAD^+/NADH$ ratio equivalent of that in ischemic hearts and restored at $NAD^+/NADH$ ratio equivalent to that of the reperfused hearts. Collectively, these data suggest that HNE metabolism is inhibited during myocardial ischemia due to NAD^+ depletion, suggesting that the compromise of mitochondrial metabolism of enals could contribute to myocardial ischemia/reperfusion injury. In vivo, it has been observed that protein–HNE and protein–acrolein adducts accumulate in ischemic hearts[194,195] and in atherosclerotic lesions.[137,152,153] We also observed that genetic ablation of aldose reductase (AR) increases atherosclerosis in mice and augments the accumulation of protein–HNE adducts in the plaques and free HNE in the plasma.[137]

13.4.4 Systemic Metabolism of Enals

Most of the metabolites of enals, including gluathionyl conjugates of aldehydes are actively extruded from the cells in which they are generated. In vivo the glutathionyl conjugates appearing in the circulation are taken up by the kidney and further processed (N-acetylated) and excreted in the urine as mercapturic acids. Parent et al., reported that intravenous injection or oral feeding of acrolein in rats results in the excretion of multiple mercapturic acids in the urine, which include N-acetyl-S-2-carboxy-2-hydroxyetylcysteine, N-acetyl-S-3-hydroxypropylcysteine, N-acetyl-S-2-carboxyetylcysteine, and 3-hydroxypropionic acid.[196,197] We recently showed that the hydroxypropyl derivative also known as hydroxypropyl mercapturic acid (HPMA) is the major metabolite recovered in the urine of mice exposed to acrolein or cyclophosphamide.[97] The 3-hydroxy-1-methylpropylmercapturic acid (HMPMA) and 2-carboxyl-1-methylethylmercpturic acid (CMEMA) are the two major metabolites detected in the urine of rats subcutaneously injected with crotonaldehyde.[198] Scherer et al., observed that the abundance of HMPMA was 5-fold higher in the urine of smokers than the control subjects.[199] In that study, only a moderate increase in the concentration of CMEMA was observed in the urine of smokers. Similarly, 1,4 dihydroxynonene mercapturic acid (DHN-MA) has been detected as the major urinary metabolite recovered in the urine of rats exposed to HNE.[200] DHN-MA has also been detected as a physiological component of human and rat urine.[200] It has been suggested that DHN-MA concentration in the urine could be used as a biomarker of oxidative stress in vivo.[201] Methods for the synthesis, quantification and characterization of enals have recently been reviewed elsewhere, and thus, will not be detailed here.

13.5 Environmental Aldehydes: Detection and Quantitation

13.5.1 Measurements of Aldehydes in Air, Water and Food

Measurements of carbonyl compounds such as aldehydes in the urban and indoor environments have received considerable attention due to their potential impact on human health. Atmospheric carbonyl compounds are known as precursors of photochemical smog and increase ozone concentrations. Also, aldehyde exposures are highly associated with adverse health effects causing cardiovascular diseases (long-term exposure), respiratory irritation, impaired lung function, bronchitis and inflammation of pneumonia (short-term exposure).[202] Occupational exposure (TWA) and RfC guidelines for common gas-phase aldehydes are presented in Table 13.1.

A variety of measurement techniques has been applied to quantify atmospheric carbonyl compounds. Because aldehydes are primarily produced from incomplete combustion of organic materials (direct production) and photo-oxidation of atmospheric hydrocarbons (indirect production) and decomposed through photochemical-oxidation reaction in the presence of hydroxyl radicals, the lifetime of gas-phase aldehydes is reported as from a few hours to days in ambient air.[203] Thus, there is a need for accurate measurement of aldehydes as a fundamental step for making an assessment of potential health effects and influence on the environment. Similarly, there is a need to characterize currently available aldehyde measurement techniques including sampling and analytical methods for better understanding of individual method's properties that should be critically considered in selecting measurement methods. Otson and Fellin summarized many different techniques for determination of airborne aldehydes through careful literature reviews.[204] Herein, we highlight a few common certified methods of the US EPA, OSHA and NIOSH for quantification of aldehydes in air, water and food samples, and describe aspects of their physical principles and application.

13.5.1.1 Measurement of Airborne Aldehydes

a. *Sampling methods:* One of the most common techniques for measurement of carbonyl compounds is the derivatization with 2,4-dinitrophenylhydrazine (DNPH), and is the basis for US EPA Methods TO-5 and TO-11. These methods have been employed to measure various atmospheric aldehydes including formaldehyde, acetaldehyde, and acrolein, *etc.* The California Air Resources Board Method 430 and EPA Method TO-5 use impingers or cartridges for air sampling, whereas EPA Method TO-11 utilizes adsorption tubes (silica gel tubes). In impinger samplers (EPA TO-5), airborne aldehydes react with DNPH as collected air is bubbled through an acidified DNPH solution. In the EPA Method TO-11, adsorption tubes allow collected airborne aldehydes to be adsorbed onto the DNPH-coated material. Cartridges coated with DNPH are also frequently used in the air-sampling device of EPA Method TO-5. Likewise, the NIOSH Method 2016 also employs a cartridge containing silica gel coated with DNPH to collect gas-phase aldehydes and subsequently analyze samples using an HPLC/UV (360 nm) endpoint technique. The NIOSH Method 2541 uses a solid sorbent tube for sampling of aldehydes but employs a GC/FID technique for sample analysis. The NIOSH Method 3500 uses both a filter (1-μm PTFE membrane) and two impingers containing sodium bisulfite solution to collect atmospheric aldehydes, and visible absorption spectrometry is applied for aldehyde analyses. Among NIOSH-approved methods for aldehyde quantification, the NIOSH Method 3500 is the most sensitive method for formaldehyde measurement, but oxidizable organic compounds cause interference during analysis.[205] Of the OSHA-approved methods, adsorbent tubes and diffusive samplers (passive samplers) are used. In the application of adsorbent tube (OSHA Method 52), airborne aldehyde samples pass through the sampling tubes containing XAD-2 adsorbent that is coated with 2-hydroxymethylpiperidine (HMP). The collected air samples are desorbed with solvent (*e.g.*, toluene) and then analyzed by gas chromatography (GC) using a nitrogen selective detector. The OSHA Method 52 is very similar to NIOSH Method 2541 but uses smaller sampling tubes. Diffusive or passive samplers (OSHA Method 1007) collect air samples by exposing Supelco DSD-DNPH Diffusive Sampling Device (DSD-DNPH) to air where adsorption of gas-phase aldehydes onto the sorbent media occurs *via* diffusion as opposed to pulling the air through the media with a pump. This sampling technique is discreet and used for long-term sampling. Adsorbed aldehydes are extracted with acetonitrile and analyzed by HPLC with UV detection.

Besides the officially certified methods described above (US EPA, NIOSH, and OSHA methods), an Airscan Monitor is used for determination of atmospheric aldehydes. In this device, aldehydes are collected on a glass-fiber filter treated with 2,4-DNPH and phosphoric acid. The sample filter is extracted with acetonitrile and analyzed by a HPLC/ UV system. A Foxboro Miran 1A analyzer is used to measure formaldehyde concentrations in air. This device applies the absorption of nondispersive infrared (IR) energy to measure the concentration of atmospheric aldehydes. Additionally, as an on-site sampling technique, a "detector tube" is used for "quick and rough" determinations of aldehyde levels, and thus, the reliable detection limit is relatively high (*e.g.*, range 20–200 ppb) depending on the manufacturer (*e.g.*, Matheson–Kitagawa, AUER/MSA, Gastec Incorporation). Because detector tubes are grab samplers (short or instantaneous samplers), they are often utilized for detection of high levels of aldehydes in buildings (*e.g.*, leaks, spills), but obviously these are not suitable for determination of long-term levels (*i.e.*, time-weighted average, TWA) of aldehydes. Commercially available samplers for collection of gas-phase aldehydes are presented (Figure 13.7). A comparative study to test collection efficiency of cartridge and impinger for several carbonyl compounds (formaldehyde, acetaldehyde, 2-butanone, cyclohexanone, and biacetyl, 20–250 ppb levels) was reported by Grosjean and Grosjean (1996).[234] According to the results, the cartridge sampler ($>94\%$) showed higher collection efficiency than the impinger collector ($>84\%$) for formaldehyde, but

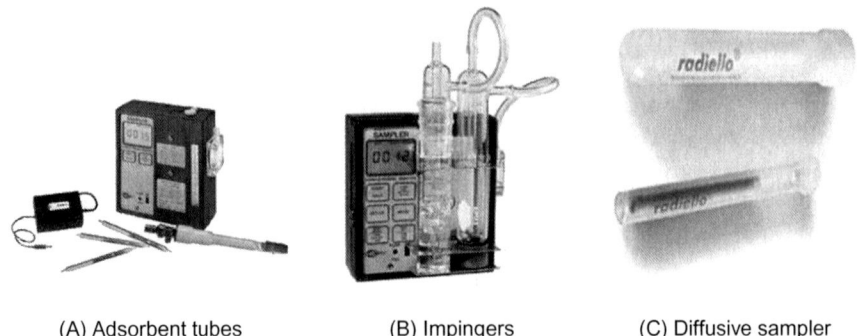

(A) Adsorbent tubes   (B) Impingers   (C) Diffusive sampler

Figure 13.7 Samplers for collection of airborne aldehydes. (A) Adsorbent tubes are DNPH-coated silica gel tubes (commercially available from SKC Inc., Fullerton, CA, USA) and is used for the EPA Method TO-11 application. (B) Impingers (available from SKC Inc.) contain DNPH solution and are used for EPA Method TO-5. (C) Passive diffusive samplers (size: 60 mm L×4.8 or 5.8 mm D, available from Sigma-Aldrich, St. Louis, MO, USA) are eligible for both ambient and indoor air sampling.

the collection efficiency for the other carbonyls tested was almost identical (approximately 100%).

b. *Analytical methods*: Many different analytical techniques are employed for determination of airborne aldehydes through direct and indirect methods. The most frequently used protocol for aldehyde analysis is DNPH derivatization (Figure 13.8A) using an HPLC/UV (360 nm) detector. At this wavelength, the derivatized aldehyde has a maximum absorbance and minimal interference. As representative techniques using a DNPH derivatization method, the EPA Method TO-11 and TO-5 utilize an HPLC/UV system. These two EPA analytical techniques are very similar, but EPA Method TO-11 is developed from modification of the EPA Method TO-5 to reinforce the reaction between collected aldehydes and DNPH coating materials by addition of a strong acid catalyst. Thus, the EPA TO-11 method provides stable and colored hydrazone derivatives and enables the measurement of many different carbonyl compounds, as shown in Figure 13.8B. In the process of aldehyde derivatization, ozone is often reported as a primary interference because ozone also can consume DNPH reagent and degrade aldehyde derivatives (Kleindienst *et al.*, 1998). In order to identify interference, GC/MS can be used as a confirmatory method. Additionally, because ketones also react with DNPH and produce derivatives, ketones are separated chromatographically from the aldehyde derivatives.

The DNPH carbonyl derivatives are also analyzed using gas chromatography (GC). As an example, NIOSH Method 2541 uses a GC technique where

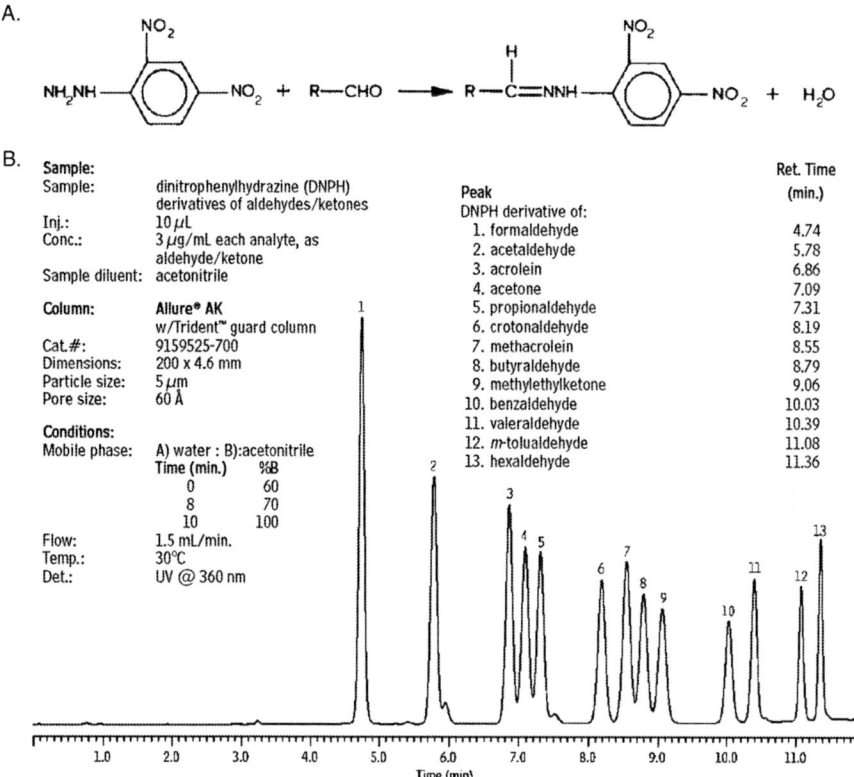

Figure 13.8 Derivatization of aldehydes with DNPH. (A) Collected aldehydes react with 2,4-DNPH solution and produce hydrazones that are analyzed by a HPLC/UV technique for determination of aldehydes. (B) Chromatogram of carbonyl compounds using HPLC/UV technique. Individual aldehydes have a different retention time (qualitative analysis) and the peak area (quantitative analysis) is proportional to the collected amount of aldehydes. (Source: Restek Corporation, http://www.restek.com)

the oxazolidine derivatives of aldehydes are analyzed by a flame ionization detector (FID) with little interference reported. This technique measures carbonyl compounds from 0.1–20 ppm in ~10 L air samples.[206] In this NIOSH method, a nitrogen-specific detector (NSD) can be used to obtain better sensitivity or a GC/MS/MS instrument will likewise enhance detection sensitivity. Besides chromatography, a capillary electrophoresis (CE) method, one of the simplest and fastest analysis techniques for measurement of aldehydes in air, has been employed for formaldehyde, acetaldehyde, propionaldehyde, and acrolein measurements.[207] This technique relies on the reaction of aldehydes with 4-hydrazinobenzoic acid (HBA) while quantifying hydrazones using a UV (maximum absorbance at 280–290 nm) detector. In other words, an electropherogram of HBA derivatives is used for determination of aldehydes. More

Table 13.7 Methods for airborne aldehyde measurement.

Sampling device	Analysis	Reference
Cartridge/Impinger	HPLC/UV	EPA TO-5 & NIOSH 2016
Filter/Impinger coupling	Visible absorption spectrometry	NIOSH 3500
Adsorbent tube	HPLC/UV & GC/FID and GC/N_2 selective detector	EPA TO-11 & NIOSH 2541 & OSHA 52
Diffusive (passive) sampler	LC/UV	OSHA 1007
Airscan Monitor	HPLC/UV	Dosimeter
Miran 1A	Nondispersive infrared (IR)	Short-term (15 min) measurement
Detector Tube	Colorimetric (length-of-stain)	Instantaneous measurement

recently, application of a mist chamber decreased the limit of detection for difficult to measure ambient acrolein and other unsaturated aldehydes.[23] Commonly used measurement methods for gas-phase aldehydes are summarized in Table 13.7 and Figure 13.8.

Aldehydes are released into the air (both ambient and indoor) through a variety of different pathways such as combustion of fossil fuel, exhaust gases from automobiles, biomass burning, industrial manufacturing processes, and cigarette smoke, *etc*. The properties and nature of atmospheric aldehydes are highly influenced by local emission sources, as well as by meteorological parameters, but the primary sources of aldehydes in the indoor environment are building materials and consumer products. Aldehyde levels in the different environments are presented in Tables 13.1–13.3. A critical element for the evaluation of the health effects of aldehyde exposures and for setting of permissible exposure limits is the proper quantification of airborne aldehyde levels. To this end, new methods for measurement of aldehydes in complex real-world airborne settings such as part of particulate matter are being developed and used to better quantitate and understand aldehyde exposure.[208] Moreover, the choice of proper sampling and endpoint analysis technique are important determinants of meeting these goals.

13.5.1.2 Aqueous Aldehydes

a. *Sampling methods:* Collection of water for aqueous aldehyde analysis is typically based on grab sampling. Glass bottles are commonly used for water-sample collection and the bottles and caps should be cleaned and rinsed with distilled water before sampling. If collected samples are not properly kept until analysis, chemical decomposition of aldehyde can occur resulting in underestimation of aldehyde concentrations, and thus, collected water samples are preserved in glass or polyethylene jars in a refrigerator (4 °C) in the dark to prevent degradation of aldehydes.

Similarly, the water sample volume to be collected will vary dependent on sampling sites, regimes, sample pretreatment methods as well as sample analysis methods. Also, the sampling frequency and number of samples or subsamples are determined based on the consideration of the properties of sample water sources.

b. *Analytical methods:* EPA Method 8315 provides an analytical protocol for determination of aqueous carbonyl compounds using a HPLC/UV technique that is already described in EPA Method TO-5. In addition, EPA Method 556 is used for measurement of aldehydes in water, where o-(2,3,4,5,6-pentafluorobenzyl)-hydroxylamine (PFBHA) makes oxime derivatives. These derivatives are extracted into hexane and can be analyzed by an electron capture detector (ECD) that provides high detection sensitivity for aldehyde compounds (*e.g.*, 27 different compounds) as shown in Figure 13.9. As in other GC methods, a head space-gas chromatography/mass spectrometry (HS-GC/MS) is used for measurement of aldehydes in tap water, commercial mineral water[209] and household products.[210] In this method, water-soluble aldehydes are treated with PFBHA and then the head space is analyzed by GC/MS. No interference is reported in this technique. High solubility of aldehydes such as formaldehyde in water leads to accumulation in the natural environment. Also, several aldehydes in drinking water can originate from the water-purification process or unsuitable treatment of water containers. Aqueous aldehydes identified in various samples are presented with concentrations and measurement techniques in Table 13.3.

13.5.1.3 Aldehydes in Food

a. *Sample preparation and extraction*: Sample preparation and extraction methods for aldehydes in food are highly related with the food type and analysis method. Basically, food materials are milled or crushed under liquid-nitrogen cooling process to avoid the transformation of food content such as oxidation of fatty acids,[211] and aldehydes in foods are extracted by solvents[212] or steam distillation.[213,214]

b. *Analytical methods*: Like air and water samples, aldehydes in foods are commonly measured through formation of DNPH-derived hydrazones that are analyzed by the HPLC-UV technique. Besides chromatography, a spectrophotometric method has been used to determine aldehydes in foodstuffs. Granados *et al.*, compared the spectrophotometric and HPLC techniques to measure furanic aldehydes from commercial wine distillates.[215] In that study, aldehydes are directly injected and analyzed by both spectrophotometry and HPLC after filtering (Millipore membranes), and no significant difference was found in quantified levels. The American Standards for Testing and Materials (ASTM E 2313-08 Method) provides spectrophotometric

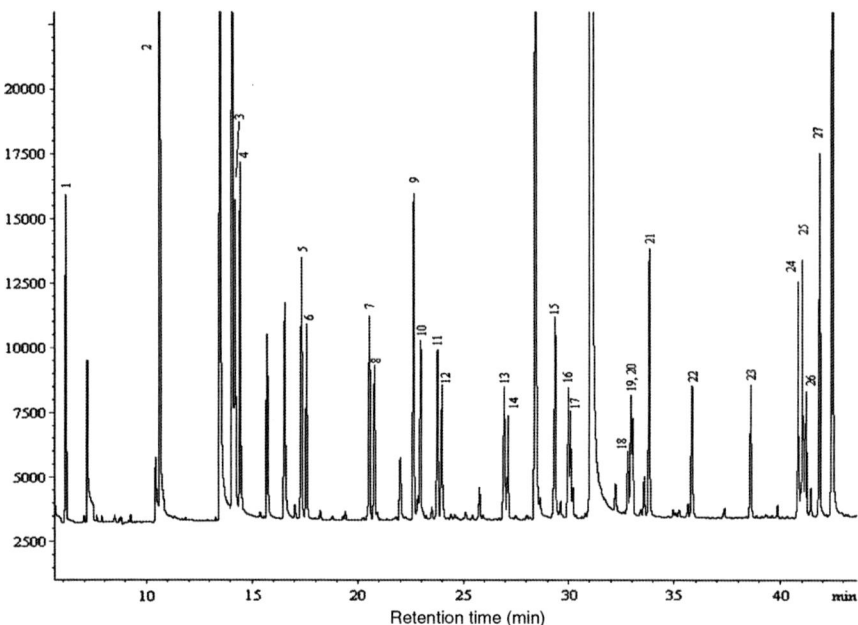

Figure 13.9 Analysis of aldehydes by gas chromatography/electron capture detector (GC/ECD) technique. Analytical conditions are DB-5 MS column, injector temp. 220 °C, head pressure 15 psi, detector temp. 300 °C, splitless injection, and 1 min split delay. Temperature program is 50 °C for 1 min, 4 °C /min to 220 °C, 20 °C/min to 250 °C, and hold at 250 °C for 10 min. Peaks(id number) are internal standard, dibromopropane, (1), formaldehyde (2), acetaldehyde (3–4), propanal (5–6), butanal (7–8), crotonaldehyde (9–10), pentanal (11–12), hexanal (13–14), cyclohexanone (15), heptanal (16–17), surrogate (trifluoro-acetophenone, 18), octanal (19-20), benzaldehyde (21), nonanal (22), decanal (23), glyoxal (24–25), methylglyoxal (26–27). Aldehydes have isomer peaks for correct identification of analyte, and the isomer ratio should be within 50% of the ratio observed in standards for valid determination. If one peak of the isomeric pair is missing, the identification is unreliable. By means of this process, interference can be measured, which is the strength of the GC/ECD method. (Source; EPA Method 556, www.epa.gov/ogwdw000/methods/pdfs/methods/met556.pdf).

determination method for aldehydes in mono-, di-, and triethylene glycol (MEG, DEG and TEG). Varlet *et al.*, used a colorimetric method for formaldehyde measurement in food using a specific absorption of spectrometry (580 nm), but the authors reported that a colorimetric technique may not be suitable for measurement of many other carbonyl compounds.[214] GC methods have also been used to analyze the derivatization of aldehydes in the form of benzyloxime[216] or thiazolidine derivatives.[212] The benzyloxime and thiazolidine derivatives are analyzed by a nitrogen phosphorus detector (NPD) and a flame photometric detector (FPD), respectively, in these GC methods. The GC-FPD is considered a proper technique for measurement of both saturated and unsaturated aldehydes in food. Also, the GC/MS technique can be applied to identify the structures of the aldehyde derivatives,[217] and we recently reported levels of acrolein in rodent chow using PFBHA-based derivatization using headspace detection by GC-MS.[118] Aldehyde levels identified in various food materials are presented in Table 13.4, and a recent publication from our laboratory summarizes acrolein levels reported in a wide variety of foods.[5]

13.5.2 Problems and Pitfalls of Aldehyde Measurements in Air, Water and Foods

It is evident that no single method is suitable for the concurrent determination of saturated and unsaturated aldehydes from environmental samples. For airborne aldehyde measurements, a number of studies have focused on detection of formaldehyde, acetaldehyde, propionaldehyde and acrolein,[207,218–221] and occasionally trace levels of other aldehydes including benzaldehyde, crotonaldehyde, and butyraldehyde are investigated.[222,223] For quantification of aqueous aldehydes from drinking water, many other studies also determined formaldehyde and acetaldehyde concentrations as major aqueous aldehydes,[209,224–226] but the levels of unsaturated aqueous aldehydes have not been sufficiently discussed. In the measurement of aldehydes from foods, the identified unsaturated aldehydes varied with food materials and analytical approaches.[227,228] In short, depending on sample types, sampling and analytical methods, a wide variety of aldehydes have been selectively detected and quantitatively analyzed, yet no single technique satisfies for the various saturated and unsaturated aldehyde measurements.

A derivatization technique using DNPH has been considered as one of the most appropriate methods for aldehyde analysis from air and water samples because this method provides relatively good sample stability and sensitivity.[229] However, interfering compounds such as ozone can affect the formation of aldehyde derivative resulting in misrepresentation of aldehyde amounts,[230] as was already discussed in this section. Also, the interference can contribute to potentially high background levels of formaldehyde and acetaldehyde in the field blank samples.[204] In order to identify the interference, the GC/ECD method is often deployed.

Direct reading analyzers or detector tubes are quick and simple airborne aldehyde measurement methods, but have less sensitivity than chromatographic instruments (HPLC or GC), and are not useful for simultaneous measurement of several aldehydes. Passive samplers are inexpensive and easy to apply, but offer limited detection sensitivity as well.

Because airborne organic compounds can be carried by particles through gas adsorption on particle surfaces and release the gas into air over time,[231] particulate matter (PM) containing gas-phase aldehydes would be an environmental issue. Past studies[232,233] covered particle-associated aldehyde measurements, and Grosjean showed that the magnitude of formaldehyde and acetaldehyde amounts in PM was negligible (1%) by comparison with the vapor-phase aldehyde levels (99%) in ambient air.[233]

Although several organizations (*i.e.* ACGIH, OSHA and NIOSH) have established exposure threshold levels for the toxicity of some aldehydes, the direct application of exposure guidelines on environmental aldehyde measurements is not simple. Due to a short-period of aldehyde persistence in air and aquatic environments as well as complex monitoring issues of inhaled aldehydes in the body, the chronic health effect of aldehydes have not been fully understood with regard to environmental measurement data. In summary, it is essential to develop an integrative aldehyde-monitoring system with high sensitivity for both environmental and biological sample analyses that supports reliable evaluation of long- and short-term cardiovascular effects of aldehyde exposure.

13.6 Conclusions and Future Directions

13.6.1 Environmental and Endogenous Aldehydes: Shared Biological Pathways?

There is ample evidence indicating that, regardless of the source, the α,β-unsaturated aldehydes, such as acrolein and HNE, will form protein–aldehyde adducts with protein lysine, histidine and cysteine residues, glutathione, and when in proximity activate the TRPA1 receptor. Similarly, these protein–aldehyde adducts could serve as the basis for initiating cellular ER stress and the integrated and multifaceted unfolded protein response (UPR). Although such a scenario is consistent with aldehyde involvement in cardiovascular pathology of diabetes, there is scant evidence that such events, however biologically plausible, are operative in environmental aldehyde exposure – even in the setting of tobacco-smoke exposure – although this would be an ideal setting to begin such an inquiry. Moreover, it is hypothetically realistic to propose that environmental aldehydes as well as other pollutants trigger formation of endogenous aldehydes *via* increased oxidative stress and subsequent lipid peroxidation. Overall, this process could be termed "aldehyde-induced aldehyde formation", with the sequence of injury and pathology (outlined above as 'ER-stress' and 'UPR') then being set in motion by endogenous aldehydes and protein–aldehyde adduction. The general applicability of this paradigm will need to be tested in multiple exposure settings.

13.6.2 Aldehyde Metabolism as a Modifier of Aldehyde Effects *in Vivo*

Several models of cardiovascular pathology involve generation of aldehydes at high levels and in specific cardiovascular locations due to selective expression of essential enzymes required for metabolism and aldehyde generation. Such examples/models, including vinyl chloride, 1,3-butadiene, cyclophosphamide and allylamide, have served to illuminate the exquisite potential for aldehydes to wreak havoc in cardiovascular cells. Classically, cyclophosphamide toxicity in bladder, kidney, heart and bone marrow has been wholly or partly attributed to acrolein – one of the 2 major metabolites of CY. While these examples serve as reminders of the potential for cardiovascular specific targeting of bioactivation and aldehyde formation, we have long hypothesized that aldehyde metabolism to more nontoxic products is an important protective process in the cardiovascular organs. Subsequently, this has been demonstrated many times that enzymes of the cardiovascular tissues/organs can rapidly convert aldehydes into less toxic molecules although not always biologically inactive molecules, as is the case for glutathionyl conjugates as mentioned in Section 13.3.1.2. Continued testing of the contribution of specific enzymes (*e.g.*, GSTs, AKRs, P450s, ALDHs) in the protection of individual cardiovascular organs (*e.g.*, endothelium, myocytes) is required to interrogate more fully the nature of aldehyde metabolism as a modulator of aldehyde actions. By taking advantage of molecular approaches to regulate gene expression (either increased or decreased) in target tissues (*e.g.*, Cre-Lox systems), genetic models will provide more focused testing of the role of these enzymes, which some whole body null systems do not.

13.6.3 Potential Approaches to Reduce and Mitigate Aldehyde Exposure

Environmental aldehydes, such as acrolein, have been measured and appear to be present even in polluted cities at relatively low levels (<1 ppb). In the past, these levels were summarily written off as "too low" to effectively induce any pathology but recent modeling of chronic acrolein exposure levels indicates that pulmonary risk could exist at environmentally relevant levels (see Woodruff *et al.*, 2007).[21] Such theoretical modeling has not been done for aldehyde exposures and cardiovascular disease, in part, because there are limited data sets for chronic aldehyde exposure in animals using well-developed cardiovascular metrics and disease models. Thus, measurement of appropriate cardiovascular endpoints in validated animal models is needed. In any case, without an emphasis on environmental aldehydes as potential agents of cardiovascular (and pulmonary) disease there will be little impetus for funding in the future to support such activities. Yet, we have provided compelling animal model data that aldehydes in foods and beverages could in fact be quantitatively more important for total aldehyde exposure than what is respired, which could prove to be more of a vexing target of regulation than air pollution.[119]

Because aldehyde metabolism appears to be an important determinant of aldehyde toxicity, providing methods and understanding mechanisms by which induction or overexpression of aldehyde-metabolizing enzymes protect cardiovascular tissues is an important need for future studies. Perhaps alterations in the level of these enzymes could be a way to mitigate the deleterious effects of aldehyde exposure. Recent investigations suggest that transcription of many of the genes coding for aldehyde-metabolizing enzymes can be induced by plant-derived compounds such as diallyl di- and trisulfides in garlic and dithiol-3-thione (D3T) in cruciferous vegetables, and that induction results in protection against subsequent aldehyde challenge. Preclinical and translational studies that establish a safe and efficacious way to boost cardiovascular expression of aldehyde-metabolizing enzymes (*e.g.*, GSTs by D3T) will be necessary to fully implement what appears to be a promising application of current research. Moreover, deciphering the mechanisms by which induction of aldehyde-metabolizing enzymes leads to cell protection will invariably provide important clues for further developing potential therapeutic targets that can be used to enhance cardiovascular protection against environmental (and perhaps endogenous) aldehyde stress. Similarly, understanding how susceptibility factors (*e.g.*, obesity, diabetes) enhance cardiovascular sensitivity to aldehydes and other pollutants will be necessary to help raise awareness of this concern.

Finally, reducing dependence on combustion-driven economies and switching to alternative energy sources and new technologies will dramatically reduce aldehyde generation, although some alternative fuels (*e.g.*, biofuels) could reduce particulate matter (PM) at the expense of forming more aldehydes. All of these issues will necessarily need to be studied in the coming decade.

Acknowledgments

The authors thank B. Awe, D. Bolanowski, S. Clausi, D. Mosley, E. Werkman and D. Young for expert technical assistance. This work was supported in part by NIH grants (ES17260, ES11594, ES11860, HL55477, HL59378, HL89380, HL95593 and RR 24489), HEI ACES 06-2 Ancillary Study, the Department of Defense (DOD) and by the US EPA.

Abbreviations

ABC, ATP-binding cassette; ALDH, aldehyde dehydrogense; AP-1, activator protein-1; AR, aldose reductase; ARE, antioxidant response element; COX2, cyclooxygenase-2; CY, cyclophosphamide; CYP, cytochromes P450; D3T, dithiol-3-thione; EDHF, endothelial-derived hyperpolarizing factor; eNOS, endothelial nitric oxide synthase; ER stress, endoplasmic reticulum stress; ERK, extracellular-regulated kinase; GSH, reduced glutathione; GST, glutathione *S*-transferase; HNE, 4-hydroxy-*trans*-2-nonenal; HRV, heart-rate variability; HUVEC, human umbilical vein endothelial cell; JNK, c-jun NH_2-terminal kinase; iNOS, inducible nitric-oxide synthase; L-NAME, L-nitroarginine methylester; MAPK, mitogen-activated protein kinase; MDR-1, multi-drug resistance protein-1; MPT, mitochondrial permeability transition; Nrf2, nuclear regulatory factor-2; NF-κB, nuclear factor- κB; PKC, protein kinase C; PM, particulate matter; PTPB1, protein tyrosine phosphatase B1; SAPK, stress-activated protein kinase; SSAO, semicarbazide-sensitive amine oxidase;

TR, thioredoxin reductase; UPR, unfolded protein response; VSMC, vascular smooth muscle cells.

References

1. A. Bhatnagar, Cardiovascular pathophysiology of environmental pollutants, *Am. J. Physiol Heart Circ. Physiol.*, 2004, **286**, H479–H485.
2. A. Bhatnagar, Environmental cardiology: studying mechanistic links between pollution and heart disease, *Circulation Res.*, 2006, **99**, 692–705.
3. V. J. Feron, H. P. Til, F. de Vrijer, R. A. Woutersen, F. R. Cassee and P. J. van Bladeren, Aldehydes: occurrence, carcinogenic potential, mechanism of action and risk assessment, *Mutat. Res.*, 1991, **259**, 363–385.
4. National Research Council, *Formaldehyde and other aldehydes*, National Academy Press, Washington D.C., 1981.
5. G. W. Wang, Y. Guo, T. M. Vondriska, J. Zhang, S. Zhang, L. L. Tsai, N. C. Zong, R. Bolli, A. Bhatnagar and S. D. Prabhu, Acrolein consumption exacerbates myocardial ischemic injury and blocks nitric oxide-induced PKCepsilon signaling and cardioprotection, *J. Mol. Cell Cardiol.*, 2008, **44**, 1016–1022.
6. A. Blair, P. Stewart, M. O'Berg, W. Gaffey, J. Walrath, J. Ward, R. Bales, S. Kaplan and D. Cubit, Mortality among industrial workers exposed to formaldehyde, *J. Natl. Cancer Inst.*, 1986, **76**, 1071–1084.
7. J. Walrath and J. F. Fraumeni Jr, Cancer and other causes of death among embalmers, *Cancer Res.*, 1984, **44**, 4638–4641.
8. R. J. Levine, D. A. Andjelkovich and L. K. Shaw, The mortality of Ontario undertakers and a review of formaldehyde-related mortality studies, *J. Occup. Med.*, 1984, **26**, 740–746.
9. E. Guberan and L. Raymond, Mortality and cancer incidence in the perfumery and flavour industry of Geneva, *Br. J. Ind. Med.*, 1985, **42**, 240–245.
10. L. Alfredsson, N. Hammar and C. Hogstedt, Incidence of myocardial infarction and mortality from specific causes among bus drivers in Sweden, *Int. J. Epidemiol.*, 1993, **22**, 57–61.
11. J. Barnoya and S. A. Glantz, Cardiovascular effects of secondhand smoke: nearly as large as smoking, *Circulation*, 2005, **111**, 2684–2698.
12. S. A. Glantz and W. W. Parmley, Passive smoking and heart disease. Mechanisms and risk, *JAMA*, 1995, **273**, 1047–1053.
13. J. C. Chow, J. G. Watson, J. L. Mauderly, D. L. Costa, R. E. Wyzga, S. Vedal, G. M. Hidy, S. L. Altshuler, D. Marrack, J. M. Heuss, G. T. Wolff, C. A. Pope III and D. W. Dockery, Health effects of fine particulate air pollution: lines that connect, *J. Air Waste Manag. Assoc.*, 2006, **56**, 1368–1380.
14. A. J. Cohen, A. H. Ross, B. Ostro, K. D. Pandey, M. Krzyzanowski, N. Kunzli, K. Gutschmidt, A. Pope, I. Romieu, J. M. Samet and K. Smith, The global burden of disease due to outdoor air pollution, *J. Toxicol. Environ. Health A*, 2005, **68**, 1301–1307.

15. D. W. Dockery, C. A. Pope III, X. Xu, J. D. Spengler, J. H. Ware, M. E. Fay, B. G. Ferris Jr and F. E. Speizer, An association between air pollution and mortality in six US cities, *N. Engl. J. Med.*, 1993, **329**, 1753–1759.
16. C. A. Pope III, Mortality effects of longer term exposures to fine particulate air pollution: review of recent epidemiological evidence, *Inhal. Toxicol.*, 2007, **19**(Suppl 1), 33–38.
17. C. A. Pope III, R. T. Burnett, G. D. Thurston, M. J. Thun, E. E. Calle, D. Krewski and J. J. Godleski, Cardiovascular mortality and long-term exposure to particulate air pollution: epidemiological evidence of general pathophysiological pathways of disease, *Circulation*, 2004, **109**, 71–77.
18. C. A. Pope III and D. W. Dockery, Health effects of fine particulate air pollution: lines that connect, *J. Air Waste Manag. Assoc.*, 2006, **56**, 709–742.
19. D. J. Conklin, P. Haberzettl, R. A. Prough and A. Bhatnagar, Glutathione-S-transferase P protects against endothelial dysfunction induced by exposure to tobacco smoke, *Am. J. Physiol Heart Circ. Physiol.*, 2009, **296**, H1586–H1597.
20. M. Riediker, R. Williams, R. Devlin, T. Griggs and P. Bromberg, Exposure to particulate matter, volatile organic compounds, and other air pollutants inside patrol cars, *Environ. Sci. Technol.*, 2003, **37**, 2084–2093.
21. T. J. Woodruff, E. M. Wells, E. W. Holt, D. E. Burgin and D. A. Axelrad, Estimating risk from ambient concentrations of acrolein across the United States, *Environ. Health Perspect.*, 2007, **115**, 410–415.
22. H. Destaillats, R. S. Spaulding and M. J. Charles, Ambient air measurement of acrolein and other carbonyls at the Oakland-San Francisco Bay Bridge toll plaza, *Environ. Sci. Technol.*, 2002, **36**, 2227–2235.
23. V. Y. Seaman, M. J. Charles and T. M. Cahill, A sensitive method for the quantification of acrolein and other volatile carbonyls in ambient air, *Anal. Chem.*, 2006, **78**, 2405–2412.
24. D. L. Costa, R. S. Kutzman, J. R. Lehmann and R. T. Drew, Altered lung function and structure in the rat after subchronic exposure to acrolein, *Am. Rev. Respir Dis.*, 1986, **133**, 286–291.
25. J. M. Hyvelin, E. Roux, M. C. Prevost, J. P. Savineau and R. Marthan, Cellular mechanisms of acrolein-induced alteration in calcium signaling in airway smooth muscle, *Toxicol. Appl. Pharmacol.*, 2000, **164**, 176–183.
26. E. Roux, J. M. Hyvelin, J. P. Savineau and R. Marthan, Calcium signaling in airway smooth muscle cells is altered by *in vitro* exposure to the aldehyde acrolein, *Am. J. Respir. Cell Mol. Biol.*, 1998, **19**, 437–444.
27. A. Ben-Jebria, Y. Crozet, M. L. Eskew, B. L. Rudeen and J. S. Ultman, Acrolein-induced smooth muscle hyperresponsiveness and eicosanoid release in excised ferret tracheae, *Toxicol. Appl. Pharmacol.*, 1995, **135**, 35–44.
28. D. M. Bautista, S. E. Jordt, T. Nikai, P. R. Tsuruda, A. J. Read, J. Poblete, E. N. Yamoah, A. I. Basbaum and D. Julius, TRPA1 mediates the inflammatory actions of environmental irritants and proalgesic agents, *Cell*, 2006, **124**, 1269–1282.

29. E. Andre, B. Campi, S. Materazzi, M. Trevisani, S. Amadesi, D. Massi, C. Creminon, N. Vaksman, R. Nassini, M. Civelli, P. G. Baraldi, D. P. Poole, N. W. Bunnett, P. Geppetti and R. Patacchini, Cigarette smoke-induced neurogenic inflammation is mediated by alpha,beta-unsaturated aldehydes and the TRPA1 receptor in rodents, *J. Clin. Invest.*, 2008, **118**, 2574–2582.
30. M. Trevisani, J. Siemens, S. Materazzi, D. M. Bautista, R. Nassini, B. Campi, N. Imamachi, E. Andre, R. Patacchini, G. S. Cottrell, R. Gatti, A. I. Basbaum, N. W. Bunnett, D. Julius and P. Geppetti, 4-Hydroxynonenal, an endogenous aldehyde, causes pain and neurogenic inflammation through activation of the irritant receptor TRPA1, *Proc. Natl. Acad. Sci. USA*, 2007, **104**, 13519–13524.
31. B. F. Bessac, M. Sivula, C. A. von Hehn, J. Escalera, L. Cohn and S. E. Jordt, TRPA1 is a major oxidant sensor in murine airway sensory neurons, *J. Clin. Invest.*, 2008, **118**, 1899–1910.
32. S. A. Simon and W. Liedtke, How irritating: the role of TRPA1 in sensing cigarette smoke and aerogenic oxidants in the airways, *J. Clin. Invest.*, 2008, **118**, 2383–2386.
33. G. M. Matanoski, C. Santos-Burgoa and L. Schwartz, Mortality of a cohort of workers in the styrene-butadiene polymer manufacturing industry (1943–1982), *Environ. Health Perspect.*, 1990, **86**, 107–117.
34. A. Penn and C. A. Snyder, Butadiene inhalation accelerates arteriosclerotic plaque development in cockerels, *Toxicology*, 1996, **113**, 351–354.
35. A. Penn and C. A. Snyder, 1,3 Butadiene, a vapor phase component of environmental tobacco smoke, accelerates arteriosclerotic plaque development, *Circulation*, 1996, **93**, 552–557.
36. R. J. Duescher and A. A. Elfarra, 1,3-Butadiene oxidation by human myeloperoxidase. Role of chloride ion in catalysis of divergent pathways, *J. Biol. Chem.*, 1992, **267**, 19859–19865.
37. A. A. Elfarra, R. J. Duescher and C. M. Pasch, Mechanisms of 1,3-butadiene oxidations to butadiene monoxide and crotonaldehyde by mouse liver microsomes and chloroperoxidase, *Arch. Biochem. Biophys.*, 1991, **286**, 244–251.
38. F. R. Fontaine, R. A. Dunlop, D. R. Petersen and P. C. Burcham, Oxidative bioactivation of crotyl alcohol to the toxic endogenous aldehyde crotonaldehyde: association of protein carbonylation with toxicity in mouse hepatocytes, *Chem. Res. Toxicol.*, 2002, **15**, 1051–1058.
39. A. Laplanche, F. Clavel-Chapelon, J. C. Contassot and C. Lanouziere, Exposure to vinyl chloride monomer: results of a cohort study after a seven year follow up. The French VCM Group, *Br. J. Ind. Med.*, 1992, **49**, 134–137.
40. C. L. Dannaher, C. H. Tamburro and L. T. Yam, Occupational carcinogenesis: the Louisville experience with vinyl chloride-associated hepatic angiosarcoma, *Am. J. Med.*, 1981, **70**, 279–287.
41. H. P. Fortwengler Jr, D. Jones, E. Espinosa and C. H. Tamburro, Evidence for endothelial cell origin of vinyl chloride-induced hepatic angiosarcoma, *Gastroenterology*, 1981, **80**, 1415–1419.

42. C. H. Tamburro, Relationship of vinyl monomers and liver cancers: angiosarcoma and hepatocellular carcinoma, *Semin. Liver Dis.*, 1984, **4**, 158–169.
43. N. Brock, J. Stekar, J. Pohl, U. Niemeyer and G. Scheffler, Acrolein, the causative factor of urotoxic side-effects of cyclophosphamide, ifosfamide, trofosfamide and sufosfamide, *Arzneimittelforschung.*, 1979, **29**, 659–661.
44. J. Drimal, J. Zurova-Nedelcevova, V. Knezl, R. Sotnikova and J. Navarova, Cardiovascular toxicity of the first line cancer chemotherapeutic agents: doxorubicin, cyclophosphamide, streptozotocin and bevacizumab, *Neuro. Endocrinol. Lett.*, 2006, **27**(Suppl 2), 176–179.
45. M. Gunther, E. Wagner and M. Ogris, Acrolein: unwanted side product or contribution to antiangiogenic properties of metronomic cyclophosphamide therapy?, *J. Cell Mol. Med.*, 2008, **12**, 2704–2716.
46. P. Morandi, P. A. Ruffini, G. M. Benvenuto, R. Raimondi and V. Fosser, Cardiac toxicity of high-dose chemotherapy, *Bone Marrow Transplant.*, 2005, **35**, 323–334.
47. M. A. Ratcliffe and A. A. Dawson, The effects of cardiotoxic chemotherapy on blood pressure in patients with lymphoma, *Clin. Lab. Haematol.*, 1998, **20**, 353–356.
48. I. Taniguchi, Clinical significance of cyclophosphamide-induced cardiotoxicity, *Intern. Med.*, 2005, **44**, 89–90.
49. R. Aplenc, J. Blanco, W. Leisenring, S. Davies, M. Relling, L. Robison, C. Sklar, M. Stovall and S. Bhatia, Polymorphisms in candidate genes in patients with congestive heart failure (CHF) after childhood cancer: A Report from the Childhood Cancer Survivor Study (CCSS), *Am. Soc. Clin. Oncol.*, 2006.
50. P. J. Boor and R. M. Hysmith, Allylamine cardiovascular toxicity, *Toxicology*, 1987, **44**, 129–145.
51. S. Awasthi and P. J. Boor, Semicarbazide protection from *in vivo* oxidant injury of vascular tissue by allylamine, *Toxicol. Lett.*, 1993, **66**, 157–163.
52. D. J. Conklin, A. Bhatnagar, H. R. Cowley, G. H. Johnson, R. J. Wiechmann, L. M. Sayre, M. B. Trent and P. J. Boor, Acrolein generation stimulates hypercontraction in isolated human blood vessels, *Toxicol. Appl. Pharmacol.*, 2006, **217**, 277–288.
53. D. J. Conklin, C. L. Boyce, M. B. Trent and P. J. Boor, Amine metabolism: a novel path to coronary artery vasospasm, *Toxicol. Appl. Pharmacol.*, 2001, **175**, 149–159.
54. J. L. Sklar, P. G. Anderson and P. J. Boor, Allylamine and acrolein toxicity in perfused rat hearts, *Toxicol. Appl. Pharmacol.*, 1991, **107**, 535–544.
55. D. J. Conklin, H. R. Cowley, R. J. Wiechmann, G. H. Johnson, M. B. Trent and P. J. Boor, Vasoactive effects of methylamine in isolated human blood vessels: role of semicarbazide-sensitive amine oxidase, formaldehyde, and hydrogen peroxide, *Am. J. Physiol. Heart Circ. Physiol.*, 2004, **286**, H667–H676.
56. P. H. Yu and D. M. Zuo, Oxidative deamination of methylamine by semicarbazide-sensitive amine oxidase leads to cytotoxic damage in

endothelial cells. Possible consequences for diabetes, *Diabetes*, 1993, **42**, 594–603.
57. D. J. Conklin, M. B. Trent and P. J. Boor, The role of plasma semicarbazide-sensitive amine oxidase in allylamine and beta-aminopropionitrile cardiovascular toxicity: mechanisms of myocardial protection and aortic medial injury in rats, *Toxicology*, 1999, **138**, 137–154.
58. B. Gong and P. J. Boor, The role of amine oxidases in xenobiotic metabolism, *Expert. Opin. Drug Metab. Toxicol.*, 2006, **2**, 559–571.
59. S. A. Langford, M. B. Trent, A. Balakumaran and P. J. Boor, Developmental vasculotoxicity associated with inhibition of semicarbazide-sensitive amine oxidase, *Toxicol. Appl. Pharmacol.*, 1999, **155**, 237–244.
60. P. J. Boor, S. A. Langford, I. G. Sipes, C. A. McQueen and A. G. Gandolfi, in *Comprehensive Toxicology*, Elsevier Science, New York, edn, 1997, pp. 309–332.
61. B. Gong, M. B. Trent, D. Srivastava and P. J. Boor, Chemical-induced, nonlethal, developmental model of dissecting aortic aneurysm, *Birth Defects Res.*, 2006, **76**, 29–38.
62. Y. S. Park, J. Kim, Y. Misonou, R. Takamiya, M. Takahashi, M. R. Freeman and N. Taniguchi, Acrolein induces cyclooxygenase-2 and prostaglandin production in human umbilical vein endothelial cells: roles of p38 MAP kinase, *Arterioscler. Thromb. Vasc. Biol.*, 2007, **27**, 1319–1325.
63. Y. Misonou, M. Takahashi, Y. S. Park, M. Asahi, Y. Miyamoto, H. Sakiyama, X. Cheng and N. Taniguchi, Acrolein induces Hsp72 *via* both PKCdelta/JNK and calcium signaling pathways in human umbilical vein endothelial cells, *Free Radic. Res.*, 2005, **39**, 507–512.
64. K. Ranganna, Z. Yousefipour, R. Nasif, F. M. Yatsu, S. G. Milton and B. E. Hayes, Acrolein activates mitogen-activated protein kinase signal-transduction pathways in rat vascular smooth muscle cells, *Mol. Cell Biochem.*, 2002, **240**, 83–98.
65. H. Zhang and H. J. Forman, Acrolein induces heme oxygenase-1 through PKC-delta and PI3K in human bronchial epithelial cells, *Am. J. Respir. Cell Mol. Biol.*, 2008, **38**, 483–490.
66. A. Tanel and D. A. Averill-Bates, P38 and ERK mitogen-activated protein kinases mediate acrolein-induced apoptosis in Chinese hamster ovary cells, *Cell Signal*, 2007, **19**, 968–977.
67. M. U. Dianzani, 4-hydroxynonenal from pathology to physiology 4, *Mol. Aspects Med.*, 2003, **24**, 263–272.
68. M. U. Dianzani, 4-Hydroxynonenal and cell signalling, *Free Radic. Res.*, 1998, **28**, 553–560.
69. G. Robino, E. Zamara, E. Novo, M. U. Dianzani and M. Parola, 4-Hydroxy-2,3-alkenals as signal molecules modulating proliferative and adaptative cell responses 3, *Biofactors*, 2001, **15**, 103–106.
70. M. Parola, G. Robino, F. Marra, M. Pinzani, G. Bellomo, G. Leonarduzzi, P. Chiarugi, S. Camandola, G. Poli, G. Waeg, P. Gentilini and M. U. Dianzani, HNE interacts directly with JNK isoforms in human hepatic stellate cells 7, *J. Clin. Invest.*, 1998, **102**, 1942–1950.

71. K. V. Ramana, A. A. Fadl, R. Tammali, A. B. Reddy, A. K. Chopra and S. K. Srivastava, Aldose reductase mediates the lipopolysaccharide-induced release of inflammatory mediators in RAW264.7 murine macrophages, *J. Biol. Chem.*, 2006, **281**, 33019–33029.
72. M. Spite, L. Summers, T. F. Porter, S. Srivastava, A. Bhatnagar and C. N. Serhan, Resolvin D1 controls inflammation initiated by glutathione-lipid conjugates formed during oxidative stress, *Br. J. Pharmacol.*, 2009, **158**, 1062–1073.
73. J. D. Adams Jr and L. K. Klaidman, Acrolein-induced oxygen radical formation, *Free Radic. Biol. Med.*, 1993, **15**, 187–193.
74. U. Dittberner, G. Eisenbrand and H. Zankl, Genotoxic effects of the alpha, beta-unsaturated aldehydes 2-trans-butenal, 2-trans-hexenal and 2-trans, 6-cis-nonadienal, *Mutat. Res.*, 1995, **335**, 259–265.
75. J. J. Horvath, C. M. Witmer and G. Witz, Nephrotoxicity of the 1:1 acrolein-glutathione adduct in the rat, *Toxicol. Appl. Pharmacol.*, 1992, **117**, 200–207.
76. K. V. Ramana, A. Bhatnagar, S. Srivastava, U. C. Yadav, S. Awasthi, Y. C. Awasthi and S. K. Srivastava, Mitogenic responses of vascular smooth muscle cells to lipid peroxidation-derived aldehyde 4-hydroxy-trans-2-nonenal (HNE): role of aldose reductase-catalyzed reduction of the HNE-glutathione conjugates in regulating cell growth, *J. Biol. Chem.*, 2006, **281**, 17652–17660.
77. M. M. Warnke, E. Wanigasekara, S. S. Singhal, J. Singhal, S. Awasthi and D. W. Armstrong, The determination of glutathione-4-hydroxynonenal (GSHNE), E-4-hydroxynonenal (HNE), and E-1-hydroxynon-2-en-4-one (HNO) in mouse liver tissue by LC-ESI-MS, *Anal. Bioanal. Chem.*, 2008, **392**, 1325–1333.
78. C. F. Mueller, J. D. Widder, J. S. McNally, L. McCann, D. P. Jones and D. G. Harrison, The role of the multidrug resistance protein-1 in modulation of endothelial cell oxidative stress, *Circulation Res.*, 2005, **97**, 637–644.
79. J. D. Widder, T. J. Guzik, C. F. Mueller, R. E. Clempus, H. H. Schmidt, S. I. Dikalov, K. K. Griendling, D. P. Jones and D. G. Harrison, Role of the multidrug resistance protein-1 in hypertension and vascular dysfunction caused by angiotensin II, *Arterioscler. Thromb. Vasc. Biol.*, 2007, **27**, 762–768.
80. B. F. Bessac and S. E. Jordt, Breathtaking TRP channels: TRPA1 and TRPV1 in airway chemosensation and reflex control, *Physiology (Bethesda)*, 2008, **23**, 360–370.
81. S. E. Jordt and B. E. Ehrlich, TRP channels in disease, *Subcell. Biochem.*, 2007, **45**, 253–271.
82. S. Earley, A. L. Gonzales and R. Crnich, Endothelium-Dependent Cerebral Artery Dilation Mediated by TRPA1 and Ca2+-activated K+ channels, *Circulation Res.*, 2009, **104**(8), 987–994.
83. C. Gratzke, T. Streng, E. Waldkirch, K. Sigl, C. Stief, K. E. Andersson and P. Hedlund, Transient Receptor Potential A1 (TRPA1) activity in the human urethra-evidence for a functional role for TRPA1 in the outflow region, *Eur. Urol.*, 2008, **55**, 696–704.

84. F. A. Kullmann, M. A. Shah, L. A. Birder and W. C. de Groat, Functional TRP and ASIC-like channels in cultured urothelial cells from the rat, *Am. J. Physiol. Renal Physiol.*, 2009, **296**(4), F892–F901.
85. S. O. Awe, A. S. Adeagbo, S. E. D'Souza, A. Bhatnagar and D. J. Conklin, Acrolein induces vasodilatation of rodent mesenteric bed *via* an EDHF-dependent mechanism, *Toxicol. Appl. Pharmacol.*, 2006, **217**, 266–276.
86. D. S. Celermajer, M. R. Adams, P. Clarkson, J. Robinson, R. McCredie, A. Donald and J. E. Deanfield, Passive smoking and impaired endothelium-dependent arterial dilatation in healthy young adults, *N. Engl. J. Med.*, 1996, **334**, 150–154.
87. M. R. Law and N. J. Wald, Environmental tobacco smoke and ischemic heart disease, *Prog. Cardiovasc. Dis.*, 2003, **46**, 31–38.
88. R. Puranik and D. S. Celermajer, Smoking and endothelial function, *Prog. Cardiovasc. Dis.*, 2003, **45**, 443–458.
89. T. Raupach, K. Schafer, S. Konstantinides and S. Andreas, Secondhand smoke as an acute threat for the cardiovascular system: a change in paradigm, *Eur. Heart J.*., 2006, **27**, 386–392.
90. J. Torok, A. Gvozdjakova, J. Kucharska, I. Balazovjech, S. Kysela, F. Simko and J. Gvozdjak, Passive smoking impairs endothelium-dependent relaxation of isolated rabbit arteries, *Physiol. Res.*, 2000, **49**, 135–141.
91. Z. Yang, C. M. Harrison, G. C. Chuang and S. W. Ballinger, The role of tobacco smoke induced mitochondrial damage in vascular dysfunction and atherosclerosis, *Mutat. Res.*, 2007, **621**, 61–74.
92. B. Q. Zhu, Y. P. Sun, R. E. Sievers, S. A. Glantz, W. W. Parmley and C. L. Wolfe, Exposure to environmental tobacco smoke increases myocardial infarct size in rats, *Circulation*, 1994, **89**, 1282–1290.
93. B. Q. Zhu, Y. P. Sun, R. E. Sievers, W. M. Isenberg, S. A. Glantz and W. W. Parmley, Passive smoking increases experimental atherosclerosis in cholesterol-fed rabbits, *J. Am. Coll. Cardiol.*, 1993, **21**, 225–232.
94. S. Srivastava, D. J. Conklin, S. Q. Liu, N. Prakash, P. J. Boor, S. K. Srivastava and A. Bhatnagar, Identification of biochemical pathways for the metabolism of oxidized low-density lipoprotein derived aldehyde-4-hydroxy trans-2-nonenal in vascular smooth muscle cells, *Atherosclerosis*, 2001, **158**, 339–350.
95. S. Srivastava, S. Q. Liu, D. J. Conklin, A. Zacarias, S. K. Srivastava and A. Bhatnagar, Involvement of aldose reductase in the metabolism of atherogenic aldehydes, *Chemico-Biol. Interact.*, 2001, **130–132**, 563–571.
96. D. Conklin, R. Prough and A. Bhatanagar, Aldehyde metabolism in the cardiovascular system, *Mol. Biosyst.*, 2007, **3**, 136–150.
97. D. J. Conklin, P. Haberzettl, J. F. Lesgards, R. A. Prough, S. Srivastava and A. Bhatnagar, Increased sensitivity of glutathione S-transferase P-null mice to cyclophosphamide-induced urinary bladder toxicity, *J. Pharmacol. Exp. Ther.*, 2009, **331**, 456–469.
98. V. Adler and M. R. Pincus, Effector peptides from glutathione-S-transferase-pi affect the activation of jun by jun-N-terminal kinase, *Ann. Clin. Lab Sci.*, 2004, **34**, 35–46.

99. V. Adler, Z. Yin, S. Y. Fuchs, M. Benezra, L. Rosario, K. D. Tew, M. R. Pincus, M. Sardana, C. J. Henderson, C. R. Wolf, R. J. Davis and Z. Ronai, Regulation of JNK signaling by GSTp, *EMBO J*, 1999, **18**, 1321–1334.
100. T. E. O'Toole, Y. T. Zheng, J. Hellmann, D. J. Conklin, O. Barski and A. Bhatnagar, Acrolein activates matrix metalloproteinases by increasing reactive oxygen species in macrophages, *Toxicol. Appl. Pharmacol.*, 2009, **236**, 194–201.
101. H. S. Deshmukh, C. Shaver, L. M. Case, M. Dietsch, S. C. Wesselkamper, W. D. Hardie, T. R. Korfhagen, M. Corradi, J. A. Nadel, M. T. Borchers and G. D. Leikauf, Acrolein-activated matrix metalloproteinase 9 contributes to persistent mucin production, *Am. J. Respir. Cell Mol. Biol.*, 2008, **38**, 446–454.
102. C. E. Kim, S. J. Lee, K. W. Seo, H. M. Park, J. W. Yun, J. U. Bae, S. S. Bae and C. D. Kim, Acrolein increases 5-lipoxygenase expression in murine macrophages through activation of ERK pathway, *Toxicol. Appl. Pharmacol.*, 2010, **245**, 76–82.
103. S. J. Lee, C. E. Kim, M. R. Yun, K. W. Seo, H. M. Park, J. W. Yun, H. K. Shin, S. S. Bae and C. D. Kim, 4-Hydroxynonenal enhances MMP-9 production in murine macrophages *via* 5-lipoxygenase-mediated activation of ERK and p38 MAPK, *Toxicol. Appl. Pharmacol.*, 2010, **242**, 191–198.
104. A. Helgadottir, A. Manolescu, G. Thorleifsson, S. Gretarsdottir, H. Jonsdottir, U. Thorsteinsdottir, N. J. Samani, G. Gudmundsson, S. F. Grant, G. Thorgeirsson, S. Sveinbjornsdottir, E. M. Valdimarsson, S. E. Matthiasson, H. Johannsson, O. Gudmundsdottir, M. E. Gurney, J. Sainz, M. Thorhallsdottir, M. Andresdottir, M. L. Frigge, E. J. Topol, A. Kong, V. Gudnason, H. Hakonarson, J. R. Gulcher and K. Stefansson, The gene encoding 5-lipoxygenase activating protein confers risk of myocardial infarction and stroke, *Nat. Genet.*, 2004, **36**, 233–239.
105. K. A. Berry, P. M. Henson and R. C. Murphy, Effects of acrolein on leukotriene biosynthesis in human neutrophils, *Chem. Res. Toxicol.*, 2008, **21**, 2424–2432.
106. P. A. Kirkham, G. Spooner, C. Ffoulkes-Jones and R. Calvez, Cigarette smoke triggers macrophage adhesion and activation: role of lipid peroxidation products and scavenger receptor, *Free Radic. Biol. Med.*, 2003, **35**, 697–710.
107. N. Ambalavanan, W. F. Carlo, A. Bulger, J. Shi and J. B. Philips 3rd, Effect of cigarette-smoke extract on neonatal porcine vascular smooth muscle cells, *Toxicol. Appl. Pharmacol.*, 2001, **170**, 130–136.
108. F. Facchinetti, F. Amadei, P. Geppetti, F. Tarantini, S. C. Di, A. Dragotto, P. M. Gigli, S. Catinella, M. Civelli and R. Patacchini, Alpha,beta-unsaturated aldehydes in cigarette smoke release inflammatory mediators from human macrophages, *Am. J. Respir. Cell Mol. Biol.*, 2007, **37**, 617–623.
109. T. Mio, D. J. Romberger, A. B. Thompson, R. A. Robbins, A. Heires and S. I. Rennard, Cigarette smoke induces interleukin-8 release from

human bronchial epithelial cells, *Am. J. Respir. Crit. Care Med.*, 1997, **155**, 1770–1776.
110. G. Valacchi, E. Pagnin, A. Phung, M. Nardini, B. C. Schock, C. E. Cross and A. van der Vliet, Inhibition of NFκB activation and IL-8 expression in human bronchial epithelial cells by acrolein, *Antioxid. Redox Signal*, 2005, **7**, 25–31.
111. C. Lambert, J. McCue, M. Portas, Y. Ouyang, J. Li, T. G. Rosano, A. Lazis and B. M. Freed, Acrolein in cigarette smoke inhibits T-cell responses, *J. Allergy Clin. Immunol.*, 2005, **116**, 916–922.
112. C. Lambert, J. Li, K. Jonscher, T. C. Yang, P. Reigan, M. Quintana, J. Harvey and B. M. Freed, Acrolein inhibits cytokine gene expression by alkylating cysteine and arginine residues in the NF-κB1 DNA binding domain, *J. Biol. Chem.*, 2007, **282**, 19666–19675.
113. Y. F. Xia, B. Q. Ye, Y. D. Li, J. G. Wang, X. J. He, X. Lin, X. Yao, D. Ma, A. Slungaard, R. P. Hebbel, N. S. Key and J. G. Geng, Andrographolide attenuates inflammation by inhibition of NF-κB activation through covalent modification of reduced cysteine 62 of p50, *J. Immunol.*, 2004, **173**, 4207–4217.
114. D. I. Kasahara, M. E. Poynter, Z. Othman, D. Hemenway and A. van der Vliet, Acrolein inhalation suppresses lipopolysaccharide-induced inflammatory cytokine production but does not affect acute airways neutrophilia, *J. Immunol.*, 2008, **181**, 736–745.
115. B. C. Liao, C. W. Hsieh, Y. C. Liu, T. T. Tzeng, Y. W. Sun and B. S. Wung, Cinnamaldehyde inhibits the tumor necrosis factor-alpha-induced expression of cell-adhesion molecules in endothelial cells by suppressing NF-κB activation: effects upon IκB and Nrf2, *Toxicol. Appl. Pharmacol.*, 2008, **229**, 161–171.
116. L. K. Chao, K. F. Hua, H. Y. Hsu, S. S. Cheng, I. F. Lin, C. J. Chen, S. T. Chen and S. T. Chang, Cinnamaldehyde inhibits proinflammatory cytokines secretion from monocytes/macrophages through suppression of intracellular signaling, *Food Chem. Toxicol.*, 2008, **46**, 220–231.
117. Y. C. Liu, C. W. Hsieh, C. C. Wu and B. S. Wung, Chalcone inhibits the activation of NF-κB and STAT3 in endothelial cells *via* endogenous electrophile, *Life Sci.*, 2007, **80**, 1420–1430.
118. D. J. Conklin, O. A. Barski, J. F. Lesgards, P. Juvan, T. Rezen, D. Rozman, R. A. Prough, E. Vladykovskaya, S. Liu, S. Srivastava and A. Bhatnagar, Acrolein consumption induces systemic dyslipidemia and lipoprotein modification, *Toxicol. Appl. Pharmacol.*, 2010, **243**, 1–12.
119. D. J. Conklin, O. A. Barski, J. F. Lesgards, P. Juvan, T. Rezen, D. Rozman, R. A. Prough, E. Vladykovskaya, S. Liu, S. Srivastava and A. Bhatnagar, Acrolein consumption induces systemic dyslipidemia and lipoprotein modification, *Toxicol. Appl. Pharmacol.*, 2010, **243**, 1–12.
120. S. D. Sithu, S. Srivastava, M. A. Siddiqui, E. Vladykovskaya, D. W. Riggs, D. J. Conklin, P. Haberzettl, T. O'Toole, A. Bhatnagar and S. E. D'Souza, Exposure to acrolein by inhalation causes platelet activation, *Toxicol. Appl. Pharmacol.*, 2010, (in press).

121. M. M. Rahman and I. Laher, Structural and functional alteration of blood vessels caused by cigarette smoking: an overview of molecular mechanisms, *Curr. Vasc. Pharmacol.*, 2007, **5**, 276–292.
122. O. C. Burghuber, C. Punzengruber, H. Sinzinger, P. Haber and K. Silberbauer, Platelet sensitivity to prostacyclin in smokers and non-smokers, *Chest*, 1986, **90**, 34–38.
123. J. W. Davis, L. Shelton, I. S. Watanabe and J. Arnold, Passive smoking affects endothelium and platelets, *Arch. Intern. Med.*, 1989, **149**, 386–389.
124. J. H. Ware, Particulate air pollution and mortality--clearing the air, *N. Engl. J. Med.*, 2000, **343**, 1798–1799.
125. G. Hoek, B. Brunekreef, P. Fischer and J. van Wijnen, The association between air pollution and heart failure, arrhythmia, embolism, thrombosis, and other cardiovascular causes of death in a time series study, *Epidemiology*, 2001, **12**, 355–357.
126. J. Denton, D. A. Lane, L. Thunberg, A. M. Slater and U. Lindahl, Binding of platelet factor 4 to heparin oligosaccharides, *Biochem. J.*, 1983, **209**, 455–460.
127. A. Gugliucci, Antithrombin activity is inhibited by acrolein and homocysteine thiolactone: Protection by cysteine, *Life Sci.*, 2008, **82**, 413–418.
128. I. Martinez-Martinez, A. Ordonez, J. A. Guerrero, S. Pedersen, A. Minano, R. Teruel, L. Velazquez, S. R. Kristensen, V. Vicente and J. Corral, Effects of acrolein, a natural occurring aldehyde, on the anticoagulant serpin antithrombin, *FEBS Lett.*, 2009, **583**, 3165–3170.
129. J. E. Freedman and J. Loscalzo, Platelet-monocyte aggregates: bridging thrombosis and inflammation, *Circulation*, 2002, **105**, 2130–2132.
130. A. D. Michelson, M. R. Barnard, L. A. Krueger, C. R. Valeri and M. I. Furman, Circulating monocyte-platelet aggregates are a more sensitive marker of *in vivo* platelet activation than platelet surface P-selectin: studies in baboons, human coronary intervention, and human acute myocardial infarction, *Circulation*, 2001, **104**, 1533–1537.
131. J. Sarma, C. A. Laan, S. Alam, A. Jha, K. A. Fox and I. Dransfield, Increased platelet binding to circulating monocytes in acute coronary syndromes, *Circulation*, 2002, **105**, 2166–2171.
132. R. Ruckerl, R. P. Phipps, A. Schneider, M. Frampton, J. Cyrys, G. Oberdorster, H. E. Wichmann and A. Peters, Ultrafine particles and platelet activation in patients with coronary heart disease – results from a prospective panel study, *Part. Fibre Toxicol.*, 2007, **4**, 1.
133. A. J. Lucking, M. Lundback, N. L. Mills, D. Faratian, S. L. Barath, J. Pourazar, F. R. Cassee, K. Donaldson, N. A. Boon, J. J. Badimon, T. Sandstrom, A. Blomberg and D. E. Newby, Diesel exhaust inhalation increases thrombus formation in man, *Eur. Heart J.*, 2008, **29**, 3043–3051.
134. S. A. Harding, J. Sarma, D. H. Josephs, N. L. Cruden, J. N. Din, P. J. Twomey, K. A. Fox and D. E. Newby, Upregulation of the CD40/CD40 ligand dyad and platelet-monocyte aggregation in cigarette smokers, *Circulation*, 2004, **109**, 1926–1929.

135. G. M. Mutlu, D. Green, A. Bellmeyer, C. M. Baker, Z. Burgess, N. Rajamannan, J. W. Christman, N. Foiles, D. W. Kamp, A. J. Ghio, N. S. Chandel, D. A. Dean, J. I. Sznajder and G. R. Budinger, Ambient particulate matter accelerates coagulation *via* an IL-6-dependent pathway, *J. Clin. Invest.*, 2007, **117**, 2952–2961.
136. Y. Takajo, H. Ikeda, N. Haramaki, T. Murohara and T. Imaizumi, Augmented oxidative stress of platelets in chronic smokers. Mechanisms of impaired platelet-derived nitric oxide bioactivity and augmented platelet aggregability, *J. Am. Coll. Cardiol.*, 2001, **38**, 1320–1327.
137. S. Srivastava, E. Vladykovskaya, O. A. Barski, M. Spite, K. Kaiserova, qJ. M. Petrash, S. S. Chung, G. Hunt, B. Dawn and A. Bhatnagar, Aldose reductase protects against early atherosclerotic lesion formation in apolipoprotein E-null mice, *Circ. Res.*, 2009, **105**, 793–802.
138. S. P. Baba, O. A. Barski, Y. Ahmed, T. E. O'Toole, D. J. Conklin, A. Bhatnagar and S. Srivastava, Reductive metabolism of AGE precursors: a metabolic route for preventing AGE accumulation in cardiovascular tissue, *Diabetes*, 2009, **58**, 2486–2497.
139. C. G. Gairola, M. L. Drawdy, A. E. Block and A. Daugherty, Sidestream cigarette smoke accelerates atherogenesis in apolipoprotein E–/– mice, *Atherosclerosis*, 2001, **156**, 49–55.
140. B. Shao, D. O'Brien K, T. O. McDonald, X. Fu, J. F. Oram, K. Uchida and J. W. Heinecke, Acrolein modifies apolipoprotein A-I in the human artery wall, *Ann. N. Y. Acad. Sci.*, 2005, **1043**, 396–403.
141. S. Srivastava, Ramana, K.V., Srivastava, S.K., D'Souza, S.E., and Bhatnagar, A., in *Aldo-Keto Reductases and Toxicant Metabolism*, ed. T. M. Penning, Petras, Mark. J., American Chemical Society, 2004, pp. 49–64.
142. S. Massberg, K. Brand, S. Gruner, S. Page, E. Muller, I. Muller, W. Bergmeier, T. Richter, M. Lorenz, I. Konrad, B. Nieswandt and M. Gawaz, A critical role of platelet adhesion in the initiation of atherosclerotic lesion formation, *J. Exp. Med.*, 2002, **196**, 887–896.
143. S. Massberg, C. Schulz and M. Gawaz, Role of platelets in the pathophysiology of acute coronary syndrome, *Semin. Vasc. Med.*, 2003, **3**, 147–162.
144. D. Manka, R. G. Collins, K. Ley, A. L. Beaudet and I. J. Sarembock, Absence of p-selectin, but not intercellular adhesion molecule-1, attenuates neointimal growth after arterial injury in apolipoprotein e-deficient mice, *Circulation*, 2001, **103**, 1000–1005.
145. R. G. Collins, R. Velji, N. V. Guevara, M. J. Hicks, L. Chan and A. L. Beaudet, P-Selectin or intercellular adhesion molecule (ICAM)-1 deficiency substantially protects against atherosclerosis in apolipoprotein E-deficient mice, *J. Exp. Med.*, 2000, **191**, 189–194.
146. H. S. Kruth, Platelet-mediated cholesterol accumulation in cultured aortic smooth muscle cells, *Science*, 1985, **227**, 1243–1245.
147. B. Fuhrman, G. J. Brook and M. Aviram, Activated platelets secrete a protein-like factor that stimulates oxidized-LDL receptor activity in macrophages, *J. Lipid Res.*, 1991, **32**, 1113–1123.

148. G. Yu, A. H. Rux, P. Ma, K. Bdeir and B. S. Sachais, Endothelial expression of E-selectin is induced by the platelet-specific chemokine platelet factor 4 through LRP in an NF-κB-dependent manner, *Blood*, 2005, **105**, 3545–3551.
149. T. Nassar, B. S. Sachais, S. Akkawi, M. A. Kowalska, K. Bdeir, E. Leitersdorf, E. Hiss, L. Ziporen, M. Aviram, D. Cines, M. Poncz and A. A. Higazi, Platelet factor 4 enhances the binding of oxidized low-density lipoprotein to vascular wall cells, *J. Biol. Chem.*, 2003, **278**, 6187–6193.
150. B. S. Sachais, T. Turrentine, J. M. Dawicki McKenna, A. H. Rux, D. Rader and M. A. Kowalska, Elimination of platelet factor 4 (PF4) from platelets reduces atherosclerosis in C57Bl/6 and apoE−/− mice, *Thromb. Haemost.*, 2007, **98**, 1108–1113.
151. M. Gawaz, F. J. Neumann, T. Dickfeld, W. Koch, K. L. Laugwitz, H. Adelsberger, K. Langenbrink, S. Page, D. Neumeier, A. Schomig and K. Brand, Activated platelets induce monocyte chemotactic protein-1 secretion and surface expression of intercellular adhesion molecule-1 on endothelial cells, *Circulation*, 1998, **98**, 1164–1171.
152. K. Uchida, M. Kanematsu, Y. Morimitsu, T. Osawa, N. Noguchi and E. Niki, Acrolein is a product of lipid peroxidation reaction. Formation of free acrolein and its conjugate with lysine residues in oxidized low density lipoproteins, *J. Biol. Chem.*, 1998, **273**, 16058–16066.
153. K. Uchida, M. Kanematsu, K. Sakai, T. Matsuda, N. Hattori, Y. Mizuno, D. Suzuki, T. Miyata, N. Noguchi, E. Niki and T. Osawa, Protein-bound acrolein: potential markers for oxidative stress, *Proc. Natl. Acad. Sci. USA*, 1998, **95**, 4882–4887.
154. N. Tanaka, S. Tajima, A. Ishibashi, K. Uchida and T. Shigematsu, Immunohistochemical detection of lipid peroxidation products, protein-bound acrolein and 4-hydroxynonenal protein adducts, in actinic elastosis of photodamaged skin, *Arch. Dermatol. Res.*, 2001, **293**, 363–367.
155. N. Y. Calingasan, K. Uchida and G. E. Gibson, Protein-bound acrolein: a novel marker of oxidative stress in Alzheimer's disease, *J. Neurochem.*, 1999, **72**, 751–756.
156. M. A. Lovell, C. Xie and W. R. Markesbery, Acrolein is increased in Alzheimer's disease brain and is toxic to primary hippocampal cultures, *Neurobiol. Aging*, 2001, **22**, 187–194.
157. M. Yoshida, K. Higashi, L. Jin, Y. Machi, T. Suzuki, A. Masuda, N. Dohmae, A. Suganami, Y. Tamura, K. Nishimura, T. Toida, H. Tomitori, K. Kashiwagi and K. Igarashi, Identification of acrolein-conjugated protein in plasma of patients with brain infarction, *Biochem. Biophys. Res. Commun.*, 2009.
158. K. Higashi, M. Yoshida, A. Igarashi, K. Ito, Y. Wada, S. Murakami, D. Kobayashi, M. Nakano, M. Sohda, T. Nakajima, I. Narita, T. Toida, K. Kashiwagi and K. Igarashi, Intense correlation between protein-conjugated acrolein and primary Sjogren's syndrome, *Clin. Chim. Acta.*, 2009, **411**(5-6), 359–363.

159. K. Sakata, K. Kashiwagi, S. Sharmin, S. Ueda and K. Igarashi, Acrolein produced from polyamines as one of the uraemic toxins, *Biochem. Soc. Trans.*, 2003, **31**, 371–374.
160. K. Sakata, K. Kashiwagi, S. Sharmin, S. Ueda, Y. Irie, N. Murotani and K. Igarashi, Increase in putrescine, amine oxidase, and acrolein in plasma of renal failure patients, *Biochem. Biophys. Res. Commun.*, 2003, **305**, 143–149.
161. N. Uesugi, N. Sakata, M. Nangaku, M. Abe, S. Horiuchi, S. Hisano and H. Iwasaki, Possible mechanism for medial smooth muscle cell injury in diabetic nephropathy: glycoxidation-mediated local complement activation, *Am. J. Kidney Dis.*, 2004, **44**, 224–238.
162. M. Daimon, K. Sugiyama, W. Kameda, T. Saitoh, T. Oizumi, A. Hirata, H. Yamaguchi, H. Ohnuma, M. Igarashi and T. Kato, Increased urinary levels of pentosidine, pyrraline and acrolein adduct in type-2 diabetes, *Endocr. J.*, 2003, **50**, 61–67.
163. D. Suzuki and T. Miyata, Carbonyl stress in the pathogenesis of diabetic nephropathy, *Intern. Med.*, 1999, **38**, 309–314.
164. D. Suzuki, T. Miyata, N. Saotome, K. Horie, R. Inagi, Y. Yasuda, K. Uchida, Y. Izuhara, M. Yagame, H. Sakai and K. Kurokawa, Immunohistochemical evidence for an increased oxidative stress and carbonyl modification of proteins in diabetic glomerular lesions, *J. Am. Soc. Nephrol.*, 1999, **10**, 822–832.
165. X. Zhang, Y. Lai, D. R. McCance, K. Uchida, D. M. McDonald, T. A. Gardiner, A. W. Stitt and T. M. Curtis, Evaluation of N^ε-(3-formyl-3,4-dehydropiperidino)lysine as a novel biomarker for the severity of diabetic retinopathy, *Diabetologia*, 2008, **51**, 1723–1730.
166. K. N. Thakore, J. C. Gan and G. A. Ansari, Rapid plasma clearance of albumin-acrolein adduct in rats, *Toxicol. Lett.*, 1994, **71**, 27–37.
167. T. Ishii, T. Yamada, T. Mori, S. Kumazawa, K. Uchida and T. Nakayama, Characterization of acrolein-induced protein crosslinks, *Free Radic. Res.*, 2007, **41**, 1253–1260.
168. I. Dalle-Donne, M. Carini, G. Vistoli, L. Gamberoni, D. Giustarini, R. Colombo, R. Maffei Facino, R. Rossi, A. Milzani and G. Aldini, Actin Cys374 as a nucleophilic target of alpha,beta-unsaturated aldehydes, *Free Radic. Biol. Med.*, 2007, **42**, 583–598.
169. A. Furuhata, T. Ishii, S. Kumazawa, T. Yamada, T. Nakayama and K. Uchida, N(epsilon)-(3-methylpyridinium)lysine, a major antigenic adduct generated in acrolein-modified protein, *J. Biol. Chem.*, 2003, **278**, 48658–48665.
170. P. C. Burcham, F. R. Fontaine, L. M. Kaminskas, D. R. Petersen and S. M. Pyke, Protein adduct-trapping by hydrazinophthalazine drugs: mechanisms of cytoprotection against acrolein-mediated toxicity, *Mol. Pharmacol.*, 2004, **65**, 655–664.
171. P. C. Burcham and S. M. Pyke, Hydralazine inhibits rapid acrolein-induced protein oligomerization: role of aldehyde scavenging and adduct trapping in crosslink blocking and cytoprotection, *Mol. Pharmacol.*, 2006, **69**, 1056–1065.

172. L. M. Kaminskas, S. M. Pyke and P. C. Burcham, Reactivity of hydrazinophthalazine drugs with the lipid peroxidation products acrolein and crotonaldehyde, *Org. Biomol. Chem.*, 2004, **2**, 2578–2584.
173. L. M. Kaminskas, S. M. Pyke and P. C. Burcham, Strong protein adduct trapping accompanies abolition of acrolein-mediated hepatotoxicity by hydralazine in mice, *J. Pharmacol. Exp. Ther.*, 2004, **310**, 1003–1010.
174. D. L. Carbone, J. A. Doorn, Z. Kiebler and D. R. Petersen, Cysteine modification by lipid peroxidation products inhibits protein disulfide isomerase, *Chem. Res. Toxicol.*, 2005, **18**, 1324–1331.
175. J. D. West and L. J. Marnett, Alterations in gene expression induced by the lipid peroxidation product, 4-hydroxy-2-nonenal, *Chem. Res. Toxicol.*, 2005, **18**, 1642–1653.
176. A. Hengstermann and T. Muller, Endoplasmic reticulum stress induced by aqueous extracts of cigarette smoke in 3T3 cells activates the unfolded-protein-response-dependent PERK pathway of cell survival, *Free Radic. Biol. Med.*, 2008, **44**, 1097–1107.
177. P. Haberzettl, E. Vladykovskaya, S. Srivastava and A. Bhatnagar, Role of endoplasmic reticulum stress in acrolein-induced endothelial activation, *Toxicol. Appl. Pharmacol.*, 2009, **234**, 14–24.
178. H. Esterbauer, R. J. Schaur and H. Zollner, Chemistry and biochemistry of 4-hydroxynonenal, malonaldehyde and related aldehydes, *Free Radic. Biol. Med.*, 1991, **11**, 81–128.
179. S. Srivastava, S. J. Watowich, J. M. Petrash, S. K. Srivastava and A. Bhatnagar, Structural and kinetic determinants of aldehyde reduction by aldose reductase, *Biochemistry*, 1999, **38**, 42–54.
180. R. A. Dick, M. K. Kwak, T. R. Sutter and T. W. Kensler, Antioxidative function and substrate specificity of NAD(P)H-dependent alkenal/one oxidoreductase. A new role for leukotriene B4 12-hydroxydehydrogenase/15-oxoprostaglandin 13-reductase, *J. Biol. Chem.*, 2001, **276**, 40803–40810.
181. S. Srivastava, A. Chandra, N. H. Ansari, S. K. Srivastava and A. Bhatnagar, Identification of cardiac oxidoreductase(s) involved in the metabolism of the lipid peroxidation-derived aldehyde-4-hydroxynonenal, *Biochem. J.*, 1998, **329**, 469–475.
182. L. Zhong, Z. Liu, R. Yan, S. Johnson, Y. Zhao, X. Fang and D. Cao, Aldo-keto reductase family 1 B10 protein detoxifies dietary and lipid-derived alpha, beta-unsaturated carbonyls at physiological levels, *Biochem. Biophys. Res. Commun.*, 2009, **387**, 245–250.
183. L. E. Rikans, The oxidation of acrolein by rat liver aldehyde dehydrogenases. Relation to allyl alcohol hepatotoxicity, *Drug Metab. Dispos.*, 1987, **15**, 356–362.
184. D. Y. Mitchell and D. R. Petersen, Inhibition of rat liver aldehyde dehydrogenases by acrolein, *Drug Metab. Dispos.*, 1988, **16**, 37–42.
185. D. Y. Mitchell and D. R. Petersen, Oxidation of aldehydic products of lipid peroxidation by rat liver microsomal aldehyde dehydrogenase, *Arch. Biochem. Biophys.*, 1989, **269**, 11–17.

186. D. Y. Mitchell and D. R. Petersen, Metabolism of the glutathione–acrolein adduct, S-(2-aldehydo-ethyl)glutathione, by rat liver alcohol and aldehyde dehydrogenase, *J. Pharmacol. Exp. Ther.*, 1989, **251**, 193–198.
187. N. S. Kolb, L. A. Hunsaker and D. L. Vander Jagt, Aldose reductase-catalyzed reduction of acrolein: implications in cyclophosphamide toxicity, *Mol. Pharmacol.*, 1994, **45**, 797–801.
188. E. Dicker and A. I. Cederbaum, Inhibition of the oxidation of acetaldehyde and formaldehyde by hepatocytes and mitochondria by crotonaldehyde, *Arch. Biochem. Biophys.*, 1984, **234**, 187–196.
189. U. H. Danielson, H. Esterbauer and B. Mannervik, Structure–activity relationships of 4-hydroxyalkenals in the conjugation catalysed by mammalian glutathione transferases, *Biochem. J.*, 1987, **247**, 707–713.
190. S. Srivastava, K. V. Ramana, R. Tammali, S. K. Srivastava and A. Bhatnagar, Contribution of aldose reductase to diabetic hyperproliferation of vascular smooth muscle cells, *Diabetes*, 2006, **55**, 901–910.
191. S. E. Spycher, S. Tabataba-Vakili, V. B. O'Donnell, L. Palomba and A. Azzi, Aldose reductase induction: a novel response to oxidative stress of smooth muscle cells, *FASEB J.*, 1997, **11**, 181–188.
192. S. Srivastava, A. Chandra, L. F. Wang, W. E. Seifert Jr, B. B. Dague, N. H. Ansari, S. K. Srivastava and A. Bhatnagar, Metabolism of the lipid peroxidation product, 4-hydroxy-trans-2-nonenal, in isolated perfused rat heart, *J. Biol. Chem.*, 1998, **273**, 10893–10900.
193. B. G. Hill, S. O. Awe, E. Vladykovskaya, Y. Ahmed, S. Q. Liu, A. Bhatnagar and S. Srivastava, Myocardial ischaemia inhibits mitochondrial metabolism of 4-hydroxy-trans-2-nonenal, *Biochem. J.*, 2009, **417**, 513–524.
194. P. Eaton, J. M. Li, D. J. Hearse and M. J. Shattock, Formation of 4-hydroxy-2-nonenal-modified proteins in ischemic rat heart, *Am. J. Physiol.*, 1999, **276**, H935–H943.
195. N. Vasilyev, T. Williams, M. L. Brennan, S. Unzek, X. Zhou, J. W. Heinecke, D. R. Spitz, E. J. Topol, S. L. Hazen and M. S. Penn, Myeloperoxidase-generated oxidants modulate left ventricular remodeling but not infarct size after myocardial infarction, *Circulation*, 2005, **112**, 2812–2820.
196. R. A. Parent, H. E. Caravello and D. E. Sharp, Metabolism and distribution of [2,3-14C]acrolein in Sprague–Dawley rats, *J. Appl. Toxicol.*, 1996, **16**, 449–457.
197. R. A. Parent, D. E. Paust, M. K. Schrimpf, R. E. Talaat, R. A. Doane, H. E. Caravello, S. J. Lee and D. E. Sharp, Metabolism and distribution of [2,3-14C]acrolein in Sprague–Dawley rats. II. Identification of urinary and fecal metabolites, *Toxicol. Sci.*, 1998, **43**, 110–120.
198. J. M. Gray and E. A. Barnsley, The metabolism of crotyl phosphate, crotyl alcohol and crotonaldehyde, *Xenobiotica*, 1971, **1**, 55–67.
199. G. Scherer, M. Urban, H. W. Hagedorn, S. Feng, R. D. Kinser, M. Sarkar, Q. Liang and H. J. Roethig, Determination of two

mercapturic acids related to crotonaldehyde in human urine: influence of smoking, *Hum. Exp. Toxicol.*, 2007, **26**, 37–47.
200. J. Alary, F. Bravais, J. P. Cravedi, L. Debrauwer, D. Rao and G. Bories, Mercapturic acid conjugates as urinary end metabolites of the lipid peroxidation product 4-hydroxy-2-nonenal in the rat, *Chem. Res. Toxicol.*, 1995, **8**, 34–39.
201. H. C. Kuiper, C. L. Miranda, J. D. Sowell and J. F. Stevens, Mercapturic acid conjugates of 4-hydroxy-2-nonenal and 4-oxo-2-nonenal metabolites are *in vivo* markers of oxidative stress, *J. Biol. Chem.*, 2008, **283**, 17131–17138.
202. ACGIH, in *American Conference of Governmental Industrial Hygienists*, Documentation of the threshold limit values and biological exposure indices, 6th ed., Cincinnati, OH, 1991.
203. P. H. Howard, *Handbook of Environmental Fate and Exposure Data for Organic Chemicals, Volume 1 – Large Production and Priority Pollutants*, Lewis, Chelsea, MI, 1989.
204. R. Otson and P. Fellin, A review of techniques for measurement of airborne aldehydes, *Sci. Total Environ.*, 1988, **77**, 95–131.
205. NIOSH, National Institute for Ocupational Safety and Health. DHHS (NIOSH) Publication No. 94–113, Cincinnati, OH, edn, 1994b.
206. NIOSH, National Institute for Occupational Safety and Health, DHHS (NIOSH) Publication No. 94–113, Cincinnati, OH, edn, 1994a.
207. E. A. Pereira, M. O. Rezende and M. F. Tavares, Analysis of low molecular weight aldehydes in air samples by capillary electrophoresis after derivatization with 4-hydrazinobenzoic acid, *J. Sep. Sci.*, 2004, **27**, 28–32.
208. X. Rao, R. Kobayashi, R. White-Morris, R. Spaulding, P. Frazey and M. J. Charles, GC/ITMS measurement of carbonyls and multifunctional carbonyls in $PM_{2.5}$ particles emitted from motor vehicles, *J. AOAC Int.*, 2001, **84**, 699–705.
209. N. Sugaya, T. Nakagawa, K. Sakurai, M. Morita and S. Onodera, Analysis of Aldehydes in Water by Head Space-GC/MS., *J. Health Sci.*, 2001, **47**, 21–27.
210. N. Sugaya, K. Sakurai, T. Nakagawa, N. Onda, S. Onodera, M. Morita and M. Tezuka, Development of a headspace GC/MS analysis for carbonyl compounds (aldehydes and ketones) in household products after derivatization with o-(2,3,4,5,6-pentafluorobenzyl)hydroxylamine, *Anal. Sci.*, 2004, **20**, 865–870.
211. J. Schultheiss, D. Jensen and R. Galensa, Determination of aldehydes in food by high-performance liquid chromatography with biosensor coupling and micromembrane suppressors, *J. Chromatogr. A*, 2000, **880**, 233–242.
212. H. Kataoka, A. Sumida, N. Nishihata and M. Makita, Determination of aliphatic aldehydes as their thiazolidine derivatives in foods by gas chromatography with flame photometric detection, *J. Chromatogr. A*, 1995, **709**, 303–311.

213. R. Teranishi, E. L. Murphy and T. R. Mon, Steam distillation-solvent extraction recovery of volatiles from fats and oils, *J. Agric. Food Chem.*, 1977, **25**, 464–466.
214. V. Varlet, C. Prost and T. Serot, Volatile aldehydes in smoked fish: Analysis methods, occurence and mechanisms of formation, *Food Chem.*, 2007, **105**, 1536–1556.
215. J. Q. Granados, M. V. Mir, H. L. G. Serrana and M. C. L. Martínez, Comparison of spectrophotometric and chromatographic methods of determination of furanic aldehydes in wine distillates, *Food Chem.*, 1995, **52**, 203–208.
216. A. Yasuhara and T. Shibamoto, Quantitative Analysis of Volatile Aldehydes Formed from Various Kinds of Fish Flesh during Heat Treatment, *J. Agric. Food Chem.*, 1995, **43**, 94–97.
217. E. K. Long, I. Smoliakova, A. Honzatko and M. J. Picklo Sr, Structural characterization of alpha,beta-unsaturated aldehydes by GC/MS is dependent upon ionization method, *Lipids*, 2008, **43**, 765–774.
218. K. Kuwata, M. Uebori, H. Yamasaki, Y. Kuge and Y. Kiso, Determination of aliphatic aldehydes in air by liquid chromatography, *Anal. Chem*, 1983, **55**, 2013–2016.
219. D. L. MacIntosh, S. A. Zimmer-Dauphinee, R. O. Manning and P. L. Williams, Aldehyde concentrations in ambient air of coastal Georgia, USA., *Environ. Monitor. Assess.*, 2000, **63**, 409–429.
220. M. Ongwandee, R. Moonrinta, S. Panyametheekul, C. Tangbanluekal and G. Morrison, Concentrations and strengths of formaldehyde and acetaldehyde in office buildings in Bangkok, Thailand, *Indoor Built Environ.*, 2009, **18**, 569–575.
221. H. Schlitt, Impinger sampling coupled to high-performance liquid chromatography by a modified autoinjector interface, *J. Chromatogr. A*, 1997, **762**, 187–192.
222. H. Destaillats, R. S. Spaulding and M. J. Charles, Ambient air measurement of acrolein and other carbonyls at the Oakland–San Francisco Bay Bridge Toll Plaza, *Environ. Sci. Technol.*, 2002, **36**, 2227–2235.
223. M. Possanzini, V. D. Palo, M. Petricca, R. Fratarcangeli and D. Brocco, Measurements of lower carbonyls in Rome ambient air, *Atmos. Environ.*, 1996, **30**, 3757–3764.
224. A. Asthana, D. Bose, S. Kulshrestha, S. P. Pathak, S. K. Sanghi and W. T. Kok, Determination of aldehydes in water samples by capillary electrophoresis after derivatization with hydrazino benzene sulfonic acid, *Chromatographia*, 1998, **48**, 807–810.
225. C. -F. Tsai, H. -W. Shiau, S. -C. Lee and S. -S. Chou, Determination of low-molecule-weight aldehydes in packed drinking water by high performance liquid chromatography, *J. Food Drug Anal.*, 2003, **11**, 46–52.
226. M. Mutsuga, Y. Kawamura, Y. Sugita-Konishi, Y. Hara-Kudo, K. Takatori and K. Tanamoto, Migration of formaldehyde and acetaldehyde into mineral water in Polyethyleneterephthalate (PET) bottles, *Food Add. Contamin.*, 2006, **23**, 212–218.

227. H. Kataoka, A. Sumida, N. Nishihata and M. Makita, Determination of aliphatic aldehydes as their thiazolidine derivatives in foods by gas chromatography with flame photometric detection, *J. Chromatogr. A*, 1995, **709**, 303–311.
228. J. Schultheiss, D. Jensen and R. Galensa, Determination of aldehydes in food by high-performance liquid chromatography with biosensor coupling and micromembrane suppressors, *J. Chromatogr. A*, 2000, **880**, 233–242.
229. R. Otson and P. Fellin, A review of techniques for measurement of airborne aldehydes., *Sci. Total Environ.*, 1988, **77**, 95–131.
230. T. E. Kleindienst, E. W. Corse and F. T. Blanchard, Evaluation of the performance of DNPH-coated silica gel and C_{18} cartridges in the measurement of formaldehyde in the presence and absence of ozone, *Environ. Sci. Technol.*, 1998, **32**, 124–130.
231. J. Lee and Y. Zhang, Evaluation of gas emissions from animal building dusts using a cylindrical convective chamber, *Biosyst. Eng.*, 2008, **99**, 403–411.
232. J. Schnelle-Kreis, I. Gebefügi, G. Welzl, T. Jaensch and A. Kettrup, Occurrence of particle-associated polycyclic aromatic compounds in ambient air of the city of Munich, *Atmos. Environ.*, 2001, **35**, 71–81.
233. D. Grosjean, Formaldehyde and other carbonyls in Los Angeles ambient air, *Environ. Sci. Technol.*, 1982, **16**, 254–262.
234. E. Grosjean and D. Grosjean, Carbonyl collection efficiency of the DNPH-coated C_{18} cartridge in dry air and in humid air, *Environ. Sci. Technol.*, 1996, **30**, 859–863.
235. R. Saiki, K. Nishimura, I. Ishii, T. Omura, S. Okuyama, K. Kashiwagi and K. Igarashi, Intense correlation between brain infarction and protein-conjugated acrolein, *Stroke*, 2009, **40**, 3356–3361.

Subject Index

AAC (abdominal aorta calcium) 223
AAI (ankle-arm index) 106, 108–9, 113–14, 115, 223, 236
abdominal aorta calcium (AAC) 223
ABI (ankle-brachial index) 106, 108–9, 113–14, 115, 223, 236
accumulation mode 77, 78
acetaldehyde 306, 308, 309, 311
acetylation, proteins, day–night cycle 16
acetylcholine 239, 240, 323–7
acquired long QT sundrome (acLQTS) 279–80
acrolein 52
   acrolein–protein adducts 188, 189, 332–8
   allylamine 314, 322
   ambient air levels 305, 307–10, 313
   aorta 317, 321, 322–3, 325
   cardiotoxicity studies 187–91, 301–53
   cyclophosphamide 302, 314, 317
   dyslipidemia/vascular effects 321–7
   11,12-EET-induced dilation effects 325, 327, 328–9
   endoplasmic reticulum stress 302–3, 319–20, 336–8
   endothelial cell function 307, 316–17, 319–20
   in vitro kinetics 338–9
   MAP kinase activation 315
   matrix metalloproteinase (MMP)-9 effects 318
   metabolism kinetics 338–41
   platelet activation 327, 329–30
   structure 306
   tobacco smoke 52, 307, 313, 327, 333, 337
   urban ambient air 307, 309, 310
ACS (American Cancer Society) study 161, 184
activated cells, spill-over 85, 87–8
activating transcription factor 6 (ATF6) 336, 337
activity, physical 32, 43, 46–9
activity regulation, aldehyde-metabolising enzymes 339
acute dyslipidemia 321
acute myocardial infarction (AMI) 81, 221, 235
acute phase reactants 88, 224
adhesion molecules
   CD18/CD54 130
   ICAM-1 148, 202, 207, 258, 320, 321
   macrophage adhesion 319
   PM air pollution 225
   UFP air pollution 202
   VCAM-1 148, 150, 320
adiposity, type-2 diabetes 143
administrative data sources 144–6
ADRB2 (β2-adrenergic receptor) 9
adsorbent tube samplers 343, 344, 346
adverse cardiovascular events 17, 19, 20, 235
   see also stroke
aerosol particles *see* particulate matter (PM) air pollution

African populations 9, 10, 15
Agatston scoring system 107–8
aging
　see also elderly people
　CVD death risk 3, 7, 10–12
　Normative Aging Study 146, 149
air
　see also gaseous pollutants;
　　particulate matter (PM) air
　　pollution; traffic-related air
　　pollution
　aldehydes 305, 307–10, 313, 341,
　　342–4, 345, 349–50
Aitken mode 77, 78, 199
albumin 333–4
alcohol, moderate consumption 43,
　44–5
aldehydes 301–53
　see also acrolein
　airborne 305, 307–10, 313, 341,
　　342–4, 345, 349–50
　aldehyde-induced aldehyde
　　formation 350
　　see also endoplasmic reticulum
　　　stress; unfolded protein
　　　response
　aqueous 305, 311, 341, 346–7
　chromatography 344, 345, 347,
　　348–9
　definition 301–4, 306
　detection 341–50
　enals, vascular toxicity 321–7
　endoplasmic reticulum
　　stress 302–3, 319–20, 336–8
　epidemiological studies 305, 307,
　　310–11, 313
　exposure assessment 305, 308–10,
　　311, 312, 313–14
　foods 301, 305, 312, 341, 347–50
　mechanisms of toxic action
　　302–3
　metabolism 338–41, 350–1
　occupational exposure 302, 307,
　　313
　signalling mechanisms 315–38
　structures 301, 306

tobacco smoke 52, 307, 313,
　316–20, 327, 329, 337
α,β-unsaturated see enals
xenobiotic metabolism 313–14
aldo-ketoreductases 338
aldose reductase (AR) 330, 340–1
allylamine 314, 322
altitude effects 24–6
American Cancer Society study 80,
　83, 221, 222
ancestral genes 6, 9–10
angiogenesis, metals toxicity 288–9
angiotensin II 89, 122, 126
angiotensinogen gene 9, 10, 15
animal studies
　aldehydes
　　atherosclerosis 330–2
　　cardiotoxicity studies 187–91,
　　　317, 321–41
　　enals, vascular effects 321–7
　atherosclerosis 4–6, 203–6, 330–2
　autonomic tone/function
　　studies 123–4
　carbon nanotubes 259–61
　diabetes, PM air pollution 148–9,
　　150
　diesel-engine emissions 242–6
　dyslipidemia 321–7
　enals cardiovascular
　　metabolism 340
　fullerenes 258
　high-fat diet, atherosclerosis 223–4
　hypertension/vascular
　　toxicity 122–7
　platelet activation/
　　thrombosis 327, 329–30
　quantum dots 261
　seasonal variation 20–1
　UFP, atherogenesis 203–6
ankle-arm index (AAI) 106, 108–9,
　113–14, 115, 223, 236
ankle-brachial index (ABI) 106,
　108–9, 113–14, 115, 223, 236
ANS see autonomic nervous system
anti-inflammatory actions 320–1
antimony 280

Subject Index

antioxidant response elements (ARE) 277
aorta
  acrolein 317, 321, 322–3, 325
  aneurysm/dissection causes 2
  atherosclerosis
    animal studies 203, 204, 205, 206
    calcium 108, 113, 223, 106
    plaque formation 46, 223, 203–6
  carbon nanotubes 259–61
  distensibility, Doppler assessment 109
  enals, acrolein 317, 321, 322–3, 325
  NAD(P)H oxidase/Rho-kinase 1 126
AP-1-dependent gene transcription 315, 316
apes 4–6, 8
APHEA2 study 79, 82, 162
APHENA study 79
apoE gene
  ε2/ε3/ε4 alleles 8–9
    beneficial effects 10
    ε3 most common variant 8, 36
    ε4 maladaptive in humans 8, 36
  mismatch hypothesis 8, 10
  selection in Africa 15
  transgenerational environmental effects 36
apoE$^{-/-}$ (apolipoprotein E-null) mice 242, 243–4, 258, 262–3, 330–2
apolipoprotein A 331
aqueous aldehydes 305, 311, 341, 346–7
AR (aldose reductase) 330, 340–1
ARE (antioxidant response elements) 277
ARIC (Atherosclerosis Risk in Communities) Study 31, 146
aromatic carbonyls 306, 320–1
arrhythmias 86, 87, 89, 146, 279–81
arsenic
  penta-/tri-valent 274
  toxicity 276–7, 278, 279–80, 282–3, 286–7, 288–9
  type-2 diabetes 150
arteriosclerosis 281–2, 283
  see also atherosclerosis
ATF6 (activating transcription factor 6) 336, 337
atherosclerosis
  see also plaque
  accompanying changes 1
  air pollution
    diesel-engine emissions 243–4
    epidemiological studies 105–17
    manufactured nanoparticles 255–6
    particulate matter 86, 223–4, 263
  1,3-butadiene exposure 313
  cadmium toxicity 283–4
  definition/characteristics 1, 106
  development steps 220
  ischemic heart disease 220, 221
  metals toxicity 263, 281–2
  protein–aldehyde adducts 330–2
Atherosclerosis Risk in Communities (ARIC) Study 31, 146
ATP7A/B P-type ATPases 274
Augsburg, MI Registry study 81
Australia 162, 181–3
autonomic nervous system (ANS)
  see also heart-rate variability
  PM air pollution 132
    heart failure 186
    imbalance induction 85, 87, 89
    stroke 166
    tone/function 123–4, 129–30

B-cell destruction, type-1 diabetes 150
B-type natriuretic peptide (BNP) 180
BALF (bronchoalveolar lavage fluid) 263
BASIC (Brain Attack Surveillance in Corpus Christi) Project 163
bed rest 48
behavioural contagion 33–5

benzaldehyde 306, 308, 310
bioavailability 209, 290
biomarkers 148, 240–1
bisphenol A 150–1
Blackfoot disease 283
blood cell counts 130
blood pressure
  see also hypertension
  PM air pollution 86, 121–34
    animal studies 125–7
    integrated human studies 131–2
    vascular toxicity 121–34
  seasonal variation 22–3
BNP (B-type natriuretic peptide) 180
bone marrow 130
brachial-artery diameter 147
bradykinin 239
Brain Attack Surveillance in Corpus Christi (BASIC) Project 163
breast-feeding 41
bronchoalveolar lavage fluid (BALF) 263
built environment 14, 32–3, 49
1,3-butadiene 313–14
butanal 308, 310, 312

c-Jun N-terminal kinases (JNK) 315, 317, 336, 337
C-reactive protein (CRP) 130, 148, 166, 186, 224
C57BL/6 mice 329, 331, 333
CAC (coronary artery calcium) 106, 107–8, 112, 113, 115, 223, 236–7
cadmium 278, 279, 283–4, 289
  hypertension 285–6
  ZIP8 transport protein 275
CAFE (Corporate Average Fuel Economy) program 235, 236
calcium
  acute influx in HUVEC, fullerenes 258
  aortic calcium 106, 108, 113, 223
  coronary artery calcium 106, 107–8, 112, 113, 115
    ischemic heart disease 223
    roadway proximity 236–7
  myofibrillar calcium ion transients 188
calcium channels
  aldehyde effects 302–3, 316–18
  cardiomyocyte action potentials 238
  diesel-engine emissions 239, 279–81
Canada 144
cancer, smoking relationship 50
CAP see concentrated air particles
capillary electrophoresis 332–4, 344, 345
capsazapine 123
car use 32
  see also traffic-related air pollution
carbohydrate consumption 58
carbon monoxide
  diesel/gasoline engines comparison 238, 245
  heart failure 180, 183, 185
  US vehicular emission trends 235, 236
carbon nanotubes 259–61
2-carboxyl-1-methylethylmercapturic acid (CMEMA) 341
cardiac arrhythmias/dysrhythmias 86, 87, 89, 146, 279–81
cardiac chemiluminescence 123
cardiac performance, day–night cycle 17
cardioembolic stroke 163
cardiomyocytes 238
cardiomyopathies, metal-induced 278–9
cardiovascular bed, particle action 222
carotid artery calcification 236
carotid artery intima-media thickness (CIMT) 106, 107, 112, 113, 115, 209–10, 223
carotid DIRECT trial 45
case cross-over studies see cross-over studies
catalytic converters 235

Subject Index

catecholamine catabolism 288
CD18 adhesion molecule 130
CD40-ligand 225
CD54 adhesion molecule 130
CD142 tissue factor 258
cell cycle arrest 258
cell proliferation 288–9
cell subsets 130–1, 273, 274
chalcone 321
charcoal filters, smoking 319
children
   arsenic exposure,
     atherogenesis 283
   early atherosclerosis 108
   fetal reprogramming 41
   infection/inflammation 11, 224
   ischemic heart disease 224
   obesity 42
   type-1 diabetes 150
chimpanzees 4–6, 8
China 29
chloroacetaldehyde 314
cholesterol
   altitude effects on metabolism 25
   apoE gene 8–9, 36
   apoE-null mice 331
   cholesterol-lowering drugs,
     drawbacks 54
   diabetes 143
   fat consumption 44, 45
   hypercholesterolemia 41, 143
   seasonal variations 18
CHOP (CCAAT/enhancer binding protein) 336, 337
chromatography 344, 345, 347, 348–9
chronic exposure *see* long-term exposure
chronobiology, day–night cycle 16–18
CIMT (carotid artery intima-media thickness) 106, 107, 112, 113, 115, 209–10, 223
cinnamaldehyde 320–1
circadian rhythms 16–18
Clean Air Act (CAA, 2008, US) 235
climate change 37–8

clinical cardiovascular
   outcomes 84–7
clock mechanism, instrinsic 16, 17
CMEMA (2-carboxyl-1-methylethylmercapturic acid) 341
coagulation markers 224–5, 241
coarse mode particulate matter 77, 78, 79, 129, 134, 199
cobalt 279
cohort survival studies
   PM air pollution 80–1, 82, 83, 161
     traffic-related 239–42, 244, 236–7
   UFP air pollution 210
   time series studies comparison 116
cold exposure, seasonal mortality 20
combustion engine exhaust *see* traffic-related air pollution
community environment
   definition 14–15
   deprived neighbourhoods 31–3
   diet choice 45
   discrete communities, evolutionary effects 26–7
   secondhand smoke 51–2, 206, 327
   social aspects 33–5
   socioeconomic status 27, 29–33
computed tomography (CT) 106, 107–8, 112, 113, 115
concentrated airborne particles (CAP)
   animal studies 125–7
   autonomic tone/function 129
   integrated human studies 131–2
   lung-related pro-inflammatory effects 88
   systemic oxidative stress/endothelial function 123
congenital defects 2
continuity of environmental conditions 35–7
coordination complexes, metals 275
copper 274, 276, 277, 281, 282, 288
coronary arteries
   calcium 106, 107–8, 115, 223
   traffic-related air pollution 110, 112–13, 236–7

coronary arteries (*continued*)
  chronic PM exposure 223
  diesel emissions 240, 242
  population health studies 236
  subclinical atherosclerosis assessment methods 106–7
coronary artery calcium (CAC) 106, 107–8, 112, 113, 115, 223, 236–7
Corporate Average Fuel Economy (CAFE) program 235, 236
costs, stroke health-care 159
cross-over studies 81, 145, 181–3
cross-sectional studies 109–17, 223
crotonaldehyde
  cytokine inhibition 319–20
  exposure limits 308
  formation from 1,3-butadiene 302, 313–14
  levels in air 310
  structure 306
  tobacco smoke 307
culture, continuity of environmental conditions 35–7
CXCL1 (C-X-C motif ligand 1) 202
cycles of night and day 16–18
cyclophosphamide 302, 314, 317
CYP3A5 gene 9
cystitis 317
cytokines
  see also individual cytokines
  aromatic carbonyls 320–1
  PM air pollution 85, 87–8
    atherosclerosis 199
    heart failure 186
    vascular toxicity 124–5
  production inhibition effects 318–20
  UFP air pollution, atherosclerosis 207

D-dimer 241
D3T (dithiol-3-thione) 303
definition of cardiovascular disease 1–2
depressive mood, inactivity 48
detection of aldehydes 303–4, 341–50

DHN (1,4-dihydroxynonene) 340
DHN-MA (1,4-dihydroxynonene mercapturic acid) 341
diabetes
  see also type-2 diabetes
  administrative data source evidence 144–6
  air pollution effects 150
  apoE-null mice 330
  exercise effects 48
  genetic polymorphisms 6
  physiologic outcomes evidence 146–50
  stroke 167
  type-1, children 150
  vitamin D synthesis, seasonal effects 23
diesel-exhaust particles (DEP)
  atherosclerosis 200
  characterisation 238
  endothelial dysfunction 123, 127–9
  lung-related proinflammatory effects 88
  NERC systems 238
  particle filters 235
  systemic inflammation 130
  ultrafine 199, 202
diet 29, 32, 43–6
diffuse cardiac fibrosis, chimps/gorillas 5
diffusive samplers 343, 344, 346
1,4-dihydroxynonene (DHN) 340
1,4-dihydroxynonene mercapturic acid (DHN-MA) 341
dioxins 150–1
DIRECT-Carotid trial 45
discrete communities 26–7
disease pathways 85–9
distance to roadways 236–7
dithiole-3-thione (D3T) 303
dithiothreitol (DTT) assay 200
diurnal cycles 16–18
DNA binding, aldehydes 320
DNPH derivatisation 342, 344, 345, 349

Doppler assessment 106, 108–9, 113–14, 115, 223, 236
dose-response relationships, metals 273
drug delivery devices, fullerenes 257
DTT (dithiothreitol) assay 200
dysfunctional HDL 207
dyslipidemia 321–7
dysrhythmias 86, 87, 89, 146, 279–81

E-selectin 207
early life 7–8
see also children
ECG (electrocardiogram)
  ischemic heart disease 225
  long QT syndrome 279–80
  ST-segment changes 240
  T-wave alterations 242
EDHF (endothelium-derived hyperpolarising factor) 322, 323, 325, 327
EDRF (endothelium-derived relaxing factor) 323, 327
education 30
11,12-EET (epoxyeicosatrienoic acid) 325, 327, 328–9
elderly people
  infection/inflammation 19, 224
  PM air pollution 179, 182, 185, 186, 224
  UFP air pollution 211
  winter mortality 19–20
emissions *see* traffic-related air pollution
enals
  aldehyde-metabolising enzymes regulation 339
  cardiovascular metabolism 340–1
  glutathionyl conjugates 315–16
  5-lipoxygenase 318
  matrix metalloproteinase-9 318
  systemic metabolism 341
  TRPA1 receptor 316–18
  vascular toxicity 321–7
endogenous vasopressor hormones 126

endogenous/exogenous aldehyde production 304–5
endoplasmic reticulum (ER) 302–3, 319–20, 336–8
endothelial cell function
  acrolein 307, 316–17, 319–20
  atherosclerosis 224–5
  metals toxicity 273, 279, 283, 286–7, 289
  PM air pollution
    diabetes 147
    diesel emissions, human studies 239–42
    ischemic heart disease 224–5
    stroke 166
    systemic oxidative stress 122–3, 126, 127–9
    ultrafine particles, atherosclerosis 199
endothelin 1 (ET-1) 89, 225, 240–1, 242
endothelin A receptors 123, 244
endothelium-derived hyperpolarising factor (EDHF) 322, 323, 325, 327
endothelium-derived relaxing factor (EDRF) 323, 327
environment categories/types 12–15
Environmental Protection Agency (EPA) 235, 236, 246
  approved quantification methods 342, 344, 346, 347
enzyme activity regulation 339
EPA *see* Environmental Protection Agency
epidemiological studies
  air pollution
    atherosclerosis 105–17, 198, 210
    morbidity/mortality 78–84, 160–1
    relationships consistency 114–15
    traffic-related 235–7
  aldehydes 305, 307, 310–11, 313
  arsenic toxicity 286
  metals toxicity 289–90
  traffic-related air pollution 235–7

ER (endoplasmic reticulum) 302–3, 319–20, 336–8
ERK1/2 123
evolutionary effects 26–7
exercise 32, 43, 46–9
exogenous/endogenous aldehyde production 304–5
exposure assessment
    aldehydes 305, 308–10, 311, 312, 313–14
    manufactured nanoparticles 264–5
    traffic-related air pollution 237

families, obesity/smoking 33–5
famine, thrifty-gene hypothesis 37
fats *see* lipids
FDP-lysine adduct 332
FELIC (The Fate of Early Lesions in Children) study 41
Fenton reaction 276
fetal reprogramming 41
fibrinogen 19, 20, 224
fibrosis, diffuse cardiac 5
filtered diesel engine emissions 243–4
fine particles 77, 78, 177
    see also ultrafine particles (UFP)
flow-mediated dilation (FMD) 147
foods, aldehydes 301, 305, 312, 341, 347–50
forearm blood flow 106, 108–9, 113–14, 115, 223, 236, 239
formaldehyde 306, 308, 309, 310, 311
Framingham Heart Study 33
France 161–2, 163, 164
free radicals 276
    see also reactive oxygen species
friendships, obesity/smoking habits 33–5
FTO gene, obesity 34
fullerenes 257–8

G(-6)M235 variant 9
gas chromatography 343, 344, 345, 347, 348–9
gaseous pollutants 77, 177
    aldehydes 342–4, 345, 346, 348–9

diesel/gasoline engines
    comparison 238, 245
    gas–particle interactions 245
    ischemic heart disease 226
    PM vulnerability causation 246
gasoline-engine emissions 238, 242, 244–5
    see also traffic-related air pollution
gene expression regulation 23, 263
gene polymorphisms 6, 241, 247, 303, 304
genetic diversity, humans 4
genetic drift 7, 37
genetic similarity, apes/humans 4
genetic/environmental treatment integration 57
geographic location 24–6, 236–7
geographical information system (GIS) methods 236–7
Germany
    PM air pollution
        acute myocardial infarction 235
        diabetes, hospital admission rates 146
        epidemiological studies 110, 111, 112, 113–14, 210
    UFP air pollution 210
GIS (geographical information system) methods 236–7
glutathione (GSH) 338, 339
glutathione S-transferase (GST) 200, 241, 247, 303
glutathione S-transferase mu-1 (*GSTM1*) gene 149
glutathione S-transferase P (*GSTP1*) gene 241
glutathione S-transferase theta 1 (*GSTT1*) gene 303
glutathionyl conjugates 315–16, 340
GM-CSF (granulocyte macrophage colony-stimulating factor) 318, 319, 320
GNB3 gene 9
gorillas 4, 5
grandmothering 36

Subject Index

granulocyte macrophage colony-stimulating factor (GM-CSF) 318, 319, 320
gravitational load 47
GS-HNE (glutathionyl conjugate of HNE) 315–16, 340
GSH/GSSG (glutathione to oxidised glutathione) ratios 200
GST (glutathione S-transferase) 200, 247
GSTM1 (glutathione S-transferase mu-1) gene 149
GSTP1 (glutathione S-transferase P) gene 241
GSTT1 (glutathione S-transferase theta 1) gene 303

Haber–Weiss reaction 276
HAEC (human aortic endothelial cells) 259
hardwood smoke 245
Harvard Six Cities study 80, 83, 160–1, 221
HBA (4-hydrazinobenzoic acid) 345
HDL (high-density lipoprotein) levels 53, 207
Health Professionals Follow-up Study 42, 47
heart failure
  PM air pollution
    aldehyde effects 187–91
    causative mechanisms 186–7
    clinical/pathological characteristics 178–9
    hospital visits/ admissions 181–3
    long-term effects 183–5
    mortality 180–1
    motor-vehicle traffic exposure 184–5
    particulate exposure 184
    short-term effects 179–83
    signs and symptoms 180
heart rate variability (HRV) 86, 89, 123–4, 129–30, 146–7, 186
heme oxygenase 1 (HO-1) 200, 263

hemodynamic studies 131–2, 166
hemolytic effects, fullerenes 257
hemorrhagic stroke 164
hERG/IKr trafficking 280
heritability of environment 35–7
heterogeneous composition of particulates 144
hexamethonium 123
HIF-1α (hypoxia-inducible factor 1α) 278
highlanders 24–5
history 37–8, 235
HMPMA (3-hydroxy-methylpropylmercapturic acid) 341
HNE (4-hydroxy-*trans*-2-nonenal) 302, 315, 318, 331, 340
HO-1 (heme oxygenase 1) 200, 263
homocysteine 277, 282
hospital admission rates
  heart failure 181–3
  PM air pollution
    diabetes 145–6
    epidemiological studies 82
    hemorrhagic stroke 164
    stroke risk 161–2
HRV (heart rate variability) 86, 89, 123–4, 129–30, 146–7, 186
HSPA8 (heat shock protein 70 kDa protein 8) 202
human aortic endothelial cells (HAEC) 259
human environment categories 12–15
human studies
  see also epidemiological studies
  manufactured nanoparticles 263
  traffic-related air pollution 239–42, 244
HUVEC (human umbilical vein endothelial cells) 257–8, 337
4-hydrazinobenzoic acid (HBA) 345
3-hydroxy-methylpropylmercapturic acid (HMPMA) 341
4-hydroxy-*trans*-2-nonenal (HNE) 302, 315, 318, 331, 340
hydroxyl radicals 276
hypercholesterolemia 41, 143

hypertension
  metals toxicity 284–8
  PM air pollution
    animal studies 125–7
    toxicity, biological
      mechanisms 132, 134
    vascular toxicity 121–34
  salt hypothesis 9–10
  type-2 diabetes 143
hypertrophy
  left ventricle 178–9, 186, 191
  right ventricle, newborns 24
hypoxia-inducible factor 1α
  (HIF-1α) 278

ICAM-1 (intercellular adhesion
  molecule-1)
  aldehydes 320, 321
  atherosclerosis 207, 331, 332
  fullerene effects 258
  PM air pollution, diabetes 148
  UFP air pollution,
    atherosclerosis 202
ICD (International Classification of
  Disease), Revisions 9/10 162
IDL (intermediate density
  lipoproteins) 8–9
IFN-γ (interferon γ) 124
IHCS (Intermountain Heart
  Collaborative Study) 81, 82
IL-6 (interleukin 6) 124, 130, 207, 263
IL-8 (interleukin 8) 319
immigrants to a new community 27,
  28
immunity, gene loci 11
immunoelectrophoresis 332–4, 344,
  345
impingers 343, 344, 346
in utero arsenic exposure 283
inactivity effects 47, 48
incidence, CVD 2
independence, risk factor effects 54
India 29
infection, childhood 11
inflammatory response
  see also proinflammatory cytokines

carbon nanotubes 260
cardiovascular bed 222
CVD significance 11, 224
manufactured nanoparticles 256
nickel nanoparticles 262–3
PM air pollution
  diabetes 130–1, 149
  diesel emissions 239
  heart failure 186
  ischemic heart disease 220, 221
  spill-over of inflammatory
    mediators 85, 87–8
  stroke 166
  ultrafine particles 198–212
α,β-unsaturated aromatic
  carbonyls 320–1
inositol-requiring enzyme 1
  (IRE) 336, 337
insulin resistance 18, 143, 150, 167
integrated genetic/environmental
  treatment approach 57
intercellular adhesion molecule 1 *see*
  ICAM-1
interferon γ (IFN-γ) 124
interleukin 6 (IL-6) 124, 130, 207,
  263
interleukin 8 (IL-8) 319
intermediate density lipoproteins
  (IDL) 8–9
intracerebral hemorrhage,
  primary 160
intrinsic clock mechanism 16, 17
ion channels 279–81, 302, 303,
  316–18
  see also calcium channels
IRE1 (inositol-requiring enzyme
  1) 336, 337
IREs (iron response elements) 274
iron 274–5, 276, 278, 281–2
IRPs (iron regulatory proteins) 274–5
ischemic heart disease
  ECG recorded 225
  endothelial cell activation 224–5
  inflammation as marker 224
  metals toxicity 281–2
  PM air pollution 220–7

Subject Index 381

components 226
long-term exposure 221–3
short-term exposure 225–6
ischemic stroke *see* stroke
Italy 180

Japan 165
jet lag 17
JNK (c-Jun N-terminal kinases) 315, 317, 336, 337

kinetic studies, aldehydes 338–9
KLH (keyhole limpet hemocyanin) 332–3, 334, 335
KORA Registry, Augsburg, Germany 146
Korea 165

L-type currents 238
lactase 36
lacunar (small-vessel) stroke 163
land use change 39
large-vessel atherothromboembolic stroke 163
latitudinal effects 22–3
LDL (low-density lipoprotein) 8, 44, 53, 200–2
lead 286
left ventricular remodelling 178–9, 186, 191, 274
leishmaniasis 280
leptin 25, 41
lethal dose, metals 272
lethal oxidative stress 277
lifestyle choices 15, 40–52
see also smoking
ligand-binding, metals 275
lipids 44–6, 276, 322
see also cholesterol
HDL (high-density lipoprotein) 53, 207
LDL (low-density lipoprotein) 8, 44, 53, 200–2
ox-LDL 202, 209
trans fatty acids 44, 46
unsaturated fat 44, 45

5-lipoxygenase, enals 318
liver, arsenic toxicity 283, 287
long QT syndrome (LQTS) 279–80
long-term exposure
particulate matter (PM) 82–3, 262–3
clinical/subclinical outcomes 84–7
atherosclerosis 223–4
heart failure 183–5
ischemic heart disease 221–3
stroke risk 165–6
versus short-term exposure 161
metals 262–3, 273, 278–9
Los Angeles, US
PM air pollution 109, 110
atherosclerosis 206, 211
diabetes 145
ischemic heart disease 222, 223
UFP air pollution 211
loss-of-function mutations 5–6
low-density lipoprotein (LDL) 8, 44, 53, 200–2
LQTS (long QT syndrome) 279–80
lung *see* pulmonary aspects
LV (left ventricular)
remodelling 178–9, 186, 191, 274
lysine dimer adducts 334, 335, 336

M-CSF (macrophage colony-stimulating factor) 201, 202
macrophage scavenger receptor Type A 319
macrophages 201, 202, 319
maladaptation 6, 7
mismatch hypothesis 6, 8–10
MALDI/TOF MS 334, 335
man-made environment *see* built environment; plastic environment
manufactured nanoparticles 253–65
cardiac toxicity 255–62
causative mechanisms for cardiovascular effects 256
characterisation requirements 264
definition 253–5, 257
direct cardiac exposure 255–6

manufactured nanoparticles (*continued*)
  exposure scenarios 264–5
  fullerenes 257–8
  future studies 263–5
  human data 263
  metallic/metallic oxide-based 261–3
  nanotoxicology 253–5
  nanotubes 259–61
  nickel nanoparticles 262–3
  potential uses 255
  quantum dots 261
  study review 256–62
MAPK (mitogen-activated protein kinases) 88, 134, 315
maternal aspects 41
matrix metalloproteinase-9 (MMP-9) 240, 244, 245, 318
MCAPS (US Medicare Pollution Study) 161, 162
MCP-1 (monocyte chemotactic protein-1) 263
measurement methods 106–9, 303–4, 341–50
mechanically-produced coarse mode PM 77, 78, 79, 129, 134, 199
Menkes disease 274
mercapturic acids 341
mercury toxicity 278, 279, 284, 287–8
MESA (Multi-Ethnic Study of Atherosclerosis) 112–13, 223
mesenteric vasculature 321–7
metabolic syndrome 129, 150–1, 241
metabolism, aldehydes 338–41, 350–1
metal-responsive-element-binding transcription factor-1 (MTF-1) 277
metals 272–90
  angiogenesis 288–9
  antimony 280
  arsenic 276–7, 278, 279–80, 282–3, 286–7, 288–9
  atherosclerosis 281–2
  cadmium 278, 279, 283–4, 285–6, 289
  cardiac arrhythmias 279–81
  cardiac pathogenic actions 278–9
  copper 274, 276, 277, 281, 282, 288
  dose-response relationships 273
  exposure routes 273
  hypertension 284–8
  iron 274–5, 276, 278, 281–2
  ischemic diseases 281–2
  lead 286
  lethal dose 272
  mechanisms of action 275–8
  mercury 278, 279, 284, 287–8
  metal-oxide-based nanoparticles 254, 260, 261–2
  metal-responsive transcription regulators 277
  nanoparticles 260, 261–2
  nickel 246, 262–3, 280, 281
  zinc 275, 280, 281, 284
MI Registry in Augsburg study 81
1,4-Michael addition 332, 333, 338
Michael adducts 334, 335
microgravity effects, space travel 47
migrants to new communities 27, 28
minimally-modified low-density lipoprotein 202
mismatch hypothesis 6, 8–10
mitochondria 200, 340–1
mitogen-activated protein kinases (MAPK) 88, 134, 315
mmLDL (minimally-modified low-density lipoprotein) 202
MMP-9 (matrix metalloproteinase-9) 240, 244, 245
monocyte chemotactic protein-1 (MCP-1) 263
monocytes 201, 202
morbidity 78–84, 165–6
mortality
  cause comparison data 2
  CVD 2, 3
  different communities, big differences 28–9
  heart failure 178, 180–1
  PM air pollution 76
    diabetes 144–5

Subject Index

epidemiological studies 78–84
heart failure 185
long-term exposure 165–6
smoking/PM 40
stroke 159, 160–1, 165–6
seasonal variations 19–20
motor-vehicle traffic *see* traffic-related air pollution
mountain sickness 24, 25
MPO (myeloperoxidase) 122
MTF-1 (metal-responsive-element-binding transcription factor-1) 277
Multi-Ethnic Study of Atherosclerosis 210
MWCNT (multiwalled carbon nanotubes) 259–61
myeloperoxidase (MPO) 122
myocardium
  infarction 81, 221, 235
  PM air pollution, ischemia 128, 220–7
  vulnerability 40–1, 187–91
myofibrillar calcium ion transients 188

NAD^+ levels 16
NADH 339
NAD(P)H oxidase
  metals toxicity 276
    arsenic 283, 287
    hypertension 285
  PM air pollution, hypertension 126
  upregulation 122
nanoparticles
  manufactured 253–65
    carbon nanotubes 259–61
    cardiac toxicity 255–62
    definition 253–5, 257
    fullerenes 257–8
    metallic/metallic oxide-based 261–2
    quantum dots 261
National Environmental Respiratory Center (NERC) 238

National Morbidity, Mortality, and Air Pollution Study (NMMAPS) 79, 82, 116
natural environment
  altitude effects 24–6
  CVD in relation to 15–26
  day–night cycle 16–18
  definition 13, 14
  seasonal variation 18–21
  sunlight exposure/vitamin D 21–4
natural selection 3, 7, 10–12
NCHS (National Center for Health Statistics) study 79
neighourhood environment 31–3, 45
NERC (National Environmental Respiratory Center) 238
Netherlands 165–6, 180
Netherlands Cohort Study on Diet and Cancer 165–6
neutrality hypothesis 7
New Zealand 162, 181–3
NF-κB (nuclear factor kappa B) 315, 316
  aldehydes 319, 320
  PM air pollution 88, 134
NHANES database 286
NHS study 81
nickel 246, 262–3, 280, 281
night–day cycles 16–18
NIOSH-approved quantification methods 342, 343, 345, 346
nitrogen oxides
  diesel-engine emissions 238, 241, 242, 243, 245
  nitrogen dioxide, heart failure 180, 183, 185
  NO synthase 242
  US vehicular emission trends 235, 236
NMMAPS (National Morbidity, Mortality, and Air Pollution Study) 79, 82, 116
nonlinear relationship, smoking/CVD 50

Normative Aging Study 146, 149
northern/southern latitudes 22
Nrf2-regulated genes 200, 277, 315
nuclear factor kappa B (NF-κB) 315, 316
  aldehydes 319, 320
  PM air pollution 88, 134
nucleation mode 77, 78, 199
Nurses' Health Study 42, 44
nutrition *see* diet

obesity
  behaviour contagion 33
  children 42
  deprived neighbourhoods 32
  FTO gene 34
  obesogenic environments 32
  PM air pollution, diabetes 149
  thrifty-gene hypothesis 36–7
  vitamin D synthesis 23
occupational aldehyde exposure 302, 307, 313
ODD (oxygen-dependent degradation) 278
OSHA-approved quantification methods 342, 343, 346
outcomes, clinical 84–7, 146–50
ox-LDL (oxidised low-density lipoprotein) 202, 209
ox-PAPC (oxidised 1-palmitoyl-2-arachidonoyl-sn-glycero-3-phosphocholine) 201, 202, 209
oxidative stress
  arsenic 277
  carbon nanotubes 260
  cardiovascular bed 222
  manufactured nanoparticles 256
  nickel nanoparticles 262–3
  PM air pollution
    diabetes 149
    heart failure 186
    pulmonary stress 85, 86, 87–9
    systemic stress 122–3
    ultrafine particles, atherosclerosis 199
  vascular toxicity 126

2-oxoglutarate-dependent prolyl-4-hydroxylases (PHD) 278
oxygen-dependent degradation (ODD) 278
ozone 150, 180

P-type ATPases ATP7A/B 274
p38 MAPK 123
p45-NFE2 related transcription factor 200
PAH (polycyclic aromatic hydrocarbons) 200
PAI1 (plasminogen activator inhibitor) 225, 241–2
PAPC (1-palmitoyl-2-arachidonyl-sn-glycero-3-phosphorylcholine) 201, 202, 209
particle filters, diesel engines 235
particulate matter (PM) air pollution 76–90
  atherosclerosis 105–17, 198–212
    epidemiological studies 105–17
    ultrafine particles 198–212
  cardiac arrhythmias 280
  cardiovascular health effects 160–1
  cardiovascular vulnerability 40, 187–91
  classification 77, 78, 198
  clinical/subclinical outcomes 84–7
  components/characterisation 77–8
  definition 177
  diabetes 143–51
  diesel engine emissions 239–46
  effects 38, 39–40
  epidemiological studies 78–84
  gas-phase aldehydes 349–50
  heart failure 177–91
  heterogeneous composition 144
  hypertension/vascular toxicity 121–34
  ischemic heart disease 76–90, 220–7
  morbidity/mortality 76, 78–84
  stroke risk 159–68
  vascular toxicity 121–34

Subject Index

PDI (protein disulfide isomerase) 336, 337
o-(2,3,4,5,6-pentafluorobenzyl)-hydroxylamine (PFBHA) 347
pentavalent antimony 280
pentavalent arsenic 274
peripheral vascular disease
  ankle-arm/brachial index 106, 108–9, 113–14, 115, 223, 236
  arsenic 283
  cadmium toxicity 283–4
  copper 282
peripheral vasoconstriction, cold exposure 20
PERK (PKR-like endoplasmic reticulum kinase) 336, 337
personal choice, smoking 49–52
personal environment 14, 15, 40–52
personalised medicine 56
PF4 (platelet factor 4) 329, 331, 332
PFBHA (*o*-(2,3,4,5,6-pentafluorobenzyl)-hydroxylamine) 347
phase II enzymes 200
PHD (2-oxoglutarate-dependent prolyl-4-hydroxylases) 278
phenylepinephrine 322–4
phenylketonuria 12
physical activity 32, 43, 46–9
physical neighbourhood 31–3
pigmentation, skin 21–2
PKC (protein kinase C) 189, 315, 316
PKR-like endoplasmic reticulum kinase (PERK) 336, 337
plaque
  formation
    aorta 46, 203–6, 223
    causative mechanisms 200–3, 207–9
    diesel-engine emissions 243–4
    long timescale 40–1
    risk factors 53–4
    smoking 53–4
  rupture
    acute myocardial infarction 221
    atherosclerosis progression 106

myocardial ischemia 220, 221
  seasonal variation 20
plasminogen activator inhibitor (PAI1) 225, 241–2
plastic environment 13, 14, 26–35
platelets
  activation, aldehydes 327, 329–30, 331
  aggregation 89
  carbon nanotubes 260
  PF4 (platelet factor 4) 329, 331, 332
  platelet–leucocyte aggregates 329
  protein–aldehyde adducts 327, 329–30
pleiotropic effects 10
PM *see* particulate matter
polycyclic aromatic hydrocarbons (PAH) 200
polymorphisms 6, 241, 247, 303, 304
population studies *see* epidemiological studies
positive pleiotropy 10
potassium channels 279–80
PPARG gene 6, 10
predation threat 37
prenatal period 41
prevalence
  CVD 2–3
  diabetes 143
  heart failure 178
prevention strategies 55–6
primary intracerebral hemorrhage 160
proinflammatory cytokines
  see also individual cytokines
  PM air pollution
    atherosclerosis 199
    heart failure 186
    vascular toxicity 124–5
  UFP air pollution, atherosclerosis 207
proinflammatory pathways 200–3
propanal (propionaldehyde) 306, 308, 309, 310, 311, 312

protective mechanisms 303, 304, 351–2
protein disulfide isomerase (PDI) 336, 337
protein kinase C (PKC) 189, 315, 316
protein synthesis, day–night cycle 16, 17
protein–aldehyde adducts 188, 189, 332–8
  acrolein–protein adducts 188, 189, 332–8
  atherosclerosis 330–2
  formation/detection 332–5
  signalling pathways 332–8
  thrombosis/platelet activation 327, 329–30
  toxicity 302–3
prothrombotic effects see thrombosis
PSAS (French National Program on Air Pollution Health Effects) 161
pulmonary aspects
  deposition 254, 256, 259–60
  inflammation 124–5, 242
  oxidative stress, PM air pollution 85, 87–9, 199
  particle translocation, myocardial ischemia 220, 222
  respiratory diseases 76, 182, 238
  UFP retention, atherosclerosis 208
pyrrhic hypothesis 8, 11

quantitation, aldehydes 341–50
quantum dots 261

rare metals 273
rare-variant-common disease hypothesis 7
reactive oxygen species (ROS) see also oxidative stress
  PM air pollution 88
    atherosclerosis 199, 200
    hypertension 121, 122, 134
  signalling, metals toxicity 276
  UFP air pollution, atherosclerosis 202
redox potential 77, 78, 208–9

remodelling, left ventricle 178–9, 186, 191, 274
renin–angiotensin system 22–3
reproductive selection pressure 3, 7–8
respiratory diseases 76, 182, 238
retinoic acid x-receptor 23
Rho/Rho-kinase pathway 126, 134
rhymthic natural environmental changes 16–24
right ventricular hypertrophy 24
risk factors
  environmental CVD 52–5
  independence of effects 54
  lifestyle choices 43
  major/secondary, CVD 52, 53
  modification by other risk factors 53
  PM synergistic interactions 121
  socioeconomic status relationship 30–1
rMSSD (root mean square of successive differences) 123, 124
roadway proximity measurement 236–7

salt regulation 9–10, 44
sampling methods 342–4, 346–7
saturated fat 44
scavenger receptors (SRA) 319
SCCN1A gene 9
schistosomiasis 280
SCN (suprachiasmatic nucleus) 16
SDCCA (simulated downwind coal-combustion atmosphere) 245
SDNN (standard deviation of normal-to-normal beats) 146
seasonal variation 18–24
secondary organic aerosol (SOA) 245
secondhand smoke 51–2, 206, 327
selection pressure 3, 7–8, 46–7
selenium 284
semicarbazide-sensitive amine oxidase (SSAO) 314
serum leptin, altitude 25
shift workers 17

Subject Index

short-term PM exposure
  heart failure 179–83
  ischemic heart disease 225–6
  PM air pollution 78, 81, 82
    clinical/subclinical outcomes 84–7
    stroke risk 161–2, 164–5
    versus long-term studies 161
shunting of fluid 243
sidestream tobacco smoke 51–2, 206
signalling pathways
  aldehyde exposure 315–38
    cardiovascular effects 318–32
    enals 315–18
    protein–aldehyde adducts role 332–8
  ROS-mediated, metals toxicity 276
  systemic oxidative stress/endothelial function 123, 134
simulated downwind coal-combustion atmosphere (SDCCA) 245
single-nucleotide polymorphisms (SNPs) 303, 304
single-walled carbon nanotubes (SWCNT) 259–61
Sirt1 deacetylase 16
size fractions 77, 78, 198
skin pigmentation 21–2
small-vessel (lacunar) stroke 163
smoking
  abstinence 43
  aldehydes in tobacco smoke 51, 307, 313, 316–20, 327, 329, 333, 337
  behaviour contagion 33
  cancer relationship 50
  charcoal filters 319
  independent CVD risk factor 50
  mortality 40
  personal choice 49–52
  plaque formation 53–4
  secondhand smoke 51–2, 206, 327
  United States ban 29
smooth muscle cell proliferation 201, 202, 315, 316, 340

SNPs (single nucleotide polymorphisms) 303, 304
SOA (secondary organic aerosol) 245
social effects *see* community environment
socioeconomic status 27, 29–33
sodium channels 279–80
sodium hypothesis 9–10
sodium nitroprusside 239, 240
soot particles 226
source apportionment studies 237
Southern California Particle Center (SCPC) 206
space travel, microgravity effects 47
spill-over of inflammatory mediators 85, 87–8
SRA (macrophage scavenger receptor) Type A 319
SRE (stress-response elements) 277
SREBP gene transcription 45
SSAO (semicarbazide-sensitive amine oxidase) 314
ST-segment changes 240
standard deviation of normal-to-normal beats (SDNN) 146
STAT3 tyrosine phosphorylation 321
statins 224
statistical independence 54
stress *see* oxidative stress
stress-repsonse elements (SRE) 277
stroke
  PM air pollution 159–68
    cardiovascular effects 160–1
    causative mechanisms 166–7
    hemorrhagic stroke 164
    ischemic stroke 163–4
    long-term exposure 165–6
    mortality 164–6
    public health problem 159–60
    short-term exposure 161–2, 164–5
      cerebrovascular mortality effects 164–5
    transient ischemic attack 163–4
  subtypes 163
  WHO definition 159

structures, aldehydes 301, 306
subarachnoid hemorrhage 160
subclinical atherosclerosis 106–9
subclinical cardiovascular
   outcomes 84–7
subendothelial space 201, 202, 331
sulfate, diabetes 150
sulfur dioxide 180
sunlight exposure, vitamin D 21–4
suprachiasmatic nucleus (SCN) 16
surface-to-volume ratio 254
susceptibility factors 84, 114, 117,
   167, 289
SWCNT (single-walled carbon
   nanotubes) 259–61
sympathetic nervous system see
   autonomic nervous system
symptomatic treatments,
   drawbacks 54
synergistic interactions 121, 202, 206
synthetic chemicals, pollution 38
systemic aspects
   circulation, PM translocation 85,
      87, 88, 132, 134
   inflammation 124–5, 130–1,
      149–50, 186
   oxidative stress 122–3, 127–9, 256

temperature, seasonal mortality 20
tetrahydrobiopterin 122
thoracic particles 77, 78, 129, 177
thrifty-gene hypothesis (Neel) 6,
   36–7
thrombosis
   carbon nanotubes 260
   PM air pollution 86, 89
      diesel-engine emissions 241
      myocardial ischemia 220, 221
      roadway proximity 236
   protein–aldehyde adducts 327,
      329–30
   seasonal variations 18
TIA (transient ischemic
   attack) 163–4
time series studies 79, 116
tissue distribution 333–5

tissue plasminogen activator 19, 128,
   239
titanium dioxide 254, 261–2
TNF-α (tumour necrosis factor-α)
   331
   aldehydes 319
   atherosclerosis 207
   systemic inflammation 124, 239
      PM toxicity 130
tobacco smoke see smoking
toll-like receptors (TLR) 88
traffic-related air pollution
   see also air pollution; particulate
      matter (PM) air pollution
   chemistry 237–9
   epidemiological studies 109–15,
      235–7
   exposure assessment 237
   future research areas 246–7
   heart failure 185
   human studies 239–42, 244
   ischemic heart disease 222, 226
      long-term exposure 223–4, 226
   overview 234
   three air pollution
      mechanisms 234
   toxicology research
      findings 239–46
   United States 235, 236
trans fatty acids 44, 46
transcription regulators, metal-
   responsive 277
transgenerational environmental
   effects 36
transient ischemic attack
   (TIA) 163–4
transient receptor potential (TRP)
   channels 302
translocation of particles
   cardiovascular bed 222
   manufactured nanoparticles 254
   PM air pollution
      heart failure 186
      toxic effects 85, 87, 88, 132, 134
   UFP air pollution,
      atherosclerosis 209

Subject Index

transport, copper/iron 274-5
transport-related pollution *see* traffic-related air pollution
triglyceride-rich lipoproteins 8
trivalent antimony 280
trivalent arsenic 274
TRP (transient receptor potential) membrane channels 302, 316-18, 350
TRPV1 (Transient Potential Vanilloid 1) receptor 123, 124
tumour angiogenesis, metals toxicity 288
tumour necrosis factor-α (TNF-α) 331
  atherosclerosis 207
  systemic inflammation 124, 239
  PM toxicity 130
tunnel pollution 246
type-1 diabetes, children 150
type-2 diabetes
  air pollution 143-51
  PM air pollution 149, 150
  UFP air pollution 208
  Western diet 44

ultrafine particles (UFP)
  see also manufactured nanoparticles
  atherosclerosis 198-212
    pro-oxidative potential 199-200
    proinflammatory pathways 200-3
  autonomic tone/function 129-30
  bioavailability 209
  characteristics 77, 78
  diabetes 150
  epidemiological studies 83-4
  greater lung retention 208
  larger particle number 208
  larger redox active compounds content 208-9
  similarity to manufactured nanoparticles 256
  systemic oxidative stress/endothelial function 123

vascular toxicity 134, 177
unfolded protein response (UPR) 302, 303, 336-8, 350
United States
  ACS (American Cancer Society) study 161, 184
  approved measurement methods 342-6
  carbohydrate consumption increase 58
  Clean Air Act (CAA, 2008) 235
  deaths due to CVD 3
  EPA (Environmental Protection Agency) 235, 236, 246
    approved quantification methods 342, 344, 346, 347
  MCAPS (US Medicare Pollution Study) 161, 162
  NCHS (National Center for Health Statistics) study 79
  NERC (National Environmental Respiratory Center) 238
  NIOSH-approved quantification methods 342, 343, 345, 346
  OSHA-approved quantification methods 342, 343, 346
  PM air pollution
    diabetes 145
    epidemiological studies 80, 83, 109, 110, 111, 113
    heart failure studies 181
    ischemic heart disease 222-3
    traffic-related 235, 236
    UFP air pollution 203, 206, 207, 211
  SCPC (Southern California Particle Center) 206
  smoking ban 29
α,β-unsaturated aldehydes *see* enals
α,β-unsaturated aromatic carbonyls 320-1
unsaturated fat 44, 45
upper respiratory tract infection 19
UPR (unfolded protein response) 302, 303, 336-8, 350

urbanisation *see* built environment
urinary bladder cystitis 317
UVB radiation 21–4

VACES (Versatile Aerosol Concentration Enrichment System) 206
vanadium 280, 281
vascular toxicity
  enals 321–7
  PM air pollution 121–34
    diesel engine emissions 239–42
    human studies 127–32, 239–42
    ultrafine particles 200–3
    vasoconstrictive effects 89, 244, 258
  vasodilation 323–7
vasomotor function 258
vasopressor hormones 126
VCAM-1 (vascular adhesion molecule-1) 148, 150, 320
vehicle emissions *see* traffic-related air pollution
venous circulation 242–3
ventricular remodelling 24, 179, 186, 191, 274
Versatile Aerosol Concentration Enrichment System (VACES) 206
vinyl chloride 314
virus infection 277, 279
vitamin D 21–4, 25
vitamin D receptor (VDR) 23
vitamin D3 synthesis 21
von Willebrand factor 19, 148, 225

vulnerability, cardiovascular system 40–1, 187–91

walking, urban area design 32
water, aqueous aldehydes 305, 311, 341, 346–7
Western blot 333–5
Western diet 44
WHI (Women's Health Initiative) study 81, 83, 147, 161, 165
white blood cells 130
WHO (World Health Organisation) 2–3, 49, 76, 159
Wilson's disease 274
winter mortality 19–20, 21–4
women
  causes of death worldwide 2
  early atherosclerosis 108
  obesity/physical inactivity 47
  PM air pollution 114, 280
  pregnancy 41
Women's Health Initiative (WHI) study 81, 83, 147, 161, 165
World Health Organisation (WHO) 2–3, 49, 76, 159

XBP-1 (X-box binding protein) 336, 337
xenobiotic metabolism 313–14

zeitgebers (environmental cues) 16
zinc 275, 280, 281, 284
ZIP8 transport protein 275, 284